Bioprocess Engineering

Bioprocess Engineering

Edited by **Edgardo Turner**

SYRAWOOD
PUBLISHING HOUSE

New York

Published by Syrawood Publishing House,
750 Third Avenue, 9th Floor,
New York, NY 10017, USA
www.syrawoodpublishinghouse.com

Bioprocess Engineering
Edited by Edgardo Turner

© 2016 Syrawood Publishing House

International Standard Book Number: 978-1-68286-195-0 (Hardback)

Printed in the United States of America.

Contents

Preface

Bioprocess engineering is an emerging field of study under the discipline of chemical engineering that focuses on creating useful designs for developing products like pharmaceuticals, polymers, etc. using biological substances. This book elucidates the concepts and innovative models around prospective developments with respect to bioprocess engineering. It includes some of the vital topics such as biomolecular engineering, producing enzymes, fermentation technology, etc. The various studies that are constantly contributing towards advancing technologies and evolution of this field are examined in detail. Scientists and students actively engaged in this field will find this book full of crucial and unexplored concepts.

The researches compiled throughout the book are authentic and of high quality, combining several disciplines and from very diverse regions from around the world. Drawing on the contributions of many researchers from diverse countries, the book's objective is to provide the readers with the latest achievements in the area of research. This book will surely be a source of knowledge to all interested and researching the field.

In the end, I would like to express my deep sense of gratitude to all the authors for meeting the set deadlines in completing and submitting their research chapters. I would also like to thank the publisher for the support offered to us throughout the course of the book. Finally, I extend my sincere thanks to my family for being a constant source of inspiration and encouragement.

Editor

Timeline of Gas Production under Anaerobic Conditions

Bernadette E Teleky* and Mugur C Bălan

Department of Mechanical Engineering, Technical University of Cluj-Napoca, Romania

Abstract

For the study of biogas production by anaerobic digestion of lignocellulose substrate, it was designed an experiment in which the quantity of gas released by the anaerobic hydrogen producing bacteria was measured. The experiment aimed to highlight the differences occurred in the quantity and composition of the gas provided in 5 consecutive experiments conducted in similar conditions. The experimental data were processed and it was analysed the timeline evolution of the gas production rate of which kinetic was numerically modelled by several three-parameter logistic functions. Even being a preliminary study, it is innovative because it combines experiment with simulation and it proved two mathematical models, based on Gompertz and logistic functions. It was found that the anaerobic digestion process with hydrogen production has a sigmoid dependence on time.

Keywords: Anaerobic digestion; Gas production; Lignocellulose biomass; Numerical modelling; Logistic functions

Introduction

The anaerobic digestion of biomass can be considered one of the most promising technologies for renewable energy production. It can provide biogas with high energy potential and thus it represents a possible alternative to the conventional fuels like coal, wood, petroleum, natural gas and others.

Hydrogen production using agricultural residues and wastes [1] like cheese whey [2,3], rice straw [4], wheat straw [5,6], manure [7], municipal waste [8], and food waste [9] has received special attention currently.

Biomass rich in lignocellulose, like energy crops, agricultural and forest management residues [10] and municipal wastes is a versatile renewable energy source [11]. It can replace fossil fuels in power and heat generation and natural gas in the production of chemicals. Additionally it can also serve as a gaseous fuel [12].

Biomass with high lignocellulose content, like wheat straw and other agricultural residues have a very compact crystalline structure, thus they are hardly degradable [13] and conversion to hydrogen through anaerobic fermentation is very hard [14]. Because wheat straw residues are widespread all over the world and it could be gained at low cost, its conversion to hydrogen is intensively studied [1,10], and could be considered quite a good substrate for hydrogen production [15].

Straw is one of the major crop residues in Europe that could be used for the production of biogas [16]. Wheat straw generally consists of around (7.4-8.2)% lignin, 46% cellulose, 33.7% hemicellulose, and forasmuch lignin physically shield the cellulose and hemicellulose parts and it can be hardly degraded microbiologically under anoxic conditions in engineered systems [17]. Because straw has a high carbon-nitrogen ratio (C/N) and low levels of trace elements it limits the activity and growth of microbes. In order to stimulate the anaerobic digestion, different pre-treatment methods can be used [18,19].

Mathematical models can be used to explain effects of the various components and to perform different behavioural predictions [20]. Many studies use mathematical models to study the microorganism growth, cumulative hydrogen production [21]. Between the many kinetic models of the gas production, are well known Gompertz [22,23] or logistic [20,24].

The aim of this study was to model, from quantitative point of view, the gas production through anaerobic fermentation of straw with hydrogen producing microorganisms. A series of five experiments was conducted and in each one the production of hydrogen was recorded. A series of alternative models for the timeline gas production were fitted and the best alternatives were selected.

Material and Methods

Samples gathering

Sediments samples, as source of digesting microorganisms, were taken from Lake Héviz (Hungary) at the temperature of 29°C and at the pH of (7.1-7.2) in April 18, 2011. In the sampling process were considered two different areas of the lake: one was located in a foreshore zone vegetated with plants from the Typhaceae family being labelled "Typha" and the other one was located in the middle of the lake being labelled "Sediment". The samples were preserved at -20°C until the beginning of the experiments.

Design of experiment

Each experiment of anaerobic digestion with gas production was conducted with the same quantity of straw as substrate (250 mg), the same quantity of culture media (25 ml), and the same number of microorganisms (10^6). The process of preparation for all the components introduced in the culture bottles is presented in Figure 1.

The straw substrate was prepared by chopping followed by sterilisation. The culture media was a DMSZ 640 modified and specially prepared with the composition indicated in Table 1.

The Cysteine is added as a source of sulphur and seeks to reduce the traces of oxygen. NaOH is added to adjust the pH to the value of 7.3. In order to remove the oxygen, this media was boiled in the microwave and

---

*Corresponding author: Bernadette E Teleky, Department of Mechanical Engineering, Technical University of Cluj-Napoca, Romania

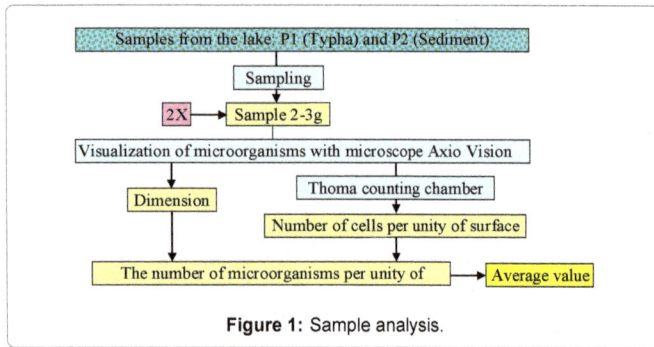

Figure 1: Sample analysis.

Components	Modified DSMZ 640
Distilled water [ml]	1000.00
NH_4Cl [g]	0.90
NaCl [g]	0.90
$MgCl_2 \cdot 6H_2O$ [g]	0.40
KH_2PO_4 [g]	0.75
K_2HPO_4 [g]	1.50
Peptone [g]	1.00
Yeast extract [g]	0.50
Trace element solution SL-10 (DSM 320) [ml]	1.00
$FeCl_3 \cdot 6H_2O$ [mg]	2.50
Straw [g]	0.25
NaOH [g]	0.50
Cystein – $HCl \cdot H_2O$ [g]	0.75
Resazurin [mg]	0.50

Table 1: Modified DSMZ 640 culture media.

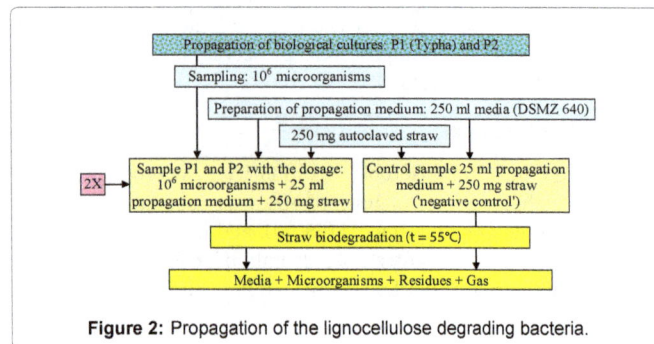

Figure 2: Propagation of the lignocellulose degrading bacteria.

Model	Rational	Gompertz	Freundlich	Hill	Weibull	Logistic
Eq.	$\dfrac{a+b \cdot x}{c+x}$	$\dfrac{e^{-e^{-b(x-c)}}}{a^{-1}}$	$a \cdot x^{b \cdot x^{-c}}$	$\dfrac{a \cdot x^d}{c^d + x^d}$	$\dfrac{1 - e^{-(b \cdot (x-c))^d}}{a^{-1}}$	$\dfrac{a}{1 + e^{-b \cdot (x-c)}}$

Table 2: Pre-selected alternatives for fit of the hydrogen production.

cooled down with nitrogen. In the anaerobic box it was measured and closed with butyl robber stopper and for sterilizing it was autoclaved for 30 minutes, at 121°C and at 2 bars.

The microorganisms were counted using an Axio Vision 4.8 microscope and a Thoma counting chamber. The preparation procedure was repeated two times for both Typha and Sediment.

Negative controls were made for control, to see that without the microorganisms from the Lake Hévíz, it produced negligible quantities of gas, and the cultures were not contaminated throughout the experiments.

The anaerobic digestion took place in each experiment, in the culture bottles at the temperature of 55°C. Each experiment was continued until the exhausting of the substrate by the bacteria, revealed by the severe reduction of gas production. Each experiment was continued for (21–30) days. The components of gas production were hydrogen and CO_2. Methane was not detected. Experiments succeeded each other following the scheme presented in Figure 2.

For this study were used the results obtained in 5 consecutive experiments.

Data processing

Some functions were considered to describe the time influence on the cumulative gas production throughout the experiments. Indirectly, the time is embedding the influence growing of microorganisms. The first two functions taken into account were Gompertz [25] and logistic [26].

Like in similar studies [27-29], a series of other possible alternatives of fit: Rational, Freundlich, Hill and Weibull were pre-selected for investigation from more than 100 available models, using FindGraph Software [30]. The pre-selected models, considered in the study, are presented in Table 2.

In all models, the variable x was assigned to the day of the observation in all the considered models and each model estimate the cumulative gas production, resulted from the experiments. The considered alternative models to fit were investigated to identify the adequate ones. The evaluation criterion for the evaluated models was the reliable confidences in their parameters. Into the analysis were considered independently 20 data sets, representing the data recorded for each of the 5 experiments, corresponding to the four bottles of culture: two bottles with Typha and to two bottles with Sediment. The expectance for correct modelling was that no more than 1/20=5% can have parameters with confidence not cancelling them.

Results and Discussion

Gas production

The results obtained from the 5 consecutive experiments are presented in Figure 3. The figured gas production is representing the normalized volume of gas, taking into account the temperature, the pressure and the volumes of gas, extracted at each measurement.

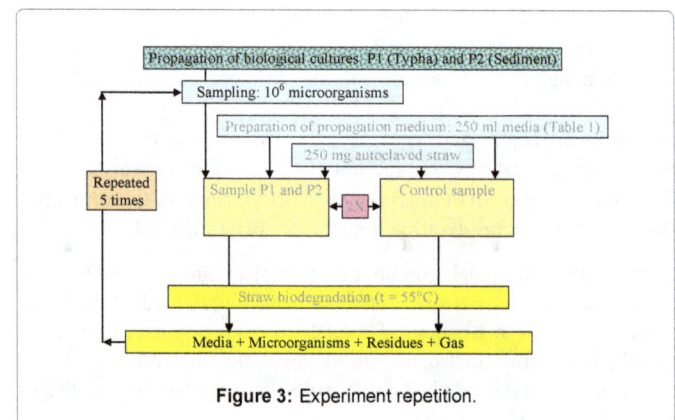

Figure 3: Experiment repetition.

The highest gas production was observed for the cultures Typha "a" and Sediment "a", while the lowest gas production was observed for the culture Typha "b".

As it was expected, in the Negative controls culture bottles, no changes were observed during the entire period of all the experiments: the culture medium remained clear and the wheat straw substrate remained unmodified. The gas productions in the Negative controls culture bottles were insignificant, of $(0.1-0.2)$ ml_N/ml until the day 10 of each experiment, when gas production completely stopped.

Evaluation of models

Four of the pre-selected models failed to correctly estimate the gas production. In Table 3 are presented failure fits for these alternatives. From all the evaluated models, only Gompertz and logistic, accomplished the selection criteria. For these models, the estimated of parameters, together with their relative confidence intervals are given in Table 4.

The presented results show that parameters of the model might change from one experiment to another. This inconsistency could be due to the slight differences in the quantity of substrate, in the quantity of culture media and/or in the number of microorganisms participating to the experiments.

Between the Gompertz and logistic functions are some differences [31]:

- The Gompertz model is symmetric;

- The logistic model is asymmetric.

Both models are used quite frequently in modelling of the experimental data as mentioned in [32-34]. When comparing the two models it should be taken into account that:

- The Gompertz model has a double exponential dependence, which is not the expected dependence for a kinetic of a chemical process.

- When the width of the confidence interval, given in Table 4, is averaged, one may find that the average of the relative width of the confidence intervals of the parameters for Gompertz model is 22.26% while for Logistic model is 20.27%, providing a better estimation for the parameters of the digestion process.

Considering the logistic model more accurate than the Gompertz model, as follows are presented detailed results of the estimations with logistic model. Anyway further investigations, also with others models, can be performed, to search for even more accurate estimation models.

The Logistic function model

The total gas production during the experiments, estimated by the logistic model as function of time, is presented in Figure 4. The parameters of the Logistic function are those presented in Table 4.

For all the 5 experiments and for all cultures, the gas production was almost continuous in the first part and then stagnated in the last

Model	Rational	Freundlich	Hill	Weibull
Failure 1	"Ta2"	"Ta2"	"Ta4"	"Ta2"
a =	0.6 ± 529%	.1 ± 122%	3.5 ± 296%	1.1 ± 7%
b =	1.8 ± 33%	1.2 ± 96%	61 ± 617%	.06 ± 75%
c =	17 ± 86%	0.15 ± 103%	0.8 ± 75%	-6 ± 191%
d=				2.3 ± 94%
Failure 2	"Ta3"	"Ta3"	"Tb3"	"Ta5"
a =	-.3 ± 848%	0.1 ± 135%	1.3 ± 117%	1.3 ± 2.5%
b =	1.9 ± 31%	1.6 ± 92%	14 ± 285%	0.03 ± 118%
c =	9.8 ± 97%	0.2 ± 67%	0.8 ± 81%	-23 ± 170%
d =				5 ± 123%

Each of these models failed to accomplish selection criteria
(already 2/20 unreliable estimates of the parameters)

Table 3: Failure fits for some of the pre-selected alternatives.

Model	Gompertz			Logistic		
Eq.	$a \cdot e^{-e^{-b\cdot(x-c)}}$			$a/(1+e^{-b\cdot(x-c)})$		
Estimates	a	b	c	a	B	c
Ta1	1.1 ± 5.2%	0.25 ± 27%	2.8 ± 26%	1.1 ± 6.2%	0.34 ± 37%	4.4 ± 23%
Ta2	1.1 ± 7.2%	0.16 ± 24%	5.2 ± 17%	1.1 ± 5.4%	0.24 ± 21%	7.4 ± 12%
Ta3	1.3 ± 5.7%	0.23 ± 23%	4.1 ± 15%	1.2 ± 3.5%	0.35 ± 18%	5.7 ± 8.9%
Ta4	1.2 ± 28%	0.14 ± 66%	5.5 ± 51%	1.1 ± 25%	0.20 ± 62%	8.0 ± 47%
Ta5	1.3 ± 4.3%	0.16 ± 17%	5.4 ± 11%	1.3 ± 2.5%	0.23 ± 11%	7.8 ± 5.9%
Tb1	1.2 ± 3.7%	0.33 ± 23%	2.6 ± 20%	1.2 ± 4.9%	0.46 ± 34%	3.8 ± 20%
Tb2	0.79 ± 19%	0.09 ± 43%	5.4 ± 44%	0.73 ± 12%	0.15 ± 31%	8.5 ± 27%
Tb3	0.78 ± 9.6%	0.18 ± 34%	3.7 ± 26%	0.75 ± 5.6%	0.27 ± 24%	5.7 ± 15%
Tb4	0.79 ± 38%	0.12 ± 69%	7.5 ± 55%	0.72 ± 30%	0.19 ± 59%	10 ± 45%
Tb5	0.64 ± 3.6%	0.26 ± 24%	3.3 ± 20%	0.63 ± 3.6%	0.35 ± 27%	4.9 ± 14%
Sa1	1.1 ± 4.3%	0.26 ± 23%	3.1 ± 19%	1.1 ± 5.4%	0.36 ± 33%	4.6 ± 19%
Sa2	1.2 ± 5.3%	0.29 ± 33%	2.3 ± 34%	1.2 ± 5.6%	0.4 ± 6%	3.7 ± 26%
Sa3	1.0 ± 2.8%	0.36 ± 15%	3.0 ± 11%	1.0 ± 3.1%	0.51 ± 18%	4.2 ± 9.7%
Sa4	0.88 ± 13%	0.29 ± 61%	5.7 ± 23%	0.83 ± 11%	0.62 ± 79%	6.6 ± 19%
Sa5	1.0 ± 3.5%	0.22 ± 19%	4.9 ± 12%	0.99 ± 2%	0.32 ± 12%	6.8 ± 5.5%
Sb1	1.1 ± 5.5%	0.3 ± 32%	2.7 ± 28%	1.1 ± 6.4%	0.41 ± 43%	4.1 ± 25%
Sb2	1.2 ± 5.3%	0.26 ± 30%	2.7 ± 30%	1.2 ± 6%	0.36 ± 37%	4.2 ± 25%
Sb3	1.2 ± 4.2%	0.35 ± 24%	2.6 ± 19%	1.2 ± 5.4%	0.48 ± 33%	3.8 ± 19%
Sb4	1.1 ± 7.2%	0.39 ± 40%	3.8 ± 21%	1.1 ± 7.2%	0.62 ± 43%	4.9 ± 18%
Sb5	1.2 ± 3.2%	0.35 ± 26%	2.7 ± 22%	1.2 ± 3.6%	0.47 ± 32%	3.8 ± 18%

Table 4: Estimation of the parameters (a, b, c, d) along with their confidence.

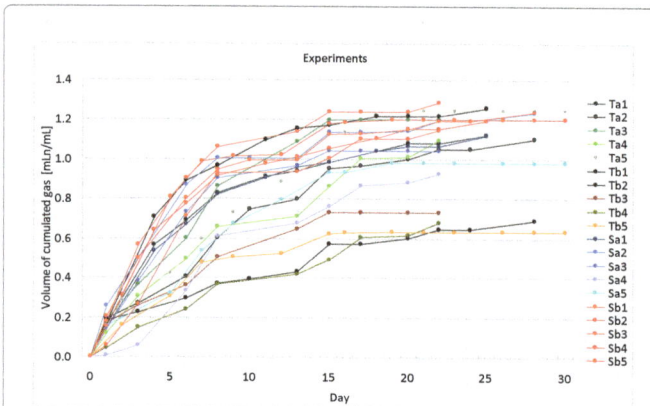

Figure 4: Volume of cumulated gas through the five repetitions of the experiment.

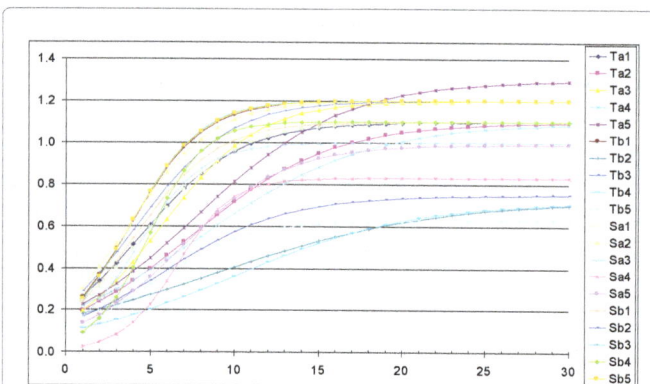

Figure 5: Digestion modelled by Logistic function as are estimated for the experiments.

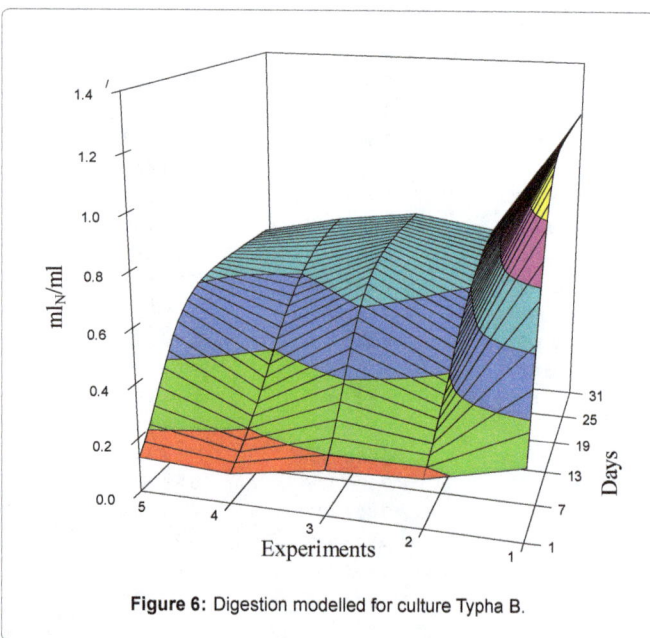

Figure 6: Digestion modelled for culture Typha B.

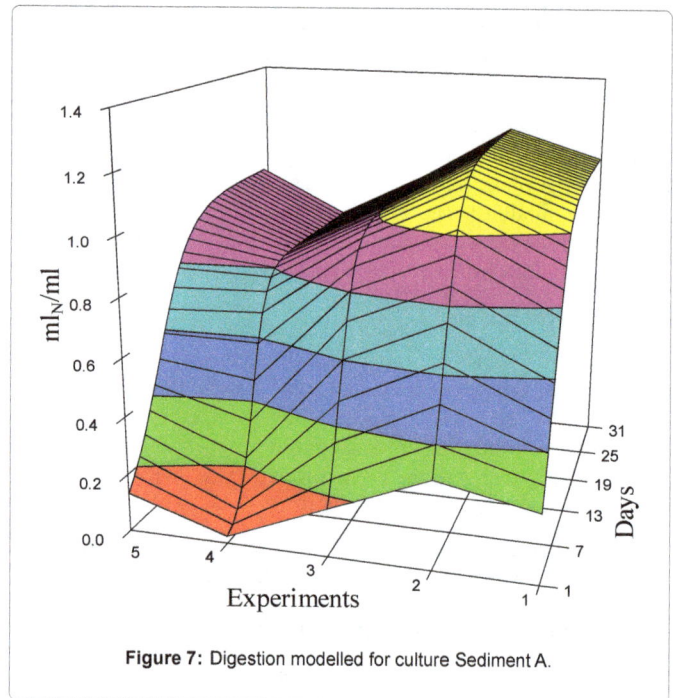

Figure 7: Digestion modelled for culture Sediment A.

period of experiments. This behaviour explain the sigmoid or "S" shape of the presented curves and simultaneous the sigmoid character of gas production dependence of time. Similar character of gas production in anaerobic digestion of biomass was reported in several references [35-37].

As it can be observed, there are some significant differences between experiments:

- The shapes of the dependences may be related with the accessible surface of the straw to the microorganisms;

- The total quantity of the produced gas may be related with the total bioactive content of the straw.

The results of the modelling with the Logistic model are presented for each of the anaerobic digestion for each of the two types of cultures: Typha and Sediment and for each of the two culture bottles used in each experiment:

- The results for culture Typha "a" are presented in Figure 5;

- The results for culture Typha "b" are presented in Figure 6;

- The results for culture Sediment "a" are presented in Figure 7;

- The results for culture Sediment "b" are presented in Figure 8.

The synthetic analysis of the data presented in Figures 5-8, is presented in Table 5.

As it can be seen in the table, the culture "Typha a" had a gas production between 0.8–1.25 ml_N/ml, with the percentage of hydrogen in the produced gas was between 22.21%-34.06%, leading to a hydrogen production of 0.18-0.43 ml_N/ml. The lowest gas production corresponded to the culture "Typha b" with 0.63–1.26 ml_N/ml, with the percentage of hydrogen in the produced gas was between 12.83%-22.91%, leading to a hydrogen production of 0.08-0.29 ml_N/ml. The duration of the experiments was between 22–33 days for all cultures. The moment of stagnation was different and usually started at days 13, 15, or 18.

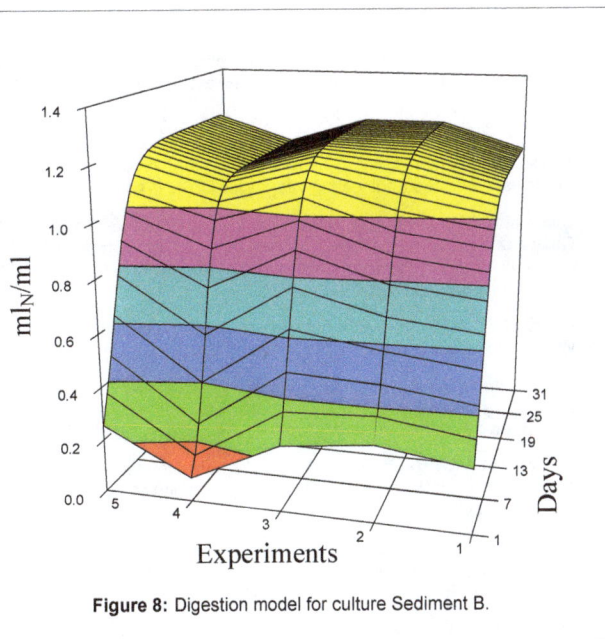

Figure 8: Digestion model for culture Sediment B.

Experiment	Total gas production [ml$_N$/ml]	Duration of experiment [days]	Moment of stagnation [day]	Percentage of hydrogen production [%]	Hydrogen production [ml$_N$/ml]
Typha 'a"	0.80-1.25	22-33	15-33	22.21-34.06	0.18-0.43
Typha 'b"	0.63-1.26	22-33	13-33	12.83-22.91	0.08-0.29
Sediment "a"	0.70-1.23	22-33	15-33	24.31-28.67	0.17-0.35
Sediment "b"	0.53-1.28	22-33	18-33	2.11-25.65	0.01-0.33

sTable 5: Main characteristics of the gas production in the experiments.

It can be observed that hydrogen production obtained from the experiments and from correlated modelling, are in good agreement with similar values reported in the literature [38-40], but are under the values reported by [15,41-43].

Conclusion

The study presents both experimental and modelling results concerning the gas production during the anaerobic digestion of biomass with high content of lignocellulose.

The analysis of the gas production revealed the sigmoid character of the gas production dependence of time in the anaerobic digestion of biomass. It was highlighted that this character is typical for this process, as it was previously reported in literature.

It were evaluated several models for the timeline of gas production and for all the considered models was analysed the expectance for correct modelling.

Two of the 6 models considered in the study passed the selection criterion and were considered correct. The two models are Gompertz and logistic. The study could not obtain clear evidence on about which model is the best one to be used for these estimates, and requires further investigation, but it could be stated that Logistic model is more accurate than Gompertz model.

The Logistic model was used to provide detailed information and characteristics of the gas production in anaerobic digestion of biomass. It was provided synthetic information about the following: duration of experiments, total gas production, moment of stagnation and hydrogen production.

The study proved that anaerobic digestion of biomass with lignocellulose presents high potential for hydrogen production to be used in energy applications.

Acknowledgements

This study is supported by the Sectoral Operational Programme for Human Resources Development POSDRU/159/1.5/S/137516 financed from the European Social Fund and by the Romanian Government. The experimental part was supported by German Federal Environmental Foundation and Helmholtz Centre for Environmental Research - UFZ Leipzig, Germany.

References

1. Guo XM, Trably E, Latrille E, Carrère H, Steyer JP (2010) Hydrogen production from agricultural waste by dark fermentation: A review. Int J Hydrogen Energy 10: 10660-10673.

2. Frascari D, Cappelletti M, Mendes Jde S, Alberini A, Scimonelli F, et al. (2013) A kinetic study of biohydrogen production from glucose, molasses and cheese whey by suspended and attached cells of Thermotoga neapolitana. Bioresour Technol 147: 553-561.

3. Kargi F, Eren NS, Ozmihci S (2012) Bio-hydrogen production from cheese whey powder (CWP) solution: Comparison of thermophilic and mesophilic dark fermentations. Int J Hydrogen Energy 37: 8338-8342.

4. Gu Y, Chen X, Liu Z, Zhou X, Zhang Y (2014) Effect of inoculum sources on the anaerobic digestion of rice straw. Bioresour Technol 158: 149-155.

5. Motte JC, Escudié R, Beaufils N, Steyer JP, Bernet N, et al. (2013) Morphological structures of wheat straw strongly impacts its anaerobic digestion. Industrial Crops and Products 52: 695-701.

6. Nkemka VN, Murto M (2013) Biogas production from wheat straw in batch and UASB reactors: the roles of pretreatment and seaweed hydrolysate as a co-substrate. Bioresour Technol 128: 164-172.

7. Wang KS, Chen JH, Huang YH, Huang SL (2012) Integrated Taguchi method and response surface methodology to confirm hydrogen production by anaerobic fermentation of cow manure. Int J Hydrogen Energy 38: 45-53.

8. De Gioannis G, Muntoni A, Polettini A, Pomi R (2013) A review of dark fermentative hydrogen production from biodegradable municipal waste fractions. Waste Manag 33: 1345-1361.

9. Kobayashi T, Xu KQ, Li YY, Inamori Y (2012) Effect of sludge recirculation on characteristics of hydrogen production in a two-stage hydrogen–methane fermentation process treating food wastes. Int J Hydrogen Energy 37: 5602-5611.

10. Kaparaju P, Serrano M, Angelidaki I (2009) Effect of reactor configuration on biogas production from wheat straw hydrolysate. Bioresour Technol 100: 6317-6323.

11. Sims REH (2004) Bioenergy options for a cleaner environment: in developed and developing countries. Elsevier, pp. 198.

12. Mosier N, Wyman C, Dale B, Elander R, Lee YY, et al. (2005) Features of promising technologies for pretreatment of lignocellulosic biomass. Bioresour Technol 96: 673-686.

13. Fan YT, Zhang YH, Zhang SF, Hou HW, Ren BZ (2006) Efficient conversion of wheat straw wastes into biohydrogen gas by cow dung compost. Bioresour Technol 97: 500-505.

14. Cao GL, Guo WQ, Wang AJ, Zhao L, Xu CJ, et al. (2012) Enhanced cellulosic hydrogen production from lime-treated cornstalk wastes using thermophilic anaerobic microflora. Int J Hydrogen Energy 37: 13161-13166.

15. Nasirian N, Almassi M, Minaei S, Widmann R (2011) Development of a method for biohydrogen production from wheat straw by dark fermentation. Int J Hydrogen Energy 36: 411-420.

16. Kim SM, Dale BE (2003) Global potential bioethanol production from wasted crops and crop residues. Biomass and Bioenergy 26: 361- 375.

17. Monlau F, Barakat A, Trably E, Dumas C, Steyer JP, et al. (2011) Lignocellulosic materials into biohydrogen and biomethane: impact of structural features and pre-treatment. Crit Rev Environ Sci Technol 43: 260-322.

18. Hendriks AT, Zeeman G (2009) Pretreatments to enhance the digestibility of lignocellulosic biomass. Bioresour Technol 100: 10-18.

19. Taherzadeh MJ, Karimi K (2008) Pretreatment of lignocellulosic wastes to improve ethanol and biogas production: a review. Int J Mol Sci 9: 1621-1651.

20. Wang J, Wan W (2008) Effect on temperature on fermentative hydrogen production by mixed cultures. Int J Hydrogen Energy 33: 5392-5397.

21. Argun H, Kargi F, Kapdan IK, Oztekin R (2008) Batch dark fermentation of powdered what starch to hydrogen gas: Effects of the initial substrate and biomass concentrations. Int J Hydrogen Energy 33: 6109-6115.

22. Kirkwood TB (2015) Deciphering death: a commentary on Gompertz (1825) 'On the nature of the function expressive of the law of human mortality, and on a new mode of determining the value of life contingencies'. Philos Trans R Soc Lond B Biol Sci 370.

23. Sagnak R, Kargi F, Kapdan IK (2011) Bio-hydrogen production from acid hydrolyzed waste ground wheat by dark fermentation. Int J Hydrogen Energy 36: 12803-12809.

24. Ghatak MD, Mahanta P (2014) Comparison of kinetic models for biogas production rate from saw dust. IJRET 3: 248-254.

25. Waliszewski P, Konarski J (2005) A mystery of the Gompertz Function. In: Losa, GA, Merlini D, Nonnenmacher TF, Weibel ER (eds.) Fractals in Biology and Medicine. Birkhäuser, Switzerland, pp. 277-286.

26. Vieira S, Hoffmann R (1977) Comparison of the logistic and the Gompertz growth functions considering additive and multiplicative error terms. Appl Statist 26: 143-148.

27. Boni MR, Sbaffoni S, Tuccinardi L, Viotti P (2013) Development and calibration of a model for biohydrogen production from organic waste. Waste Manag 33: 1128-1135.

28. Chen WH, Chen SY, Khanal SK, Sung S (2006) Kinetic study of biological hydrogen production by anaerobic fermentation. Int J Hydrogen Energy 31: 2170-2178.

29. Mu Y, Yu HQ, Wang G (2007) A kinetic approach to anaerobic hydrogen-producing process. Water Res 41: 1152-1160.

30. http://www.uniphiz.com/

31. Nguimkeu P (2014) A simple selection test between the Gompertz and Logistic growth models. Technological Forecasting and Social Change 88: 98-105.

32. Trappey CV, Wu HY (2008) An evaluation of the time-varying extended logistic, simple logistic, and Gompertz models for forecasting short product lifecycles. Advanced Engineering Informatics 22: 421- 430.

33. Chowdhury BR, Chakraborty R, Chaudhuri UR (2007) Validity of modified Gompertz and Logistic models in predicting cell growth of Pediococcus acidilactici H during the production of bacteriocin pediocin AcH. J Food Eng 80: 1171-1176.

34. Wilson DL (1994) The analysis of survival (mortality) data: fitting Gompertz, Weibull, and logistic functions. Mech Ageing Dev 74: 15-33.

35. Bah H, Zhang W, Wu S, Qi D, Kizito S, et al. (2014) Evaluation of batch anaerobic co-digestion of palm pressed fiber and cattle manure under mesophilic conditions. Waste Manag 34:1984-1991.

36. Krishania M, Vijay VK, Chandra R (2013) Methane fermentation and kinetics of wheat straw pretreated substrates co-digested with cattle manure in batch assay. Energy 57: 359-367.

37. Xu F, Wang ZW, Tang L, Li Y (2014) A mass diffusion-based interpretation of the effect of total solids content on solid-state anaerobic digestion of cellulosic biomass. Bioresour Technol 167: 178-185.

38. Aguilar MAR, Fdez-Güelfo LA, Álvarez-Gallego CJ, García LIR (2013) Effect of HRT on hydrogen production and organic matter solubilization in acidogenic digestion of OFMSW. Chem Eng J 219: 443-449.

39. Hernanández M, Rodríguez M (2013) Hydrogen production by anaerobic digestion of pig manure: Effect of operating conditions. Renewable Energy 53: 187-192.

40. Xiao L, Deng Z, Fung KY, Ng KM (2013) Biohydrogen generation from anaerobic digestion of food waste. Int J Hydrogen Energy 38: 13907-13913.

41. Zhu H, Stadnyk A, Béland M, Seto P (2008) Co-production of hydrogen and methane from potato waste using a two-stage anaerobic digestion process. Bioresour Technol 99: 5078-5084.

42. Wang W, Xie L, Luo G, Zhou Q (2013) Enhanced fermentative hydrogen production from cassava stillage by co-digestion: The effects of different co-substrates. Int J Hydrogen Energy 38: 6980-6988.

43. Liu H, Fang HH (2003) Hydrogen production from waste water by acidogenic granular sludge. Water Sci Technol 47: 153-158.

Effect of Different Impellers and Baffles on Aerobic Stirred Tank Fermenter using Computational Fluid Dynamics

Sharma Alok[1] and Genitha Immanuel[2]

[1]Food Technology (Food Process Engineering), SHIATS, Allahabad, India
[2]Department of Food Process Engineering, SHIATS, Allahabad, India

Abstract

The present research was carried out with objective to study the effect of different types of Impellers and Baffles on mixing and to examine the correlation between mixing time and mass transfer of Aerobic Stirred Tank Fermenter. An Aerobic Stirred Tank fermenter was assembled with different Impellers and Baffles alternately. The four different Impellers used were Rushton Impeller, Marine Impeller, A320 Impeller and HE3 Impeller while walled and un walled baffles were used in combination with these Impellers. The Fermenter was assembled with Resistance Temperature Detector, pH Probe and Pressure Gauge. Tachometer was used to calculate the Mixing Time of Fermenter. Volumetric Mass Transfer Coefficient was experimentally determined and calculated by the respective formulae. MATLAB was used for Mathematical Modelling of CFD, while Autodesk Simulation CFD and ANSYS FLUENT were used to generate the simulations. Turbulence lengths and Trailing Vortices were used to understand flow patterns inside the fermenter created by Impellers and Baffles. The k_La values suggested that Rushton Impeller was the most efficient among all. Turbulent Kinetic Energy and Turbulent Dissipation Rate were to understand the mixing efficiency of every Impeller.

Keywords: CFD; Aerobic stirred tank fermenter; Turbulent kinetic energy; Turbulent; Dissipation rate

Introduction

Fermentation is the conversion of carbohydrates to alcohols and carbon dioxide or organic acids using desirable microorganisms. The fermentation is carried out in specially designed and engineered device or system called fermenter [1,2]. The class of fermenters varies in a wide range according to the requirement of process to be carried out. They have the following functions: homogenization, suspension of solids, dispersion of gas-liquid mixtures, aeration of liquid and heat exchange. It is provided with a baffle and a rotating stirrer attached either at the top or at the bottom of the fermenter [3].

The typical decision variables are type, size, and location impellers and baffles. These determine the hydrodynamic pattern in the reactor, which in turn influence mixing time, mass and heat transfer coefficients, shear rates etc. [4-7]. The conventional fermentation is carried out in a batch mode. Since stirred tank reactors are commonly used for batch processes with slight modifications, these reactors are simple in design and easier to operate. The batch stirred tanks generally suffer due to their low volumetric productivity. The Stirred tank reactor offers excellent mixing and reasonably good mass transfer rates. The cost of operation is lower and the reactors can be used with a variety of microbial species [8,9].

A research team led by Chaim Weizmann in Great Britain during the First World War developed a process for the production of acetone by a deep liquid fermentation using *Clostridium acetobutylicum* which led to the eventual use of the first truly large scale aseptic fermentation vessels [10,11]. The fermenters consisted of large cylindrical tanks with air introduced at the base via networks of perforated pipes. Later, mechanical impellers were used to increase the mixing rate and to breakup and disperse the air bubbles. Baffles on the walls of vessels prevented a vortex formation in fermentation broth [12]. In 1932, a system was introduced in which aeration tubes were provided with water and steam for cleaning and sterilizing. Construction work on the first large scale plant to produce penicillin by deep fermentation was started in September 1943, at Terre Haute in the United States of America, building steel fermenters with working volumes of 54,000 dm^3 [13].

In food processing industries, the fermentation is very important for various products e.g. alcoholic beverages, organic acids (acetic acid, citric acid etc.). Today's fermenters are more sophisticated as reflected by the various monitoring and control facilities with computer interfere for a more efficient fermentation process while the hardware remains the same [14]. The main function of a fermenter is to provide a controlled environment for the growth of microorganisms to get the desired product as like adequate aeration and agitation, low power consumption, temperature and pH control, minimum labor requirement and economy. In aerobic fermenters the dissolution of oxygen to the fermentation broth is important for the efficient operation of reactor [15,16]. Traditionally, stirred tank design is driven by the oxygen transfer capability needed to achieve cell growth. However, design methodologies available for stirred tank fermenters are insufficient and many times contains errors [17].

The most difficult part of the design is matching the fermenter capability to the oxygen demand of the fermentation culture. Some general guidelines have been offered on how to improve mass transfer in stirred tank reactors. In addition some correlations have been formed to provide predictions on stirred tank performance. However, the guidelines offered do not provide information on how different aspects of the tank (i.e. impeller and baffle geometry) specifically effect oxygen transfer in stirred tanks. The correlations offered do not provide

*Corresponding author: Sharma Alok, M. Tech, Food Technology (Food Process Engineering), SHIATS, Allahabad, India

a wide enough range of tank sizes, power inputs or gas flow rates to be useful to more than just a handful of people. In addition the efficacy of different baffle and impeller types in STRs were assessed. This was accomplished through four key areas. First, empirical studies were used to quantify the mass transfer capabilities of several different reactors; second, Computational Fluid Dynamics (CFD) was used to assess the impact of certain baffle and impeller geometries; third, correction schemes were developed and applied to the experimental data; and fourth, dimensionless correlations were created to act as a guide for future production scale fermenter design [18-22].

The CFD analysis of this research work was based on Turbulence Modelling of the fermenter. Turbulence modelling is construction and use of a model to predict the effects of turbulence [23]. Among these Reynolds-averaged Navier Stokes equation and κ-ε models were mainly used for the mathematical modelling of fermenter. These equations provided the outcome of the modelling of fermenter and the experimental data were evaluated accordingly. In light of above discussion, a study on the effect of different Impellers and Baffles was undertaken with the following objectives

1. To study the effect of different types of impeller and baffle on mixing of aerobic stirred tank fermenter.

2. To examine the possibility of a correlation between mixing time and mass transfer in aerobic stirred tank fermenter.

Material and Methods

The present study was carried out in Faculty of Agriculture Engineering, Indira Gandhi Agriculture University, Raipur. The experimental plan of present study is presented in Table 1. The materials required for this study were Aerobic Stirred Tank F e r m e n t e r, Impellers (i.e. Rushton Impeller, Marine Impeller, A320 Impeller and HE3 Impeller), Baffles (Walled and Unwalled)

Design calculations

The most popular range of ratio of height to diameter is 1:1 to 3:1 for stirred tank fermenters. In this project the ration of height (h) is diameter (D) is taken as 2:1. The diameter of the vessel was 0.50 m and the height of vessel was 1.00 m.

Hence, the calculated volume of the fermenter was 0.1964 m³ or 196.35 L. The working volume of fermenter was calculated to be 75% of the total volume. The impeller diameter must be one-third to the vessel diameter. So, the impeller diameter was calculated as 0.167 m. The pitch between impeller and vessel bottom was taken equal to impeller diameter i.e. 0.167 m. The length of impeller blade must be one-fourth of impeller diameter and calculation showed it 0.042 m. The width of impeller must be one-fifth of impeller diameter, after calculating it was 0.035 m [24-26].

Assembling of impellers and baffles

This step involved the assembling all parts of the fermenter. The vessel body was already fitted with the baffles and discharge port. The drive motor was installed on the cover plate of fermenter using J-bolts and oil spill. J-bolts were used to support and hold the motor while the oil spill was used to ensure that there is no leakage or air passage from the entry point of drive shaft [27-30]. A coupling was used to facilitate the changing of impeller assembled with drive shaft. A port was provided for feed inlet at the cover plate. Similarly, a port was made to facilitate the air inlet in the fermenter. A port was made to fit the thermometer inside the fermenter. A pressure gauge was installed along

with air emergency air outlet. O rings and gaskets were used at all the ports to avoid any kind of air leakage and contamination inside and outside of the fermenter. A spray ball was fitted at the inside-top of the fermenter to facilitate the cleaning and sterilization of fermenter.

Processing

The processing was done by running the fermenter with cane juice. The cane juice was collected from the local market. 18L of cane juice was used as feed for the fermenter. After filling the fermenter with cane juice, the fermenter was closed tightly (air tight). Then it was allowed to run, during this process the required data was obtained and the processing was regularly monitored. The temperature was about 37°C [31].

Performance evaluation

The performance evaluation of aerobic stirred tank fermenter was done by using various software packages i.e. MATLAB 8.01, Autodesk Simulation CFD and ANSYS FLUENT. The mathematical modelling was done by using MATLAB 8.01. It is the fourth generation programming language tool. Mathematical equations like Reynolds averaged Navier Stokes equation were programmed with proper inputs, which finally calculated the output of the process. Turbulence modelling method of Computational Fluid Dynamics was used for mathematical modelling. The models used were κ-ε model, κ-ω model, Shear Stress Transport model (SST) and Scale Adaptive Simulation SST (SAS-SST). All the data generated by these models were implemented in Autodesk Simulation CFD and ANSYS FLUENT software packages [32-34]. These software packages showed the results in various simulations. Simulation was preferred because it becomes very easy to understand the complex conclusion of CFD equations. These simulations were showing various coloured patterns of the fermenter performance. Results were obtained for both impellers i.e. Rushton turbine and Paddle impeller. In the k-ε model the turbulent kinetic energy k and its rate of dissipation ε are obtained from the transport equations.

Fluent used two approaches to modelling this "near-wall" region. The wall function approach bridges the viscosity-affected region between the wall and the fully turbulent region. The turbulence length-scales used in this study are given for the k-ε and SSG-RSM model and for the SST and SAS-SST models [35].

Experimental determination of volumetric mass transfer coefficient

For this method first the water in the tank is deoxygenated by sparging nitrogen until the Dissolved Oxygen (DO) in the tank reaches below 10% of the saturation level. Then air is reintroduced into the tank through the sparge at a known mass flow rate while the DO is monitored over time. This is monitored until the oxygen reaches close to 85% of the saturation level.

$$\frac{dC_{AL}}{dt} = k_L a(\overline{C_{AL}} - C_{AL})$$

C_{AL} is the dissolved oxygen concentration in percentage of saturation, t is time, C_{AL} is the final DO concentration and C_{AL1} and C_{AL2} are the DO concentrations at times t_1 and t_2, respectively [36,37].

For the configurations outlined in Table 1 steady-state method was used to give $k_L a$ values which serve as a quantitative comparison of the tanks. The volumetric mass transfer coefficient was determined at several points throughout the tank to give a volume-averaged mass transfer coefficient for each configuration. This data was used to empirically derive the dimensionless correlations. It also assisted in assessing the mass transfer capabilities of specific impellers and baffles [38].

Results and Discussions

During this research work the effect of different types of impellers and baffles on mixing of aerobic stirred tank fermenter were studied.

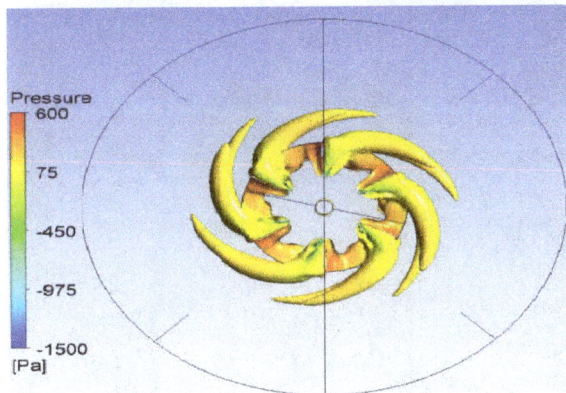

Figure 1: Simulation of Trailing Vortices by k-ε model.

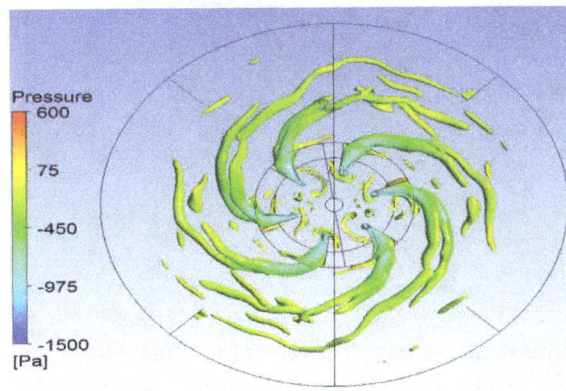

Figure 2: Simulation of Trailing Vortices by SAS SST.

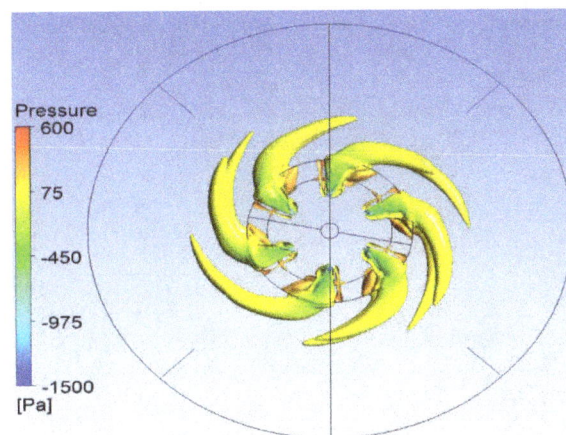

Figure 3: Simulation of Trailing Vortices by SST model.

This study involved the use of Computational Fluid Dynamics software packages i.e. Autodesk Simulation CFD and ANSYS FLUENT to obtain the simulations of the fluid mixing in Aerobic Stirred Tank Fermenter. The mathematical modelling was done by turbulence modelling and solved using MATLAB software package. The mixing time and mass transfer were also correlated to examine the performance of Aerobic Stirred Tank Fermenter.

Effect of different types of Impellers and Baffles on Mixing

First, the steady-state flow field was calculated and visually displayed to identify "dead zones" where the fluid was not moving or mixing very well. The pictures of the flow field gave information on how each impeller moves fluid through the tank. This aided in determining the effectiveness of impellers in their mixing capability. The second output from the CFD is a mixing time for each configuration [39]. After the steady-state formulation was calculated, the simulation was changed to a transient formation and a tracer fluid was introduced into the tank. The volume fraction of tracer fluid was monitored at several locations in the tank, according to Plate, and the mixing time was calculated as the time when 90% of homogeneity was reached. The effect of different Impellers and Baffles on Mixing has been studied and reviewed by Nurtuno T [37].

Trailing Vortices

The vortical structures in a flow were visualized in a number of different ways. Here the swirling strength has been used, based on the computation of the Eigen values of the velocity gradient tensor. A threshold value of 0.1 was found to be a good compromise between missing structures if the value was too high and masking the structures if the value was too low.

Unsteady two-equation models, such as k-ε and SST, were found for excessively damping turbulence so that any detail of the turbulent structure (even on the larger scales) cannot be resolved directly. This is reflected by simulations generated using Autodesk Simulation CFD in Figure1 for k-ε model, Figure 2 for SAS SST model, and Figure 3 for SST model, which show that these models predict very small and hence dissipative trailing vortices with no secondary vortex motion apparent. Nevertheless, both the k-ε and SST models predict the appearance of the pair of vortices, one vortex above and one vortex below the disk of the impeller that originate from behind each blade and trail out into the bulk of the flow [40-42].

Turbulence Length

In the k-ε and SST models, all turbulence scales are modelled through Reynolds-averaging, and hence a relatively large length-scale of turbulence was calculated using, as can be seen in Figure 4 and Figure 5. The length-scales of turbulence of up to 3 mm predicted by the k-ε and SST models in the region of the impeller are similar to the experimental turbulence length-scales [43]. For the SAS-SST model, some details of the turbulence structures -the larger scale structures-can be directly resolved, while Reynolds averaging accounts for the smaller-scale turbulence structures. Figure 6 show that the SAS-SST model predicts longer trailing vortices and secondary vortex motions. The turbulence length-scales predicted by the SAS-SST model are much shorter than those predicted by the k-ε and SST models, because the larger scale turbulence is now being directly resolved through the "LES" content of the model [44,45].

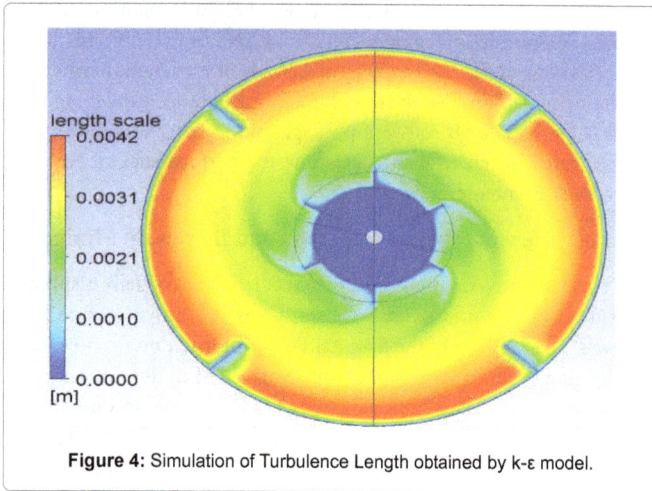

Figure 4: Simulation of Turbulence Length obtained by k-ε model.

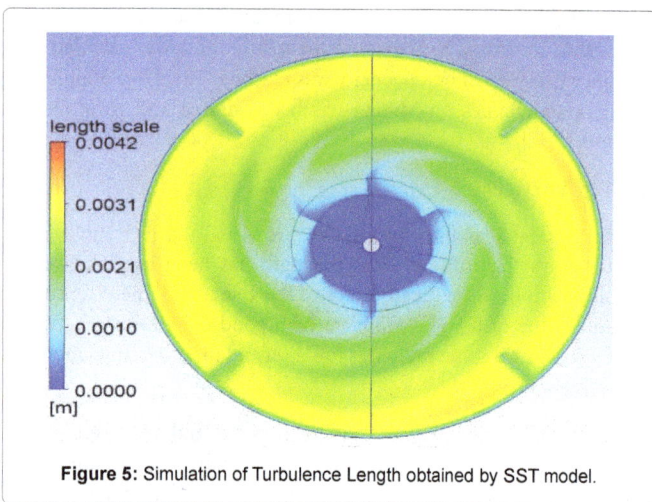

Figure 5: Simulation of Turbulence Length obtained by SST model.

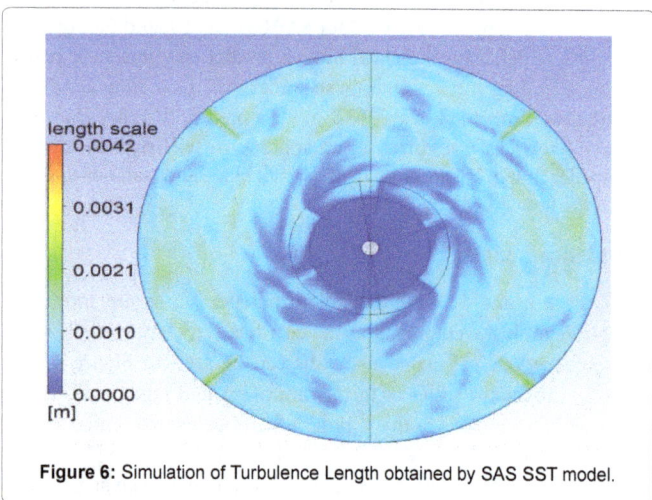

Figure 6: Simulation of Turbulence Length obtained by SAS SST model.

CFD results of impellers and baffles obtained from ansys fluent

The CFD results calculated in this study can be used to give a better understanding of mixing in stirred tanks, and how certain aspects of the tank produce better mixing. These pictures are a slice of the mid-plane of the tank and the arrows represent the direction of flow. The different

colors of arrows represent faster moving fluid, where the length of the arrows represents the direction of the fluid moving at that point. Where the arrows are longer, the fluid is moving more in line with the mid-plane of the tank; where the arrows are shorter they are moving more perpendicular to the mid-plane of the tank. Figure 7 shows CFD

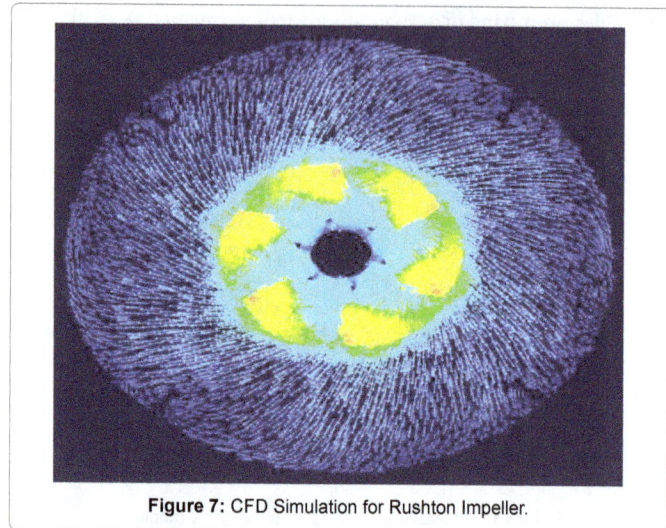

Figure 7: CFD Simulation for Rushton Impeller.

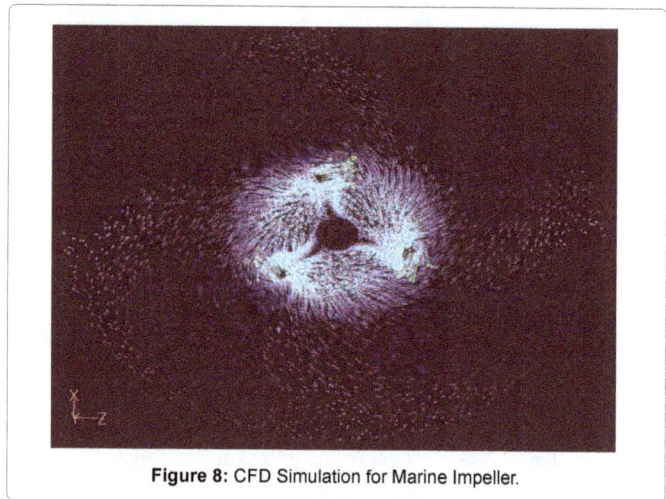

Figure 8: CFD Simulation for Marine Impeller.

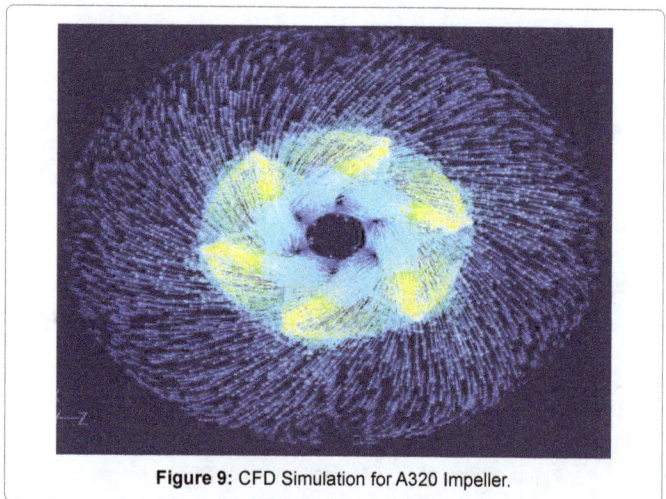

Figure 9: CFD Simulation for A320 Impeller.

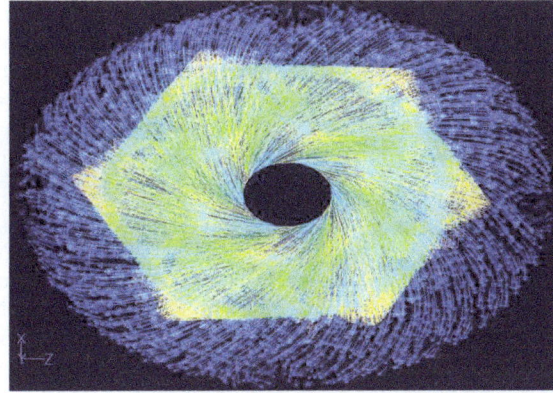

Figure 10: CFD Simulation for HE3 Impeller.

S.No.	Impeller	Baffle Configuration 1	Baffle Configuration 2	Experimental Determination of $k_L a$	CFD Steady State Calculation	CFD Mixing Time Calculation
1	Rushton Impeller	Walled	Un walled	Yes	Yes	Yes
2	Marine Impeller	Walled	Un walled	Yes	Yes	Yes
3	A320 Impeller	Walled	Un walled	Yes	Yes	Yes
4	HE3 Impeller	Walled	Un walled	Yes	Yes	Yes

Table 1: Impeller Tank Configuration and in which studies they were used.

Impellers	T_m (sec)	$k_L a$ at 40 lpm	$k_L a$ at 70 lpm
Rushton Impeller	5.5	135	141
Marine Impeller	7.9	72	78
A320 Impeller	10.5	85	90
HE3 Impeller	11.5	65	70

Table 2: Average $k_L a$ values for different Impellers at Gas Flow Rate of 40 lpm and 70 lpm.

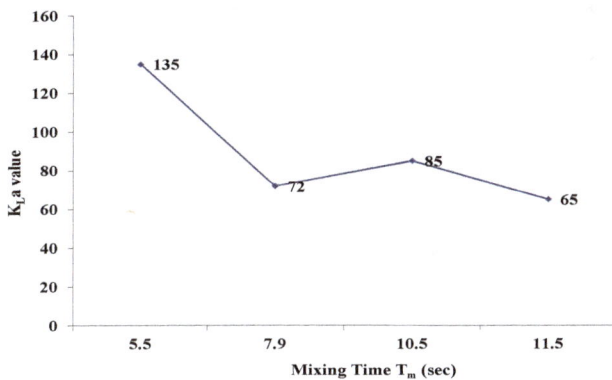

Figure 11: Average $K_L a$ values for different Impellers at Gas Flow Rate of 40 lpm.

Simulation for Rushton Impeller. Similarly, Figure 8, Figure 9 and Figure 10 show CFD Simulation for Marine Impeller, A320 Impeller and HE3 Impeller respectively [46-48].

Correlation between Mixing Time Tm (sec.) and Volumetric Mass Transfer Rate $k_L a$ (1/h)

The mixing times obtained in this study were compared to the experimental $k_L a$ values to examine the possibility of a correlation. Table 2 shows the average $k_L a$ values for different Impellers at the Gas

Flow Rate of 40 and 70 liters per minute (lpm). The figures show the plot of $k_L a$ of each impeller versus the mixing time at the two flow rates [49,50]. Figure 11 and Figure 12 shows that non uniform $k_L a$ values were observed for all four Impellers with respect to Mixing Time (T_m) at the Gas Flow Rate of 40 lpm and 70 lpm. The correlation between Mixing Time and Volumetric Mass Transfer Rate were found to be more or less similar to those mentioned by Byung-Hwan Um (2007) (Table 3).

As seen in these figures the mixing times do not correlate to the experimentally determined $k_L a$ values. Due to this lack of correlation, gas flow rates of 20 and 10 liters per minute (lpm) were tested for all four Impellers. These additional tests are outlined in Table 3 and plotted in Figure 13 and Figure 14 [51-54].

Effect of Impellers and Baffles on Turbulent Kinetic Energy and Turbulent Dissipation Rate with respect to Mixing Time

The values of Turbulent Kinetic Energy k (m^2/s^2) and Turbulent Dissipation Rate ε (m^2/s^3) were experimentally determined. The distribution of turbulent kinetic energy and dissipation rates as shown in Figures 15, 16, 17, 18, 19, 20, 21 and 22 are characteristic of the reactor geometry [55]. The turbulent k and ε predicted by the various viscosity suspensions with the maximum values are found in the discharge region and a surrounding zone of relatively high turbulent kinetic energy. As expected, relatively high dissipation rates were found near the impellers. The values of k are close to zero with low dissipation rates elsewhere [56].

Figure 15 and 19 shows that there was uniform increase in Turbulent Kinetic Energy (k) and Turbulent Dissipation Rate (ε) with respect to Mixing Time in case of Rushton Impeller [57]. But, Figure 16, 17, 18, 20, 21 and 22 shows that there was fluctuations in Turbulent Kinetic Energy (k) and Turbulent Dissipation Rate (ε) with respect to Mixing Time in case of Marine Impeller, A320 Impeller and HE3 Impeller.

Since, the Turbulent Kinetic Energy (k) and Turbulent Dissipation

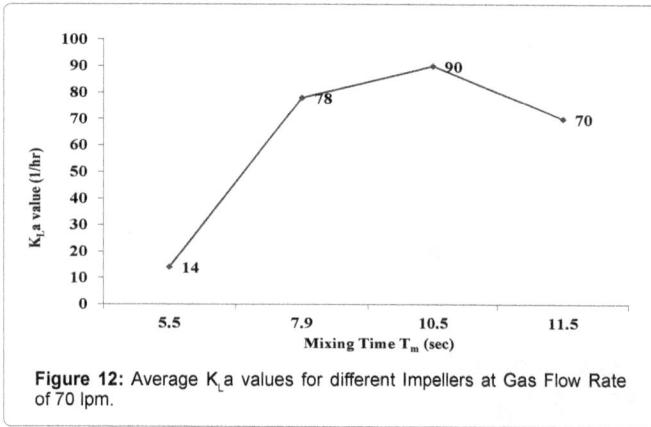

Figure 12: Average K_La values for different Impellers at Gas Flow Rate of 70 lpm.

Impellers	T_m(sec)	k_La at 20 lpm	k_La at 10 lpm
Rushton Impeller	16.0	56	79
Marine Impeller	14.7	67	87
A320 Impeller	10.2	63	93
HE3 Impeller	9.1	71	112

Table 3: Additional k_La Testing performed at Gas Flow Rate of 20 lpm and 10 lpm.

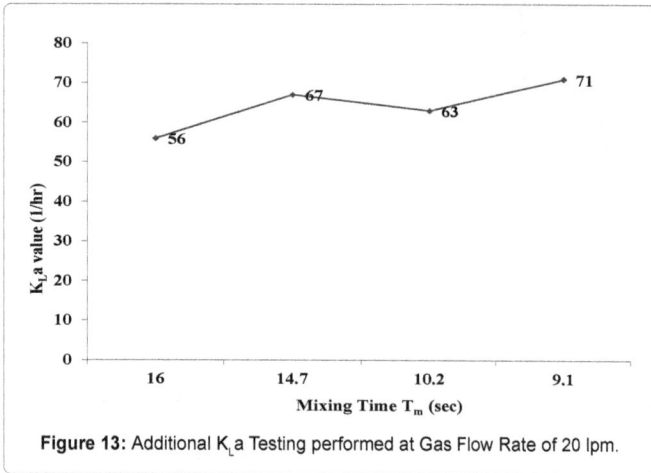

Figure 13: Additional K_La Testing performed at Gas Flow Rate of 20 lpm.

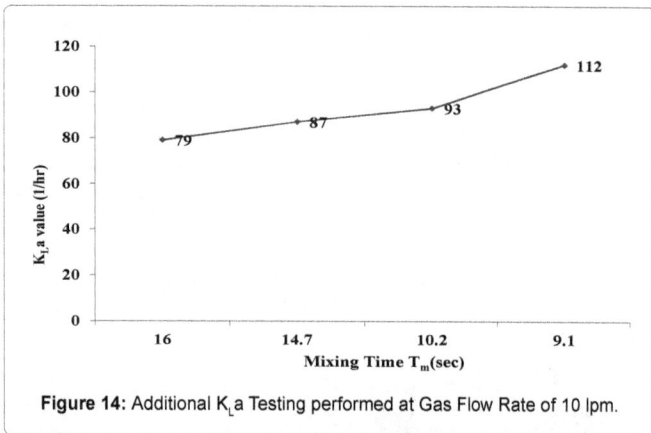

Figure 14: Additional K_La Testing performed at Gas Flow Rate of 10 lpm.

These data also indicated that Rushton Impeller was indeed more efficient for Mass Transfer in Aerobic Stirred Tank Fermenter when used with walled baffles instead of unwalled baffles [35]. The other three impellers i.e. Marine Impeller, A320 Impeller and HE3 Impeller were not as much efficient as Rushton Impeller to provide efficient Mass Transfer in Aerobic Stirred Tank Fermenter with walled as well as unwalled baffle. The configuration of different impellers with baffles is mentioned in Table 1 [17].

Conclusion

Flow pattern calculations for potential operating conditions of Rushton six blade Impeller, Marine Impeller, A320 Impeller and HE3 Impeller in the ellipsoidal bottom tank have been performed to assess mixing behavior. The trailing vortices and turbulence length modeled by k-ε model, SST model and SAS SST model were used to understand

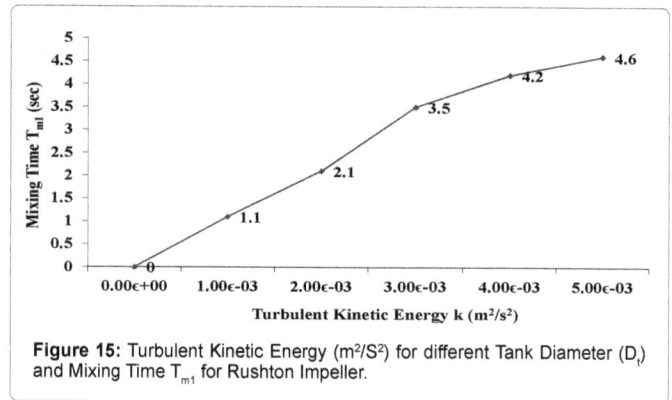

Figure 15: Turbulent Kinetic Energy (m²/S²) for different Tank Diameter (D_t) and Mixing Time T_{m1} for Rushton Impeller.

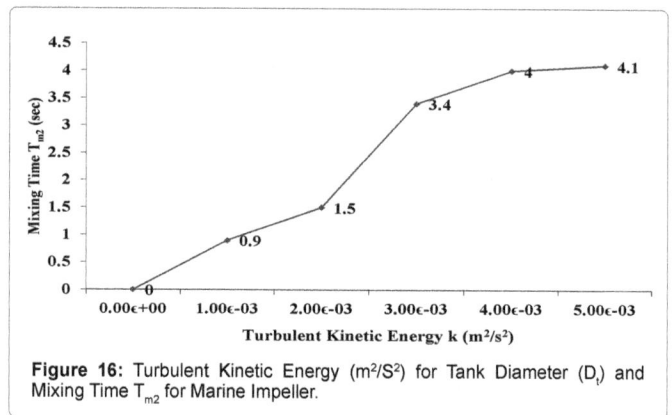

Figure 16: Turbulent Kinetic Energy (m²/S²) for Tank Diameter (D_t) and Mixing Time T_{m2} for Marine Impeller.

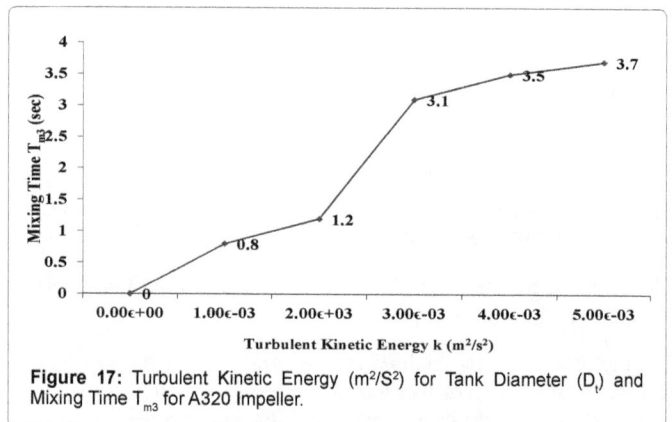

Figure 17: Turbulent Kinetic Energy (m²/S²) for Tank Diameter (D_t) and Mixing Time T_{m3} for A320 Impeller.

Rate (ε) were used to represent Volumetric Mass Transfer Rate. The data from these figures indicated that Mass Transfer was very efficient whenever the Rushton Impeller was used while Marine Impeller, A320 Impeller and HE3 Impeller were somewhere inefficient as compared to latter [58].

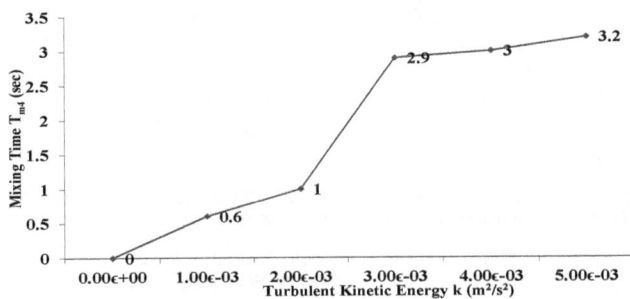

Figure 18: Turbulent Kinetic Energy (m²/S²) for Tank Diameter (D$_t$) and Mixing Time T$_{m4}$ for HE3 Impeller.

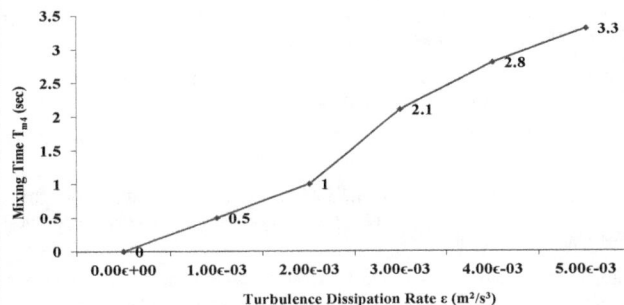

Figure 19: Turbulent Dissipation Rate (m²/S²) for Tank Diameter (D$_t$) and Mixing Time T$_{m1}$ for Rushton Impeller.

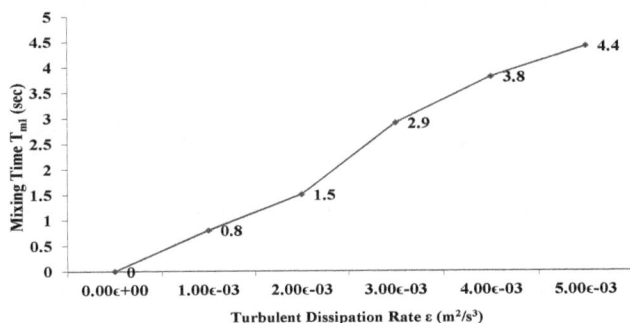

Figure 20: Turbulent Dissipation Rate (m²/S³) for Tank Diameter (D$_t$) and Mixing Time T$_{m2}$ for Marine Impeller.

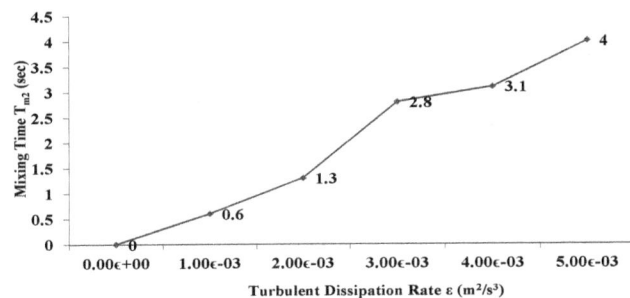

Figure 21: Turbulent Dissipation Rate (m²/S³) for different Tank Diameter (D$_t$) and Mixing Time T$_{m3}$ for A320 Impeller.

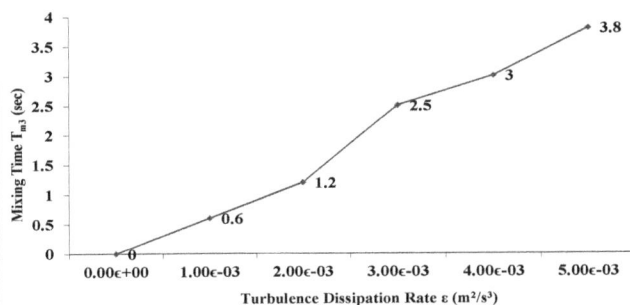

Figure 22: Turbulent Dissipation Rate (m²/S³) for different Tank Diameter (t$_t$) and Mixing Time T$_{m4}$ for HE3 Impeller.

the effect of Impeller and Baffles. Among these Impellers Rushton Impeller was the most efficient. The modeling results were used determine acceptable agitator speeds and tank liquid levels to ensure suspension of solid particles deposited during high solid fermentation.

A few important observations with regard to the effect of fluid viscosity on fermentation suspension in the laminar flow regime have been made in this work. The main interest was axial and mixed-flow pattern of impellers since they are the most important considered for viscous suspension mixing. It was found that the axial flow component for these impellers was suppressed on the bottom of the tank, such that overall flow was predominantly radial. Specifically, this relatively weak distribution of axial velocities at the bottom of the tank may cause the solid particles to stay around the bottom of the tank. This condition becomes more significant with increased solid concentration.

The simulation shows that there is a potential for slow flow or stagnant fluid between the bottom of tank and the fermentor wall and also above the top impeller. In an aerobic fermentation, both of these regions could become depleted of oxygen. High shear rates and energy dissipation rates could be found near both impellers. In all of fermentations, high shear and energy dissipation regions could deactivate the microorganism.

From the experimental k$_L$a studies we see that for a 197 liter tank with a 2:1 height to diameter ratio, 6-blade Rushton impeller used with the walled baffles creates the best conditions for mass transfer. The axial flow impellers so not seem to have any advantage, or disadvantage over the radial flow impellers. Numerically calculated mixing times do not correlate with mass transfer for the gas flow rates used by this STR and thus can only be used to give information on mixing.

References

1. Ali HKhQ, Zulkali MMD (2011) Design Aspects of Bioreactors for Solid-State Fermentation: A Review.Chem Biochem Eng 25: 255-266.

2. Ameur H, Bouzit M, Helmaoui M (2011)Numerical study of fluid flow and power consumption in a stirred vessel with a Scaba 6SRGT impeller. Chemical and Process Engineering 32: 351-366.

3. Aubin J, Fletcher DF,Xuereb C (2004) Modeling Turbulent Flow in Stirred Tanks with Computational Fluid Dynamics: The Influence of the Modeling Approach, Turbulence Model & Numerical Scheme.Experimental Thermal and Fluid Science 28: 431-445.

4. Bakker Andre (2006) Fermenter Specific Modeling Issues.

5. Benz GT (2003) Optimize Power Consumption in Aerobic Fermenters. CEP Magazine, 23-28.

6. Benz GT (2004) Sizing Impellers for Agitated Aerobic Fermenters. CEP Magazine, 18-24.

7. Ariffin MAB (2010) Design, Construction and Commissioning of Bioreactor Experimental Rig to Measure K_La of Oxygen in Newtonian Fluid. Project Report. Universiti Malaysia Pahang.

8. Blakerough N (2012) Fundamentals of Fermenter Design, Department of Chemical Engineering, University of Birmingham, UK.

9. Elqotbi M, Montastruc L, Vlaev SD, Nikov I (2004) CFD simulation of gluconic acid production in a stirred gas-liquid fermenter.

10. Bourdichon F, Casaregola S, Farrokh C, Frisvad JC, Gerds ML, et al. (2012) Food fermentations: microorganisms with technological beneficial use. Int J Food Microbiol 154: 87-97.

11. Breedveld P, Couenne F, Jallut C, Maschke B, Tayakout M (2003) Model of a Continuous Stirred Tank Reactor using Bond Graph Formalism. Biotechnol Bioeng 1009-1105.

12. Chain FB, Paladino S, Callow DS, Ugolini F, Vander Sluis J (1952) Studies on aeration. I. Bull World Health Organ 6: 73-97.

13. Charles Marvin, Wilson Jack (1994) Fermentation Design.

14. Chisti Y, Moo-Young M (1993) Aeration and mixing in vortex fermenters. J Chem Technol Biotechnol 58: 331-336.

15. Dale MC, Zhou C (1995) Design of a Pilot Scale Continuous Stirred Ethanol Reactor Separation with Solvent Absorption & Extractive Distillation. CoFE 1-6.

16. Angelique D, Delvigne F, Collignon ML, Crine M, Thonart P, et al. (2010) Development of compartment model based on CFD simulation for description of mixing in biorectors. Biotechnol Agron Soc Environ 14: 517-522.

17. Dutta R (2008) Fundamentals of Biochemical Engineering, Amity Institute of Biotechnology.

18. Egedy A, Marcus DS, Dominique A (2011) Application of Models with Different Complexity for a Stirred Tank Reactor. Hungarian Journal of Industrial Chemistry 39: 335-339.

19. Ein-Mozaffari F, Upreti SR (2010) Investigation of mixing in shear thinning fluids using computational fluid dynamics. Intech Open.

20. Gaden EL Jr (2000) Fermentation process kinetics. Reprinted from Journal of Biochemical Microbiological Technology and Engineering VOl. 1, No. 4 Pages 413-29 (1959). Biotechnol Bioeng 67: 629-635.

21. Gill NK, Appleton M, Baganz F, Lye GJ (2008) Quantification of power consumption and oxygen transfer characteristics of a stirred miniature bioreactor for predictive fermentation scale-up. Biotechnol Bioeng 100: 1144-1155.

22. Gimbun J, Radiah ABD, Chuah TG (2004) Bioreactor design via spreadsheet--a study on the monosodium glutamate (MSG) process. Journal of Food Engineering 64: 277-283.

23. Guha D, Dudukovic MP, Ramachandran PA, Mehta S, Alvare J (2004) CFD based Compartmental Modelling of Single Phase Stirred Tank Reactors. AIChE Journal 52: 1836-1846.

24. Hristov HT, Mann R, Lossev V, Vlaev SD (2004) A Simplified CFD for Three-dimensional Analysis of Fluid Mixing, Mass Transfer and Bioreaction in a Fermenter Equipped with Triple Novel Geometry Impellers. Food and Bioproducts Processing 82: 21-34.

25. Hugo A, Mork M, Grislingas A (2003) Stirred Tank Reactor, Reactor Technology.

26. Karcz J, Siciarz R, Bielka I (2004) Gas Hold-Up in a Reactor with Dual System of Impellers. Chem Pap 58: 404-409.

27. Karimi A, Golbabaei F, Mehrnia MR, Neghab M, Mohammad K, et al. (2013) Oxygen mass transfer in a stirred tank bioreactor using different impeller configurations for environmental purposes. Iranian J Environ Health Sci Eng 10:6.

28. Kaushal P, Sharma HK (2012) Concept of Computational Fluid Dynamics (CFD) and its Applications in Food Processing Equipment Design. J Food Process Technol 3:138.

29. Kokoo R, Narataruksa P, Pana-Suppamassadu K, Tungkamani S (2008) Determination of optimum rotational speed of heterogeneous catalytic reactor using computational fluid dynamic. Songlanakarin J Sci Technol 755-760.

30. Kozma L, Nyeste L, Szentirmai A (2006) Optimization Problems of Fermenter Aeration-Agitation System. Hungarian Journal of Industrial Chemistry 34: 35-39.

31. Kunik S, Mudroncik D, Kopcek M, Stremy M (2007) Virtual and Analogue Model of a Continuous Stirred Tank Reactor. 16th Int. Conference Process Control 1-4.

32. Le Roux JMW, Purchas K, Nell B (1986) Refrigeration requirement for Precooling & Fermentation Control in Wine Making. S Afr J Enol Vitic 7: 6-13.

33. McNeil B, Harvey LM (2008) Practical Fermentation Technology. John Wiley & Sons, USA.

34. Moilaneu P, Laakkonen M, Aittamaa J (2005) Modelling fermenters with CFD. European Symposium on Computer Aided Process Engineering-15.

35. Montante G, Coroneo M, Francesconi JA, Paglianti A, Magelli F (2012) CFD Modelling of a Novel Stirred reactor for the Bioproduction of Hydrogen. 14th European Conference on Mixing 311-316.

36. Nikhil TR (2007) Software Design of Stimulating Microbial Process in Bioreactor. Environmental Informatics Archive 5.

37. Nurtono T, Nirwana WOC, Anwar N, Nia SM, Widjaja A, et al. (2012) A CFD Study into a Hydrodynamic Factor that affects a Biohydrogen Production Process in a Stirred Tank Reactor. Procedia Engineering 50: 232–245

38. Ochieng Aoyi, Onyango Maurice S (2010). CFD Simulation of Solid Suspension in Stirred Tanks.

39. Patarinska T, Trenev V, Popova S (2010) Software Sensors Design for a Class of Aerobic Fermentation Processes. International Journal of Bioautomation 99-118.

40. Ranade VV, Tayalia Y, Krishnan H, (2002) CFD Predictions of Flow near Impeller Blades in Baffled Stirred Vessels. Chemical Engineering Communications 189: 895-922.

41. Ranade VV, Perrard M, Le SN, Xuereb C, Bertrand J (2002) Trailing vortices of Rushton turbines. Chemical Engineering Research and Design 79: 3-12.

42. Saarela U, Kauko L, Esko J (2003) Modeling of a Fed Batch Fermentation Process.

43. Sahlin Peter (1999) Fermentation as a Method of Food Processing. Production of Organic Acids pH-development & Microbial Growth in Fermenting Cereals.

44. Vanea SM, Lena BE, Rolland M (2012) Numerical CFD Simulation of a Batch Stirred Tank Reactor with Stationary Catalytic Basket. Chemical Engineering Journal 207: 596-606.

45. Madhavi SV, Ranade VV (2012) Computational Fluid Dynamics Modelling of Solid Suspension in Stirred Tanks. Current Science 102: 1539-1551.

46. Shukla S, Nayak D, Khan HS (1998) Numerical prediction of flow fields in baffled stirred vessels: a comparison of alternative modeling approaches. Chemical Engineering Science 53: 3653-3684.

47. Srinophakun T, Jessada J (2000) Computational Fluid Dynamics for Mixing Behaviour in Bakers' Yeast Fermentation.

48. Stanbury PF, Hall S, Whittakar A (1999) Principles of Fermentation Technology. (2nd edition).

49. Sungkorn R, Derksen JJ, Khinast JG (2012) Modelling of Aerated Stirred Tanks with Shear thinning Power Law Liqiud. International Journal of Heat and Fluid Flow 36: 153-166.

50. UdayaBhaskar RR, Goplakrishnan S, Ramaswamy E (2002) CFD Analysis of Turbulence Effect on Reaction in Stirred Tank Reactor.

51. Vamsi KK, Boraste A, Jhadav A (2009) Construction & Standardization of a Bioreactor for Production of Alkaline Protease from Bacillus licheniformis. International Journal of Microbiology Research 1: 32-44.

52. Verschuren IL (2001) Feed Stream Mixing in Stirred Tank Reactor. Int of Biochem Eng 44: 56-72.

53. Walter PH (2005) Fermentation based manufacturing processes & fermentation products. GMFFTL: 1-50.

54. Wan TEL, Kumar P, Samyudia Y (2011) Computational Fluid Dynamics of Mixing in Aerated Bioreactors. International Conference on Biology, Environment & Chemistry.

55. Warfvinge P (2012) Reactor Calculation, The continuous Flow, Stirred Tank Reactor CSTR, Department of Chemical Engineering, Lund University.

56. Lin X1, Liu H, Zhu F, Wei X, Li Q, et al. (2012) Enhancement of biodesulfurization by Pseudomonas delafieldii in a ceramic microsparging aeration system. See comment in PubMed Commons below Biotechnol Lett 34: 1029-1032.

57. Zdenka P, Roman P (2009) Modeling & Simulation of Dry Anaerobic Fermentation. 24 European Conference on Modelling& Simulation.

58. Zhou Z, Mirac N, Iminia F, Goniya J (2013) Experimental Study & CFD Simulation of Mass Transfer Characteristics of a Gas-Induced Pulsating Flow Bubble Column. Chem Biochem Eng 27: 167-175.

3

Fructose-Mediated Elevation of Hydrogen Production in Glucose Tolerant Mutant of *Synechocystis* Sp. Strain PCC 6803 under the Dark Anaerobic Nitrate-Free Condition

Adipa Chongsuksantikul*, Kazuhiro Asami, Shiro Yoshikawa and Kazuhisa Ohtaguchi

Department of Chemical Engineering, Tokyo Institute of Technology, Japan

Abstract

Fructose is a potential additive that elevates a supply of electrons to hydrogenase for hydrogen production in cyanobacteria. A series of dark anaerobic hydrogen production experiments was performed to evaluate the validity of this assumption, in which fructose from 0 to 110 mmol/L was added to HEPES buffer solution on which a glucose tolerant mutant of unicellular cyanobacterium *Synechocystis* sp. strain PCC 6803 (GT strain) was incubated. Despite the reported knowledge that fructose was an inhibitor for photoheterotrophic growth of *Synechocystis* cells, GT strain assimilated fructose and represented a limited heterotrophic growth on fructose in HEPES buffer solution under dark anaerobic condition. The initial hydrogen production rate that was 0.025 mmol/L h in run without fructose increased to 0.0917 mmol/L h in run with 60-83 mmol/L fructose. The associated increase in the initial amount of endogenous glucose from 0.22 mmol/L in run without fructose to 0.36 mmol/L in run with 50 mmol/L fructose was observed. Fructose released the complete suppression of hydrogen production by nitrate. This work presents the first experimental evidence that cells of GT strain are able to assimilate fructose for cell growth in dark anaerobic condition. Our results show that hydrogen production in *Synechocystis* sp. strain can be significantly elevated by a proper addition of fructose to dark anaerobic HEPES buffer solution.

Keywords: Dark; Fructose; Hydrogen; Lactate; NiFe-hydrogenase; Nitrate-free; *Synecocystis* sp. strain PCC 6803

Introduction

Cyanobacteria are capable of carbon dioxide fixation under the light, utilizing water as a primary electron donor, and generating free oxygen, ATP and low potential reductive compounds NADH, NADPH and $FADH_2$ for metabolism. A number of cyanobacteria express an extremely oxygen-labile bidirectional NiFe-hydrogenase activity that, under the absence of oxygen and highly reductive condition, serves to provide a terminal electron sink and reduce excess protons and NAD(P) H by evolving hydrogen [1]. Stresses caused by the nitrate or sulfate limitations are effective to trigger hydrogen production in a unicellular non-nitrogen-fixing cyanobacterium *Synechocystis* sp. strain PCC 6803 [2,3]. Eliminating nitrate from medium constituents is established to be a key for dark hydrogen production, which is supported by an observation that a deletion mutant of the genes encoding both nitrate reductase and nitrite reductase significantly enhanced production [4]. Inhibitor addition to block of a certain electron sink pathway that is competitive to NiFe-hydrogenase is applied to increase hydrogen production [5].

Fermentation of the endogenous glucose in the form of glycogen that is accumulated during photosynthesis is shown to be accompanied by the dark anaerobic hydrogen production in a unicellular cyanobacterium *Gloeocapsa alpicola* CALU 743 [6]. Dark anaerobic incubation of the glucose tolerant mutant of *Synechocystis* sp. strain PCC 6803 (hereafter GT strain) in the nitrate-free and glucose-supplemented HEPES (4-(2-hydroxyethyl)-1-piperazineethanesulfonic acid)) buffer solution results in an eminent increase in hydrogen production on bidirectional NiFe-hydrogenase [7]. Fructose is an isomer of glucose. Heterotrophic growth of the nitrogen-fixing filamentous cyanobacteria *Anabaena* species on fructose under the light and dark was previously reported [8], however, there are remarkably few data available on the dark heterotrophic growth and the dark hydrogen production of GT strain on fructose.

Fructose, along with glucose and sucrose, are three key sugars for photolithotrophs. Fructose is an extremely soluble ketohexose that exists in a cyclized form hemiacetal fructofuranose in solution. The fructofuranose has the anomeric carbon directly bonding to two oxygen atoms in the forms of a free OH group and an OR group. This anomeric carbon is susceptible to be oxidized via a redox reaction; hence ketohexose fructose is classified into a reducing sugar. If hydrogen production on NiFe-hydrogenase is the cellular response against redox unbalance, then a possibility exists that the presence of fructose in cellular environment affects hydrogen production [9].

Exogenous glucose and fructose are known to be transported through facilitated diffusion *via* a glucose-fructose permease (the product of the *glc*P transport gene) [10]. Previous works show that *Synechocystis* sp. strain PCC6 714 transports glucose via permease with K_m of about 0.4 mmol/L [11] and fructose inhibits the uptake of glucose slightly [12]. The K_m for fructose consumption of *Anabaena variabilis* is reported to be approximately 0.16 mmol/L that does not change in the light versus dark [13]. Although sensing mechanism in GT strain for glucose is not defined, growth arrested cells are known to respond to glucose at the level of post-translational modifications or modulation of enzyme activities and phosphorylate the transported glucose to glucose 6-phosphate on glucokinase [14]. Glucose 6-phosohate is

*Corresponding author: Adipa Chongsuksantikul, Department of Chemical Engineering, Tokyo Institute of Technology, Tokyo 2-12-1 Ookayama, Meguro-Ku, Tokyo 152-8552, Japan

further decomposed mainly in the Oxidative Pentose Phosphate (OPP) pathway [15]. Fructose inhibits photoautotrophic growth of a culture of *Synechocystis* sp. strain PCC 6714 but, cannot result in a toxic effect on the chemoheterotrophic growth of the strain on 5 mmol/L glucose in the light [14]. Genome sequence of *Synechocystis* sp. strain PCC 6803 represents a potential-coding regions for fructokinase (Qry match, 18.4%) in ORF slr1448 [16]. Glycogen accumulation in a heterocystous cyanobacterium *Anabaena azollae* is reported to be increased by fructose [17]. A recent work on higher plant *Arabidopsis thaliana* suggests that fructose is a signaling molecule regulating growth and development and that fructokinase plays a putative regulatory role in fructose signaling [18]. However fructose signal transaction pathway is yet unsolved.

In the present paper we report results of experiments which were designed to provide additional information on the fructose supported hydrogen production in *Synechocystis* sp. strain PCC 6803-GT under the dark anaerobic condition with or without nitrate.

Materials and Methods

Strain, cell preparation and hydrogen production

A glucose tolerant mutant of *Synechocystis* sp. strain PCC 6803 (hereafter GT strain) was supplied by Professor Y. Hihara, Department of Biochemistry and Molecular Biology, Saitama University, Japan. For cell preparation, cells were grown photoautotrophically for 3 days in BG-11 medium (initial pH 7.8) at 34°C, aerated by 6 % CO_2 in air and illuminated by fluorescent lamps at 100 μmol photons/m² photosynthetic photon flux density (PPFD). Utilizing the culture with OD_{730} at 3, logarithmically growing cells were collected by centrifugation at 1000 g and 25°C for 10 min. The cell pellets were first washed by 50 mmol/L HEPES buffer plus 18 mmol/L $NaNO_3$ and then washed again by pure HEPES (4-(2-hydroxyethyl)-1-piperazineethanesulfonic acid) buffer (pH 7.8). The washed cells were collected by centrifugation. Cells were re-suspended and incubated in 10 mL of 50 mmol/L HEPES buffer (pH 7.8) solutions without or with 18 mmol/L $NaNO_3$, without or with 5, 25, 50, 60, 70 and 110 mmol/L fructose and with 18 mmol/L $NaNO_3$ plus 50 mmol/L fructose in 32 mL glass test tubes with butyl rubber caps. The initial dry cell weight concentration was set at 2 g/L (OD_{730}=5.4) and cell suspension was purged with nitrogen gas for a few minutes to remove oxygen molecules. Incubation was carried out under the dark anaerobic conditions with shaking at 145 rpm in a reciprocating shaker. The reciprocating distance was 40 mm and the horizontal angle was about 30°. Dark incubation in this study was slightly modified with a daily pulse of dim light. All experiments were carried out in duplicate.

Measurement of hydrogen production

The amount of molecular hydrogen in the gas phase of the closed 32 mL test tube was measured utilizing a gas chromatograph equipped with a molecular sieve column and a thermal conductivity detector (GC-320, GL science Inc.; column, Molecular sieve 13X; column temperature, 37°C; injector temperature, 45°C; detector temperature 80°C) with nitrogen gas as the carrier gas.

Measurement of fructose consumption and metabolic product production

The amount of unreacted fructose in cell suspensions was analyzed by a high performance liquid chromatography (HPLC) (Shimadzu LC-10AD and RID-6A) with a 6.0φ×150 mm column SZ5532 (Showa Denko Co.). Column temperature was set at 50°C. Mobile phase was 70% acetonitrile. Mobile phase flow rate was 0.7 mL/min. A 10 μL of supernatant sample was injected by using a 10 μL syringe. Metabolites in supernatants were analyzed by HPLC (JASCO PU-2080 Plus and UV-2075 Plus). Mobile phase was 18 mmol/L KH_2PO_4, pH of which was adjusted at 2.3 with H_3PO_4. Flow rate of mobile phase was 0.7 mL/min. Column temperature was 30°C. A 10 μL of supernatant sample was injected by using 10 μL syringes. The wave length of 210 nm was used for analyzing lactate and acetate concentrations.

Measurement of endogenous glucose accumulation and consumption

For glucose analysis, defined dry weight cell pellets were collected by centrifugation of defined volume of cell suspension at 25°C, 3000 rpm for 10 min and washing 3 times by deionized water to eliminate extracellular carbon sources. From defined amount of cell pellets, the endogenous glucose was obtained by extracting and decomposing glycogen with 50 μL of 6 N HCl at 80°C for 30 min. The amount of endogenous glucose in glycogen per culture volume was determined by applying Glucose CII Test Wako (Wako Pure Chemical Ind., Ltd.). The almost 100% conversion of glycogen to glucose by this method is confirmed by employing glycogen assay kit containing amyloglucosidase on glycogen (data not shown).

Measurement of dry cell weight

The dry cell weight per culture volume was determined according to previous report [3].

Results and Discussions

Figure 1 compares the initial hydrogen production rates of dark anaerobically incubated cells of *Synechocystis* sp. strain PCC 6803-GT in HEPES buffer solutions at different conditions. The number of moles of hydrogen per culture volume versus time data of those four runs gives linear relation in the first 48 h (data not shown), hence the initial hydrogen production rates are calculated by the amount of hydrogen per culture volume at 48 h, divided by 48 h. The hydrogen production data of the runs without fructose support the reported

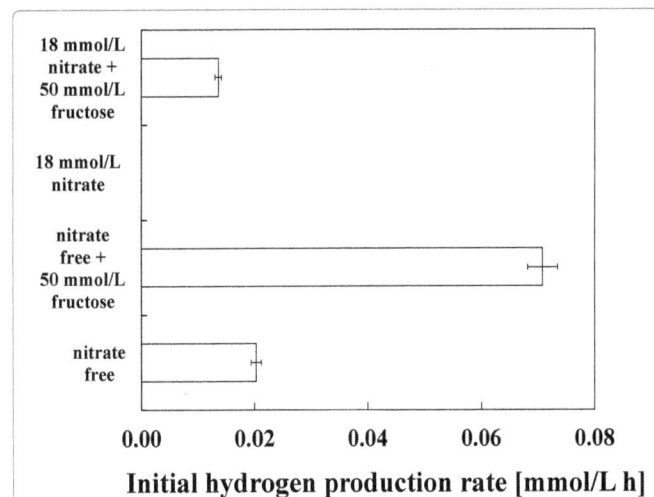

Figure 1: Effect of nitrate and fructose on the initial hydrogen production rate in *Synechocystis* sp. PCC 6803-GT. The hydrogen production was evaluated by the measurements of the amount of hydrogen per culture volume after 48 h incubation of cells in HEPES buffer solutions (a) without nitrate or fructose, (b) without nitrate and with 50 mmol/L fructose, (c) with 18 mmol/L $NaNO_3$ and without fructose, and (d) with 50 mmol/L fructose and 18 mmol/L $NaNO_3$. All experiments were carried out in duplicate.

conclusion that hydrogen production is inhibited by nitrate because of the NAD(P)H competition between nitrate assimilation and hydrogen production [4]. The result of run without nitrate shows that fructose elevates dark anaerobic hydrogen productions in GT strain. The initial hydrogen production rate in run with fructose is 3.5 times that without fructose. The positive effect of fructose on the dark anaerobic hydrogen production is found to overcome the inhibitory effect of nitrate. This observation appears to be the first description on the dark anaerobic hydrogen production in GT strain in the presence of nitrate (Figure 1).

Figure 2 shows the time courses of the number of moles of hydrogen per culture volume (y_{H2}). Hydrogen production starts right after inoculation, irrespective of varying c_{S0}. The time course of hydrogen production rate (r_{H2}=dy_{H2}/dt) is studied. The initial hydrogen production rates (r_{H20}) for runs with c_{S0} at 0, 5, 25, 50, 60 and 70 mmol/L are 0.025, 0.031, 0.059, 0.091, 0.093 and 0.094 mmol/L h, respectively. The addition of 70 mmol/L fructose results in 3.7-fold increase in the initial hydrogen production rate from run without fructose. The 48 h dark anaerobic incubation of *Synechocystis* cells in HEPES buffer solution without fructose and N-deprivation condition in this work generated 0.927 μmol hydrogen from (7) (10^9) cells, which was acceptable when compared to 2.37 μmol hydrogen from (7) (10^9) cells shown in previous report (OD$_{730}$=0.1 equal to (1.4)(10^8) cells/mL) [5]. This minor difference can be explained by the difference of temperature, gas phase composition and growth phase of *Synechocstis* cells. The r_{H2} is not time-varied and same as the r_{H20} at 0.0927 mmol/L h when c_{S0} is no less than 50 mmol/L. The y_{H2} versus time data for runs with 50, 60 and 70 mmol/L fructose are fitted by the relation:

$$y_{H2} = r_{H2,0}t \tag{1}$$

The straight broken line in Figure 2 fits well with observed values of runs with c_{S0} at 50, 60 and 70 mmol/L. The number of moles of hydrogen per culture volume at 120 h of run without fructose was 11.1 mmol/L. When fructose concentration was further increased, the observed y_{H2} versus time data were fitted well with Equation (1) for runs with c_{S0} at no more than 83 mmol/L. The rate of hydrogen production in GT strain is found to be highest and extremely stable against the progress of hydrogen production when initial fructose concentration is set at between 50 and 83 mmol/L. The upper limit of fructose concentration for the stable hydrogen production was observed at 110 mmol/L, where cell weight concentration decreased with time and the initial hydrogen production rate dropped significantly (data not shown). This cell death represented that GT strain could not survive under the solution of high fructose concentration.

The y_{H2} versus time data for run without fructose is fitted by the relation:

$$y_{H2} = \frac{r_{H2,0}}{k}\{1-\exp(-kt)\} \tag{2}$$

in which k is the rate constant for the deactivation of hydrogen production. The smooth broken line in Figure 2 shows calculated y_{H2} with the parameters r_{H20}, 0.025 mmol/L h; and k, 0.0151 1/h, which shows a fairly good fit to observed y_{H2}. The y_{H2} at 120 h of run without fructose is 1.7 mmol/L. The hydrogen production with 50-83 mmol/L fructose is found to be 6.54 times that without fructose. The long dashed short dashed line in Figure 2 is the calculated time course of y_{H2} utilizing Equation (2) with the reference parameters r_{H20}, 0.0405 mmol/L h and k, 0.0186 1/h, which were obtained from the results of experiments with late-logarithmically growing cells. [9]. The present observed y_{H2} of run without fructose and with logarithmically growing cells is lower than that on the long dashed short dashed line.

The difference between broken lines (present work) and long dashed short dashed line (previous work) is highly suggestive of the inoculum-growth-phase dependency of hydrogen production.

The result of run with 5 mmol/L fructose shows that effect of fructose is not obvious in the first 72 h, while it becomes apparent after 72 h. The y_{H2} versus time data is on the curve of Equation (2) before 72 h and appears to approach to the straight line of Equation (1) after 72 h. Interestingly, hydrogen production rate of runs with 5 and 25 mmol/L fructose became almost similar to that of runs with 50-70 mmol/L fructose.

The result of run with 25 mmol/L fructose shows that effect of fructose is obvious right after inoculation, however $r_{H2,0}$ of Equation (1) is lower than 0.0927 mmol/L that fits the result of run with c_{S0} no less than 50 mmol/L. After 72 h, r_{H2} of this run gradually increases with time.

Of greatest interest is the fructose-mediated activation and stabilization of hydrogen production after 72 h. Fructose is found to elevate the number of moles of hydrogen per culture volume at 120 h by increasing the initial hydrogen production rate, and stabilizing it after 72 h (Figure 2).

To elucidate relations of hydrogen production to fructose utilization, acid evolution, endogenous glycogen metabolism and dry cell weight production, culture variables concerning these mechanisms are totally monitored. For runs with fructose, run with 25 mmol/L fructose is selected as a representative run because time courses of culture variables of runs with 5, 50, 60 and 70 mmol/L fructose qualitatively resembled those of run with 25 mmol/L fructose. Figure 3 compares the time courses of culture variables measured in run (a) without or (b) with 25 mmol/L fructose in nitrate-free HEPES buffer solutions under the dark anaerobic conditions. In the absence of fructose, lactate concentration was extremely low and reached only 0.2 mmol/L at 120 h, whereas, in the presence of fructose, lactate synthesis continuously occurred during 120 h. The lactate production rate after 48 h was similar to the hydrogen production rate. The level of lactate concentration is seen to be the same order of magnitude of number of moles of hydrogen per culture volume. Increase in the lactate

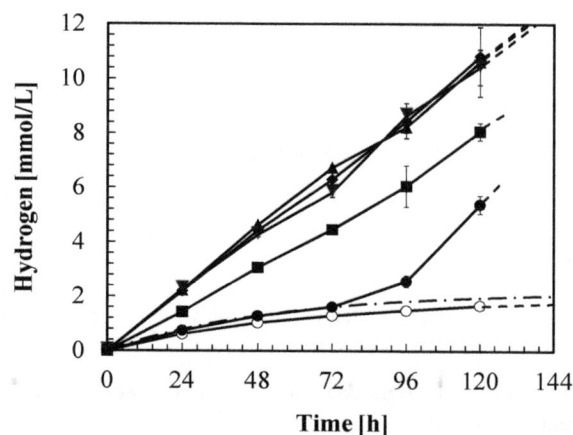

Figure 2: Time courses of number of moles of hydrogen per culture volume. Cells were incubated under the dark anaerobic condition in nitrate-free HEPES buffer solution supplemented with 0(○), 5(●), 25(■), 50(♦), 60(▲) or 70 (▼) mmol/L fructose. All experiments were carried out in duplicate.

concentration in the presence of fructose suggests that exogenous fructose is an additional carbon source for lactate production. The amount of intracellular lactate was not detectable level.

Different from lactate, acetate concentration rose to a maximum level at 72 h and then decreased. The highest concentration of acetate is 0.2 mmol/L, hence the time course of acetate concentration is omitted from Figure 3.

The time course data of dry cell weight concentration versus time of run without or with 25 mmol/L fructose shows that the dry cell weight concentration in run without fructose decreases slightly with time, whereas it in run with fructose slightly increases in the first 24 h and in the period between 72 and 120 h. An interesting feature of Figure 3b is a limited heterotrophic growth of GT strain on fructose under the dark anaerobic condition. This research is carried out to find the condition to elevate hydrogen production; hence cells were incubated not in BG-11 medium but in HEPES buffer solution that contained only HEPES and KOH for pH adjustment. Fructose assimilation is seen in the first 24 h and in the final period between 72 and 120 h. If $(-r_{S0}=-dc_S/dt)$ is the initial fructose consumption rate, the $(-r_{S0})$ of run with 25 mmol/L fructose is 0.174 mmol/L h. A cessation of fructose assimilation is seen between 24 and 72 h. The unreacted fructose concentration in this period is 23.9 ± 0.56 mmol/L. The combination of the curves for fructose concentration and dry cell weight concentration versus time in Figure 3b eminently demonstrates the limited heterotrophic growth on fructose in the first 24 h and in the period between 72 and 120 h.

This discontinuous heterotrophic growth on fructose suggests that GT strain appear to respond on cell cycle events to the initial shift of cellular environment to dark anaerobic condition with fructose.

The number of moles of endogenous glucose per culture volume of inoculum cells was fixed at 0.16 mmol/L by collecting the cells from the photoautotrophic culture of OD_{730} at 3. The number of moles of endogenous glucose per culture volume for run without or with 25 mmol/L fructose increased right after dark incubation, reached the maximum value at 24 h $(c_{G,max})$ of 0.22 mmol/L in run without fructose and 0.31 mmol/L in run with 25 mmol/L fructose, rapidly decreased from 24 h to 48 h, and then approached to the stable level after 48 h. The stable level observed at 120 h is higher in run with 25 mmol/L fructose than in run without fructose. The $c_{G,max}$ increases 1.40-fold from run without fructose by adding 25 mmol/L fructose. For run without fructose, it is difficult to consider any carbon source for this glycogen synthesis other than endogenous carbonaceous compounds such as amino acids in protein, fatty acids in membrane lipid and derivatives of poly-β-hydroxybutyrate (PHB). Turnover of cell constituting materials is known to activate many NAD(P)H-generating enzymes. If r_{G0} and $r_{G0,0}$ are the initial rates of the accumulation of endogenous glucose and that in the absence of fructose, and if $r_{GS,0}$ is an increment of r_{G0} from $r_{G0,0}$, then the experimental data of run with 25 mmol/L fructose results that r_{G0}, $r_{G0,0}$ and $r_{GS,0}$ are 0.00625, 0.0025, 0.00375 mmol/L h, respectively.

Glycogen in terms of endogenous glucose of runs without and

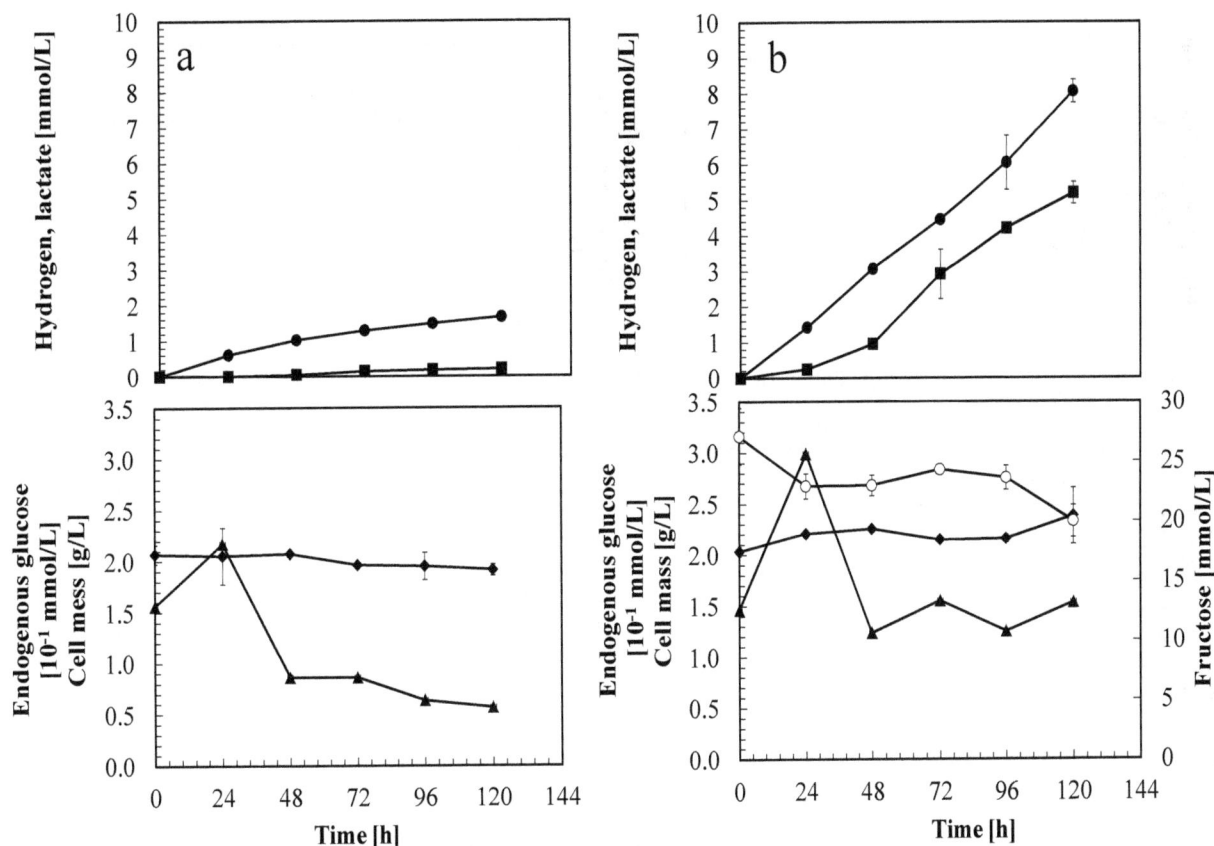

Figure 3: Time courses of the number of moles of hydrogen per culture volume (●), the concentrations of lactate (■), dry cell weight (♦) and unreacted fructose (○), and the number of moles of endogenous glucose per culture volume (▲). Cells were incubated under the dark anaerobic nitrate-free HEPES buffer solutions supplemented without (a) or with 25 mmol/L fructose (b). All experiments were carried out in duplicate.

with 25 mmol/L fructose is drastically decomposed between 24 and 48 h. This drop is from 0.22 mmol/L to 0.09 mmol/L in the run without fructose and from 0.31 mmol/L to 0.15 mmol/L in the run with 25 mmol/L fructose. The assimilated amount of endogenous glucose per culture volume of run with fructose is comparable with that without fructose. Glycolysis is found to be independent from the presence of exogenous fructose and to proceed with the characteristic reaction rate of 0.00604 mmol/L h and for the degradable glycogen amount of 0.145 mmol/L. When the degradable glycogen is assimilated, glycolysis is terminated.

Glycogen decomposition is known to generate NADH. The hydrogen production rate between 24 h and 48 h in run with 25 mmol/L fructose is 0.0604 mmol/L h that is 2.39 times that in run without fructose (Figure 2), hence the hydrogen production rate after 24 h is found to be independent from the amount of NADH generated by glycolysis. Considering that the amount of endogenous glucose per culture volume is increased by increasing initial fructose concentration and that the amount of assimilated endogenous glucose per culture volume after 24 h is not affected by initial fructose concentration, it is found that the amount of endogenous glucose per culture volume after 24 h is always higher in run with fructose than in run without fructose. Similar to the time course of the number of moles of endogenous glucose per culture volume, that per dry cell weight rises to a maximum at 24 h and then decreases (data not shown) (Figure 3).

Figure 4 shows the effect of change in initial fructose concentration

on culture characteristic variables observed in runs without and with 5, 25, 50, 60 and 70 mmol/L fructose. Culture parameters states at 0 h and 120 h are shown by subscript 0 and 1 respectively. Fructose increased the initial hydrogen production rate and the total increase in number of molecules of hydrogen per culture volume (Figure 4a).

The plot of initial fructose consumption rate ($-r_{s_0}$) versus initial fructose concentration (c_{s_0}) results the K_m for fructose consumption of 12.5 mmol/L and the maximum consumption rate ($-r_{s_0}$)$_{max}$ of 1.07 mmol/L h. Broken line in Figure 4b is the ($-r_{s_0}$) calculated by the Michaelis-Menten kinetics, which fits well with observed consumption rate. The K_m of GT strain for fructose in HEPES buffer solution is found to be greater than that reported for *Anabaena variabilis* [13]. Low affinity of fructose to fructokinase in our experiment appears to be caused by limitation of ATP that is imperative for activation of fructokinase, since HEPES buffer solution lacks any phosphate ions.

The activated production of hydrogen in the presence of 50-83 mmol/L fructose is found to be accompanied by the elevated assimilation of fructose (Figure 4b), the elevated initial synthesis of glycogen (Figure 4c), followed by the assimilation of endogenous glucose after 24 h and the elevated production of lactate (Figure 4d). Because of the insufficiency of required elements for normal cell growth, small portion of assimilated fructose is directed to cell growth (Figure 4b). When cells are exposed to fructose in buffer solution, the initial glycogen synthesis is activated, while the subsequent glycogen decomposition was not affected by initial concentration of fructose

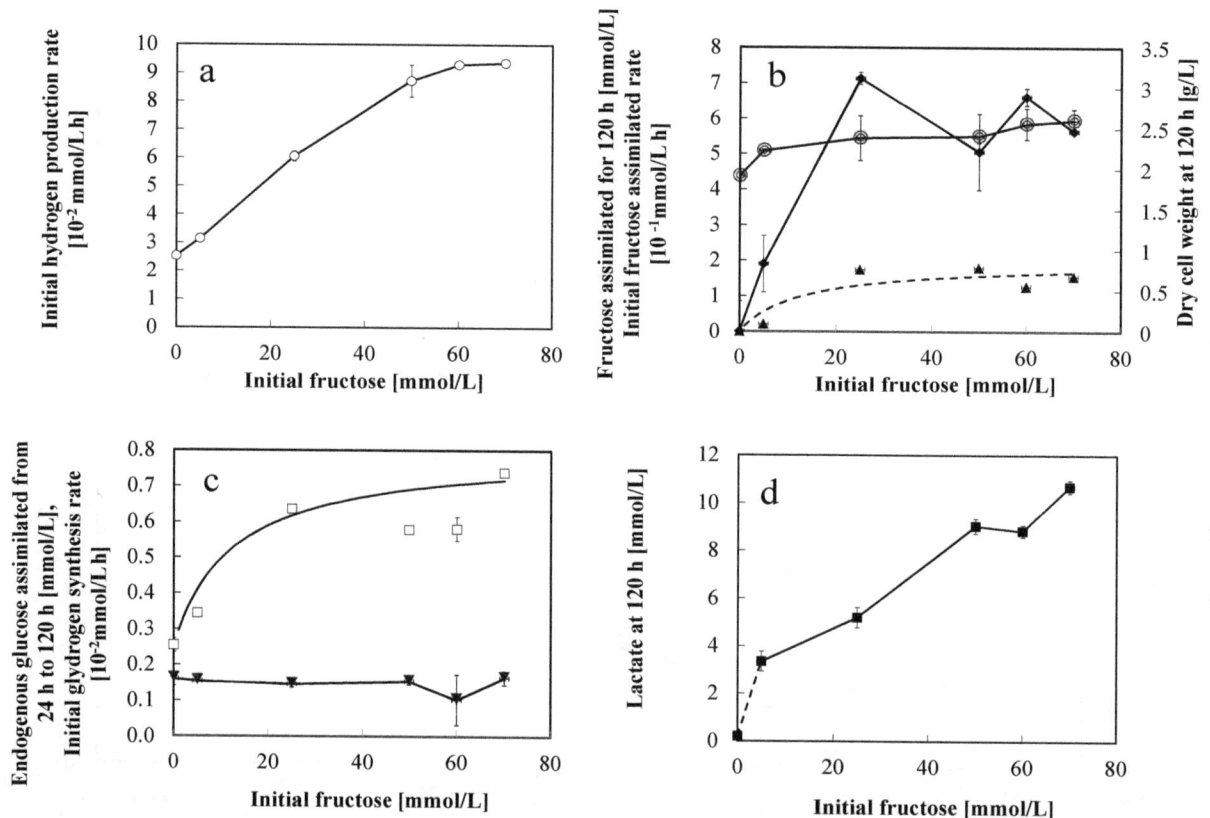

Figure 4: Relation between the initial fructose concentration and the characteristic variables of hydrogen production in *Synechocystis* sp. strain PCC 6803-GT under the dark anaerobic nitrate-free condition. Characteristic variables are the initial hydrogen production rate (○), the number of moles of assimilated fructose per culture volume for 120 h (◎), the initial rate of fructose consumption that is observed from 0 h to 24 h (▲), the dry cell weight concentration at 120 h (◎), the initial endogenous glucose accumulation rate (□), the number of moles of assimilated endogenous glucose per culture volume from 24 to 120 h (▼) and the lactate concentration at 120 h (■).

(Figure 4c).

In the absence of fructose, endogenous glucose is synthesized in the first 24 h with the rate of 0.0025 mmol/L h utilizing carbons from turnover mechanism, since the yield of endogenous glucose from fructose to endogenous glucose ($=r_{G0}/(-r_{S0})$) is calculated as 0.0279 ± 0.0146, which is quite low. Fructose is not mainly turned to endogenous glucose.

The time course of the number of moles of endogenous glucose per culture volume in Figure 3 shows that endogenous glucose consumed drastically between 24 and 48 h. The assimilated amount of endogenous glucose from 24-120 h are not affected by initial fructose concentration, hence the amount of endogenous glucose per culture volume is found to be always higher in the presence of fructose than in the absence of fructose (Figure 4).

The number of moles of hydrogen per culture volume at 120 h is plotted against the number of moles of endogenous glucose per culture volume at 120 h in Figure 5a. The number of moles of hydrogen per culture volume at 120 h correlates well with the number of moles of endogenous glucose per culture volume at 120 h. Figure 5b also shows the relation between the r_{G0} and the number of moles of hydrogen per culture volume at 120 h. Good correlation is seen in this plot. To keep high level in the amount of endogenous glucose per culture volume appears to be of prime importance to result high hydrogen production in GT strain (Figure 5).

The assimilation of endogenous glucose is obvious between 24 h and 48 h, hence the instantaneous fractional yield of hydrogen on endogenous glucose $Y_{H2/G}$ that is defined by the amount of hydrogen produced per amount of endogenous glucose utilized in this period is evaluated and plotted in Figure 6. Glucose 6-phosphate is an intermediate both in gluconeogenesis pathway and glycolysis pathway. If glucose 6-phosphate is catabolized in the Oxidative Pentose Phosphate (OPP) pathway and if both NADPH and ATP are required in this pathway, theoretical conversion of 1 mole glucose 6-phosphate to 5/3 mole pyruvate is accompanied by the generation of 2 mole NADPH and 5/3 mole NADH. When 1 mole pyruvate is completely converted to 3 mole carbon dioxide, this reaction yields 4 mole NADH and 1 mole $FADH_2$. If all of NAD(P)H generated in these pathways are directed to NiFe-hydrogenase, potential yields of hydrogen in the pathway from glucose 6-phosphate to pyruvate and from glucose 6-phosphae to carbon dioxide are estimated as 3.67 and 12, respectively. The yield of hydrogen for run without fructose in Figure 6 is slightly lower than the theoretical yield from glucose 6-phosphate to pyruvate and that for runs with 50-70 mmol/L fructose is much higher than the theoretical yield from glucose-6-phosphate to carbon dioxide. Our observation suggests that NAD(P)H for hydrogen production are supplied not only from glycogen degradation but also from turnover of proteins, membrane lipid and PHB. This turnover appears to be accompanied by glycolysis. The observed lactate level was high at high initial fructose concentration. High concentration fructose appears to activate NiFe-hydrogenase and lactate dehydrogenase to discharge excess reductants generated in the incomplete tricarboxylic acid (TCA) cycle [13] and in the pathways for turnover of above components.

Two additional experiments were done in which GT strain was added to the dark anaerobic HEPES buffer solution supplemented with 5 mmol/L $FeCl_2$ and that with 5 mmol/L ascorbic acid, in order to observe if the exposure of cells to reductive compounds is related to hydrogen production. When autoclaved, Fe^{2+} (green) solution turned to Fe^{3+} (red brown) due to unstable form of Fe^{2+}. There were

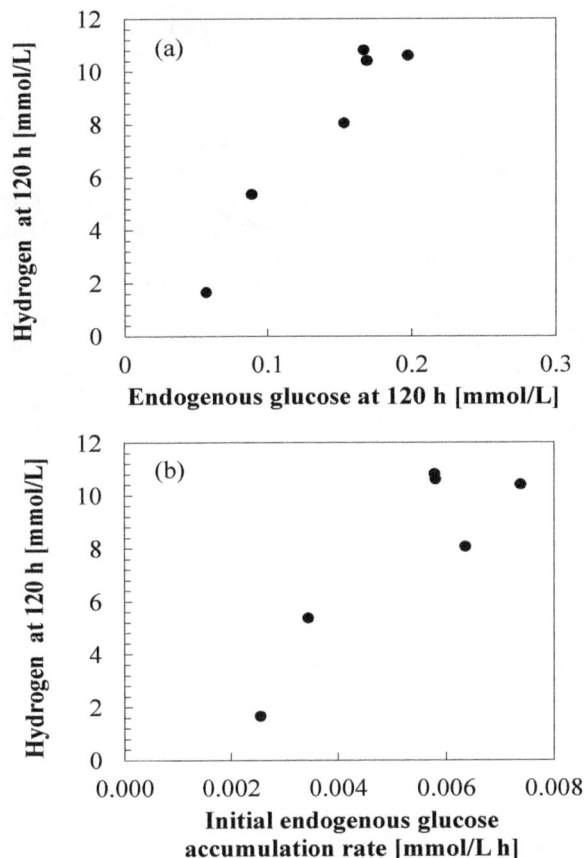

Figure 5: Relation between the number of moles of hydrogen per culture volume at 120 h versus (a) endogenous glucose at 120 h and (b) the initial endogenous glucose accumulation rate. R^2 were 0.93 and 0.79 for (a) and (b), respectively.

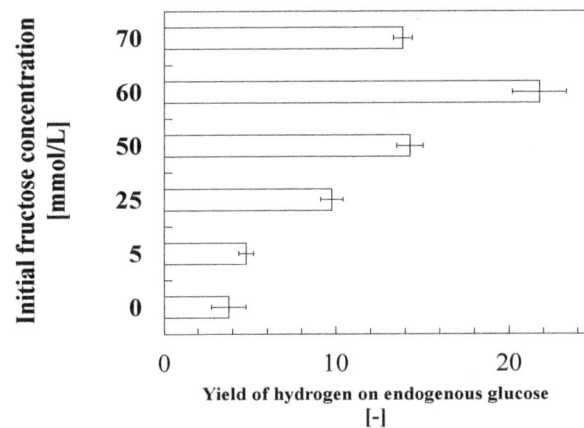

Figure 6: Relation between initial concentration of fructose and yield of hydrogen on endogenous glucose in the period between 24 h and 48 h.

no improvements of hydrogen production in the presence of those reductive compounds (data not shown). The positive effect of fructose on hydrogen production in our result suggests that fructose transport through cellular membrane is of prime importance to elevate hydrogen production.

Previously, we reported the elevation of dark anaerobic production of hydrogen in GT strain by adding glucose in HEPES buffer solution [7]. Hydrogen production in cells under glucose fermentation is different from that in fructose fermentation because cells extensively utilize glucose for cell division and thus, more cell population results in more hydrogen production (data not shown). In fructose under dark anaerobic condition, cells utilize fructose for endogenous glucose buildup rather than limited cell growth. This endogenous glucose accumulation is followed by the endogenous glucose break down by glycolysis that triggers many intracellular associated turnover reactions against cell constituting components. Cells gain more electrons through glycolysis and associated turnover reactions. This is partially supported by the increase in lactate concentration in runs with fructose since lactate is one of key carbon terminal compounds. Since fructose inhibited photoheterotrophic growth, our observation suggests that fructose might suppress an important pathway or mechanism related to light mode of life. Interestingly, this growth inhibition was not observed under dark anaerobic mode of life. Suppression of specific mechanism by fructose might be a key for redirection of more electrons to hydrogenase enzyme (Figure 6).

Conclusions

This work has demonstrated that fructose is a carbon source for a heterotrophic growth of a glucose tolerant mutant of cyanobacterium *Synechocystis* sp. strain PCC 6803 in the dark anaerobic HEPES buffer solution. Growth was observed but limited because only HEPES buffer with fructose was utilized as a medium. This observation differs from previous works that showed that fructose inhibited photoheterotrophic cell growth. Right after dark incubation, cells consumed fructose for endogenous glucose accumulation and dry cell weight production. Affinity to fructose was extremely low. At the end of dark incubation, the number of moles of endogenous glucose per culture volume correlates well with the number of moles of hydrogen per culture volume. To increase hydrogen production, a shift-up of the number of moles of endogenous glucose per culture volume is found to be of utmost importance. Our result shows that addition of fructose into HEPES buffer solution is effective to increase the number of moles of endogenous glucose per culture volume. Further experiments are necessary before the roles of fructose and endogenous glucose for hydrogen production are known.

References

1. Ducat DC, Sachdeva G, Silver PA (2011) Rewiring hydrogenase-dependent redox circuits in cyanobacteria. Proc Natl Acad Sci U S A 108: 3941-3946.

2. Antal TK, Oliveira P, Lindblad P (2006) The bidirectional hydrogenase in the cyanobacterium *Synechocystis* sp. strain PCC 6803. International Journal of Hydrogen Energy 31: 1439-1444.

3. Yamamoto T, Asami K, Ohtaguchi K (2012) Anaerobic production of hydrogen in the dark by *Synechocystis* sp. strain PCC6803: effect of photosynthesis media for cell preparation. J Biochem Tech 3: 344-348.

4. Baebprasert W, Jantaro S, Khetkorn W, Lindblad P, Incharoensakdi A (2011) Increased H_2 production in the cyanobacterium *Synechocystis* sp. strain PCC 6803 by redirecting the electron supply via genetic engineering of the nitrate assimilation pathway. Metab Eng 13: 610-616.

5. Burrows EH, Chaplen FW, Ely RL (2011) Effects of selected electron transport chain inhibitors on 24-h hydrogen production by *Synechocystis* sp. PCC 6803. Bioresour Technol 102: 3062-3070.

6. Troshina O, Serebryakova L, Sheremetieva M, Lindblad P (2002) Production of H_2 by the unicellular cyanobacterium *Gloeocapsa alpicola* CALU 743 during fermentation. International Journal of Hydrogen Energy 27: 1283-1289.

7. Yamamoto T, Chongsuksantikul A, Asami K, Ohtaguchi K (2012) Anaerobic production of hydrogen in the dark by *Synechocystis* sp. strain PCC 6803 supplemented with D-glucose. J Biochem Tech 4: 464-468.

8. Ungerer JL, Pratte BS, Thiel T (2008) Regulation of fructose transport and its effect on fructose toxicity in *Anabaena* spp. J Bacteriol 190: 8115-8125.

9. Chongsuksantikul A, Asami K, Yoshikawa S, Ohtaguchi K (2014) Hydrogen production by anaerobic dark metabolism in *Synechocystis* sp. strain PCC6803-GT: effect of monosaccharide in nitrate free solution. J Biochem Tech 5: 685-692.

10. Joset F, Buehou T, Zhang CC, Jeanjean R (1988) Physiological and genetic analysis of the glucose-fructose permeation system in two *Synechocystis* species. Archives of Microbiology 149: 417-421.

11. Beauclerk AA, Smith AJ (1978) Transport of D-glucose and 3-O-methyl-D-glucose in the cyanobacteria Aphanocapsa 6714 and Nostoc strain Mac. Eur J Biochem 82: 187-197.

12. Flores E, Schmetterer G (1986) Interaction of fructose with the glucose permease of the cyanobacterium *Synechocystis* sp. strain PCC 6803. J Bacteriol 166: 693-696.

13. Haury JF, Spiller H (1981) Fructose uptake and influence on growth of and nitrogen fixation by *Anabaena variabilis*. J Bacteriol 147: 227-235.

14. Kahlon S, Beeri K, Ohkawa H, Hihara Y, Murik O, et al. (2006) A putative sensor kinase, Hik31, is involved in the response of *Synechocystis* sp. strain PCC 6803 to the presence of glucose. Microbiology 152: 647-655.

15. Ungerer JL, Pratte BS, Thiel T (2008) Regulation of fructose transport and its effect on fructose toxicity in *Anabaena* spp. J Bacteriol 190: 8115-8125.

16. Kaneko T, Sato S, Kotani H, Tanaka A, Asamizu E, et al. (1996) Sequence analysis of the genome of the unicellular cyanobacterium *Synechocystis* sp. strain PCC6803. II. Sequence determination of the entire genome and assignment of potential protein-coding regions (supplement). DNA Res 3: 185-209.

17. Rozen A, Arad H, Schonfeld M, Tel-Or E (1986) Fructose supports glycogen accumulation, heterocysts differentiation, N2 fixation and growth of the isolated cyanobiont *Anabaena azollae*. Archives of Microbiology 145: 187-190.

18. Cho YH, Yoo SD (2011) Signaling role of fructose mediated by FINS1/FBP in *Arabidopsis thaliana*. PLoS Genet 7: e1001263.

19. Zhang S, Bryant DA (2011) The tricarboxylic acid cycle in cyanobacteria. Science 334: 1551-1553.t

4

A Highly Stable Biocatalyst Obtained from Covalent Immobilization of a Non-Commercial Cysteine Phytoprotease

Walter David Obregón[1], José Sebastián Cisneros[1], Florencia Ceccacci[1] and Evelina Quiroga[2*]

[1]Research Laboratory Vegetable Proteins (LIPROVE), Faculty of Sciences, National University of La Plata (UNLP), 47 and 115 s / N, La Plata (B1900AVW), Argentina
[2]Membranes and Biomaterials Laboratory, Institute of Applied Physics (INFAP) -CONICET, National University of San Luis (UNSL), Almirante Brown 907, San Luis (D5700HHW), Argentina

Abstract

In this work, *araujiain* (enzymatic preparation obtained from the latex of *Araujia hortorum* fruits) was successfully immobilized on glyoxyl-agarose via multipoint covalent attachment. Thus, good efficiency of immobilization and high operational stability of immobilized enzyme were obtained. The activity of *araujiain* at alkaline pH was significantly improved after immobilization. In addition, immobilized *araujiain* also showed high activity and good stability, without significant loss in its activity, at temperatures between 37 and 60°C and in the presence of immiscible organic solvents. Immobilized *araujiain* also showed good performance in a mixture of 50% ethyl acetate in buffer, used for peptide synthesis, with better results than when the free enzyme was used as catalyst. These results indicate that immobilized *araujiain* via multipoint covalent attachment can be highly stabilized and this method might be used for practical applications of *araujiain* in hydrolytic and synthetic processes.

Keywords: *Araujiain*; Protease; Enzyme technology; Immobilized enzymes; Enzymatic stabilization

Introduction

Proteases are a family of enzymes that play a prominent role in plant physiology. These enzymes are catalysts of processes such as storage of proteins during seed germination, activation of proenzymes, and degradation of defective proteins [1,2]. Due to their good solubility and stability, proteases are widely used in medicine and industry. In particular, alkaline proteases have applications in leather processing [3], laundry detergents [4], production of protein hydrolysates [5] and food processing [6-8]. The recent application of proteases to the production of certain peptides and peptides derivatives has received great attention as a viable alternative to the chemical approach [9-13]. This wide range of applications insures proteases the first place in the world market of enzymes.

In the last decades, proteolytic plant enzymes have received special attention due to their property of being active in a wide range of temperature and pH values [13-15]. Although the market of commercial proteases includes a high number of proteases with high proteolytic activity and available at a low cost, there is a need to discover new plant sources of more active and more specific proteases.

As the number of practical applications increase, the requirement for bulk amounts of enzyme becomes a limiting factor. Therefore, immobilization is considered to be one possible way to allow the reuse of the enzyme [16-19]. In addition, because immobilization induces rigidification of the enzyme, avoids enzyme aggregation and autolysis, provides considerable stability towards temperature variations and organic solvents and generates a more suitable environment, a high enzymatic stabilization may be achieved [20,21].

Immobilization on solid carriers is perhaps the most used strategy to improve the operational stability of biocatalysts [20,22]. Among the immobilization methods available, multipoint covalent attachment is the most effective in terms of thermal stabilization [23], although gel-entrapped has also been reported to provides thermal stabilization [24,25].

Glyoxyl or glutaraldehyde activated supports have been proven to be quite efficient in increasing tertiary enzyme stability via multipoint covalent attachment, since the formation of additional covalent bonds

increases the rigidity of the immobilized enzyme [23]. Immobilization on glyoxyl-carriers occurs at alkaline pH, via the area on the proteins surface richest in lysines. The main advantage of this protocol is the high stability usually achieved.

In this contribution, an immobilized and highly stabilized biocatalyst of *araujiain* on glyoxyl-agarose via multipoint covalent attachment was accomplished. This enzymatic preparation consisting on three papain-like cysteine peptidases was obtained from the latex of *Araujia hortorum* fruits (a climbing plant that grows in Brazil, Paraguay, Uruguay and Argentina) [26-28]. *Araujiain* has been demonstrated to be a successful biocatalyst for the synthesis of amide bonds in aqueous-organic media [9]. N,N-dimethylformamide (with low water content) and mixtures of the Tris-HCl buffer (0.1 M, pH 8.5) and hexane, ethyl acetate or propanone in 50:50 ratio were selected to perform the peptide synthesis catalyzed by *araujiain*. The maximum conversion (35%) was obtained using ethyl acetate as organic medium [9]. In later work, *araujiain* was immobilized on diverse supports: i) deposited onto polyamide *araujiain* demonstrated to be good catalyst for the condensation of coded and non- coded Cbz-amino acids and amines such as amino alcohols and amino acetals in acetonitrile containing 1% (v/v) water [13]; ii) the use of titanium dioxide as support for the immobilization of *araujiain* led to an immobilized biocatalyst with a high protein concentration but with partial deactivation with respect to the native enzyme [29]; iii) entrapped within alginate beads *araujiain* showed good performance in the peptide synthesis using aqueous-organic media and a secondary structure with a high α-helical character was responsible for the

*Corresponding author: Evelina Quiroga, Membranes and Biomaterials Laboratory, Institute of Applied Physics (INFAP) -CONICET, National University of San Luis (UNSL), Almirante Brown 907, San Luis (D5700HHW), Argentina

highest activity of entrapped *araujiain* [25,30]. Although, it has been proved the remarkable enzymatic activity of *araujiain* immobilized on diverse supports, there are no studies in the literature reporting the multipoint covalent immobilization of *araujiain*.

Because of the high biotransformation potential of *araujiain*, the present study is a natural continuation of our previous work. The main objective was to improve catalytic performance and stability of *araujiain* through multipoint covalent immobilization on glyoxyl-agarose for its application to the peptide synthesis in organic medium. The results were compared with those obtained with free *araujiain*. To our knowledge, this is the first report dealing with the immobilization of *araujiain* on glyoxyl-agarose and its use in the biocatalysis.

Material and Methods

Materials

Synthetic substrates L-pyroglutamyl-L-phenylalanyl-L-leucine *p*-nitroanilide (PFLNA), L-phenylalanine methyl ester (Phe-OMe·HCl), *N*-(benzyloxycarbonyl)-L-alanine (Z-Ala) and *N*-(benzyloxycarbonyl)-L-alanine *p*-nitrophenyl ester (ZAla-*p*No) were supplied by Bachem (California, USA). Bovine Serum Albumin (BSA) and casein from bovine milk were purchased from Sigma Chemical Co. The rest of the chemicals used in this work were of analytical grade and solvents (hexane and ethyl acetate) were of HPLC grade.

Proteolytic extract preparation

Araujiain is a proteolytic extract obtained from the latex of fruits of *Araujia hortorum* Fourn. (*Asclepiadaceae*) containing papain-like cysteine proteases. The latex obtained by superficial incisions of fruits, gathered in 0.1 M citrate-phosphate buffer (pH 6.5) containing 5mM EDTA and cysteine as preservatives, was centrifuged at $16,000 \times g$ (30 min at 4°C) in order to discard gums and other insoluble materials. The supernatant was then ultracentrifuged ($100,000 \times g$ (60 min at 4°C)) and this new supernatant, called *araujiain*, was fractionated and conserved at -20°C for further studies. Cation exchange chromatography of *araujiain* reveals the presence of three fractions with proteolytic activity, according to previously described studies by Priolo and Obregón [26,28]. The protein content was estimated according to Bradford's assay using BSA as standard [31]. Whereas that the proteolytic activity of *araujiain* was determined using casein as substrate in Tris-HCl buffer (0.1 M, pH 8) at 37°C. Caseinolytic activity was expressed as an arbitrary enzymatic unit (Ucas), according to Priolo et al. [32].

Immobilization of *araujiain*

Araujiain was immobilized on glyoxyl-agarose gel via multipoint covalent attachment. Glyoxyl-agarose gel was prepared by etherification of agarose 10 BLC (Hispanagar) with glycidol and further oxidation of the resulting glyceryl-agarose gel with periodates [33]. The immobilization process was carried out according to the protocol used by Guisan [23] with modifications [34]. For this purpose, 10 g of the activated support were suspended in 25 mL of 0.1 M sodium bicarbonate buffer (pH 10) containing an appropriate amount of *araujiain* so as to get a ratio of 35 mg enzyme/g carrier. The obtained suspension was gently stirred at 25°C during 20 h. Aliquots of the supernatant and whole suspension were withdrawn at different times and the catalytic activity was measure as described below. The protein content of *araujiain* solution was determined before and after its contact with the support. The difference in the protein content of the

araujiain solution (before and after of immobilization) was considered as the amount of the immobilized enzyme. Finally, derivatives were reduced for 30 min at 25°C adding 0.025 M sodium borohydride ($NaBH_4$). The gel was washed with distilled water and 0.1 M Tris-HCl buffer (pH 8.0).

Hydrolytic activity assays of immobilized araujiain

In order to make more accurate the study of the effect of immobilization on the enzyme activity, the hydrolytic activity of immobilized *araujiain* was measured using a specific chromogenic substrate for papain-like thiol proteases (L-pyroglutamyl-L-phenylalanyl-L-leucine *p*-nitroanilide (PFLNA)). The assay was carried out according to Obregón [15]: 0.1 mL of biocatalyst (0.12 g immobilized *araujiain*/mL) was added to 1.5 mL of phosphate buffer (0.1 M, pH 6.5) containing 0.3 M KCl, 10 mM EDTA, 3 mM DTT (Dithiothreitol) and 0.18 ml of 4 mM PFLNA. For comparison, an adequate dilution of free *araujiain* was also tested under same conditions. The liberation of *p*-nitro aniline from PFLNA by hydrolysis of the thiol protease was estimated spectrophotometrically at 410 nm during 3 min of reaction at 37°C in an orbital shaker at 160 rpm. One unit of protease activity (IU) was defined as the amount of protease which liberated one micromole of *p*-nitro aniline per minute in the assay conditions. These assays were carried out by triplicate, and Standard Deviation (SD) was calculated.

Effect of pH and temperature

The effect of pH on activity of free and immobilized *araujiain* was studied in the pH range from 5 to 11, after 3 min of reaction at 37°C, using PFLNA as substrate in 25 mM sodium salts of "Good" buffers (MES, MOPS, TAPS, AMPSO and CAPS) [35].

The effect of temperature was tested by measuring free and immobilized *araujian* activity at temperatures varying between 27 and 70°C, after 3 min of reaction in 0.1 M phosphate buffer pH 6.5 and using PFLNA as substrate. Additionally, thermal stability was evaluated as residual activity (%) under standard assay conditions, after incubation the enzyme in absence of substrate between 5 and 120 min at 37, 45, 55, 65 and 75°C in 0.1 M phosphate buffer pH 6.5. The enzyme activity prior to incubation was taken as 100 % at each assayed temperature. All reported results were the average values of three replicates for each experimental condition. Variation coefficients (($Sd\ Mean^{-1}$) 100) of reported values were less than 2.5% for activity assays, calculated in each case from triplicate results.

Effect of organic solvents on hydrolytic activity of immobilized *araujiain*

The immobilized enzyme (2.1 IU/g carrier) suspended in 0.1 M phosphate buffer pH 6.5 was incubated for 4 h at 37°C under controlled stirring, with an organic phase (hexane or ethyl acetate) at 1:1 ratio. Samples were withdrawn at desired time intervals to test the hydrolytic activity using PFLNA as substrate.

Peptide synthesis catalyzed by immobilized *araujiain*

Enzymatic peptide synthesis catalyzed by immobilized *araujiain* was carried out in an aqueous-organic medium formed by Tris-HCl buffer (0.1 M, pH 8.5) and ethyl acetate at 1:1 ratio, according to conditions established in previous studies [9,25]. The condensation reaction was initiated by mixing the aqueous phase containing the immobilized enzyme (2.1 IU/g carrier) and 4mM Phe-OMe as amino component, with the organic phase containing the carboxylic component (Z-Ala or Z-Ala-*p*No). The reaction was conducted at 37°C in stopper flask under stirring at 160 rpm during 24 h. At time intervals, aliquots were analyzed by HPLC. Simultaneously, blanks with identical composition but without the enzyme and with only the enzyme in the organic system, were analyzed. Gilson HPLC System (Model 712) equipped with a C-18 Luna (5 μm), 250mm × 4.60mm column (Phenomenex) was used. The eluent was a mixture of acetonitrile and water (1:1 ratio), containing 0.1% (v/v) trifluoroacetic acid (TFA), at flow rate of 0.8 mL/min (254 nm at 25°C).

Results and Discussion

Process of immobilization

The general conditions for the immobilization of *araujiain* on glyoxyl-agarose were selected in function of i) the immobilization of other enzymes on glyoxyl-agarose; ii) previous knowledge on the preparation and characterization of the proteolytic extract from *A. hortorum* (*Asclepiadaceae*) fruit latex, and iii) the non-deactivating conditions for the proteases present in the enzymatic extract: the use of high pH values does not affect the performance of *araujiain* [13,26,28,36]. Since biocatalysts are designed to perform their activities in aqueous medium and in organic media, the pH value is important, since it is well established that enzymes "remember" the pH of the last aqueous solution in which they were dissolved and, after water removal, their ionization state remains unchanged (i.e., the optimum to display its maximum activity) [37].

The proteolytic extract obtained from latex of *A. hortorum* fruits, named *araujiain*, was characterized in terms of its protein content and specific activity. The total protein content of *araujiain* was 5.4 mg protein/mL and the specific activity was 12.8 Ucas/mg protein (68.4 Ucas/mL *araujiain*) when casein was used as substrate. Previous studies of Priolo and Obregón [9] reveal the presence of three fractions with proteolytic activity contained in this enzymatic extract [26,28]. Such fractions (called *araujiain* hI, hII and hIII) have synergistic activity and show better performance at different pH and temperatures in the proteolytic extract that when each working alone and isolated. Although the immobilization of the enzymatic extract obtained in a laboratory containing several enzymes is not an ideal situation, numerous researches have been carried out using this mixture of enzymes [18,38-41]. The properties observed using these crude preparations will be the average of the whole mixture of enzymes able to catalyze the studied reactions.

According to established conditions for the *araujiain* immobilization, a high percentage of the immobilized enzyme was achieved (~96%) (Figure 1). Although 5 h were enough to achieve the maximum amount of immobilized *araujiain*, an additional time of 20 h was used to increase a multipoint attachment. It is well known that long contact periods between enzyme and active support allow the formation of more stable enzyme-support bonds thereby promoting an increase in the stability of the enzyme [42]. After such time, the highest immobilization yield (defined as the ratio between the activity of the immobilized enzyme and the activity of the free enzyme) was obtained (~80%; corresponding to 2.1 IU/g carrier).

Immobilized enzymes are preferred as they can be recycled, resulting in lower production costs. The operational stability of immobilized *araujiain* was determined for thirty consecutive cycles. Each cycle was defined as the number of enzymatic reactions carried out using PFLNA as substrate, as described in the Material and methods section. Up to twenty six consecutive cycles, no loss of enzymatic activity

was observed (Figure 2). Thereafter, a decrease in activity (< 10%) between 27th and 30th cycles was observed. Thus, the immobilized enzyme could be reused for several cycles without substantial loss of activity retaining more than 90% of its initial activity. The decrease (<10 %) in enzyme activity during repeated use might be due to the frailty associated to agarose structure and the behavior of agarose gels under stirring [43].

As it was demonstrated in previous studies, when *araujiain* is immobilized onto TiO$_2$ its amidasic activity is drastically reduced with the number of uses (the immobilized catalyst retained ~20% of its initial activity after five cycles) [29]. When *araujiain* is entrapped into alginate beads, 78% of the initial enzymatic activity is maintained after twenty cycles [25]. Thus, it was demonstrated better operational stability unlike to previous case or some examples in the literature where some entrapped enzymes in alginate show a minor possibility of reuse [25]. Comparing these results with those obtained when agarose was used as support, it can be concluded that the formation of *araujiain*-agarose covalent bonds improved the performance of the immobilized enzyme and allowed its use without loss of activity in successive cycles of hydrolytic reactions.

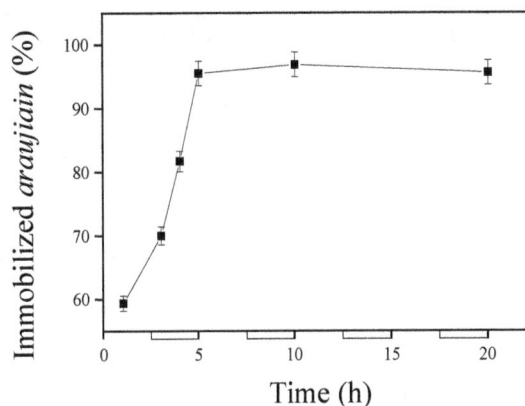

Figure 1: Time course of immobilized araujiain on activated agarose. Immobilized araujiain (%) corresponds to the difference in the protein content of araujiain determined before and after of immobilization.

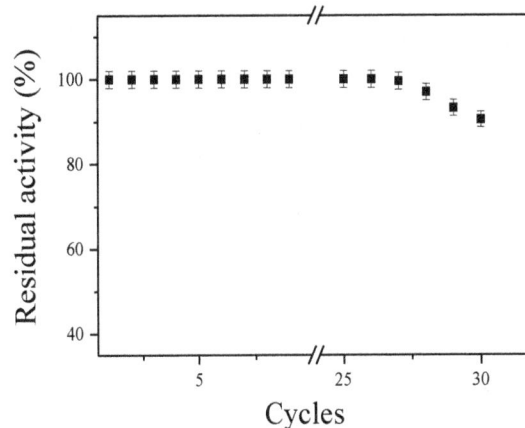

Figure 2: Residual activity (%) of immobilized araujiain after thirty cycles of the hydrolytic reaction. Each cycle was defined as the number of enzymatic reactions carried out during 3 min at 37°C and using PFLNA as substrate.

Effect of pH and temperature

The effect of pH on the activity of free and immobilized *araujiain* was analyzed in the pH range from 5 to 11. The curve of free *araujiain* peaked at pH 7, whereas immobilized *araujiain* shifted at the optimal pH of about 2 units toward the high pH side (from 7 for free *araujiain* to 9 for immobilized *araujiain*) (Figure 3). The residual activity of the free enzyme decreased as pH increased. In contrast, the activity of the enzyme against alkaline pH was significantly improved upon immobilization, probably due to the stabilization of enzyme molecules resulting from the multipoint attachment on the surface of agarose. At lower pH ranges, immobilized *araujiain* showed lower activities than free *araujiain* (Figure 3). A change in pH would affect the intramolecular hydrogen bonding leading to a distorted conformation that would reduce the activity of the enzyme. The conformation of the free enzyme would be more favorable in the low pH range [44].

To analyze the effect of temperature, *araujiain* immobilized on glyoxyl-agarose was subjected to a range of temperatures from 27 to 70°C (Figure 4). The free enzyme was simultaneously tested at those temperatures and then compared with the immobilized form. The optimum temperature for the free enzyme was 37°C. After the immobilization process, a shift in such temperature was observed and immobilized *araujiain* exhibited highest activity at 45°C (Figure 4a). As it can be observed, the immobilized enzyme was able to retain a high activity between 30 and 60°C. Hydrophobic interactions and other secondary interactions of immobilized *araujiain* might impair the conformational flexibility, needing higher temperatures for the enzyme molecule to reorganize and attain a proper conformation to keep its reactivity [24].

The thermal stability of both free and immobilized *araujiain* was also analyzed. As it can be observed in figure 4b, the free enzyme was stable at temperatures ranging between 37 and 55°C, after 120 min of incubation. When the temperature was higher, a drastic inactivation was observed. After the same time of incubation, immobilized *araujiain* was activated, showing a high hydrolytic activity at 55°C, whereas it was able to retain 70% and 40% of initial activity at 65 and 75°C, respectively (Figure 4b). High activity and good stability, without significant loss in its activity, was observed at the temperatures studied. It is important to highlight that *araujiain* immobilized on a matrix of glyoxyl-agarose showed similar performance to that observed when *araujiain* is entrapped in alginate beads [25].

Enzyme thermal inactivation is the consequence of the breaking of the intermolecular forces responsible for maintaining the three-dimensional structure, leading to a reduction in its catalytic capacity [45]. The higher stability of immobilized *araujiain* upon heating may be ascribed to the stabilizing effects of the covalent and secondary interactions between the enzyme and the support [46]. As it is well known, the activation of a support with agents as glycidol generates high concentration of aldehyde groups on the support surface [18,47]. Aldehyde groups in the support and amine groups in the enzyme from lysine residues are a good choice to achieve multipoint attachment and, therefore, to obtain highly thermo-stable enzyme derivatives [48,49]. Thermal stability upon immobilization is the result of molecular rigidity and the creation of a protected microenvironment.

Effect of immiscible organic solvents on the hydrolytic activity of immobilized *araujiain*

The use of organic solvents as reaction media can thus expand the

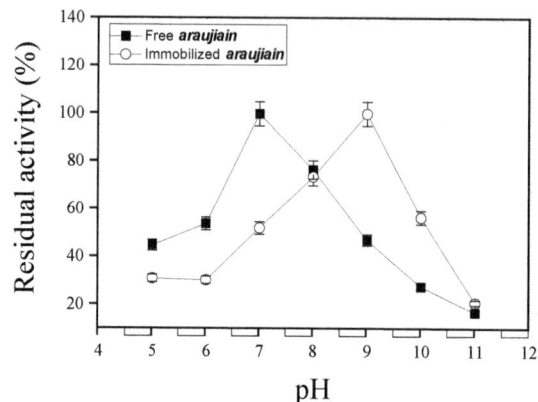

Figure 3: Effect of pH on the activity of free and immobilized araujiain, measured in the pH range from 5 to 11, after 3 min of reaction at 37°C and using PFLNA as substrate.

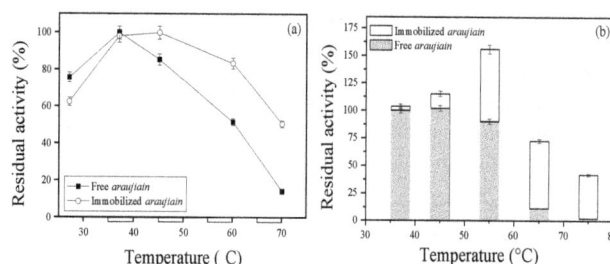

Figure 4: Effect of temperature on the (a) activity (measured at temperatures from 27 to 70°C, after 3 min of reaction) and (b) stability (after 120 min of incubation at temperatures from 37 to 75°C) of free and immobilized araujiain.

repertoire of enzyme-catalyzed transformations [50]. Nevertheless, organic solvents may affect the conformational stability of biocatalysts because these may interact with the hydration layer essential for activity and proper folding, or may alter the protein structure by direct interactions with protein solvation sites [51]. It has been previously demonstrated that free *araujiain* is not inactivated in aqueous-immiscible organic systems and that it has higher activity in these systems than in buffer [36,52,53]. The partition of the organic solvents into the aqueous phase reduces the autolysis degree and produces a considerable activation of *araujiain* [52]. In this opportunity and taking into account such considerations, hexane and ethyl acetate were selected to study the organic solvent effect on the hydrolytic activity of immobilized *araujiain* in aqueous-organic mixtures (0.1 M Tris-HCl buffer (pH 8.5) and water-immiscible organic solvents at 50:50) as a function of time. This was expressed as a percentage of the initial activity in buffer. High activity and stability were observed in both aqueous-organic media (Figure 5). Nevertheless, the highest activity of immobilized *araujiain* was observed in 50% (v/v) hexane. High activity as well as a good stability of the immobilized preparation in organic solvents is one important feature to consider as it allows the use of the enzyme as catalyzer of reactions which are not possible in aqueous phase.

Application of immobilized *araujiain* to biocatalysis

Araujiain immobilized on glyoxyl-agarose was used as catalyzer for the synthesis of a precursor of a bitter dipeptide of commercial interest for the food industry (Z-Ala-Phe-OMe). The synthesis was performed in a mixture of Tris-HCl buffer and ethyl acetate (1:1 ratio). Such medium of reaction was selected on the basis of the results previously obtained when the reaction was catalyzed by free *araujiain* [9]. Although high hydrolytic activity of free *araujiain* is observed in 50% (v/v) hexane, the enzyme is not able to synthesize Z-Ala-Phe- OMe in such medium.

Enzymatic peptide synthesis can be achieved by two different approaches thermodynamically controlled synthesis or kinetically controlled synthesis. In the thermodynamic approach, the acyl donor (an N-terminally protected amino acid) reacts with an acceptor (nucleophile), resulting in the formation of an amide bond. Thermodynamically controlled synthesis is slow and the thermodynamic equilibrium must be shifted towards the synthetic direction by means of product precipitation, water withdrawal, or organic solvent addition [54,55]. In contrast, the kinetic approach involves a C-terminally activated acyl donor that reacts with the nucleophile to give the product in high yields in generally shorter reaction times than the thermodynamic approach [55].

For the kinetically controlled peptide bond formation, Z-Ala-*p*No was used as the carboxylic component. The highest concentration of the dipeptide Z-Ala-Phe-OMe (2.23 mM) corresponding to the maximum product conversion (Xp: 57.5%) was obtained after 1 h of reaction (Figure 6a). The parameter Xp, which represents the amount of product obtained as a function of the limiting reagent, was defined as the relation between the dipeptide concentration (mM) and the initial concentration (mM) of the limiting reagent (Phe-OMe) [9]. The Xp value here reported was 1.67-fold higher than that obtained when the free enzyme was used as catalyst under the same reaction conditions [9]. In addition, *araujiain* immobilized onto agarose showed higher performance than those observed when it was entrapped into alginate beads [25]. After 1h of reaction, a decrease in the amount of dipeptide Z-Ala-Phe-OMe was accompanied by an increase in the by-product Z-Ala-Phe, with a maximum conversion of 32% after 9 h of reaction (Figure 6a).

When the peptide synthesis was thermodynamically controlled, lower yields were obtained (Xp: 25.25%) after 24 h of reaction (Figure 6b). Nevertheless, such value of peptide conversion was 1.74-fold higher than that reported using the free enzyme under the same conditions [9]. After the same reaction time, the results showed here were similar to those obtained when the entrapped *araujiain* was used as catalyzer [25].

Conclusions

In this work, *araujiain* was successfully immobilized on glyoxyl-agarose by multipoint-attachment: i) good efficiency of immobilization and high operational stability of immobilized enzyme were obtained; ii) the activity of *araujiain* at alkaline pH was significantly improved after immobilization; iii) immobilized *araujiain* also showed high activity and good stability at temperatures between 37 and 60°C and in the presence of immiscible organic solvents; iv) immobilized *araujiain* also showed good performance in a mixture of 50% ethyl acetate in buffer, used for peptide synthesis, with better results than when the free enzyme was used as catalyst. This behavior can be related to the covalent linkage and secondary interactions (ionic and polar stabilization, hydrogen bonding, etc.) between the enzyme and the support, which enhance the stability of *araujiain* structure. Thus, considering the low cost of the enzymatic extract used as protease source, the good efficiency of the immobilization and the high operational stability of the immobilized enzyme, this method

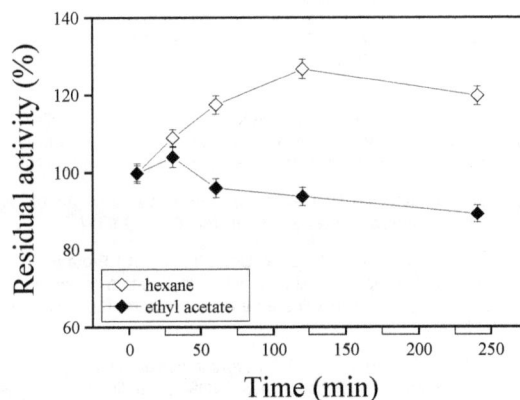

Figure 5: Effect of hexane and ethyl acetate on the hydrolytic activity of immobilized araujiain in aqueous-organic mixtures (0.1 M Tris-HCl buffer (pH 8.5) and water-immiscible organic solvents at 50:50) after 4 h of incubation at 37°C.

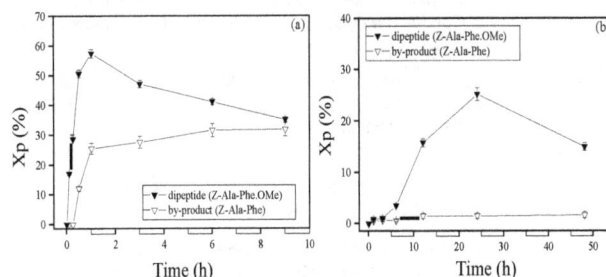

Figure 6: Time-course production of dipeptide Z-Ala-Phe-OMe and by-product (Z-Ala-Phe) in 50% (v/v) ethyl acetate, using (a) Z-Ala-pNo and (b) Z-Ala as carboxylic components and Phe-OMe as amino component. Xp represents the amount of product obtained as a function of the limiting reagent.

might be used for practical applications of *araujiain* in different applications of biotechnological interest.

Acknowledgments

The present work was supported by grants from ANPCyT (PICT 02224), CONICET (PIP 0120), PPID X/004 (UNLP) and REDES VII (number 27-52-265). W.D.O. and E.Q. are career members of CONICET; JSC hold a doctoral fellowship from CONICET.

References

1. van der Hoorn RA (2008) Plant proteases: from phenotypes to molecular mechanisms. Annu Rev Plant Biol 59: 191-223.

2. Parde VD, Aregai T, Abburi MR, Belay K, Saripalli HKRP (2012) Role of the proteolytic enzymes in the living organisms. Int J Int Sci Inn Tech Sec A 4: 32-41.

3. Dayanandana A, Kanagaraj J, Sounderraj L, Govindaraju R, Rajkumar GS (2003) Application of an alkaline protease in leather processing: an ecofriendly approach. J Clean Prod 11: 533-536.

4. Barberis S, Quiroga E, Barcia C, Liggieri C (2013) Effect of laundry detergent formulation on the performance of alkaline phytoproteases. Electron J Biotechnol 16: 1-8.

5. Rocha C, Goncalves MP, Teixeira JA (2011) Immobilization of trypsin on spent grains for whey protein hydrolysis. Process Biochem 46: 505-511.

6. González-Rábade N, Badillo-Corona JA, Aranda-Barradas JS, Oliver-Salvador Mdel C (2011) Production of plant proteases *in vivo* and *in vitro*--a review. Biotechnol Adv 29: 983-996.

7. Feijoo-Siota L, Villa TG (2011) Native and biotechnologically engineered plant proteases with industrial application. Food Bioprocess Technol 4: 1066-1088.

8. Bajaj BK, Jamwal G (2013) Thermostable alkaline protease production from *Bacillus pumilus* D-6 by using agro-residues as substrates. Adv Enzyme Res 1: 30-36.

9. Quiroga E, Priolo N, Obregón D, Marchese J, Barberis S (2008) Peptide synthesis in aqueous-organic media catalyzed by proteases from latex of *Araujia hortorum* (Asclepiadaceae) fruits. Biochem Eng J 39: 115-120.

10. Hernáiz MJ, Alcántara AR, García JI, Sinisterra JV (2010) Applied biotransformations in green solvents. Chemistry 16: 9422-9437.

11. Viswanathan K, Omorebokhae R, Li G, Gross RA (2010) Protease-catalyzed oligomerization of hydrophobic amino acid ethyl esters in homogeneous reaction media using l-phenylalanine as a model system. Biomacromolecules 11: 2152-2160.

12. Xu J, Jiang M, Sun H, He B (2010) An organic solvent-stable protease from organic solvent-tolerant *Bacillus cereus* WQ9-2: purification, biochemical properties, and potential application in peptide synthesis. Bioresour Technol 101: 7991-7994.

13. Morcelle SR, Cánepa AS, Padró JM, Llerena-Suster CR, Clapés P (2013) Syntheses of dipeptide alcohols and dipeptide aldehyde precursors catalyzed by plant cysteine peptidases. J Mol Catal B: Enzym 89: 130-136.

14. Llorente BE, Obregón WD, Avilés FX, Caffini NO, Vairo-Cavalli S (2014) Use of artichoke (*Cynara scolymus*) flower extract as a substitute for bovine rennet in the manufacture of Gouda-type cheese: characterization of aspartic proteases. Food Chem 159: 55-63.

15. Obregón WD, Lufrano D, Liggieri CS, Trejo SA, Vairo-Cavalli SE, et al. (2011) Biochemical characterization, cDNA cloning, and molecular modeling of araujiain aII, a papain-like cysteine protease from *Araujia angustifolia* latex. Planta 234: 293-304.

16. Hanefeld U, Gardossi L, Magner E (2009) Understanding enzyme immobilisation. Chem Soc Rev 38: 453-468.

17. Torres-Salas P, del Monte-Martinez A, Cutiño-Avila B, Rodriguez-Colinas B, Alcalde M, et al. (2011) Immobilized biocatalysts: novel approaches and tools for binding enzymes to supports. Adv Mater 23: 5275-5282.

18. Rodrigues RC, Ortiz C, Berenguer-Murcia Á, Torres R, Fernández-Lafuente R (2013) Modifying enzyme activity and selectivity by immobilization. Chem Soc Rev 42: 6290-6307.

19. Sheldon RA, van Pelt S (2013) Enzyme immobilisation in biocatalysis: why, what and how. Chem Soc Rev 42: 6223-6235.

20. Garcia-Galan C, Berenguer-Murcia A, Fernandez-Lafuente R, Rodrigues RC (2011) Potential of different enzyme immobilization strategies to improve enzyme performance. Adv Synth Catal 353: 2885-2904.

21. Aissaoui N, Landoulsi J, Bergaoui L, Boujday S, Lambert JF (2013) Catalytic activity and thermostability of enzymes immobilized on silanized surface: influence of the crosslinking agent. Enzyme Microb Technol 52: 336-343.

22. Mateo C, Palomo JM, Fernandez-Lafuente G, Guisan JM, Fernandez-Lafuente R (2007) Improvement of enzyme activity, stability and selectivity via immobilization techniques. Enzyme Microb Technol 40: 1451-1463.

23. Guisán JM, Bastida A, Blanco RM, Fernández-Lafuente R, Garcia-Junceda E (1997) Immobilization of enzymes on glyoxyl agarose: Strategies for enzyme stabilization by multipoint attachment. Immobilization of enzymes and cells. Bickerstaff GF (editor) Humana Press, Totowa NJ, USA, pp 277-287.

24. Munjal N, Sawhney SK (2002) Stability and properties of mushroom tyrosinase entrapped in alginate, polyacrylamide and gelatin gels. Enzyme Microb Technol 30: 613-619.

25. Quiroga E, Illanes CO, Ochoa NA, Barberis S (2011) Performance improvement of *araujiain*, a cystein phytoprotease, by immobilization within calcium alginate bead. Process Biochem 46: 1029-1034.

26. Priolo N, Morcelle del Valle S, Arribére MC, López L, Caffini N (2000) Isolation and characterization of a cysteine protease from the latex of *Araujia hortorum* fruits. J Protein Chem 19: 39-49.

27. Priolo N, Arribére CM, Caffini N, Barberis S, Vázquez RN, et al. (2001) Isolation and purification of cysteine peptidases from the latex of *Araujia hortorum* fruits. Study of their esterase activities using partial least-squares (PLS) modeling. J Mol Catal B: Enzym 15: 177-189.

28. Obregón WD, Arribére MC, del Valle SM, Liggieri C, Caffini N, et al. (2001) Two new cysteine endopeptidases obtained from the latex of *Araujia hortorum* fruits. J Protein Chem 20: 317-325.

29. Llerena-Suster CR, Foresti ML, Briand LE, Morcelle SR (2009) Selective adsorption of plant cysteine peptidases onto TiO_2. Colloids Surf B Biointerfaces 72: 16-24.

30. Illanes CO, Quiroga E, Camí GE, Ochoa NA (2013) Evidence of structural changes of an enzymatic extract entrapped into alginate beads. Biochem Eng J 70: 23-28.

31. Bradford MM (1976) A rapid and sensitive method for the quantitation of microgram quantities of protein utilizing the principle of protein-dye binding. Anal Biochem 72: 248-254.

32. Priolo NS, López LMI, Arribére MC, Natalucci CL, Caffini NO (1991) New purified plant proteinases for the food industry. Acta Alimentaria 20: 189-196.

33. Blanco RM, Calvete JJ, Guisán JM (1988) Immobilization-stabilization of enzymes; variables that control the intensity of the trypsin (amine)-agarose (aldehyde) multipoint attachment. Enzyme Microb Technol 11: 353-359.

34. Obregón WD, Ghiano N, Tellechea M, Cisneros JS, Lazza CM, et al. (2012) Detection and characterization of a new metallocarboxypeptidase inhibitor from *Solanum tuberosum* cv. Desirèe using proteomic techniques. Food Chem 133: 1163-1168.

35. Good NE, Izawa S (1972) Hydrogen ion buffers. Methods Enzymol 24: 53-68.

36. Barberis S, Quiroga E, Morcelle S, Priolo N, Luco JM (2006) Study of phytoproteases stability in aqueous-organic biphasic systems using linear free energy relationships. J Mol Catal B: Enzym 38: 95-103.

37. Polizzi KM, Bommarius AS, Broering JM, Chaparro-Riggers JF (2007) Stability of biocatalysts. Curr Opin Chem Biol 11: 220-225.

38. Palomo JM, Segura RL, Mateo C, Terreni M, Guisan JM, et al. (2005) Synthesis of enantiomerically pure glycidol via a fully enantioselective lipase-catalyzed resolution. Tetrahedron: Asymmetry 16: 869-874.

39. Palomo JM, Ortiz C, Fernandez-Lorente G, Fuentes M, Guisan JM, et al. (2005) Lipase-lipase interactions as a new tool to immobilize and modulate the lipase properties. Enzyme Microb Technol 36: 447-454.

40. Masuda M, Sakurai A, Sakakibara M (2001) Effect of enzyme impurities on phenol removal by the method of polymerization and precipitation catalyzed by *Coprinus cinereus* peroxidase. Appl Microbiol Biotechnol 57: 494-499.

41. Lotti M, Grandori R, Fusetti F, Longhi S, Brocca S, et al. (1993) Cloning and analysis of *Candida cylindracea* lipase sequences. Gene 124: 45-55.

42. Bolivar JM, Wilson L, Ferrarotti SA, Guisán JM, Fernández-Lafuente R, et al. (2006) Improvement of the stability of alcohol dehydrogenase by covalent immobilization on glyoxyl-agarose. J Biotechnol 125: 85-94.

43. Mateo C, Palomo JM, Fuentes M, Betancor L, Grazu V, et al. (2006) Glyoxyl agarose: A fully inert and hydrophilic support for immobilization and high stabilization of proteins. Enzyme Microb Technol 39: 274-280.

44. Reshmi R, Sanjay G, Sugunan S (2007) Immobilization of α-amylase on zirconia: A heterogeneous biocatalyst for starch hydrolysis. Catal Commun 8: 393-399.

45. Gupta MN (1991) Thermostabilization of proteins. Biotechnol Appl Biochem 14: 1-11.

46. Le-Tien C, Millette M, Lacroix M, Mateescu MA (2004) Modified alginate matrices for the immobilization of bioactive agents. Biotechnol Appl Biochem 39: 189-198.

47. Adriano WS, Mendonça DB, Rodrigues DS, Mammarella EJ, Giordano RL (2008) Improving the properties of chitosan as support for the covalent multipoint immobilization of chymotrypsin. Biomacromolecules 9: 2170-2179.

48. Palomo JM, Fernández-Lorente G, Mateo C, Fuentes M, Guisan JM, et al. (2002) Enzymatic production of (3S,4R)-(-)-4-(4'-fluorophenyl)-6-oxo-piperidin-3- carboxylic acid using a commercial preparation from *Candida antarctica* A: the role of a contaminant esterase. Tetrahedron: Asymmetry 13: 2653-2659.

49. Mendes AA, Giordano RC, Giordano RLC, de Castro HF (2011) Immobilization and stabilization of microbial lipases by multipoint covalent attachment on aldehyde-resin affinity: Application of the biocatalysts in biodiesel synthesis. J Mol Catal B: Enzym 68: 109-115.

50. Klibanov AM (2001) Improving enzymes by using them in organic solvents. Nature 409: 241-246.

51. Yoon JH, Mckenzie D (2005) A comparison of the activities of three β-galactosidases in aqueous-organic solvent mixtures. Enzyme Microb Technol 36: 439-446.

52. Quiroga E, Priolo N, Marchese J, Barberis S (2006) Behaviour of *araujiain*, a cystein phytoprotease, in organic media with low water content. Electron J Biotechnol 9: 18-25.

53. Quiroga E, Camí G, Marchese J, Barberis S (2007) Organic solvents effect on the secondary structure of *araujiain* hI, in different media. Biochem Eng J 35: 198-202.

54. Bordusa F1 (2002) Proteases in organic synthesis. Chem Rev 102: 4817-4868.

55. Nuijens T, Cusan C, Schepers ACHM, Kruijtzer JAW, Rijkers DTS, et al. (2011) Enzymatic synthesis of activated esters and their subsequent use in enzyme-based peptide synthesis. J Mol Catal B: Enzym 71: 79-84.

5

A Solid and Robust Model for Xylitol Enzymatic Production Optimization

Branco R. F¹*, Santos J. C² and Silva S.S²

¹Federal Technological University of Paraná - Campus Pato Branco, Coordination Chemistry, Research Group in bioprocesses and food technology, Fraron, Pato Branco, Paraná, Brazil

²University of Sao Paulo - School of Engineering of Lorena, Department of Biotechnology, Group of Applied Microbiology and Bioprocess,Road of the city soccer field PO Box 116, Postal code 12602810, Lorena, SP, Brazil

Abstract

Xylitol production by fermentation process is widely studied. However, few works describes the enzymatic production of this polyalcohol. This works aims to determine a model that could explain the xylitol enzymatic production as a function of major variables in this process. For this purpose we applied an adequate statistical analysis and response surface methodology (RSM). Initially, variables were selected using a 2^{5-1} fractioned factorial design. Xylose and NADPH concentrations were chosen for the optimization experiments. In order to use the RSM, experiments according to a 2^2 factorial design with star points and triplicate in the center were carried out. The statistical analysis resulted in a quadratic model which could explain 98.6 % of the volumetric productivity in xylitol in function of xylose and NADPH concentrations. Using predicted experimental conditions of 7.0, 25°C, 1.2 mM NADPH, 0.34 M xylose, glucose 0.2 U mg⁻¹ xylose reductase and 0.2 U mg⁻¹ glucose dehydrogenase, this solid model was possible to achieve in batch reaction a xylitol volumetric productivity of 1.58 ± 0.05 g L⁻¹ h⁻¹ with stoichiometric xylose/xylitol conversion efficiency. These values are considered higher and significant in comparing with the traditional fermentation processes. Our results contribute for development of a novel and promising alternative process for xylitol production.

Keywords: Multi-enzymatic process;experimental design; Coenzyme regeneration; Oxidoreductive enzymes

Nomenclature: ANOVA: Analysis of variance; CTAB: Cetyltrimethyl ammonium bromide ; DF: Degree of freedom; a: gravity acceleration (9.8 m s⁻²); MS: Mean of squares; NAD: Nicotinamide adenine dinucleotide; NADH: Reduced nicotinamide adenine dinucleotide; NADP: Nicotinamide adenine dinucleotide phosphate; NADPH: Reduced nicotinamide adenine dinucleotide phosphate; Qp: Xylitol volumetric productivity (g L⁻¹ h⁻¹); RSM: Response surface methodology; SQ: Sum of squares;TRIS:Tris(hydroxymethyl)aminomethane; X_1: Symbol for xylose concentration codified variable; X_2: Symbol for glucose concentration codified variable; X_3: Symbol for xylose reductase load codified variable; X_4: Symbol for glucose dehydrogenase load codified variable; X_5: Symbol for NADPH concentration codified variable; xi: Independent variable (factor) codified value; Xi: Independent variable (factor) real value; X_0: Central point real value; ΔXi: Step change value

Introduction

Xylitol is an important five carbon pentahydroxylated polyols with many significant applications in food, odontological and pharmaceutical industries. It stands out as a natural sweetener that does not cause and combats dental caries; it also prevents respiratory infections, among other properties [7-9]. This compound has been currently used, beside as sweetener, in tooth paste, gum, mouthwash and nasal spray.

Xylitol is produced traditionally by chemical means and many fermentation processes are under studies. However, the lower bioconversion rates are still a challenge to establish a feasible and low cost large scale technology. Thus, enzymatic reactions are novel biotechnological alternative for traditional microbiological and chemical production processes. In xylitol case, it is believed that the enzymatic way can surpass disadvantages of the chemical and microbiological routes, such as low conversion efficiency and productivity [1-3], being this new option also appropriate to current concepts and interests of sustainability and ecosystem preservation, for example the use of sugarcane bagasse as source of xylose and glucose [4,5]. In the enzymatic process it is possible to achieve, without difficulty, high productivity with stoichiometric conversion of xylose into xylitol process which would be very difficult in the fermentative and chemical batch processes [6]. Xylitol enzymatic production consists in the direct reduction of xylose into xylitol by the enzyme xylose reductase (E.C. 1.1.1.21) assisted, in this work, by the coenzyme nicotinamide adenine dinucleotide phosphate in its reduced form (NADPH). Further, the NADP can be reduced again in a coupling enzymatic reaction in order to minimize process cost. Here we used a glucose dehydrogenase system in which glucose is oxidized to gluconic acid (gluconate) mediated by glucose dehydrogenase (E.C. 1.1.1.119) and NADP is reduced to NADPH for the xylitol enzymatic production.

However, as a new procedure firstly it is necessary to determinate and to understand the influence and interactions of many variables involved on the enzymatic process before attempting to optimize it. In these cases, the experimental design methodology, or more specifically fractional factorial designs are indicated [10,11]. The search for a representative model of the process can represent an economic gain and allows the discovery and predicting ideal process conditions. Therefore, after the screening, the optimization can be performed also by experimental design using the response surface methodology (RSM), a very accurate method to maximize the desirable response variable [12-14].

In this context, the present work had as objective to optimize the enzymatic production of xylitol in batch regime using RSM and determine an adequate model to explain xylitol enzymatic production.

*Corresponding author: Branco R. F, Universidade Tecnológica Federal do Paraná – Campus Pato Branco, Coordenação de Química, Grupo de pesquisa em tecnologia de bioprocessos e alimentos Address: Via do Conhecimento km 01, Fraron, Pato Branco, Paraná, Brazil, Postal code 85503390

It must be highlighted that there are no records for this kind of approach for xylitol enzymatic production. The work was done in two parts; the first one was the screening of variables using a 2^{5-1} fractional factorial design with triplicate at central point, being xylitol volumetric productivity (Qp, g L^{-1} h^{-1}) the major response variable. The second step was the optimization of the process using a 2^2 factorial design with star points and triplicate in the center in which the results were used for RSM. This work contributes with a consistent approach to study optimize a bioprocess.

Material and Methods

Materials

All chemicals used in this study were of the analytical grade available and were obtained from Sigma-Aldrich (São Paulo, Brazil), save the ones mentioned in the text.

Preparation of xylose reductase pre-purified extract

The xylose reductase pre-purified extract used in all assays was produced in-house. Pre-purification was required to separate xylose reductase from xylitol dehydrogenase in order to avoid xylitol consumption. The pre-purified xylose reductase extract used in the reactions was prepared in three steps: the yeast *Candida guilliermondii* FTI 20037 cultivation, cell disruption and reverse micelle technique. The first step, cell cultivation, was performed in a batch system in a BIOFLO III bioreactor of 1.25 L (New Brunswick Scientific Co. Inc., Edison, New Jersey, USA). The total fermentation media volume was 1.0 L, containing 50 g L^{-1} xylose, 3 g L^{-1} ammonium sulfate, 10 % v v^{-1} rice bran extract, 0.1 g L^{-1} calcium chloride and 0.1 % v v^{-1} antifoaming (silicone base, Adonex, São Paulo, Brazil). The pH, dissolved oxygen and temperature were kept at 5.50; 50 % and 30 °C, respectively. Initial cell concentration of *Candida guilliermondii* FTI 20037 was 1.0 g L^{-1} (dry weight) and the inoculum was prepared as described elsewhere [15]. The pH was controlled by addition of 3 M NaOH. In the second step, the crude xylose reductase extract was obtained by the microbial cells disruption under vortex agitation using glass pearls (0.5 mm diameter). The used conditions for the disruption were previously determined by [16]. Finally, the pre-purified xylose reductase extract was produced by reverse micelle technique using CTAB-reversed-micelles in isooctane, hexanol and butanol, by a two-step procedure according methodology described in our laboratory [17]. Firstly, 3.0 ml of the crude extract was mixed with an equal volume of micellar microemulsion (CTAB in isooctane/hexanol/butanol/water). This mixture was agitated on a vortex for 1 min, to obtain the equilibrium phase, and separated into two phases by centrifugation at 6570 a for 10 min (Jouan Centrifuge model 1812, Saint-Herblain, France). Afterwards, 2.0 mL of CTAB-micellar phase was mixed with 2.0 mL of fresh aqueous phase (acetate buffer 1.0 M at pH 5.5 with 1.0 M NaCl), in order to transfer xylose reductase from the micelles to this fresh aqueous which was finally collected by centrifugation (6570 a; 10 min).

Enzymes activity assays

Xylose reductase activity was determined by spectrophotometric analysis using NADPH as the detecting parameter at 25 °C and 340 nm, in a medium composed of : 350 µL Tris buffer (71 mM, pH 7.2), 50 µL NADPH (1.2 mM), 50 µL xylose (2.0 M) and 150 µL of the enzymatic extract. The glucose dehydrogenase activity was also determined by spectrophotometric analysis using the same conditions. The medium was composed of 350 µL Tris buffer (71 mM, pH 7.2), 50 µL NADP (1.2 mM), 50 µL glucose (1.5 M) and 150 µL of the enzymatic extract. Variation of the absorbance at 340 nm of the assay against a blank without enzyme was monitored for 1 min. The activity was calculated from the slope of the absorbance versus the time curve by using the apparent molar extinction coefficient of 6.22 $mmol^{-1}$ cm^{-1} for NAD(P)H. One xylose reductase or glucose dehydrogenase unit (U) was defined as the amount of enzyme catalyzing the formation or degradation of 1 µmol of NADPH per min. The volumetric activity was expressed as U mL^{-1} (extract volume).

Xylitol enzymatic production

The enzymatic reactions were carried out in a 10 mL flask and a total reaction volume of 5 mL agitated by magnetic stirrer. In all assays the temperature and pH were kept at 25° C and 7.0, respectively. The pH was controlled by addition of 0.2 M Tris. The temperature and pH were controlled using the BIOFLO III bioreactor module (New Brunswick Scientific Co. Inc., Edison, New Jersey, USA). The other experimental conditions were varied according to the 25-1 and 22 designs.

Experimental designs for model development

Two experimental designs to achieve the optimization of xylitol enzymatic production were performed. A 2^{5-1} fractional factorial design with triplicate at center point, in a total of 19 assays, was used in order to select and evaluate the effect of five independent variables: xylose concentration, glucose concentration, NADPH concentration, xylose reductase load and glucose dehydrogenase load on xylitol enzymatic production. The study range was defined in previous experiments. The next step was the process optimization experiments which were performed according to a 2^2 experimental design with star points and triplicate at center point. The factors selected for this design were xylose and NADPH concentrations, all the others factors were kept at the low level (-1) of the first design (2^{5-1}). For appropriate statistical analysis, the variables, of both experimental designs, needed to be coded according to Equation 1.

$$x_i = \frac{X_i - X_0}{\Delta X_i}$$

Eq. (1)

xi: independent variable (factor) coded value; X_i: independent variable (factor) real value, X_0 central point real value and; ΔX_i: step change value.

The range and levels investigated for the 2^{5-1} and the 2^2 designs are presented in Table 1 and 2, respectively. Qp was the dependent variable (response) considered for selection and optimization. Both designs were proposed according to literature [18,19]. The statistical calculi were performed using the software STATISTICA (version 6.0, StatSoft, U.S.A.) and DESIGN EXPERT (version 6.0, Stat-Ease, U.S.A.).

Analytical methods

Xylose and xylitol concentrations were determined by high performance liquid chromatography (HPLC) using Waters equipment (Waters, Milford, MA, USA) with a refractive index (RI) detector (model 2414, Waters, Milford, MA, U.S.A.), UV detector (model 2487, Waters, Milford, MA, U.S.A.) and a HPX-87H column (Bio-Rad, Hercules, CA, U.S.A.). The operational conditions were: temperature of 45°C, 0.005 M sulfuric acid as eluent, flow of 0.6 mL min^{-1} and injection volume of 20 µL.

Independent variables	Symbol	Range and levels		
		-1	0	1
Xylose concentration (M)	X_1	0.033	0.066	0.100
Glucose concentration (M)	X_2	0.033	0.066	0.100
Xylose reductase load (U.mL^{-1})	X_3	0.20	0.40	0.60
Glucose reductase load (U.mL^{-1})	X_4	0.20	0.40	0.60
NADPH concentration (mM)	X_5	0.20	0.40	0.60

Table 1: Study range and levels with real values of the independent variables used for the 25-1 experimental design carried out to select and evaluate variables on the performance of the batch xylitol enzymatic production.

Independent variables	Symbol	Range and levels				
		$-\sqrt{2}$	-1	0	1	$\sqrt{2}$
Xylose concentration (M)	X1	0.039	0.100	0.250	0.400	0.460
NADPH concentration (mM)	X5	0.23	0.60	1.50	2.40	2.80

Table 2: Study range and levels with real values of the independent variables used for the 22 experimental design carried out to optimize the batch xylitol enzymatic production.

Assay	Coded variable					Response
	X_1	X_2	X_3	X_4	X_5	Qp* (g L^{-1} h^{-1})
1	-1	-1	-1	-1	1	0.46
2	1	-1	-1	-1	-1	0.36
3	-1	1	-1	-1	-1	0.20
4	1	1	-1	-1	1	0.72
5	-1	-1	1	-1	-1	0.43
6	1	-1	1	-1	1	0.57
7	-1	1	1	-1	1	0.53
8	1	1	1	-1	-1	0.35
9	-1	-1	-1	1	-1	0.45
10	1	-1	-1	1	1	0.87
11	-1	1	-1	1	1	0.52
12	1	1	-1	1	-1	0.89
13	-1	-1	1	1	1	0.38
14	1	-1	1	1	-1	0.48
15	-1	1	1	1	-1	0.35
16	1	1	1	1	1	1.20
17	0	0	0	0	0	0.89
18	0	0	0	0	0	0.88
19	0	0	0	0	0	0.90

*correspondent to 6h of reaction

Table 3: 2$^{5-1}$ experimental design matrix with correspondent Qp results for each assay.

Term	Effect value	Standard error	t –value (13 DF)*	p-value
Average	0.60	0.055	12.0	<0.01
X_1	0.26	0.110	2.4	0.03
X_2	0.09	0.110	0.8	0.42
X_3	-0.02	0.110	-0.2	0.81
X_4	0.18	0.110	1.7	0.11
X_5	0.21	0.110	2.0	0.07

*DF: Degree of freedom; t - tabulated = ±1.77 for a confidence level of 90% and 13 DF.

Table 4: Effect values and significance tests for the average and main variables effects in relation to Qp for the 2$^{5-1}$ experimental design.

Results and Discussion

Screening of major variables for enzymatic xylitol production model

Amongst the variables which influence the catalytic activity of en-

zymes it were selected for the study of xylitol enzymatic production: xylose concentration, glucose concentration, NADPH concentration, xylose reductase load and glucose dehydrogenase load. The data for analysis of influence and screening of those variables was generated by experiments to a 25-1 fractioned factorial design, with triplicate at center point. The design matrix with the volumetric productivity in xylitol (Qp, g L^{-1} h^{-1}) for each assay are presented in Table 3. The results were input in the software STATISTICA (6.0), for determining the main effect values and to carry out significance tests, which are presented in Table 4.

The observed Qp values (Table 3), demonstrating that the chosen study region was adequate, since it varied between 0.20 and 1.20 g L^{-1} h^{-1} showing an adequate range. The center point triplicate evidenced that the experiments were reproducible, once the standard error was lower than 0.01. In relation to the NADPH regeneration system (glucose dehydrogenase), the results demonstrated that it worked satisfactorily with the main reaction, once it was not observed the formation of by-products (side reactions) and in all experiments the xylose-xylitol conversion efficiency was stoichiometric.

The statistical analysis demonstrated that only the effect of xylose and NADH concentrations were significant, at a 90 % confidence level (Table 4). Both effects were positive indicating that the response (Qp) benefits by the use of these factors at their superior levels (+1) the influence of these factor can be observed by the results in Table 3, specially in assays 15 and 16. In assay number 15 xylose and NADPH concentrations were used at their inferior levels (-1) as result the Qp value was low (0.35 g L^{-1} h^{-1}), however when they were used at their superior level (+1), assay number 16, it was observed the highest Qp value (1.20 g L^{-1} h^{-1}), three times higher than assay 15. This result demonstrated that the main reaction rate was elevated with the increase of the substrates concentrations, xylose and NADPH, independent of the enzyme xylose reductase amount. This fact indicates that, probably, the reaction rate could be raised if the concentrations of xylose and NADPH were further increased. Thus it was necessary, for the process optimization, a new study range for these variables since they were still in non-limiting concentrations.

Therefore, the concentrations of xylose and NADPH were the factors selected for the study of xylitol enzymatic production optimization and for the development of a mathematical model in batch, being a new study range defined in which the concentrations were increased.

Process optimization using RSM and model development

This work aimed to optimize the enzymatic production of xylitol in batch regime using RSM and to determine an empiric model.

It was selected for xylitol enzymatic production optimization and to develop a statistical model, the concentrations of xylose and NADPH. The experiments were carried out according to a 2^2 complete factorial design, with star points and triplicate at center point. The design matrix and Qp result for each assay are presented in Table 5. For effects significance tests, analysis of variance (ANOVA) of the model, regression coefficients determination and RSM the data were input in the software STATISTICA and DESIGN EXPERT.

The results presented in Table 5 demonstrated that the shift on the study range benefited productivity, since the mean QP value, compared to the first design, increased from 0.60 to 1.42 g L^{-1} h^{-1}, 2.4 times higher. The best Qp value (1.63 g L^{-1} h^{-1}) was observed at assay 10 (Table 5), which is a replicate at center point, this result indicated that this region is, probably, near the maximum for Qp.

Assay	Coded variable		Response
	X_1	X_5	Qp* (g L^{-1} h^{-1})
1	-1	-1	1.03
2	1	-1	1.40
3	-1	1	1.33
4	1	1	1.51
5	$-\sqrt{2}$	0	1.20
6	$+\sqrt{2}$	0	1.12
7	0	$-\sqrt{2}$	1.61
8	0	$+\sqrt{2}$	1.52
9	0	0	1.59
10	0	0	1.66
11	0	0	1.63

*correspondent to 6h of reaction

Table 5: 2^2 experimental design matrix with correspondent Qp results for each assay.

Term	Effect value	Standard error	t-value (2 DF)	p-value
Average	1.63	0.020	80.2	<0.01
X_1	0.12	0.012	10.0	<0.01
X_{12}	-0.12	0.015	-8.4	0.01
X_5	0.14	0.012	11.4	<0.01
X_{52}	-0.17	0.015	-11.8	<0.01
$X_1.X^5$	-0.06	0.018	-3.3	0.08

t- tabulated = ±2.92 for a 90 % confidence level and 2 DF.

Table 6: Coefficient regression values and respective significance tests.

Source of variation	SS*	DF	MS**	F - value	p – value
Regression	0.50	5	0.10	60.0	<0.01
Residues	0.01	5	0.02	-	-
Total	0.51	10	-	-	-
Lack of fit	0.006	3	0.002	1.6	0.42
Pure error	0.002	2	0.001	-	-
% of explained variation: 98.6					
% of maximum explained variation: 99.6					

* SS: sum of squares ** MS: mean of squares. F - tabulated = 5.05 for a confidence level of 95 %, 5 DF of the regression and 5 DF of the residues.

Table 7: Analysis of variance (ANOVA) and variation values of the proposed quadratic model.

After the decision to keep all the regression terms, it was initiated the RSM which involves calculi and significant analysis of a quadratic empiric model.

The regression coefficients for each term in the quadratic model with their respective significance test are presented in Table 6. The analysis of variance (ANOVA) of the quadratic model is presented in Table 7.

As can be observed in Table 6 all regression coefficients were significant and according to the ANOVA of the quadratic model (Table 7) the proposed model is adequate to explain the behavior of Qp in function of xylose concentration and NADPH concentration, in the study region, since the lack of fit was not significant (Table 7). The quadratic model (real values) is presented in Equation 2.

$$Qp = 0.20 + 4.22[Xylose] - 5.51[Xylose]^2 + 0.91[NADPH] - 0.21[NADPH]2 - 0.43[Xylose].[NADPH] \qquad Eq. (2)$$

The model could explain 98.6 % of the experimental variation (Table 6), which means that it is solid and robust. This fact can be visual-ized in Figure 1, which presents the predicted data from the model in function of the experimental data.

After the significance confirmation of the empiric model, it was performed the optimization of the process using RSM. The response surface and contour curves are presented in Figure 2. According to Figure 2a and 2b the increase in both variables (xylose and NADPH concentrations) brings benefit to productivity probably due to an increase in the main catalysis rate (xylitol production) since the substrates of the enzyme xylose reductase are increased and consequently its xylose to xylitol conversion rate, as predicted in traditional enzyme kinetic models.

The mathematic resolution of the model pointed a maximum Qp

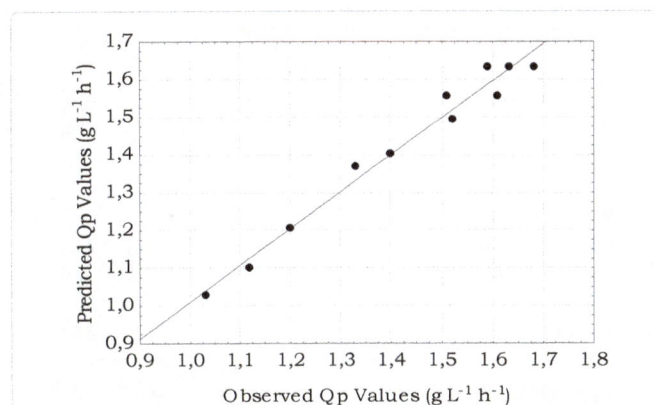

Figure 1: Observed Qp values as a function of the predicted by the quadratic model. The curve represents the model and the dots are the experimental data.

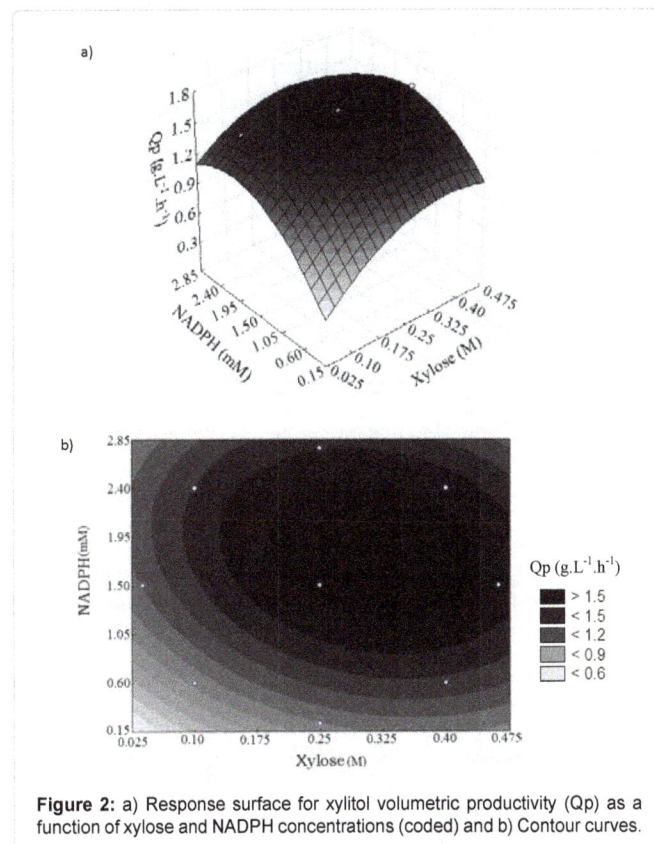

Figure 2: a) Response surface for xylitol volumetric productivity (Qp) as a function of xylose and NADPH concentrations (coded) and b) Contour curves.

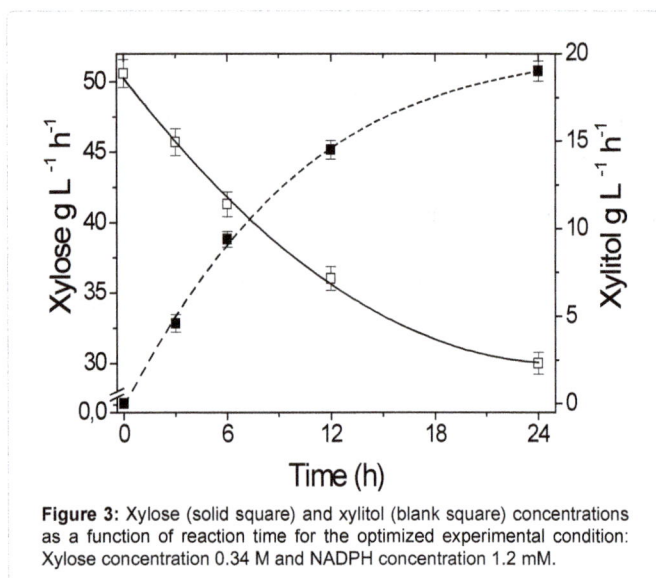

Figure 3: Xylose (solid square) and xylitol (blank square) concentrations as a function of reaction time for the optimized experimental condition: Xylose concentration 0.34 M and NADPH concentration 1.2 mM.

value of 1.68 g L^{-1} h^{-1}, referent to the coordinates 0.31 M for xylose concentration and 1.80 mM for NADPH concentration. This result surpasses the Qp values and conversion values of literature for xylitol microbial production, in batch and with synthetic media [20,21] which were 1.50 g L^{-1} h^{-1} and   71 % of xylose-xylitol conversion efficiency. However, it must be considered, even in earlier steps the process economics, in xylitol enzymatic production case the cost of the coenzyme NADPH is one of the greatest contributors for the total cost [22], hence is desirable a lower NADPH concentration since there is no important loss in productivity. In this context, in Figure 2a and 2b, it can be observed a wide region that comprehends an equal or higher Qp value of 1.60 g L^{-1} h^{-1}, confirming also the robustness of the process, which is already a superior value from mentioned literature. Therefore, this region corresponding to a Qpvalue of 1.60 g L^{-1} h^{-1} was selected for optimization. Both mathematical and graphical resolutions, given by the softwares used, demonstrated that the minimum NADPH concentration which is needed to achieve this goal is 1.20 mM, with a corresponding concentration of xylose of 0.34 M. This NADPH concentration corresponds to a 33.3 % reduction when compared to the concentration of 1.80 mM pointed by the mathematical resolution to attain the maximum Qp value, mentioned earlier, (1.68 g L^{-1} h^{-1}) and this reduction represents only in a Qp loss of only 5 % (from 1.68 to 1.60 g L^{-1} h^{-1}).

Therefore, it was carried out the model validation in those conditions, 1.20 mM NADPH concentration and 0.34 M xylose concentration, in duplicate. The experiments resulted in a Qp mean value of 1.58±0.05 g L^{-1} h^{-1}, which is inside of the confidence statistical limits for this model. Xylose and xylitol concentrations in function of reaction time for these experiments are presented in Figure 3.

Xylitol is produced in the same rate as xylose consumption until the reaction reaches 6 hours (Figure 3). From this point, these rates gradually fall and from 24 hours the production almost ceases. This phenomenon is probably due to the depletion of glucose, which deactivates the coenzyme regeneration system therefore, reducing xylitol production.

Conclusion

The results presented in this work made possible to firstly selected adequate variables from a group, xylose and NADPH concentrations, using experimental design for further optimization. The optimization step was performed successfully using RSM and experimental design. It was possible to generate an empiric quadratic model which could explain the xylitol productivity in function of the studied variables. From predicted coordinates, 1.20 mM NADPH concentration and 0.34 M xylose concentration, the model was validated and it was possible to achieve a productivity of 1.58±0.05 g L^{-1} h^{-1} with stoichiometric xylose/xylitol conversion efficiency. The results attained were higher when compared to the ones presented in literature for other xylitol production processes, in batch regime. The work also presented a potential process of xylitol enzymatic production as an alternative of traditional xylitol ways of production. Finally, it was possible to achieve a representative statistical model for xylitol enzymatic production which contributed to enlighten and improve this process.

Acknowledgements

The authors gratefully thank the financial support of the Fundação de Amparo à Pesquisa do Estado de São Paulo, FAPESP (Proc. n º 05/0280-4 and 05/02866-9).

References

1. Alanem P, Isokangas P, Gutmann K (2000) Xylitol candies in caries prevention: results of a field study in Estonian children community. Dent Oral Epidemiol 28: 218-224.

2. Gliemmo MF, Calvin AM, Tamasib O, Gerschensona LN, Camposa CA (2008) Interactions between aspartame, glucose and xylitol in aqueous systems containing potassium sorbate. LWT - Food Sci Technol 41: 611-619.

3. Makinen KK (2000) Can the pentitol–hexitol theory explain the clinical observations made with xylitol. Med Hypothe 200: 603-613.

4. Santos JC, Mussatto SI, Dragone G, Converti A, Silva SS (2005) Evaluation of porous glass and zeolite as cells carriers for xylitol production from sugarcane bagasse hydrolysate. Biochem Eng J 23: 1-9.

5. Mikkola JP, Salmi T, Villela A, Vainio H, Mäki-Arvela P,et al. (2003) Hydrogenation of xylose to xylitol on sponge nickel catalyst – a study of the process and catalyst deactivation kinetics. Brazilian J Chem Eng 20: 263-265.

6. Marton JM, Felipe MGA, Silva JBA, Pessoa Jr A (2006) Evaluation of the activated charcoals and adsorption conditions used in the treatment of sugarcane bagasse hydrolysate for xylitol production. Brazilian J Chem Eng 23: 9-10.

7. Gurpilhares DB, Hasmann FA, Pessoa Jr A, Roberto IC (2009) The behavior of key enzymes of xylose metabolism on the xylitol production by *Candida guilliermondii* grown in hemicellulosic hydrolysate. J Ind Microbiol Biotechnol 36: 87-93.

8. Sarrouh BF, Branco RF, Silva SS (2009) Production of Xylitol: Enhancement of Monosaccharide Biotechnological Production by Post-Hydrolysisof Dilute Acid Sugarcane Hydrolysate. Appl Biochem Biotechnol 153: 163-170.

9. Converti A, Torre P, Luca E, Perego P, Borghi M, et al. (2003) Continuous xylitol production from synthetic xylose solutions by *Candida guilliermondii*: Influence of pH and temperature. Eng Life Sci 3: 193-198.

10. Neto BB, Scarmínio IS, Bruns RE (2007) Como fazer experimentos: pesquisa e desenvolvimento na ciência e na indústria. UNICAMP Press.

11. Rodrigues MI, Lemma AF (2005) Planejamento e otimização de processos: uma estrategia sequencial de planejamento Casa do Pão Press.

12. Branco RF, Santos JC, Murakami LY, Mussatto SI, Dragone G, Silva SS (2007) Xylitol production in a bubble column bioreactor: Influence of the aeration rate and immobilized system concentration. Process Biochem 42: 258-262.

13. Sharma S, Malik A (2009) Application of response surface methodology (RSM) for optimization of nutrient supplementation for Cr (VI) removal by Aspergillus lentulus AML05. J Hazard Mater 164: 1198-1204.

14. Sarrouh BF, Silva SS (2010) Application of Response Surface Methodology for Optimization of Xylitol Production from Lignocellulosic Hydrolysate in a Fluidized Bed Reactor. Chem Eng Tech 33: 1481-1487.

15. Branco RF, Silva SS (2010) Contribution of Tris Buffer on Xylitol Enzymatic Production. Appl Biochem Biotechnol 162: 1558-1563.

16. Gurpilhares DB, Pessoa Jr A, Roberto IC (2006) Glucose-6-phosphate dehy-

drogenase and xylitol production by *Candida guilliermondii* FTI 20037 using statistical experimental design. Process Biochem 41: 631-637.

17. Cortez EV, Pessoa Jr A, Felipe MGA, Roberto IC, Vitolo M (2004) Optimized extraction by cetyl trimethyl ammonium bromide reversed micelles of xylose reductase and xylitol dehydrogenase from *Candida guilliermondii* homogenate. J Chromatography B 807: 47-54.

18. Box EPG, Hunter WG, Hunter JS (1978) Statistics for Experimenters: An Introduction to Design, Data Analysis, and Model Building. John Wiley & Sons Press.

19. Box EPG, Hunter WG, Hunter JS (2005) Statistics for Experimenters: Design, Innovation, and Discovery. John Wiley & Sons Press.

20. Branco RF, Santos JC, Sarrouh BF, Rivaldi JD, Pessoa Jr A, et al. (2009) Profiles of xylose reductase, xylitol dehydrogenase and xylitol production under different oxygen transfer volumetric coefficient values. J Chem Technol Biotechnol 84: 326-330.

21. Sampaio FC, Faveri D, Mantovani HC, Passos FML, Perego P, et al. (2006) Use of response surface methodology for optimization of xylitol production by the new yeast strain Debaryomyces hansenii UFV-170. J Food Eng 76: 376-386.

22. Nidetzky B, Neuhauser W, Haltrich D, Kulbe KD (1996) Continuous enzymatic synthesis of xylitol with simultaneous coenzyme regeneration in a charged membrane reactor. Biotechnol Bioeng 52: 387-396.

6

Statistical Optimization of Medium Components by Response Surface Methodology for Enhanced Production of Bacterial Cellulose by *Gluconacetobacter persimmonis*

Swati Hegde, Gururaj Bhadri, Kavita Narsapur, Shanta Koppal, Princy Oswal, Naina Turmuri, Veena Jumnal, Basavaraj Hungund*

B.V.B College of Engineering & Technology, Hubli-580031, Karnataka, India

Abstract

Bacterial cellulose has been found to be attractive for novel applications due to its material properties. An optimization of the medium used for the production of bacterial cellulose using *Gluconacetobacter persimmonis* was carried out. Plackett-Burman (PB) Design for screening of the medium constituents and a Central Composite Design (CCD) for optimization of significant factors were employed. Glucose, yeast extract and peptone were estimated as significant factors from PB design. Bacterial cellulose concentration of 1.72% (w/v) was obtained in the medium optimized using CCD method as compared to un-optimized medium that yielded 0.318% (w/v). Hence a 6 fold increase was observed after the optimization of medium was carried out. The combined effects of medium components and their optimum regions are reported in this paper.

Keywords: *Gluconacetobacter persimmonis*; Bacterial Cellulose; Plackett Burman Design; Central Composite Design

Introduction

Cellulose is the most abundant earth biopolymer, recognized as the major component of plant biomass and a representative of microbial extracellular polymers. Efficient producers of cellulose are members of acetic acid bacterium *Gluconacetobacter* (earlier known as genus *Acetobacter*). Bacterial cellulose from *Acetobacter* strains displays unique physical, chemical and mechanical properties including high crystallinity, high water holding capacity, large surface area, elasticity, mechanical strength and biocompatibility [1]. Although bacterial cellulose finds applications in several fields, productivity of cellulose production needs to be addressed so as to make it economically compatible. Hence it becomes necessary to optimize the yields of cellulose production by the use of process improvement strategies. Various workers have optimized medium constituents and process parameters for increased bacterial cellulose (BC) yield [2-6].

Optimization of processing parameters plays an important role in the development of any process owing to their impact on the economy and efficacy of the process. Designing an appropriate production medium and conditions is of crucial importance to improve the efficiency and productivity of bioactive microbial metabolites fermentation process, because it can significantly affect product concentration, yield, and the ease and cost of downstream product separation. Statistically based experimental designs have proved to be more efficient than one-at-a-time method, which is complicated and time-consuming, especially on multi-variables screening, and do not consider the complex interactions among different variables. On the other hand, statistical experimental designs provide a systematic and efficient plan for experimentation to achieve certain goals so that many factors can be simultaneously studied. Therefore, in recent years number of statistical designs were used to search the key factors rapidly from a multivariable system, such as Plackett–Burman design and response surface methodology [2,3,5,7-9] because statistical optimization not only allows quick screening of large experimental domain, but also reflects the role of each of the components and their interactions.

Plackett-Burman design allows the evaluation of (n-1) variables by n experiments; n must be multiple of 4 ex: 4,8,12 etc. Any factor not assigned to a variable can be designated as dummy variable. The incorporation of dummy variables into experiments makes it possible to estimate the variances of an effect (experimental error). The effect of dummy variables is calculated in the same way as the experimental variables. If there is no interaction and no error in measuring the response, the effect shown by dummy variable should be zero .If the effect is not equal to zero, it is assumed to be a measure of analytical error in measuring the response. This procedure will identify the important variables and allow them to be ranked in order of importance to decide which to investigate in a more detailed study to determine the optimum value to use.

Response Surface Methodology (RSM) is a collection of statistical and mathematical techniques useful for designing experiments, building models, evaluating the effects of factors, and searching optimum conditions of factors for desirable responses [5]. It also has important applications in design, development and formulation of new products as well as in improvement of existing product designs. Statistical experimental design minimizes the error in determining the effect of parameters and it shows the simultaneous, systematic, and efficient variation of all parameters. Response Surface Methodology (RSM) is an effective tool for optimizing the process condition that uses quantitative data from an appropriate experimental design to determine and simultaneously solve multivariate equations [2]. It usually involves an experimental design such as Central Composite Design (CCD) to fit a second-order polynomial by a least squares technique. An equation is used to describe the test variables, and describe the combined effect of all the test variables in the response.

*Corresponding author: Basavaraj Hungund, B.V.B College of Engineering & Technology, Hubli-580031, Karnataka, India

The objectives of the present investigation includes optimization of BC production from *G. persimmonis* GH-2 by RSM and to study the interrelationship among the media ingredients on BC yield using response surface plots. The medium ingredients were screened by Plackett-Burman (PB) Design and optimization of significant factors was done applying a Central Composite Design (CCD). We observed a 6 fold increase in cellulose yield as an optimum response of the study.

Materials and methods

Microorganism

Cellulose producing *Gluconacetobacter persimmonis* reported earlier was used in this study [10]. The culture was maintained on HS agar slants, transferred and stored at a temperature of 2oC -8oC in the refrigerator.

Culture media and growth conditions

Standard Hestrin-Schramm (HS) medium containing 2.0% D-glucose, 0.5% peptone, 0.5% yeast extract, 0.15% citrate and 0.3% disodium phosphate (pH 5.5-6.0) was used in the study [11]. A volume (100 ml) of medium in 250 ml conical flask was inoculated with bacterium and incubated at room temperature for 14 days under stationary conditions to observe the cellulose pellicles that would form at air-liquid interface.

Quantification of bacterial cellulose

Quantification of cellulose was carried-out as reported in our earlier study [10]. The cellulose pellicles obtained after 14 days of incubation were subjected to filtration to remove excess media and the pellicles were retained in flasks. A solution of 2% NaOH was added to the pellicles and boiled for 15 min. The mixture was then filtered and dried in hot air oven at 75oC for 6h and dry weight of pellicle was taken to find out the yield.

Screening of the independent variables-Placket Burman Design

The concentrations of medium components like glucose, yeast extract, peptone, citric acid and disodium phosphate were varied from high to low concentrations (Table 1) and 16 experimental trials were conducted in two batches their effects on BC production was determined . Trials for 16 sets of media with different concentrations of media components were inoculated with 2 ml of *Gluconacetobacter persimmonis* culture and were incubated at room temperature for 14 days under static condition of growth [12].

Optimization of the independent variables- Response surface methodology

To optimize carbon source (glucose), nitrogen source (yeast extract and peptone), a Central Composite Design (CCD), consisting of a set of 20 experiments with replicates at central point was conducted. A three factor 5 level CCD with 20 experiments was used. Trial for

Components	High concentration (% w/v)	Low concentration(%w/v)
Glucose (X₁)	4	2
Peptone (X₂)	1	0.5
Yeast extract (X₃)	1	0.5
Disodium phosphate (X₄)	0.6	0.3
Citric acid (X₅)	0.3	0.15

Table 1: Components of HS medium and their high and low concentrations.

20 sets of media with different concentrations of media components were inoculated with 2 ml of *Gluconacetobacter persimmonis* and were incubated at room temperature for 14 days under static condition of growth.

The following second-order polynomial equation was adopted to study the effects of variables to the response.

$$Y= \beta_o + \beta_1 X_1 + \beta_2 X_2 + \beta_3 X_3 + \beta_{11} X_{12} + \beta_{22} X_{22} + \beta_{33} X_{32} + \beta_{12} X_1 X_2 + \beta_{23} X_2 X_3 + \beta_{13} X_1 X_3 \ (1)$$

Where Y is the response (Cellulose yield, g/l), β_o is the constant term, β_1, β_2, β_3 the coefficients of linear terms and β_{11}, β_{22}, β_{33} are the coefficients of quadratic terms and β_{12}, β_{23}, β_{13} are the coefficients of cross product terms, X_1, X_2 and X_3 represent the factors glucose, yeast extract and peptone respectively.

Results and Discussion

Screening of the independent variables: Plackett-Burman design

From the Main effects plot (Figure 1) and Pareto chart (Figure 2) of standardized effects, it can be seen that nitrogen sources like peptone and yeast extract and carbon source glucose have significant effect on BC production when compared to disodium phosphate and citric acid. Yeast extract gave higher yield at its higher concentration (0.4%), yield obtained using peptone was found to be greater at its lower concentration (0.35%) and yield from glucose is highest at its higher concentration (0.35%). Hence it can be inferred that yeast extract is the best nitrogen source for cellulose production by *Gluconacetobacter persimmonis*. The analysis of yield using PB design is shown in Table 2. This is because Yeast extract contains abundant nitrogen compounds as well as many growth factors and its addition into medium might have stimulated cellulose production. In this investigation the maximum yield was 0.604% for a medium composition in which the concentrations of yeast extract and peptone were high and minimum yield was 0.0095% for a medium with low concentration of these nitrogen sources. Hence to optimize BC production it is necessary to optimize the concentrations of yeast extract, glucose and peptone using different concentrations whereas other components of the medium can be kept constant.

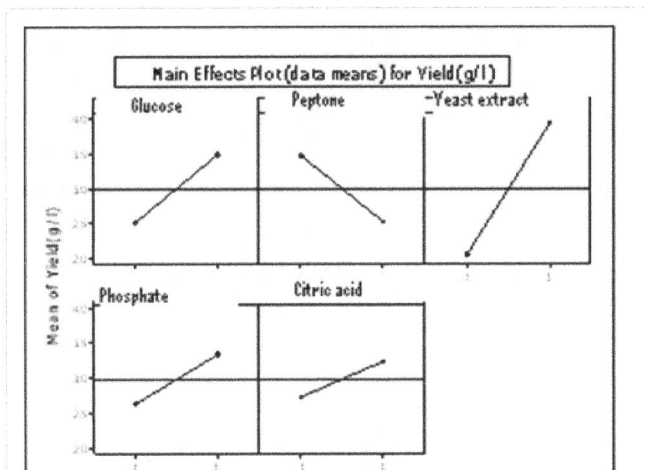

Figure 1: Main effect plot for the screening of medium components using PB design.

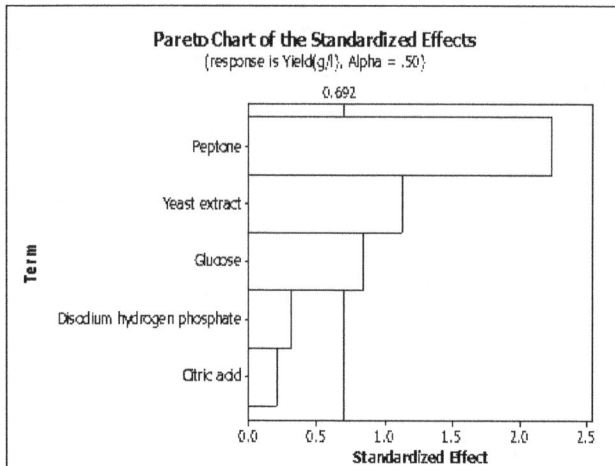

Figure 2: Pareto chart of the standardized effects showing effect of medium constituents on cellulose yield.

		Factors			
	Glucose	Peptone	Yeast	phosphate	Citric acid
$\sum$(H)	34.73	33.15	32.79	31.3	24.48
$\sum$(L)	25.10	26.68	2.042	32.027	35.35
Difference	9.63	6.47	+5.75	-0.72	-10.9
Effect	1.2035	0.8085	0.0185	-0.09	-1.36
Mean square	7.059	2.614	2.06	0.033	1.37

Table 2: Analysis of cellulose yield by *Gluconacetobacter persimmonis*.

Optimization of the independent variables: Response surface methodology

To optimize carbon (glucose) and nitrogen (yeast extract and peptone) sources, a Central Composite Design consisting of a set of 20 experiments with six replicates at central point was conducted. Table 3 shows variables and their levels for central composite design (CCD). The CCD matrix of the independent variables in coded units (experimental design) along with predicted and experimental values of response is given in Table 4. All the experiments were performed in 250 ml Erlenmeyer flask containing 100 ml of media. The quadratic model expressed by equation (2) represents cellulose yield (Y) as a function of glucose (X_1), peptone (X_2) and yeast extract (X_3).

The response equation from the above set of experiments can be written as

$$Y = 0.467 + 0.123\ X_1 + 0.01\ X_2 + 0.267\ X_3 + 0.031\ X_1^2 + 0.054\ X_2^2 + 0.319\ X_3^2 - 0.282\ X_1X_2 + 0.326\ X_1X_3 - 0.224\ X_2X_3 \quad (2)$$

The statistical significance of the polynomial equation was checked by P test and Analysis Of Variance (ANOVA) for response surface quadratic model is given in Table 5. The P value coeffiecient for X_1X_2 and X_1X_3 was 0.028 and 0.014 respectively indicating the significant interaction between two variables X_1 and X_2 as well as between X_1 and X_3.

The estimated coefficients for BC yield are presented in Table 6. The significance of each coefficient was determined by student's T-test and P values. The larger the magnitude of T-value and smaller P value, the more significant is the corresponding coefficient. This means that quadratic main effects of glucose, yeast extract and peptone are more significant. The response obtained under different combinations of variables and defined experimental design Table 2 were analyzed using the Analysis of Variance (ANOVA) appropriate to the experimental

Variable	Symbols	-1.68179	-1	0	1	1.68179
Glucose (%)	X_1	1.31821	2	3	4	4.68179
Yeast extract (%)	X_2	0.3295	0.5	0.75	1	1.1704
Peptone (%)	X_3	0.3295	0.5	0.75	1	1.1704

Table 3: Variables and their levels for CCD.

Experiment No	Glucose	Yeast extract	Peptone	Observed yield(% w/v)	Predicted yield(%w/v)
1	0.00000	0.00000	0.00000	0.4657	0.467
2	0.00000	0.00000	0.00000	0.3749	0.467
3	0.00000	0.00000	1.68179	1.6943	1.818
4	1.00000	1.00000	-1.00000	0.5189	0.354
5	1.00000	1.00000	1.00000	1.4262	1.091
6	-1.00000	-1.00000	-1.00000	0.3011	0.293
7	0.00000	-1.68179	0.00000	0.5024	0.602
8	1.00000	-1.00000	1.00000	1.0271	1.082
9	-1.68179	0.00000	0.00000	0.1493	0.348
10	0.00000	0.00000	0.00000	0.5319	0.467
11	-1.00000	-1.00000	1.00000	0.7986	0.620
12	1.00000	-1.00000	-1.00000	0.7654	0.450
13	-1.00000	1.00000	-1.00000	1.7216	1.323
14	0.00000	0.00000	-1.68179	0.5605	0.922
15	0.00000	0.00000	0.00000	0.6701	0.467
16	1.68179	0.00000	0.00000	0.4751	0.762
17	0.00000	0.00000	0.00000	0.2657	0.467
18	0.00000	1.68179	0.00000	0.2495	0.635
19	-1.00000	1.00000	1.00000	0.7846	0.756
20	0.00000	0.00000	0.00000	0.5762	0.467

Table 4: Experimental design, predicted and observed yields in CCD experiments in term of cellulose yield.

Source	df	Seq SS	Adj SS	Adj MS	F value	P
Regression	9	4.5391	4.5391	0.50434	5.22	0.08
Linear	3	1.1761	1.1761	0.39203	4.05	0.040
Square	3	1.4774	1.4774	0.49247	5.09	0.021
Interaction	3	1.8856	1.8856	0.62852	6.50	0.010
Residual error	10	0.9669	0.9669	0.09669		
Pure error	5	0.1053	0.1053	0.02105		
Total	19	5.5060				

Table 5: Analysis of variance.

Term	Coefficient	Standard error	t-value	P
Constant	0.466859	0.12682	3.681	0.004
Glucose	0.122988	0.08414	1.462	0.175
Yeast extract (YE)	0.009795	0.08414	0.116	0.910
Peptone	0.266263	0.08414	3.164	0.010
Glucose*Glucose	0.031192	0.08191	0.381	0.711
YE*YE	0.05371	0.08191	0.656	0.527
Peptone*Peptone	0.319409	0.08191	3.899	0.003
Glucose*YE	-0.281738	0.10994	-2.563	0.028
Glucose*Peptone	0.326062	0.10994	2.966	0.014
YE*Peptone	-0.223613	0.10994	-2.034	0.069

Estimated regression coefficients representing response and the variable.
S = 0.3110   R-Sq = 82.4%   R-Sq (adj) = 66.6%

Table 6: Estimated regression coefficients representing response and the variable.

design Table 4 which indicated that the sum of squares due to regression (first and second-order terms) was found to be significant ($p < 0.05$) and lack of fit was not significant. The value of coefficient of determination ($R^2 = 0.824$) suggested that the model is a good fit though not accurate. The R^2 is the proportion of variability in response values explained or accounted for by the model. The value of R^2 is always in between 0.0 and 1.0 R^2 value close to 1.0 implies that the model is accurate and predicts better response. However, model with higher R^2 value always does not mean that model is accurate. Large R^2 value also is resulted by addition of non-significant extra variables in the model. Thus, it may be possible of a model having higher R^2 value with poor prediction of response. So the term adjusted R^2 has been introduced which arranges the R^2 values for the sample size and for the number of variables in the model. Addition of insignificant model term in the model leads to decrease in adjusted R^2 value. So, the value of R^2 should be as close as that of adjusted R^2.

Parity plot (Figure 3) showed the distribution of experimental and model predicted values where data points are localized close to the diagonal line suggesting the model is adequate enough to explain cellulose production. The 3D plot (Figure 4) and their respective 2D contour plots (Figure 5) provide a visual interpretation of the interaction between two factors. In the response surface plot, glucose is held at an intermediate level and levels of peptone and yeast extract are varied from -1.5 to +1.5. It can be seen that the cellulose yield increases with increase in concentration of peptone and with decrease in concentration of yeast extract. The corresponding Contour plots indicate diferent regions of yields based on different colours. The maximum yield falls in the range 2-2.5 g/100ml as indicated by dark shaded region in the plot. Hence to optimize the levels of peptone and yeast extract, it is

necessary to carry out experiments in this region of higher yield. So the levels of peptone has to be varied between 1.4 to 1.618 and yeast extract has to be varied from -1.5 to -0.5 to optimize BC production. The graphs also show that the yield of cellulose increases with increase in concentrations of peptone and glucose. The yield is found to decrease with decrease in concentrations of glucose and peptone. The Contour plot shows different ranges of yields. To get optimum conditions the experiments have to be done in the region of maximum yield and the levels of glucose and peptone have to be varied from 1 to 1.6 and 1.5 to 1.68 respectively.

From Figure 5, it can be observed that with increase in concentration of yeast extract and with decrease in concentration of glucose, the yield of cellulose increases considerably. The contour plot indicates the levels of yeast extract and glucose that have to be used to get optimum conditions as 1 to 1.6 for yeast extract and -1.5 to -0.5 for glucose.

In the previous studies for optimization for BC production, Embuscado et. al., [2] used five-level, four factor central composite design. They found all four factors affected cellulose yield significantly from G. xylinus. This is in line with present investigation. However, Rani et. al., [5] optimized cultural conditions for BC production from G. hansenii by central composite design. They used coffee cherry husk and corn steep liquor as less expensive sources of carbon and nitrogen sources respectively. Casarica et. al., [6] brought about improvement in BC yield using poor quality horticulture substrates using Taguchi method. In the study carried-out by Bae and Shoda [3], culture conditions in a jar fermenter for BC production were optimized using Box-Behnken design, Response surface methodology was used to predict the levels of various factors.

Conclusion

This work has demonstrated the use of Central composite design by determining the conditions which are required to get optimum yield of cellulose production from G persimmonis. This methodology could therefore be successfully employed to process development where an analysis of effects and interactions of many experimental factors are required. Central composite experimental design maximizes the amount of information that can be obtained, while limiting the numbers of individual experiments required. Response curves are very helpful in visualizing the main effects and interaction of factors. Thus smaller and less time consuming experimental designs could generally suffice for the optimization of many processes. From the above Main effects plot and Pareto chart of standardized effects, it can be seen that

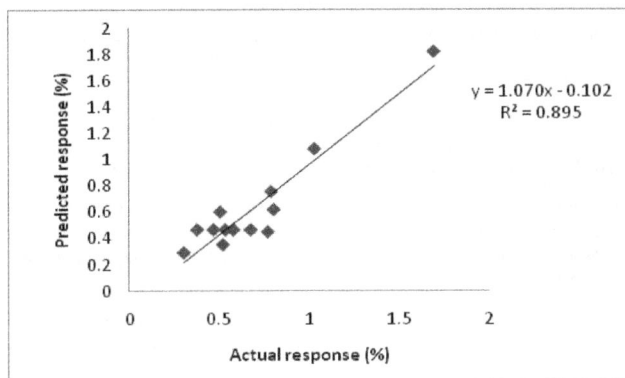

Figure 3: Parity plot showing distribution of experimental and predicted values of cellulose production (% w/v).

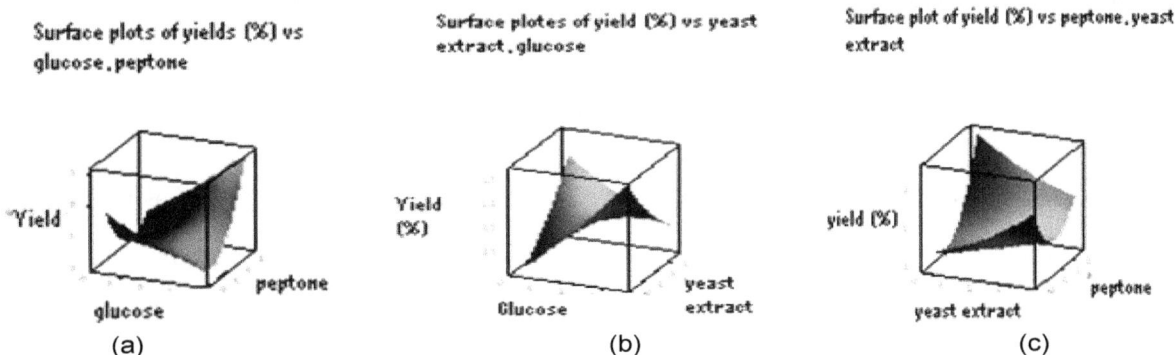

Figure 4 (a-c): Surface plot and contour plots of showing the interaction effects. While the plots show interaction of two factors, third factor is held constant at its middle value.

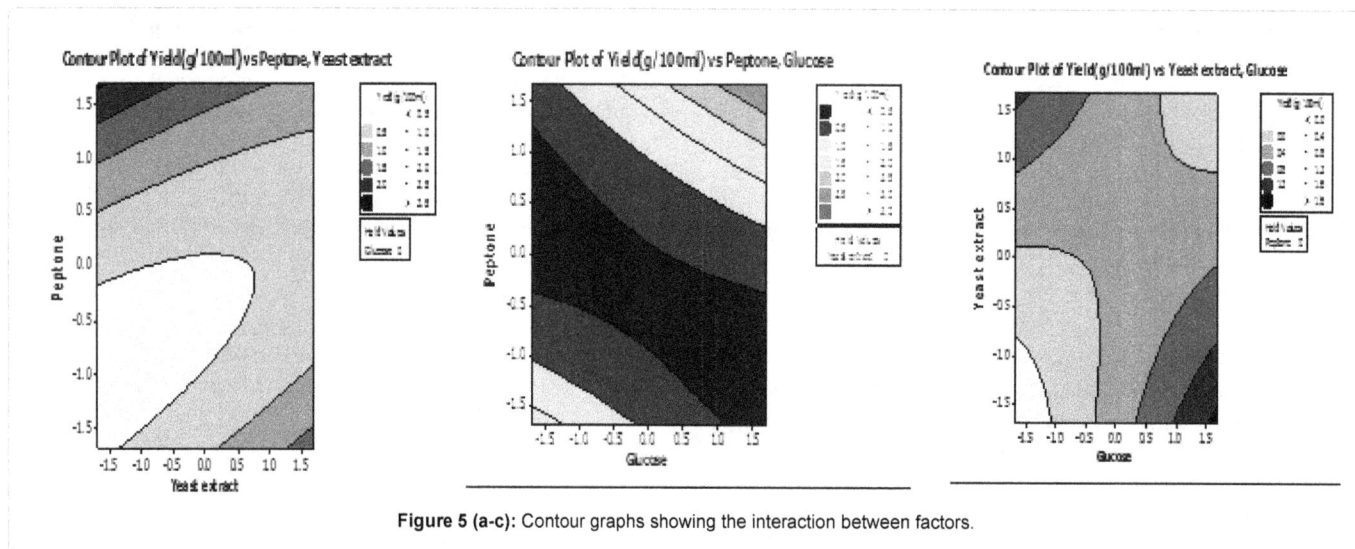

Figure 5 (a-c): Contour graphs showing the interaction between factors.

nitrogen sources like peptone and yeast extract and carbon source glucose have significant effect on BC production when compared to disodium phosphate and citric acid. Hence to optimize BC production it is necessary to optimize the concentrations of yeast extract, glucose and peptone using different concentrations whereas other components of the medium can be kept constant. Hence statistical experimental designs are powerful tools for the rapid search of key factors from a multivariable system and minimizing the error in determining the effect of parameters and the results are achieved in an economical manner.

Acknowledgement

The authors acknowledge Dr. Ashok Shettar, Principal, BVBCET for providing the facilities. The authors also acknowledge Dr. B.B Kotturshettar and Dr. V.N Gaitonde for their technical help during the studies.

References

1. Yoshinaga F, Tonouchi N, Watanabe K (1997) Research Progress in Production of Bacterial Cellulose by Aeration and Culture and its Applications as New Industrial Material. Biosci Biotechnol Biochem 61: 219-224.

2. Embuscado ME, Marks JS, BeMiller NB (1994) Bacterial Cellulose. II. Optimization of Cellulose Production by Acetobacter Xylinum through Response Surface Methodology. Food Hydrocoll 8: 419-430.

3. Bae S, Shoda M (2005) Statistical Optimization of Culture Conditions for Bacterial Cellulose Production using Box-Behnken Design. Biotechnol Bioengg 90: 20-28.

4. Hungund BS, Gupta SG (2010) Production of Bacterial Cellulose from Enterobacter Amnigenus GH-1 Isolated from Rotten Apple. World J Microb Biot 26: 1823-1828.

5. Rani Usha M, Rastogi NK, Appaiah KA (2011) Statistical Optimization of Medium Composition for Bacterial Cellulose Production by Gluconacetobacter Hanenii UAC09 using Coffee Cherry Husk Extract-an Agro Industry Waste. J Microb Biot 21: 739-745.

6. Casarica A, Campeanu G, Moscovici M, Ghiorghita A, Manea V (2011) Improvement of Bacterial Cellulose Production by Acetobacter Xylinum DSMZ-2004 on Poor Quality Horticulture Substrates using the Taguchi Method for Media Optimization. Part I. Cellulose Chem Technol 47: 61-68.

7. Vohra A, Satyanarayana T (2002) Statistical Optimization of the Medium Components by Response Surface Methodology to Enhance Phytase Production by Pichia Anomala. Process Biochem 37: 999-1004.

8. Baskar G, Renganathan (2009) Statistical Screening of Process Variables for the Production of L-Asparaginase from Cornflour by Aspergillus Terreus MTCC 1782 in Submerged Fermentation. Indian J Sci Technol 2: 45-48.

9. Vuddaraju SP, Nikku M, Chaduvulu A, Dasari VR, Donthireddy R (2010) Application of Statistical Experimental Designs for the Optimization of Medium Constituents for the Production of L-Asparginase by Serratia Marcescens. J Microbial Biocheml Technol 2: 89-94.

10. Hungund BS, Gupta SG (2010) Improved Production of Bacterial Cellulose from Gluconacetobacter Persimmonis GH-2. J Microbial Biochem Technol 2: 127-133.

11. Hestrin S, Schramm M (1954) Synthesis of Cellulose by Acetobacter Xylinum. II. Preparation of Freeze-Dried Cells Capable of Polymerizing Glucose to Cellulose. Biochem J 58: 345-352.

12. Son HJ, Heo MS, Kim YG, Lee SJ (2001) Optimization of Fermentation Conditions for the Production of Bacterial Cellulose by a Newly Isolated Acetobacter Sp.A9 in Shaking Cultures. Biotechnol Appl Biochem 33: 1-5.

7

Evaluation and Characterization of Some Egyptian *Fusarium oxysporum* Isolates for their Virulence on Tomato and PCR Detection of (SIX) Effector Genes

Mohamed E Selim[1*], EZ Khalifa[1], GA Amer[1], AA Ely-kafrawy[2] and Nehad A El-Gammal[2]

[1]*Agricultural Botany Department, Faculty of Agriculture, Menoufiya University, Egypt*
[2]*Plant Pathology research institute, Agriculture research center, Gizza, Egypt*

Abstract

The genus Fusarium is comprises of different species which are furtherer divided into small groups called *formae speciales* according to their host specificity. It is remarkable that while an isolate belonging to particular *formae speciales* is considered highly pathogenic in a certain plant species, the many other isolates belonging to the same *formae speciales* may have no harm effect or even a beneficial relation to the same host plant. Traditionally, bioassays conducted under green house conditions were used to distinguish pathogenic and non-pathogenic isolates. However, such these techniques have some limitations due to the time consuming and the inconsistency of obtained results. In contrary, molecular identification techniques are candidate to play an important role in the characterization process of Fusarium isolates as well as with many other organisms. Recently, the ability of a Fusarium isolate to infect particular plant species was found to be dependent on specific genes encoding host determining 'virulence factors' that distinguish virulent from avirulent strains. More recently, eight fungal proteins were identified from xylem sap of infected plants, encompassing the small secreted proteins called Six1, Six2, Six3, Six4, Six5, Six6, Six7 and SIX 8. In the present study, the linkage between virulence potential of fifteen pathogenic isolates in addition to one non-pathogenic isolate, Fo162 , of *Fusarium oxysporum* and the presence of secreted in xylem effector genes (SIX) was determined. The results showed that the tested isolates were varied regarding to their pathogenicity potential toward tomato plants under greenhouse conditions. On the other hand, the amplicons of SIX1, SIX5 and SIX 7 were detected with most hyper virulent isolates while no amplicons for any tested SIX genes were observed with the non-pathogenic isolate (Fo162) as well as with the most hypo virulent isolates. These results, suggested that SIX1, SIX5 and SIX6 may play a distinct role in virulence potential of these *Fusarium oxysporum* isolates toward tomato plants under Egyptian conditions.

Keywords: *Fusarium oxysporum*; Virulence; Effector genes; PCR

Introduction

Tomato (*Solanum lypersicum* Mill.) is ranked number 1 among fruits and vegetables with 14% of the total production worldwide. World production of tomato covers approximately 4 million hectares of arable land with production estimated at 100.5 million tons and valued at 5-6 billion US$ [1]. Note worthy, Egypt is among the top 10 tomato producing countries worldwide with estimated 181.000 ha of harvested area which producing about 6.4 million tons [1].

Of all vegetable crops, tomato in particular, is heavily infected with many different plant parasites and pathogens [2]. More than 200 diseases were recorded on tomato plants and were considered responsible for 70 to 95% of the annual losses of tomato production while Fusarium wilt disease is only responsible for 10 to 50% of these losses [3]. In Egypt, the losses due to tomato wilt disease reached up to 67% [4].

Many isolates of *Fusarium oxysporum* are known to be important plant pathogens that cause severe damage to many host plants [5-7]. Moreover, pathogenic strains have been grouped into host-specific forms called *formae speciales* (f.spp.), which are sometimes divided further into races based on cultivar specificity [8-10]. As a result, Fusarium genus contains approximately 120 *formae speciales* and races, based on the plant species and cultivars they infect [8,11,12]. Different isolates of a particular forma speciales are known to be varied in their virulence potential toward a certain plant species [5,12-15].

Generally, virulence characterization achieved by traditional pathogenicity bioassays under greenhouse conditions and recently through molecular approaches. Molecular identification techniques are candidate to be among the most promising useful tolls for Fusarium isolate identification as well as for many other pathogens due to the accuracy in addition to the consistency of the results [16,17].

The ability of a Fusarium isolate to infect particular plant species depends on specific genes that distinguish virulent and avirulent strains. Recently, 'virulence factors', that encoding host determining small secreted proteins, called effectors, are found to play a crucial role in Fusarium virulence potential on his hosts [18]. During the preliminary infection stage, Fusarium isolates secret a special protein and inject it into the xylem of the host plant leading to initiation of infection process. On the other hand, in resistant host, immune plant genes (i-) can recognize and counter-attack the effect of these virulence effectors [19].

In order to control this disease in tomato plants, different approaches including obtaining resistant cultivars and good understanding of pathogen population specially its virulence genes and their effects. Therefore, 15 different *Fusarium oxysporum* isolates were selected to present delta area in Egypt as a start for studding the Egyptian *Fusarium oxysporum* population.

Recently, in tomato several in planta secreted proteins have been identified with *F. oxysporum* f. sp. *lycopersici* (Sacc.). The first identified secreted in xylem protein (SIX) with *Fusarium oxysporum* f.sp. *Lycopersici* was called (Six1), which is a small cysteine-rich protein

*Corresponding author: Mohamed E Selim, Agricultural Botany Department, Faculty of Agriculture, Menoufiya University, Egypt

required for full virulence on tomato [20]. In addition, recognition of the protein by tomato plants carrying the resistance gene I-3 leads to disease resistance as mentioned by Rep et al. [21]. Therefore, Six1 is also called Avr3 to indicate its gene-for-gene relationship with the I-3 resistance gene. More recently, additional fungal proteins were identified from xylem sap of infected plants, encompassing the small secreted proteins Six2, Six3, Six4, Six5, Six6, Six7 and SIX 8 [22,23].

This study is aimed to:

1. Evaluation of virulence potential of 15 different Egyptian isolates of *Fusarium oxysporum* toward tomato using pathogenicity bioassay under Egyptian greenhouse conditions.

2. Characterization of these isolates for presence and absence of genes SIX1, SIX2, SIX3, SIX4, SIX5, SIX6, SIX7 and SIX8 using molecular technique based on PCR polymorphism analysis.

Materials and Methods

Fungal isolates

Fungal isolates of *Fusarium oxysporum* were originally isolated during 2013-2014 seasons from the cortical tissues of surface sterilized roots of naturally infected tomato plants (*Lycopersicum esculentum* Mill.) showed the typical symptoms of Fusarium wilt disease and grown in five different provinces in Egypt (Table 1). Fungal isolates which had culture and microscopic characteristics corresponding to those isolates of *Fusarium oxysporum* were cultured on Potato Dextrose Agar (PDA) media amended with 150 mg^{-1} streptomycin and 150 mg^{-1} chloramphenicol and purified using single spore colony technique. Purified isolates were reared on PDA plates and incubated at 27°C in the dark for 14 days. For each isolate, several small disks ($\approx$1 mm in diameter) were cut using a sterilized cork borer and stored in micro-bank tubes at -80°C. These initial disks were used later for preparing the inoculum for using in the next investigations.

Fungal inoculum for pathogenicity test

For preparation of the fungal inoculum for pathogenicity test in greenhouse, the initial disks of Fusarium isolates were cultured on PDA plates for 2 weeks. The mycelia and spores were scratched from the surface, suspended in water, sieved through 3 layers of cheese cloth and number of spores was counted using a Haemocytometer slide (Thoma, Germany). The concentration was adjusted using sterilized tap water. Seeds of "Super Marmande`` tomato cultivar which is susceptible to Fusarium wilt disease were sown in plastic trays filled with autoclaved beetmooth substrate (company, Egypt). After 45 days, seedlings were transplanted in plastic pots filled with 3 kg of soil and sand mixture (1:2, w/w). Two days after transplanting, tomato seedlings were injected 2 cm deep into the rhizosphere using 3 holes made around the stem base with 2×10^6 conidia suspended in 2 ml water. Five replicates were used within each Fusarium isolate. Control plants were treated only with 2 ml of tap water. Disease incidence and disease severity were recorded after 4 weeks from pathogen inoculation according to Vakalounakis and Fragkiadakis [24] on 0 to 3 visual scales:

0= no symptoms

1= light yellowing of leaves, light or moderate rot on tap root and secondary roots and crown rot

2= moderate or severe yellowing of leaves with or without wilting, stunting, sever rot on taproot and secondary roots, crown rot with or without hypocotyls rot, and vascular discoloration in the stem

3= Dead plants.

Disease severity was determined using the following formula:

$$Disease\ severity(\%) = \left[\frac{\left(\sum scale \times number\ of\ plants\ infected \right)}{Highest\ scale \times total\ number\ of\ plants} \right]$$

DNA extraction

For isolating the DNA from *Fusarium oxysporum* isolates, 1 disk of each isolate was cultured into 500 ml flask containing 200 ml of potato dextrose broth media (PDB, Defco, Germany) and incubated at 27°C at 100 rpm shaker. After 14 days, the mycelia and spores were sieved through 3 layers of cheese cloth. The mycelium was dried between 2 layers of sterilized filter papers and immediately freeze dried using lyfolization apparatus for overnight. 100 mg of the freeze dried mycelium was subjected into DNA extraction using GE Health care, illustra™, Uk kit (Table 1).

Genetic polymorphism analysis

For investigating the presence and absence of SIX genes, the DNA of 15 isolates of pathogenic *Fusarium oxysporum* isolated from tomato plants grown in five different governorates in Egypt in addition to 1 isolate of non-pathogenic *Fusarium oxysporum*, Fo162, which was kindly provided by Prof.Richard Sikora, Bonn University, Germany, was obtained using the same procedure mentioned above. DNA was diluted 10x proier to PCR amplification. PCR fragments were amplified by PCR using the 8 different primers of secreted in xylem (SIX) genes (Table 2).

The PCR mix contained 10 μL 5X Green Go Taq reaction buffer (Promega), 2 μl dNTPs (10 mM), 1 μl forward primer, 1 μl backward primer, 0.25 μl Tag polymerase (0.625 units) and 37.75 μl water and 2 μl fungal DNA as template. Polymerase Chain Reaction (PCR) procedure was performed in a Bio-RAD, C1000 thermo cycler by an initial denaturation at 95°C for 4 min, followed by 34 cycles of 95°C for 1 min, 50°C for 2 min and 72°C for 3 min. After the PCR procedure, samples were stored at 4°C. For each isolate with each SIX primer, 10 μL of the PCR product were injected into agarose gel electrophoresis to determine the presence and absence of effector avirulence gene.

Isolate number	Governorate	District	location
1	Kafer el-Sheihk	Sedi Ghazy	Sedi Nasr
2	Kafer el-Sheihk	Balteem	Sidi-Mubarak
3	Gharbya	Santa	Balkem
4	Giza	Giza	Agriculture research center
5	Kafer el-Sheigh	Balteem	Al-Shahabiya
6	Kafer el-Sheigh	Sakha	Al-Hamrawy
7	Kafer el-Sheigh	Sedi GHazy	Dukmera
8	Dakhahlia	El-Mansoura	Meet Ali
9	Dakhahlia	Shrbeen	El-Wekala
10	Dakhahlia	Talkha	Batrah
11	Dakhahlia	Meet Ghamr	Damas
12	Kafer el-Sheigh	Sedi Ghazy	Almanshiya
13	Gharbya	Qutuor	Qutur
14	Gharbya	Santa	Al-Gemmeza
15	Sharkiya	Zagazig	Al-Qinayat

Table 1: Pathogen isolates of *Fusarium oxysporum* isolated from infected tomato plants grown in different locations in Egypt and showed the typical symptoms of Fusarium wilt disease during 2013-2014 seasons.

Gel electrophoresis analysis

The agarose gel used in this analysis was prepared with 1 X Tris-Acetate EDTA-Buffer (TAE, AppliChem). 1 g agarose (Sigma) was added to 100 ml of TAE buffer and heated for 5 minutes in a microwave (MW800, Continent) at 650 watts. After cooling at approx. 50°C, 3 µL of 10mg ml^{-1} Ethidium Bromide (AppliChem) was added. This solution was poured into an electrophoresis tray and left for approx. 30 minutes until the gel had solidified. The gel was subsequently transferred to the gel electrophoresis chamber filled with 1 X TAE buffer. After transferring all samples to the wells of the gel, the electrophoresis analysis was conducted for 60 minutes at 80 Volt. An Ultraviolet transilluminator (BIO-RAD, Gel DOC™ –XR+) was used to visualize the DNA bands.

Results

Pathogenicity test

Fifteen isolates which showed the identical morphological and microscopically characters of *Fusarium oxysporum* species were isolated from infected tomato plants grown in fields located in 5 different governorates in Egypt (Table 1). To determine the virulence potential of obtained *Fusarium oxysporum* isolates, pathogenicity test was conducted under greenhouse conditions in Egypt. The results showed that all the tested isolates were pathogenic toward tomato plants and caused symptoms corresponding to the Fusarium wilt disease. Moreover, results revealed that the tested isolates were varied regarding to their aggressiveness and virulence potential under the bioassay conditions (Figure 1). In deed, the results illustrated that the highest percentage of disease severity i.e. 61%; 52%; 44%; and 41.65% were recorded within isolates 4; 5; 15; 2 respectively (Figure 1). On the other hand, isolates 8; 12 and 14 resulted in the lowest disease severity percentage on tomato plants (8,79%; 9,7% and 9,79% respectively)

Molecular identifications

Presence and absence of SIX effector genes: The linkage between virulence potential of fifteen pathogenic isolates in addition to one non-pathogenic isolate, Fo162 , of *Fusarium oxysporum* and the presence of secreted in xylem effector genes (SIX) was determined. The results showed that amplicons of SIX1 were detected with all isolates tested except isolates 8, 12, 14 and 15 in addition to the non-pathogenic isolate, Fo162, (Figure 2 and Table 3). Furthermore, no amplicons for SIX2, SIX3 and Six4 were detected among all *Fusarium oxysporum* isolates subjected to the test (Figures 2 and 3, Table 3). Similar results were also observed with SIX7 and Six8 (Figure 4). In contrarily, amplicons of Six6 were detected within all pathogenic tested isolates except isolates 1, 6, 7, 8, 11 and isolate 12 (Figure 5). Additionally, amplicons of SIX5 were detected only within isolates 2, 3, 4, 5, 7 and 15 (Figure 5, Table 3).

Discussion

Pathogenicity test was conducted under greenhouse conditions to determine the virulence potential of fifteen isolates belonging to species *Fusarium oxysporum*. The obtained results showed that all tested isolates were pathogenic on tomato plants under the bioassay conditions and can be classified into three different groups according to their virulence potential (Table 4). The first group contains the hyper-virulent isolates while the second group contains the moderate virulent isolates and the last group contains the hypo-virulent isolates (Table 4). To verify the bioassay results, the fifteen tested pathogenic isolates of *Fusarium oxysporum* in addition to one non-pathogenic Fusarium isolate, Fo162, were subjected to the polymorphic analysis based on

screening the presence and absence of the amplicons of 8 different secreted in xylem genes (SIX1, SIX2, SIX3, SIX4, SIX5, SIX6, SIX7 and SIX8). The obtained results illustrated that the amplicons of the three effectors, Six1, Six5 and Six6, were detected with most isolates of the hyper-virulent group while only two of the Six genes were detected within isolates of the moderate virulent group except isolate 3 which contain the three SIX genes (Table 4). Furthermore, No amplicons for any of the tested SIX genes were detected within any of the isolates

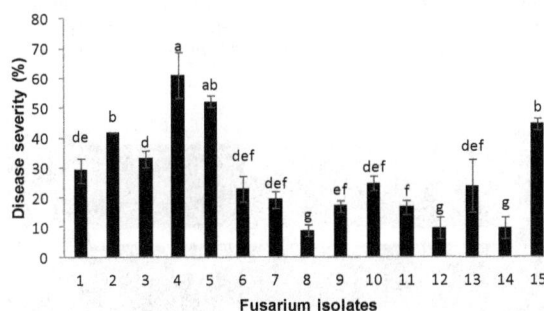

Figure 1: Disease severity recorded after 4 weeks on tomato plants inoculated with 15 different individual isolates of *Fusarium oxysporum* (1:15). Paired Means with different letters are significantly different based on Tukey test ($p \leq 0, 05$; n=5).

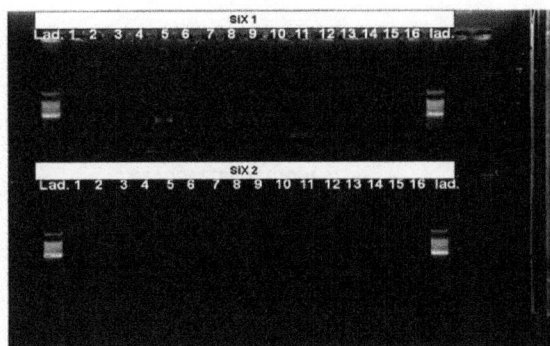

Figure 2: PCR amplicons generated by SIX 1 and SIX 2 primers using genomic DNA of 15 pathogenic isolates (lanes from 1:15) and one non-pathogenic isolate, Fo162, (lan16) of *Fusarium oxysporum*. Lad. is 1Kbp DNA ladder

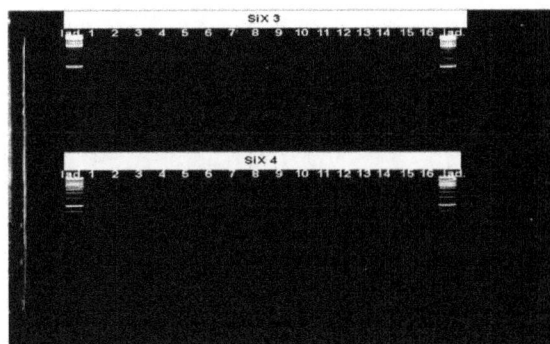

Figure 3: PCR amplicons generated by SIX 3 and SIX 4 primers using genomic DNA of 15 pathogenic isolates (lanes from 1:15) and one non-pathogenic isolate, Fo162, (lan16) of *Fusarium oxysporum*. Lad is 1Kbp DNA ladder.

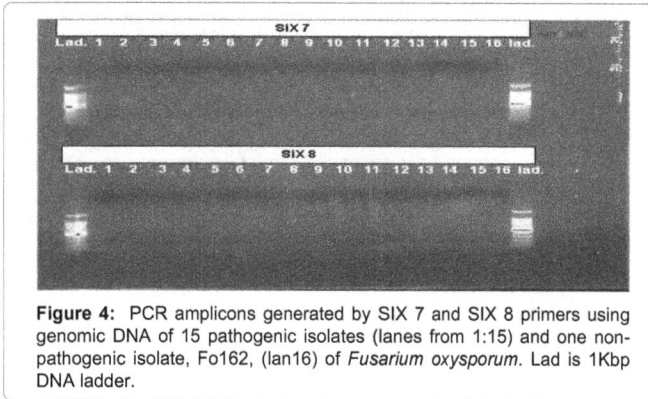

Figure 4: PCR amplicons generated by SIX 7 and SIX 8 primers using genomic DNA of 15 pathogenic isolates (lanes from 1:15) and one non-pathogenic isolate, Fo162, (lan16) of *Fusarium oxysporum*. Lad is 1Kbp DNA ladder.

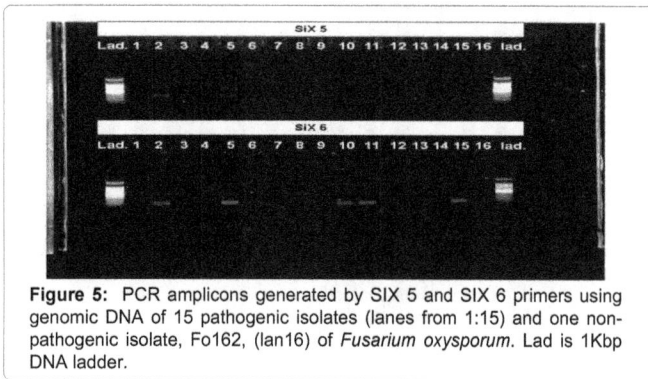

Figure 5: PCR amplicons generated by SIX 5 and SIX 6 primers using genomic DNA of 15 pathogenic isolates (lanes from 1:15) and one non-pathogenic isolate, Fo162, (lan16) of *Fusarium oxysporum*. Lad is 1Kbp DNA ladder.

belonging to the hypo-virulent group except isolate 14. Note worthy, amplicons of genes SIX2, SIX3, SIX4, SIX7 and SIX8 could not be detected within any of the tested isolates which may suggest their absence in the Egyptian isolates.

These results are in consistence with results of Lievens et al. [23] who reported that SIX1, SIX2, SIX3 and SIX5 can be used for the unambiguous identification of *F. oxysporum f. sp. lycopersici* while Six6 and SIX7 were presented in few *Fusarium oxysporum formae speciales* and suggested that these genes may play a more general role in Pathogenicity. In 2002, Rep et al. [25] mentioned that SIX1 gene encodes a protein which was detected in xylem sap of infected tomato plants and this protein is likely to play a role in host colonization at least under some conditions or in some hosts and it was clearly that SIX1 is not required for pathogenicity under the conditions of their bioassay. Furthermore, Rep et al. [21] and Houterman et al. [19] reported that two avirulence proteins of *Fusarium oxysporum*, Avr3 (SIX1) and Avr1 (SIX4), have been recently identified and we found to be required for full virulence of *Fusarium oxysporum* on tomato. Moreover, the virulence function has been established for tow of these small proteins using gene knockout experiments [26]. These are Six4 (Avr1), which is required for I and I-1-mediated resistance [27-29], and Six3 (Avr2), which is required for I-2-mediated resistance while the validity of virulence function of the other SIX genes still under investigation. In addition, Six4/Avr1 was found to suppress I-2- and I-3-mediated disease resistance [27,30,31].

In conclusion, the results obtained from present study suggest that

name	Sequence of forward primer (5'- 3')	Sequence of backward primer (5'- 3')
Six1	GTATCCCTCCGGATTTTGAGC	AATAGAGCCTGCAAAGCATG
SIX2	CAACGCCGTTTGAATAAGCA	TCTATCCGCTTTCTTCTCTC
SIX3	CCAGCCAGAAGGCCAGTTT	GGCAATTAACCACTCTGCC
SIX4	TCAGGCTTCACTTAGCATAC	GCCGACCGAAAAACCCTAA
SIX5	ACACGCTCTACTACTCTTCA	GAAAACCTCAACGCGGCAAA
SIX6	CTCTCCTGAACCATCAACTT	CAAGACCAGGTGTAGGCATT
SIX7	CATCTTTTCGCCGACTTGGT	CTTAGCACCCTTGAGTAAC
SIX8	ATGCAACCCCTACGCGTTCT	CTAGAAATTGTTTATAAACTGGAC
FO162	CGAGCAACCTCTCAGTATCAGATC	CGGAGCCTGCAAAGCTGTAC
FO162	CAAACCGGGCGAGCAA	GGAGCCTGCAAAGCTGTACAG

Table 2: Sequences of forward and backword primers of SIX1, SIX2, SIX3, SIX4, SIX5, SIX6, SIX7 and SIX8 genes in addition to specific forward and backword primers for one non-pathogenic *Fusarium oxysprum* isolate (Fo162)

isolate	SIX1	SIX2	SIX3	SIX4	SIX5	SIX6	SIX7	SIX8
1	+	-	-	-	-	-	-	-
2	+	-	-	-	+	+	-	-
3	+	-	-	-	+	+	-	-
4	+	-	-	-	+	+	-	-
5	+	-	-	-	+	+	-	-
6	+	-	-	-	-	-	-	-
7	+	-	-	-	+	-	-	-
8	-	-	-	-	-	-	-	-
9	+	-	-	-	-	+	-	-
10	+	-	-	-	-	+	-	-
11	+	-	-	-	-	-	-	-
12	-	-	-	-	-	-	-	-
13	+	-	-	-	-	+	-	-
14	-	-	-	-	-	+	-	-
15	-	-	-	-	+	+	-	-
Fo162	-	-	-	-	-	-	-	-

Table 3: PCR-screen targeting the SIX genes, SIX1, SIX2, SIX3, SIX4, SIX5, SIX6, SIX7 and SIX8 amplicons with 15 different pathogenic isolates of *Fusarium oxysporum* (1:15) and non-pathogenic isolate (Fo162).

Virulence potential	Isolate No.	Amplicons of SIX1	Amplicons of SIX5	Amplicons of SIX6
Hyper-virulent isolates	2	+	+	+
	4	+	+	+
	5	+	+	+
	15	-	+	+
Moderate isolates	1	+	-	-
	3	+	+	+
	6	+	-	-
	7	+	+	-
	9	+	-	+
	10	+	-	-
	11	+	-	-
	13	+	-	-
Hypo-virulent isolates	8	-	-	-
	12	-	-	-
	14	-	-	+

Table 4: Aggressiveness levels of 15 *Fusarium oxysporum* isolates and amplicons of SIX1, SIX5 and SIX6

the presence of SIX1, SIX5 and SIX6 genes within Egyptian isolates of *Fusarium oxysporum f. sp. lycopersici* can be used, for some extent, as a remarkable indicator for virulence potential on tomato plants under certain conditions. On the other hand, finding new tomato genotypes which contain immune genes against virulence effector genes Six1, SIX5 and SIX6 is highly recommended for using them in integrated pest management strategies in Egypt [24]. In deed, further investigations using many different pathogenic and non-pathogenic isolates belonging to various Fusarium forma speciales and representing different populations are needed to clarify the role of Fusarium virulence effector genes in pathogenicity process on different host plants.

Acknowledgments

Authors would like to thank Professor Dr. Flurin Grundler and Dr. Alexander Schouten from Bonn University, department of molecular phytomedicine, for on-going cooperation to support the research and provide the facilities necessary to achieve the present study.

References

1. Costa JM, Heuvelink E (2005) Introduction: The tomato crop and industry. Heuvelink E (Editor) Tomatoes. CAB International, UK, 1-19.

2. Sikora RA, Fernandez E (2005) Nematode parasites of vegetables. Luc, M. Sikora, R. A. And Bridge, J. (Editors). Plant parasitic nematodes in subtropical and tropical agriculture. CABI Publishing: UK, 319-392.

3. Lukyanenko AN (1991) Disease resistance in tomato. Genetic improvement of tomato 14: 99-119.

4. Lasheen HHA (2009) Control of wilt disease of tomato through application natural compost and endo-mycorrhizal fungi. Ph.D. dissertation, Mansoura University, Faculty of Agriculture, Plant Pathology department.

5. Olivain C, Alabouvette C (1997) Colonization of tomato root by a non-pathogenic strain of *Fusarium oxysporum*. New Phytol 137: 481-494.

6. Olivain C, Alabouvette C (1999) Process of tomato root colonization by a pathogenic strain of *Fusarium oxysporum* f. sp. lycopersici in comparison with a non-pathogenic strain. New Phytol 141: 497-510.

7. Olivain C, Trouvelot S, Binet M, Cordier C, Pugin A, et al. (2003) Colonization of flax roots and early physiological responses of flax cells inoculated with pathogenic and non-pathogenic strains of *Fusarium oxysporum*. Appl Environ Microb 69: 5453-5462.

8. Armstrong GM, Armstrong JK (1981) Formae speciales and races of *Fusarium oxysporum* causing wilt diseases. PE Nelson, TA Toussoun, RJ Cook (editors.) Fusarium: Diseases, biology, and taxonomy. Pennsylvania State University Press, University Park and London, 391-399.

9. Pietro AD, Madrid MP, Caracuel Z, Delgado-Jarana J, Roncero MI (2003) *Fusarium oxysporum*: exploring the molecular arsenal of a vascular wilt fungus. Mol Plant Pathol 4: 315-325.

10. Michielse CB, Rep M (2009) Pathogen profile update: *Fusarium oxysporum*. Mol Plant Pathol 10: 311-324.

11. Gordon TR, Okamoto D (1992) Population structure and the relationship between pathogenic and nonpathogenic strains of *Fusarium oxysporum*. Phytopathology 82: 73-77.

12. Alabouvette C, Edel V, Lemanceau P, Olivain C, Recorbet G, et al. (2001) Diversity and interactions among strains of *Fusarium oxysporum*: Application and biological control. Jeger MJ and Spence NJ (editors): Biotic interactions in plant-pathogen associations, CAB International, London, England, 131-157.

13. Mandeel Q, Baker R (1991) Mechanisms involved in biological control of *Fusarium* wilt of cucumber with strains of non-pathogenic *Fusarium oxysporum*. Phytopathology 81: 462-469.

14. Hallmann J, Sikora RA (1994a) Occurrence of plant parasitic nematodes and nonpathogenic species of *Fusarium* in tomato plants in Kenya and their role as mutualistic synergists for biological control of root knot nematodes. Int J Pest Manage 40: 321-325.

15. Hallmann J, Sikora RA (1994b) Influence of *Fusarium oxysporum*, a mutualistic fungal endophyte, on Meloidogyne incognita of tomato. J Plant Dis Protect 101: 475-481.

16. Edel V, Steinberg C, Avelange I, Laguerre G, Alabouvette C (1995) Comparison of three molecular methods for the characterization of *Fusarium oxysporum* strains. Phytopathology 85: 579-585.

17. Lievens B, Rep M, Thomma BP (2008) Recent developments in the molecular discrimination of formae speciales of *Fusarium oxysporum*. Pest Manag Sci 64: 781-788.

18. van der Does HC, Rep M (2007) Virulence genes and the evolution of host specificity in plant-pathogenic fungi. Mol Plant Microbe Interact 20: 1175-1182.

19. Houterman PM, Cornelissen BJ, Rep M (2008) Suppression of plant resistance gene-based immunity by a fungal effector. PLoS Pathog 4: e1000061.

20. Rep M, Meijer M, Houterman PM, van der Does H, Cornelissen BJC (2005) *Fusarium oxysporum* evades I-3-mediated resistance without altering the matching avirulence gene. Molecular Plant Microbe Interactions 18: 15-23.

21. Rep M, van der Does HC, Meijer M, van Wijk R, Houterman PM, et al. (2004) A small, cycteine-rich protein secreted by *Fusarium oxysporum* during colonization of xylem vessels is required for I-3-mediated resistance in tomato. Mol Microbiol 51: 1373-1383.

22. Houterman PM, Speijer D, Dekker HL, DE Koster CG, Cornelissen BJ, et al. (2007) The mixed xylem sap proteome of *Fusarium oxysporum*-infected tomato plants. Mol Plant Pathol 8: 215-221.

23. Lievens B, Houterman PM, Rep M (2009) Effector gene screening allows unambiguous identification of *Fusarium oxysporum* f. sp. lycopersici races and discrimination from other formae speciales. FEMS Microbiol Lett 300: 201-215.

24. Vakalounakis DJ, Fragkiadakis GA (1999) Genetic Diversity of *Fusarium oxysporum* isolates from Cucumber: Differentiation by Pathogenicity, Vegetative Compatibility, and RAPD Fingerprinting. Phytopathology 89: 161-168.

25. Rep M, Dekker HL, Vossen JH, de Boer AD, Houterman PM, et al. (2002) Mass spectrometric identification of isoforms of PR proteins in xylem sap of fungus-infected tomato. Plant Physiol 130: 904-917.

26. Houterman PM, Ma L, van Ooijen G, de Vroomen MJ, Cornelissen BJ, et al. (2009) The effector protein Avr2 of the xylem-colonizing fungus *Fusarium oxysporum* activates the tomato resistance protein I-2 intracellularly. Plant J 58: 970-978.

27. Houterman PM, Cornelissen BJ, Rep M (2008) Suppression of plant resistance gene-based immunity by a fungal effector. PLoS Pathog 4: e1000061.

28. Dababat AA, Selim ME, Saleh AA, Sikora RA (2008) Influence of *Fusarium* wilt resistant tomato cultivars on root colonization of the mutualistic endophyte *Fusarium oxysporum* strain 162 and its biological control efficacy toward the root-knot nematode *Meloidogyne incognita*. Journal of Plant Disease and Protection 115: 273-278.

29. Edel V, Steinberg C, Gautheron N, Alabouvette C (1997) Populations of Nonpathogenic *Fusarium oxysporum* Associated with Roots of Four Plant Species Compared to Soilborne Populations. Phytopathology 87: 693-697.

30. Houterman PM, Ma L, van Ooijen G, de Vroomen MJ, Cornelissen BJC, et al. (2009) The effector protein Avr2 of the xylem-colonizing fungus *Fusarium oxysporum* activates the tomato resistance protein I-2 intracellulary. The Plant J 58: 970-978.

31. Kistler HC (1997) Genetic Diversity in the Plant-Pathogenic Fungus *Fusarium oxysporum*. Phytopathology 87: 474-479.

8

A Dynamic View of Protein Phosphatase 1 Interaction with its Inhibitors

Sudhish Mishra[1], Mandal A[2], Gupta RC[1] and Mandal PK[2]*

[1]*Translational Science and Molecular Medicine, Michigan State University, Grand Rapids, MI 48202, USA*
[2]*Edward Waters College, Biology Program, Jacksonville, FL 32209, USA*

Abstract

Protein phosphatase 1 (PP1) is well known for their role in signal transduction and protein function. Together with its inhibitors 1 and 2, it regulates wide variety of cellular activities. We have amplified catalytic subunit of PP1 (PP1c) and its inhibitors 1 (I-1) and 2 (I-2) from dog cardiac mRNA by rt-PCR and cloned into bacterial expression vector pCRT7. Cloned genes were expressed in *E. coli* BL21 (DE3) pLysS by IPTG induction. Functional positive clones were identified by western blotting of bacterial lysate and polymerase chain reaction. Double transformed bacterial cells were also generated by transforming PP1c clone with either I-1 or I-2. Activity of PP1 was analyzed in whole bacterial lysate by measuring dephosphorylation of phos b. Activities of inhibitors were analyzed by their capability to inhibit PP1 from dephosphorylation of phos b. Our findings indicate differential regulation of PP1 by I-1 and I-2. Both expressed recombinant inhibitors 1 and 2 have high potency for inhibition of PP1 activity. Interestingly, I-2 co-expression caused increase in PP1c expression but no change in expression was observed when co-expressed with I-1.

Keywords: Protein phosphatase; Inhibitor; Gene expression; Cloning

Introduction

A well-controlled regulation of protein phosphatase and protein kinase is required to maintain the phosphorylation status of proteins. These proteins are responsible for modulating metabolic pathways as per their phosphorylation state. There are at least 4 types of ser-thr protein phosphatases, protein phosphatase1 (PP1), protein phosphatase 2A (PP2A), protein phosphatase 2B (PP2B (Calcineurin) and protein phosphatase 2C (PP2C), which along with multiple types of protein kinases regulate phosphorylation status of metabolically active proteins. Protein phosphatase-1 (PP1) is a major ser-thr phosphatase which comprises of a catalytic subunit and a variable regulatory subunit. Catalytic subunit is a functional unit which is directed to subcellular compartments on the basis of binding to regulatory subunit [1]. PP1 can modulate various functions like, neuronal signaling, muscle contraction, protein synthesis, glycogen metabolism, cell cycle etc. [2]. PP1 has 3 specific heat stable inhibitors-inhibitor-1, inhibitor-2, inhibitor-3 and DARPP-2. Inhibitor-1 and DARPP-2 are phosphorylation dependent and requires phosphorylation by PKA at Thr-35 and Thr-34, respectively. Inhibitor-2 does not need phosphorylation to inhibit PP1 activity. Instead, calcium channel mediated phosphorylation of Ser-43 of Inhibitor-2 relieves PP1 from its inhibition [3].

PP1 mediated phosphorylation state of many proteins are known to altered in pathological conditions including Heart failure where increased PP1 expression [4,5] caused cardiac dilation and premature death [6]. Limited gene manipulation studies with I-1 have shown promising results in normalizing cardiac function [7].

In view of these facts, a clear understanding of PP1 interactions with its inhibitors is required. We do not have enough information available about interaction of PP1 and its inhibitors and detailed knowledge will provide the information to manipulate metabolic pathways for treatment of wide spectrum of diseases wherever PP1 is involved.

In the current study, we have cloned the genes of PP1, inhibitor-1 and inhibitor-2 and expressed them in BL21 (DE3) cells by Isopropyl-D thiogalactoside (IPTG) induction. We analyzed the activity of PP1 and compared the inhibition of PP1 by inhibitor-1 and inhibitor-2. Since both inhibitors are potent inhibitors of PP1, they can be used as therapeutic agents in diseases where activity of PP1 is enhanced.

Materials and Methods

Materials

Cloning vector pcrT7-topo and bacterial expression system consist of *E. coli* BL21 (DE3) pLysS cells, mini plasmid isolation kit, IPTG and ampicillin were purchased from Invitrogen. Agar and LB media was from Bio 101 Systems (QBiogene, Carlsbad, CA).

Cloning of PP1, inhibitor-1 and Inhibitor-2

Using custom primers (Table 1), cDNA from dog heart, was amplified by PCR and cloned in to pCR-T7-topo vector according to manufacturer's instruction (Gene Bank Accession Nos. AY062037, AY063765 and EU170432). The recombinant plasmid containing gene for PP1 and inhibitors were isolated and purified. Purified recombinant plasmids were used to transform *E. coli* BL21(DE3) pLysS cells for expression studies.

Induction of gene expression

The recombinant proteins were expressed in BL21 (DE3) cells by IPTG induction according to the instructions provided by supplier (Invitrogen). Briefly, fresh LB medium was inoculated by overnight culture of cloned PP1 or inhibitors transformed in BL21 (DE3) cells. After growing the cultures for 2 hours, IPTG was added up to 1mM of final concentration. Cultures were further allowed to grow up to desired time (~4 hrs.). Bacterial cells were harvested by centrifugation and resuspended in 50 mM Tris-HCl (pH 7.4), 0.5 mM sodium EDTA (pH 7.0), 0.3 M sucrose, and protease inhibitors (0.8 mM benzamidine, 0.8 mg/l each aprotinin and leupeptin, and 0.4 g/l antipain).

PP1 activity Assay

Bacterial lysate was prepared by homogenization in the same

---

*Corresponding author: Prabir K Mandal, Professor, Biology Program, Edward Waters College, USA, Tel: 904-470-8091

Gene	Sequence	Size (bp)
Inhibitor-1	5'-CCATggAgCAAgACAACAg-3'	602
	5'-CCCAAAAgTgAAggAATAAgAA-3'	
Inhibitor-2	5'-CCAATggCggCCTCgACggCCTC-3'	697
	5'-TgAAgAACAAgAAgCAACgTACTA-3'	
PP1c	5'-gCCATgTCCgACAgCgAgAAg-3'	1122
	5'-TCCATgTTCCCCgTgACAggTg-3'	

Table 1: Primers used for polymerase chain reaction and sizes of amplified products.

buffer it was resuspended (Tris-sucrose buffer). With the use of [^{32}P] phosphorylase b as the substrate, PP activity was determined in the bacterial lysate. The assay was performed in a 60 μl aliquot that consisted of 50 mM Tris-HCl (pH 7.4), 5 mM caffeine, 0.5 mM EGTA, 0.5 mM EDTA, 50 mM β-mercaptoethanol 100 ng of aprotinin (protease inhibitor), 1 μg of the cell lysate and 550 pmol ^{32}P-phosphorylase a. The assay was initiated by adding the phos a mix and carried out for 10 min at 34°C. Incubation was rapidly stopped by addition of 20 μl of 60% TCA and 20 μl of BSA (50 mg/ml). Tubes were held in ice for 10 min and then centrifuged at 12,000 g for 5 min. After centrifugation, ^{32}P radioactivity was counted in 80 μl of clear supernatant in 5 ml of liquid scintillation fluid. PP1 activity was calculated by comparing with samples with no IPTG induction. Activity in each sample was expressed as pmol ^{32}P released/min/mg of protein.

Inhibitor activity assay

Activity of inhibitor-1 and inhibitor-2 was measured by mixing bacterial lysate of inhibitor expressions with lysate from PP1 expression before adding phos a mix to measure phosphatase activity. Briefly, 10 μl (1 μg) of PP1 and 30 μl (15 μg or 30 μg) of inhibitor-1 or inhibitor-2 was mixed in the assay tube and kept in ice for 5 min. Reaction was initiated by adding 20 μl phos a mix and was incubated at 34°C for 10 min. Reaction was terminated by adding 20 μl of 60% TCA and 20 μl of BSA (50 mg/ml). Inhibitor activity was calculated by comparing released ^{32}P with the samples without inhibitor.

Western blotting

To determine the protein expression of PP1C, Inhibitor-1 and inhibitor-2, western blotting was performed on SDS extracts of the bacterial extracts. Bacterial culture was centrifuged at 6000 rpm for 15 min. Pellet was resuspended in Tris-Sucrose homogenization buffer and homogenized for 2 × 15 sec using homogenizer. Extract was prepared by adding 1/10th volume of 20% SDS in the homogenized sample followed by incubating in boiling water for 10 min. After boiling samples were allowed to cool at room temperature and centrifuged at 12,000 × g for 15 min. Supernatant was collected in the fresh tubes and used for western blotting. Protein amount was measured according to Lowery's method [8]. Equal amount of protein was subjected to electrophoresis on 4-20% SDS-polyacrylamide gel (Bio-Rad), and the separated proteins were electrophoretically transferred to a nitrocellulose membrane [9,10]. Accuracy of the electro transfer was confirmed by staining the membrane with 0.1% ponceau S. For the immunoreaction, the nitrocellulose blot was incubated with diluted primary antibody (monoclonal or polyclonal) based on the supplier's instructions. Antibody-binding protein was visualized by autoradiography after treating the blot with horseradish peroxidase-conjugated secondary antibody (anti-mouse for PP1 and Inhibitor-2 and anti-rabbit for inhibitor-1) and enhanced chemiluminescence

color-developing reagents according to the supplier (NEN).

Co-transformation of *E. coli*

Clones of PP1c and inhibitors (either inhibitor-1 or inhibitor-2) were mixed in equal concentration of 5 ng/μl. One μl of this mixture was added into chemically competent BL21 pLysS cells and transformed by heat shock at 42°C. Transformed cells were selected on amp-LB agar plate. 10 colonies were grown in 10 ml culture and induced by IPTG for gene expression. Dual-positive clones were selected by western blot analysis for PP1 and inhibitor-2. Phosphatase activity and western blots were carried out on these samples according to the methods described in previous sections.

Results

Expression of recombinant proteins

To analyze the expression of recombinant PP1, culture was incubated for different time points (1hr-4 hrs) after induction with 0.5 mM IPTG. Both western blot and activity shows increasing trend with time (Figure 1). Similarly, cultures for inhibitor-1 and inhibitor-2 were incubated for 1-4 hrs. after adding 1 mM IPTG. Increase in time of expression after induction with 1 mM IPTG showed increasing trend in expression and PP1 inhibiting activity for both I-1 and I-2 (Figures 2 and 3).

Effect of heat on inhibitor-1 and inhibitor-2

Naturally both of these inhibitors are heat and acid stable. We tested expressed inhibitor-1 and inhibitor-2 for heat resistance. After treatment both inhibitors not only maintained their activity but also showed 10-20% enhancement in their activity (Figure 4).

Effect of co-expression

I-1 and I-2 were separately co-expressed with PP1c inside single bacterium to study their effect on each other. I-1 co-expression had no effect on amount of gene expression of PP1 or vice versa but I-2 co-expression with PP1c resulted in higher expression of both genes when compared to their individual expression (Figures 5 and 6).

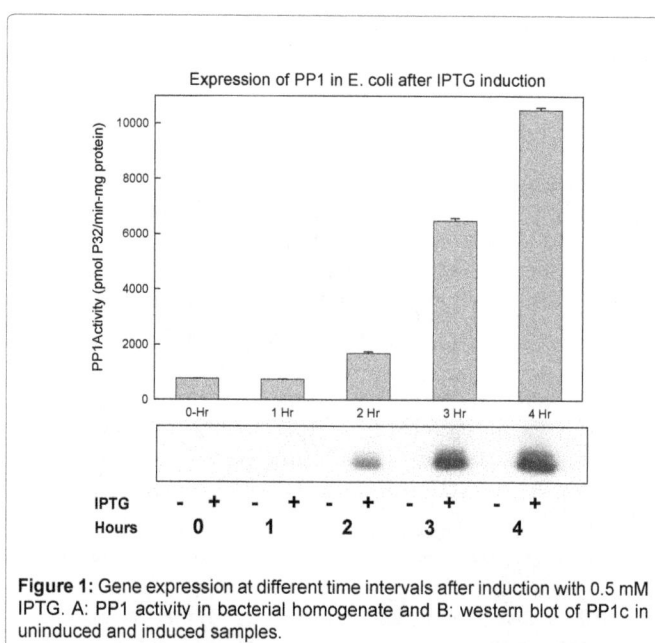

Figure 1: Gene expression at different time intervals after induction with 0.5 mM IPTG. A: PP1 activity in bacterial homogenate and B: western blot of PP1c in uninduced and induced samples.

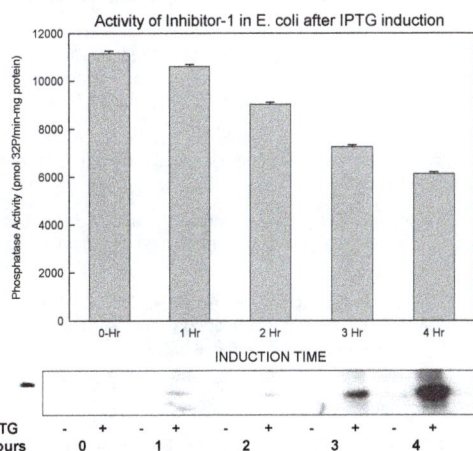

Figure 2: Gene expression at different time intervals after induction with 1.0 mM IPTG. A: PP1 activity in bacterial homogenate and B: western blot of I-1 in un-induced and induced samples.

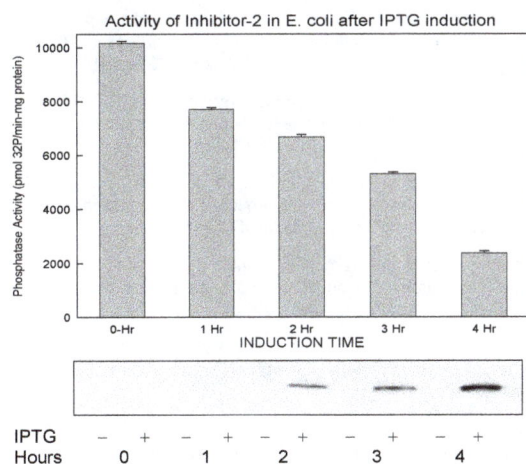

Figure 3: Gene expression at different time intervals after induction with 1.0 mM IPTG. A: PP1 activity in bacterial homogenate and B: western blot of I-2 in un-induced and induced samples.

Discussion

We have cloned and analyzed the expression of canine PP1 and inhibitors in bacterial system. The purpose of study was to characterize these genes for future gene manipulation studies. Our bacterial expression system utilizes T7 promoter which results in robust expression of mammalian genes in bacteria. We used BL21(DE3) pLysS cells for expression of canine PP1 and inhibitors genes. These cells maintain T7 RNA polymerase under suppressed conditions to minimize basal level expression. An inducer IPTG is required to relieve T7 RNA polymerase from suppression and start gene expression [11]. We analyzed the amount and activity of expressed products up to 4 hours after addition of IPTG. Samples were collected for analysis after each hour. Our results showed that gene expression is associated with time of incubation after addition of IPTG. PP1 and both inhibitors continuously increased up to 4 hrs. It indicates that the products are stable and active at least up to 4 hrs. and they can be used for further analysis.

Analysis of gene function showed maximum PP1 activity in extract of 0.5 mM IPTG induced cells in comparison to extracts prepared after incubation with other concentrations of IPTG (data not shown), thus this concentration of IPTG was used for further experiments. The decrease in activity at higher concentrations of IPTG may be due to aggregation of enzyme at very high concentrations [12,13]. Maximum activity of both inhibitors were observed at 1mM IPTG concentration (data not shown), thus selected for further experiments.

In the earlier studies several workers have reported about dependence of PP1 on Mn^{2+} [14]. There are other reports also that Mn^{2+} is required for activation of PP1 [15]. The role of Mn^{2+} in activation of PP1 is not clear. However, the PP1 described in this manuscript is Mn^{2+} independent.

PP1 inhibitors 1 & 2 are acid and heat stable and mostly exist in random coil structure [16-19]. This unique structure makes them resistant to heat but sensitive to proteases. Inhibitors acquire flexible conformation and they bind to PP1 at multiple sites with a long range distribution on the surface of the enzyme [20]. We tested the PP1 inhibition activity of both inhibitors after heat treatment and observed slight increase in PP1 inhibition after treatment. This increase in inhibition may be due to heat-induced dissociation of some interacting proteins from inhibitors. This separation might make them free and available to interact with PP1.

Co-expression of Inhibitors with PP1c gives an opportunity to study their interaction. Selective increased expression of both genes during co-expression of I-2 and PP1c indicates their unique interaction which needs to be explored. It looks like that inhibition of PP1 activity by inhibitor-2 somehow inducing more expression of PP1. This extra PP1 requires extra inhibitor-2, which induces expression of more inhibitor-2. This feedback mechanism may be the cause of increased

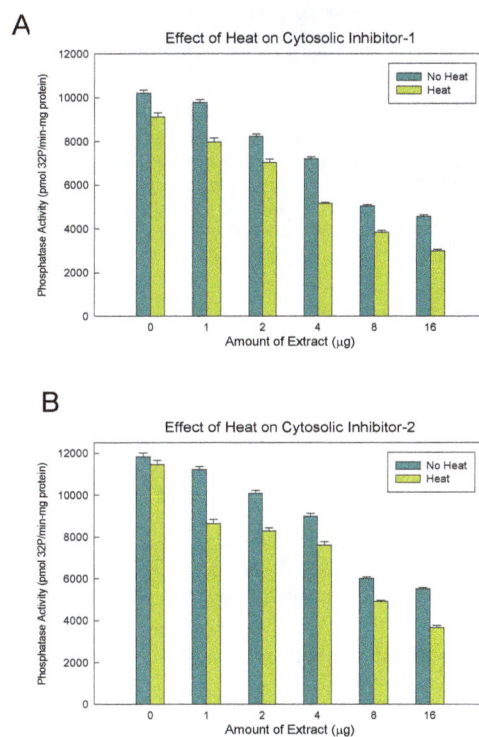

Figure 4: Effect of Heat on inhibitors. Inhibition of phosphatase activity by different amount of I-1 (A) and I-2 (B) induced bacterial extract.

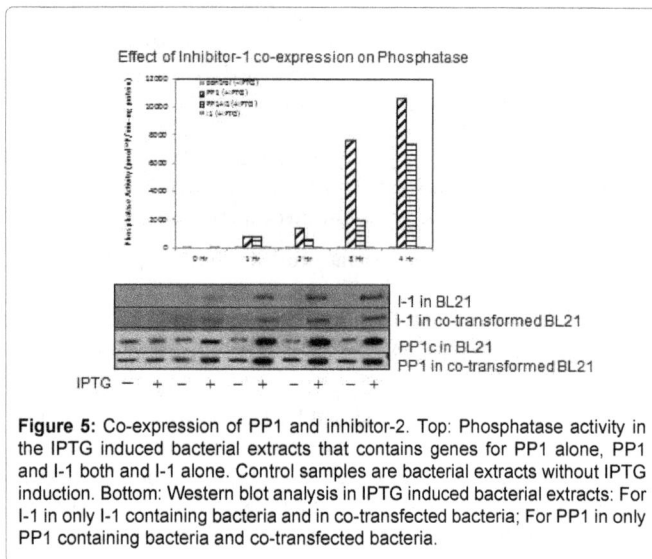

Figure 5: Co-expression of PP1 and inhibitor-2. Top: Phosphatase activity in the IPTG induced bacterial extracts that contains genes for PP1 alone, PP1 and I-1 both and I-1 alone. Control samples are bacterial extracts without IPTG induction. Bottom: Western blot analysis in IPTG induced bacterial extracts: For I-1 in only I-1 containing bacteria and in co-transfected bacteria; For PP1 in only PP1 containing bacteria and co-transfected bacteria.

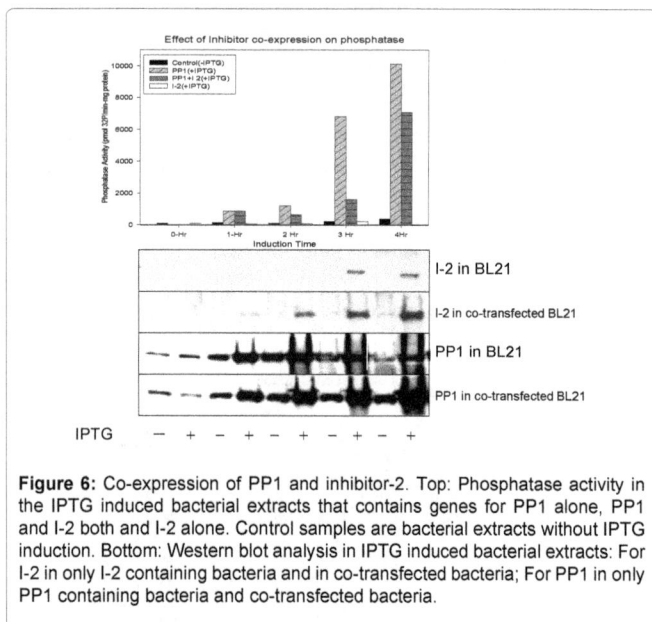

Figure 6: Co-expression of PP1 and inhibitor-2. Top: Phosphatase activity in the IPTG induced bacterial extracts that contains genes for PP1 alone, PP1 and I-2 both and I-2 alone. Control samples are bacterial extracts without IPTG induction. Bottom: Western blot analysis in IPTG induced bacterial extracts: For I-2 in only I-2 containing bacteria and in co-transfected bacteria; For PP1 in only PP1 containing bacteria and co-transfected bacteria.

expression of both genes. In our earlier study on one kidney one-clip model of hypertension, we have reported increased expression of PP1 and Inhibitor-2 [21]. This may be due to feedback mechanism proposed in the current study. One recent study on PPI-I2 complex analysis by X-ray crystallography has shown that Thr-74 phosphorylation of I-2 activates PP1 by dissociates it from metal containing active site of PP1 [22] but how they affect gene expression is not known.

In summary, phosphatase inhibition by I-1 and I-2 are very well known for long time. A detailed study about their interactions, effect on associated proteins and their role on gene expression is required. A better understanding of phosphatase-inhibitor relationship will help in better manipulation of metabolic pathways.

References

1. Cohen PT (2002) Protein phosphatase 1--targeted in many directions. J Cell Sci 115: 241-256.

2. Shenolikar S (1994) Protein serine/threonine phosphatases--new avenues for cell regulation. Annu Rev Cell Biol 10: 55-86.

3. Siddoway BA, Altimimi HF, Hou H, Petralia RS, Xu B, et al. (2013) An essential role for inhibitor-2 regulation of protein phosphatase-1 in synaptic scaling. J Neurosci 33: 11206-11211.

4. Mishra S, Gupta RC, Tiwari N, Sharov VG, Sabbah HN (2002) Molecular mechanisms of reduced sarcoplasmic reticulum Ca(2+) uptake in human failing left ventricular myocardium. J Heart Lung Transplant 21: 366-373.

5. Netticadan T, Temsah RM, Kawabata K, Dhalla NS (2000) Sarcoplasmic reticulum Ca(2+)/Calmodulin-dependent protein kinase is altered in heart failure. Circ Res 86: 596-605.

6. Carr AN, Schmidt AG, Suzuki Y, del Monte F, Sato Y, et al. (2002) Type 1 phosphatase, a negative regulator of cardiac function. Mol Cell Biol 22: 4124-4135.

7. Pritchard TJ, Kawase Y, Haghighi K, Anjak A, Cai W, et al. (2013) Active inhibitor-1 maintains protein hyper-phosphorylation in aging hearts and halts remodeling in failing hearts. PLoS One 8: e80717.

8. Laemmli UK (1970) Cleavage of structural proteins during the assembly of the head of bacteriophage T4. Nature 227: 680-685

9. Gupta RC, Mishra S, Mishima T, Goldstein S, Sabbah HN (1999) Reduced sarcoplasmic reticulum Ca(2+)-uptake and expression of phospholamban in left ventricular myocardium of dogs with heart failure. J Mol Cell Cardiol 31: 1381-1389.

10. Gupta RC, Shimoyama H, Tanimura M, Nair R, Lesch M, et al. (1997) SR Ca(2+)-ATPase activity and expression in ventricular myocardium of dogs with heart failure. Am J Physiol 273: H12-18.

11. Alexander WA, Moss B, Fuerst TR (1992) Regulated expression of foreign genes in vaccinia virus under the control of bacteriophage T7 RNA polymerase and the *Escherichia coli* lac repressor. J Virol 66: 2934-2942.

12. Zhang AJ, Bai G, Deans-Zirattu S, Browner MF, Lee EY (1992) Expression of the catalytic subunit of phosphorylase phosphatase (protein phosphatase-1) in *Escherichia coli*. J Biol Chem 267: 1484-1490.

13. Zhuo S, Clemens JC, Hakes DJ, Barford D, Dixon JE (1993) Expression, purification, crystallization, and biochemical characterization of a recombinant protein phosphatase. J Biol Chem 268: 17754-17761.

14. Silberman SR, Speth M, Nemani R, Ganapathi MK, Dombradi V, et al. (1984) Isolation and characterization of rabbit skeletal muscle protein phosphatases C-I and C-II. J Biol Chem 259: 2913-2922.

15. Brautigan DL, Ballou LM, Fischer EH (1982) Activation of skeletal muscle phosphorylase phosphatase. Effects of proteolysis and divalent cations. Biochemistry 21: 1977-1982.

16. Huang HB, Chen YC, Lee TT, Huang YC, Liu HT, et al. (2007) Structural and biochemical characterization of inhibitor-1alpha. Proteins 68: 779-788.

17. Herzig S, Neumann J (2000) Effects of serine/threonine protein phosphatases on ion channels in excitable membranes. Physiol Rev 80: 173-210.

18. Chyan CL, Tang TC, Chen Y, Liu H, Lin FM, et al. (2001) Letter to the editor: backbone 1H, 15N, and 13C resonance assignments of inhibitor-1--a protein inhibitor of protein phosphatase-1. J Biomol NMR 21: 287-288.

19. Huang HB, Chen YC, Tsai LH, Wang H, Lin FM, et al. (2000) Backbone 1H, 15N, and 13C resonance assignments of inhibitor-2 -- a protein inhibitor of protein phosphatase-1. J Biomol NMR 17: 359-360.

20. Egloff MP, Johnson DF, Moorhead G, Cohen PT, Cohen P, et al. (1997) Structural basis for the recognition of regulatory subunits by the catalytic subunit of protein phosphatase 1. EMBO J 16: 1876-1887.

21. Gupta RC, Mishra S, Yang XP, Sabbah HN (2005) Reduced inhibitor 1 and 2 activity is associated with increased protein phosphatase type 1 activity in left ventricular myocardium of one-kidney, one-clip hypertensive rats. Mol Cell Biochem 269: 49-57.

22. Cannon JF (2013) How phosphorylation activates the protein phosphatase-1 • inhibitor-2 complex. Biochim Biophys Acta 1834: 71-86.

9

Investigation of Human Mesenchymal Stromal Cells Cultured on PLGA or PLGA/Chitosan Electrospun Nanofibers

Fatemeh Ajalloueian[1,2], Moa Fransson[3], Hossein Tavanai[1], Mohammad Massumi[4], Jöns Hilborn[2], Katarina Leblanc[5], Ayyoob Arpanaei[6*] and Peetra U Magnusson[3*]

[1]Department of Textile Engineering, Center of Excellence in Applied Nanotechnology, Isfahan University of Technology, Isfahan, 84156-83111, Iran
[2]Polymer Chemistry, Department of Chemistry, Ångström Laboratory, SciLife Lab, Uppsala University, 751 21 Uppsala, Sweden
[3]Department of Immunology, Genetics and Pathology, Uppsala University, 751 85 Uppsala, Sweden
[4]Department of Physiology, University of Toronto, Toronto, ON M5S 1A8, Canada
[5]Department of Clinical Immunology and Transfusion Medicine, Karolinska Institutet and Hematology Center at Karolinska University Hospital, Stockholm, Sweden
[6]Department of Industrial and Environmental Biotechnology, National Institute of Genetic Engineering and Biotechnology, Tehran, 14965-161, Iran

Abstract

We compared the viability, proliferation, and differentiation of human Mesenchymal Stromal Cells (MSC) after culture on poly(lactic-co-glycolic acid) (PLGA) and PLGA/chitosan (PLGA/CH) hybrid scaffolds. We applied conventional and emulsion electrospinning techniques, respectively, for the fabrication of the PLGA and PLGA/CH scaffolds. Electrospinning under optimum conditions resulted in an average fiber diameter of 166 ± 33 nm for the PLGA/CH and 680 ± 175 nm for the PLGA scaffold. The difference between the tensile strength of the PLGA and PLGA/CH nanofibers was not significant, but PLGA/CH showed a significantly lower tensile modulus and elongation at break. However, it should be noted that the extensibility of the PLGA/CH was higher than that of the nanofibrous scaffolds of pure chitosan. As expected, a higher degree of hydrophilicity was seen with PLGA/CH, as compared to PLGA alone. The biocompatibility of the PLGA and PLGA/CH scaffolds was compared using MTS assay as well as analysis by scanning electron microscopy and confocal microscopy. The results showed that both scaffold types supported the viability and proliferation of human MSC, with significantly higher rates on PLGA/CH nanofibers. Nonetheless, an analysis of gene expression of MSC grown on either PLGA or PLGA/CH showed a similar differentiation pattern towards bone, nerve and adipose tissues.

Keywords: Electrospinning; PLGA; Chitosan; Synthetic-natural scaffold; Mesenchymal stromal cells; Viability; Proliferation

Introduction

Mesenchymal Stromal Cells (MSC) are considered as a heterogenous cell source for tissue engineering approaches and have received a great deal of attention during the past decade [1-3]. MSC have the capacity to replicate in an undifferentiated state, as well as the potential, when stimulated, to develop into distinct mesodermal tissues, including bone, cartilage, fat, tendon, muscle, and other lineages, both *in vitro* and *in vivo*. Although several sources of adult stem cells have been described, bone marrow is the most widely utilized as a source of autologous MSC in the clinic. These multipotent stem cells can give rise to differentiated cells found in adult tissues, have proliferative potential, and facilitate immunomodulatory effects [3,4].

One of the most exciting applications of stem cells is their potential for use in regenerative medicine [5,6]. Autologous human MSC have previously been used in combination with artificial engineered scaffolds in attempts to regenerate new tissues or organs [7]. In order to develop tissue engineered scaffolds that are suitable for preclinical and clinical applications, it is necessary to investigate the biocompatibility of the biomaterial and its capacity to support the lineage/tissue-specific differentiation of stem cells. Several studies have indicated that three-dimensional (3D) synthetic scaffolds are suitable for stem cell-based tissue engineering applications [8,9], but stem cell differentiation and organization can be influenced by the scaffold architecture. An ideal implantable scaffold is expected to recapitulate many features of the native Extracellular Matrix (ECM) in the target tissue, in addition to topographical and biochemical cues. Electrospinning is one of the most adopted techniques with the capability of fulfilling such requirements. Electrospun nano/microfibers could not only mimic the ECM 3D structural organization but also make use of a high surface area-to-volume ratio that provides cells with abundant area for attachment, migration, and proliferation [10,11].

Thus far, both synthetic biodegradable materials such as PLA (polylactic acid), PGA (polyglycolic acid), PLGA (poly(lactic-co-glycolic acid)), and PCL (polycaprolactone) and naturally derived materials such as fibrin, gelatin, chitosan, and collagen have been well used in tissue engineering and regeneration [12-16]. Synthetic materials usually suffer from poor cell-scaffold attachment interactions because of their hydrophobic structure [17], although they are biodegradable and have good mechanical properties. On the other hand, naturally derived polymers suffer from poor mechanical properties [18,19], despite their biocompatibility and favorable support for cell-scaffold interactions. Therefore, a polyblend of synthetic/natural polymers is of interest for experimental investigations involving stem cells.

In the present study, a polyblend nanofibrous scaffold of PLGA/chitosan (PLGA/CH) was produced through emulsion electrospinning [20,21], and the adhesion, growth, and differentiation properties of

*Corresponding authors: Ayyoob Arpanaei, Department of Industrial and Environmental Biotechnology, National Institute of Genetic Engineering and Biotechnology, Tehran 14965-161, Iran

Peetra U Magnusson, Department of Immunology, Genetics and Pathology, Uppsala University, 751 85 Uppsala, Sweden

human MSC seeded onto this scaffold were compared to those of PLGA electrospun scaffolds. PLGA is a biodegradable polymer with excellent mechanical properties and processability which is used in most of FDA approved therapeutic devices. PLGA electrospun scaffolds have shown mechanical stability to withstand implantation and to support the regeneration of new tissue [22-24]. However, PLGA is a synthetic polymer (with a hydrophobic nature) that suffers from a low affinity for cell attachment. It has been shown that introducing naturally derived polymers such as collagen [24], gelatin [25], or chitosan [26] to PLGA can introduce cell recognition sites and/or improvement of protein adsorption to the scaffold surface to induce better interaction between cells and scaffold. Here, chitosan has been selected as the natural polymer to enhance the biological behavior of the PLGA scaffold. This naturally occurring, biocompatible, biodegradable, and non-toxic polysaccharide is a cationic polymer with a structural similarity to various Glycosaminoglycans (GAGs) of the ECM [27]. Specific interactions between GAGs and growth factors, receptors, and adhesion proteins suggest that the bioactivity of chitosan is a result of its analogous structure to GAGs [28].

Chitosan has been widely used in research for many tissue engineering applications, especially in combination with MSC for bone [28-30] (enhancing osteogenesis) and neural tissue engineering [31,32]. Previous studies have shown that MSC cultured on non-fibrous scaffolds composed of PLGA and chitosan facilitate cell viability, proliferation, and adhesion [31-33]. Considering the key role of autologous human MSC in combination with synthetic scaffolds in regenerative medicine, our aim in this study was to evaluate whether an electrospun polyblend of PLGA/CH, when seeded with human MSC, would demonstrate cell-scaffold superior interactions to those of scaffolds electrospun from PLGA alone.

Materials and Methods

Reagents

Chitosan (MW ≈ 200,000 g/mol, degree of deacetylation ≈ 90%) and PLGA (MW 66,000-107,000 g/mol, L-lactide/glycolide, 75:25) were purchased from Sigma-Aldrich (Munich, Germany). Polyvinyl alcohol (PVA, MW ≈ 72,000 g/mol), 1,1,1,3,3,3-hexafluor-2-propanol (HFP), hexamethyldisilazane (HMDS), and glacial acetic acid were obtained from Merck KGaA (Darmstadt, Germany). All solvents were used as received without further purification. Paraformaldehyd (PFA, 1%) was purchased from Sigma (St. Louis, MO, USA) and CellTiter 96 AQueous One solution was purchased from Promega (Madison, WI, USA).

Fabrication of electrospun PLGA and PLGA/CH scaffolds

Two different scaffolds of PLGA and PLGA/CH were fabricated in the present study. A PLGA solution (16% w/v) was prepared by dissolving PLGA in HFP at 50°C, with magnetic stirring for 3 h. PLGA/CH scaffolds were electrospun from emulsions of PLGA/CH/PVA. In brief, PLGA was dissolved in HFP (16%w/v), chitosan (4% w/v) was dissolved in acetic acid (14%) at room temperature, and a PVA solution (8% w/v) was prepared by dissolving PVA (emulsifier) in double-distilled water at 90°C. The electrospinning emulsion was prepared by adding PVA and chitosan solutions to PLGA solution, followed by mixing with a magnetic stirrer. The volume ratio of the 3 solutions was 1:1:1. The final mixture was stirred at room temperature for 12 h to obtain a homogenous emulsion. Electrospinning of the prepared emulsion was carried out, with a voltage of 14-16 kV applied to the blunt needle (21-gauge) tip of the 1-mL syringe (filled with the solution/emulsion) and the grounded aluminum foil collector.

The feed rate was 1 mL/h for PLGA and 0.25 mL/h for PLGA/CH/PVA. The needle tip-to-collector distance was 15 cm for both of the scaffolds. Nanofibers collected on aluminum foil were dried overnight under vacuum and then used for the characterization and cell culture experiments. To obtain PLGA/CH nanofibers, PVA was removed from final composites of PLGA/CH/PVA using an aqueous solution containing 50% ethanol. Weight reduction measurements showed that 8 h was sufficient for the PVA removal.

Morphology and characterization of the electrospun nanofibers

The morphology of the electrospun nanofibers was studied with the help of scanning electron microscope (SEM) (Seron Technologies AIS 2100, Seron Technologies Inc., Gyeonggi-do, Korea) micrographs. The average diameter of the electrospun nanofibers was measured by applying Image J software (Image J, National Institutes of Health, USA) to the SEM micrographs. Statistical analysis was carried out using SPSS 8.0. (SPSS Inc. Chicago, IL, USA).

The hydrophilic/hydrophobic nature of the electrospun nanofibrous scaffolds was measured by sessile-drop water-contact angle measurement using a video-based optical contact angle meter (Data Physics OCA 15EC). A 6 μL drop of distilled water was placed on the sample surface. The contact angles on the left and right sides of the drop were measured by SCA software. The average of ten angles is reported for each sample.

The tensile properties of the electrospun nanofibrous scaffolds were determined under both dry and wet conditions using a tensile tester (Instron 3345, Canton, MA, USA), with a load cell capacity of 10 N. Test specimens, 10 mm in width × 60 mm in length, with a thickness of 70-80 μm, were tested at a crosshead speed of 10 mm/min and a gauge length of 20 mm, under ambient conditions. A minimum of six specimens of each individual scaffold were tested. Wet specimens were prepared by soaking the samples in Phosphate-Buffered Saline (PBS) for 8 h at 37°C.

The porosity of the PLGA and PLGA/CH fibrous scaffolds was determined using an AutoPore III mercury porosimeter (Micromeritics Instrument, Norcross, GA, USA), based on the total pore volume measured with the mercury intrusion technique.

Isolation and culture of bone marrow-derived MSC

MSC were isolated and expanded from Bone Marrow (BM) of healthy donors as previously described [34] following approval by the ethics committee at Huddinge University Hospital and thereafter cultured and utilized at Uppsala University (EPN Uppsala, Dnr2013/410). The release criteria for MSC was based on spindle shaped morphology, cell viability >95% and flow cytometry of cells with >95% positivity for CD73, CD90, CD105, HLA-ABC and <5% for CD14, CD31, CD34, CD45 and HLA-DR as previously described [34]. The cultures were negative for bacteria, fungi and Polymerase Chain Reaction (PCR)-negative for *Mycoplasma pneumoniae*. The MSC were cultured in MSC medium consisting of Dulbecco's modified Eagle's medium-low glucose (DMEM-LG), supplemented with 10% heat-inactivated fetal bovine serum (FBS from PAA Laboratories GmbH, Pasching, Austria). In this study, MSC in passages 5-8 from two different donors were used.

Mesenchymal stem cell culture on scaffolds

Prior to seeding, the nanofibers were sterilized by ultraviolet irradiation of both top and bottom surfaces in a laminar flow hood (each side for 20 min). To avoid detaching the scaffold from the glass into the medium, sterile plastic rings were used on each scaffold, and 0.5 mL of culture medium was added to each well (including scaffolds) of a 24-well plate; the plate was then incubated at 37°C overnight. The next day, the medium on the scaffold surface was removed, and MSC suspended in 100 µL of culture medium were added to each scaffold at a density of 25,000 cells/cm². Cells were incubated with the scaffolds for 2 h, and then an additional 1 mL of culture medium was added to each well. The incubation of the cells on the scaffolds was then continued as indicated.

Cell and scaffold morphology

Morphological studies (by SEM) of MSC grown on electrospun PLGA and PLGA/CH nanofibers were performed after 1, 4, and 7 days of cell culture. Cell-seeded nanofiber constructs were harvested and washed with PBS, then fixed in 2.5% glutaraldehyde for 3 h. The scaffolds were dehydrated with increasing concentrations of ethanol (30%, 50%, 70%, 90% and 100%) for 10 min each. Finally, the cell-scaffold constructs were treated with hexamethyldisilazane (HMDS; Fluka Chemical, Milwaukee, WI, USA) to allow further water extraction. The dehydrated, cell-seeded constructs were maintained in desiccators equipped with a vacuum for overnight drying. After sputter-coating with platinum, SEM was used to observe cell and scaffold morphology and cell attachment to the nanofiber scaffolds.

Metabolic activity and proliferation of MSC

Cell viability and metabolic activity in response to various substrates, Tissue Culture Polystyrene cover slips (TCP), and electrospun samples of PLGA and PLGA/CH, were measured using the MTS cytotoxicity assay (CellTiter 96°AQueous Non-Radioactive Cell Proliferation Assay, Promega, Madison, WI, USA) according to the manufacturer's instructions. The MTS substrate [3-(4,5-dimethylthiazol-2-yl)-5-(3-carboxymethoxyphenyl)-2-(4-sulfophenyl)-2H-tetrazolium] is bioreduced by the mitochondria of live cells into a soluble formazan product, which can be quantified by absorbance measurement once it is excreted into the culture medium. Cells were seeded at a density of 15,000 cells per cm², and the cell activity was evaluated during a 7-day period. On days 1, 4, and 7, the cell-seeded, electrospun samples were washed with PBS and transferred to new plates containing 500 µL of culture medium in each well. Thereafter, 100 µL of MTS solution was added to each well. Cells were maintained for an additional 4 h in a humidified incubator at 37°C with 5% CO_2. The absorbance was read at 450 nm on a Labsystems Multiskan MS plate reader (Labsystems Diagnostics Group, Vantaa, Finland), and the response was defined as (cell-seeded scaffolds A450 - mean blank)/(mean TCP 1 day A450 -mean blank) × 100%. In this way, the cell activity on each substrate at different time points was compared to that of TCP on day 1. These assays were done in triplicate, followed by calculation of the mean values and standard deviations.

Cell proliferation was determined by counting the number of DAPI-stained cell nuclei that were associated with the PLGA or PLGA/CH nanofibrous scaffolds. Cell nuclei were counterstained with 4,6-diamidino-2-phenylindole dihydrochloride (DAPI, Sigma-Aldrich; 1 µg/mL) in three separate samples (PLGA or PLGA/CH) per time point, at 1, 4 and 7 days. A minimum of five randomly selected visual fields were selected and analyzed in a Zeiss 510 confocal microscope (Carl Zeiss AG, Obercochen, Germany).

Cell tracking and confocal analysis of MSC on scaffolds

MSC grown on electrospun PLGA and PLGA/CH nanofibers were stained with fluorescent 5-chloromethylfluorescein diacetate (CMFDA cell tracker green, Molecular Probes, Life Technologies, Stockholm, Sweden) for observation by fluorescence microscopy of the cells on/in the scaffolds at 4 and 14 days. In short, cells were incubated prior to the time of examination with 5 mM of CMFDA for 60 min at 37°C in the incubator. The dye was removed after incubation, and the cells were washed with PBS and supplemented with the respective culture medium. The MSC were cultured for an additional 24 h period, then washed with PBS, fixed in 1% PFA for 1 h, mounted in mounting medium, and observed with a Zeiss 510 confocal laser scanning microscope (Carl Zeiss).

Quantitative PCR analysis of human MSC

PLGA and PLGA/CH scaffolds were rinsed with PBS and placed in 500 µL of RNALater (Life Technologies, NY, US) for 24 h. Total RNA was extracted using an RNAeasy Micro Kit according to the manufacturer's instructions (Qiagen, Holden, Germany) and transcribed into cDNA with a Superscript II Reverse Transcriptase Kit (Life Technology). Gene-specific polymerase chain reactions were analyzed using the Applied Biosystems StepOne Plus Real-Time PCR Systems machine (Life Technology) during 40 cycles with Applied Biosystems SYBR Green Supermix (Life Technology). All targets (RUNX1, NES, PPARG) were detected with QuantiTect primers (Qiagen GmbH, Hilden, Germany), and the PCR reaction was set up according to the Qiagen QuantiTect protocol. Quantitative values were obtained from the threshold cycle number, and the x-fold change in expression as compared to control samples (MSC grown on tissue-culture plastic) was calculated using the threshold cycle number method. Target genes from each individual sample were normalized against a mean threshold cycle number of the target genes SDHA and 18s. Selection of reference target genes was determined by geNorm algorithm analysis and analyzed using qbase plus software (www.biogazelle.com) according to the company's protocol.

Statistical analysis

Statistical analysis was performed using one-way analysis of variance (ANOVA) and Student's t-test to determine the statistical significance between the two means, evaluated at $p<0.05$.

Results

Characterization of the electrospun scaffold

SEM analysis of PLGA (Figure 1A) and PLGA/CH nanofibers (Figure 1B) demonstrated that PVA removal did not affect the fibrous structure of PLGA/CH nanofibers, and the scaffold kept its integrity, with an average fiber diameter of 166 ± 33 nm, as compared to a diameter of 680 ± 175 nm for the PLGA scaffold. To obtain uniform, bead-free fibers, electrospinning conditions were optimized for both the PLGA solution and PLGA/CH/PVA emulsion. It was observed that the average diameter of nanofibers in the PLGA/CH scaffold was much lower than in the PLGA scaffold.

Water contact-angle measurements of PLGA and PLGA/CH scaffolds demonstrated a large decrease from PLGA (105 ± 4.3) to PLGA/CH (27 ± 8.1) nanofibers, revealing the hydrophobic nature of the PLGA material as well as the role of chitosan in conferring hydrophilicity on the PLGA/CH scaffold.

A comparison of the tensile mechanical properties of the PLGA, PLGA/CH/PVA, and PLGA/CH nanofibers (Table 1) revealed that the addition of chitosan and PVA to PLGA led to a strong reduction in the extensibility of the scaffolds, demonstrated as an "elongation

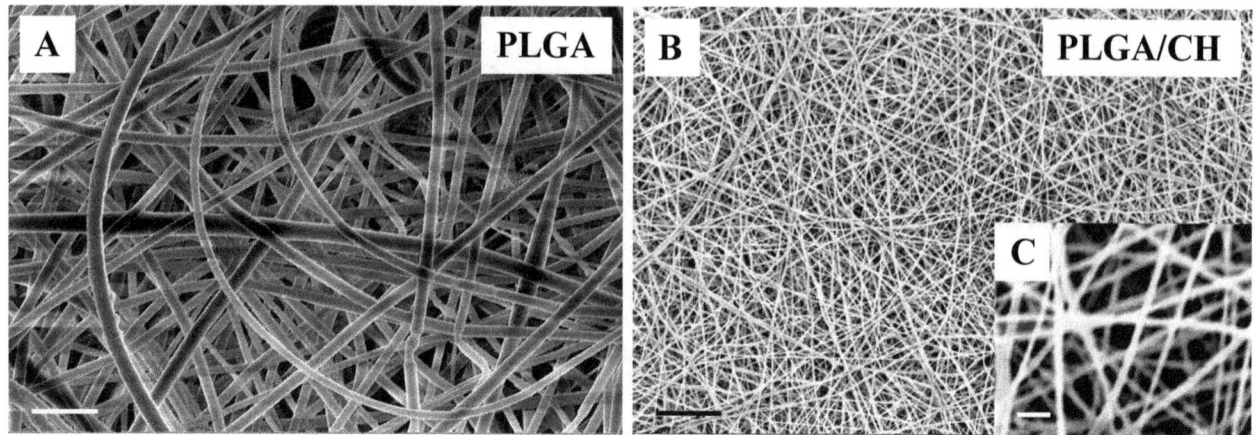

Figure 1: Magnification of the structure of electrospun **(A)** PLGA and **(B and C)** PLGA/CH nanofibers visualized by scanning electron microscopy (SEM). Scale bars represent 10 μm in A, B and 1 μm in C.

	Sample	Tensile strength (MPa)	Tensile strain (%)	Modulus (MPa)
Dry	PLGA	5.6 ± 1.10	107 ± 20	354.7 ± 57.6
	PLGA/CH/PVA	5.4 ± 0.57	14.7 ± 6.6	296 ± 32
	PLGA/CH	4.3 ± 0.68	32 ± 11	75 ± 10.8
Wet	PLGA	5.35 ± 0.80	88 ± 32	227.9 ± 57.6
	PLGA/CH/PVA	4.9 ± 0.60	12.5 ± 7.2	211 ± 24
	PLGA/CH	4.14 ± 1.05	28 ± 9.4	44 ± 18.5

Table 1: Tensile mechanical properties of PLGA, PLGA/CH/PVA, and PLGA/CH scaffolds in dry and wet state.

at break" (dry, from 107% ± 20 to 14.7% ± 6.6; wet, from 88 ± 32 to 12.5 ±7.2), whereas there were no significant differences in stiffness (Young's modulus) or tensile strength (maximum tensile stress). Also, PVA removal resulted in a sharp decrease in Young's modulus, from 296 ± 32 MPa in PLGA/CH/PVA to 75 ± 10.8 MPa in PLGA/CH under dry conditions. A similar trend was seen for wet samples, from 211 ± 24 MPa to 44 ± 18.5 MPa. However, PVA removal led to an increase in the extensibility of the hybrid structure for both dry (14.7% ± 6.6 to 32% ± 11) and wet (12.5% ± 7.2 to 28% ± 9.4) samples. The porosities of the PLGA and PLGA/CH scaffolds were calculated as 77.3% ± 4.6 and 81.7% ± 2.8, respectively.

Cell studies

MSC were cultured on PLGA and PLGA/CH scaffolds and analyzed by SEM after 1, 4 and 7 days (Figure 2). The morphology of the MSC cultured on both scaffolds changed over time. One day after seeding, the MSC had a rounded shape, in contrast to a more elongated cell shape after 4 days (Figure 2), suggesting that the seeded MSC increasingly attached to both PLGA and PLGA/CH nanofibers. Moreover, in continuous culture, the cells covered the substrate to a greater extent (day 7). Length measurements of the cells on PLGA and PLGA/CH showed increased elongation of MSC over time in culture (Figure 3).

To study the proliferation and viability of MSC cultured on the nanofibers, counting of cell nuclei by DAPI staining (Figure 4A and 4B, quantification in Figure 4C) and MTS analysis (Figure 4D) were performed after 1, 4, and 7 days of culture. Both PLGA and PLGA/CH scaffolds supported the proliferation of MSC, as shown by the higher count of DAPI-positive nuclei in PLGA/CH nanofibers (Figure 4B, white bars in Figure 4C) than in PLGA nanofibers (Figure 4A, black

bars in Figure 4C) and an increased metabolic activity in PLGA/CH scaffolds, as compared to PLGA (Figure 4D, white and black bars, respectively). Significantly higher absorbance was observed after 4 and 7 days when compared to one day of culture in both PLGA and PLGA/CH nanofibers (Figure 4D), with higher viability seen in PLGA/CH (Figure 4D, white bars) at all-time points when compared to PLGA (Figure 4D, black bars) and MSC cultured on control tissue-culture plastics (Figure 4D, gray bars).

To examine the morphological changes in MSC seeded onto PLGA and PLGA/CH scaffolds after 4 and 14 days, we cell-tracked the MSC in green (CMFDA) and analyzed them by confocal microscopy (Figure 5). In addition to an increased number of viable cells on PLGA/CH scaffolds as shown in Figure 4, we also found that the MSC elongated over time, making a connected network through interacting filopodia on both PLGA and PLGA/CH scaffolds by 14 days of culture (Figures 3 and 5).

In parallel with our morphologic analysis of MSC after 14 days, we analyzed the differentiation of MSC into different lineages, bone, neuronal and fat by qPCR. An increased expression of bone-(RUNX2), neuro-(NES) and adipose-(PPARG) related genes was detected in cells on both PLGA and PLGA/CH on day 14 when compared to MSC grown on tissue-culture plastic (Figure 6).

Discussion

Considering the recent attention that has been paid to scaffolds made from mixtures of polymers, especially synthetic-natural polyblends, we have fabricated PLGA/CH nanofibrous scaffolds and compared their behavior to that of PLGA nanofibers. Here we show that by the inclusion of two desired characteristics, the increased strength and durability of the synthetic polymer and the specific cell affinity of the natural one [18,35,36], we have produced an electrospun hybrid with enhanced physical properties and the required biological functionality to support human MSC, when compared to scaffolds made from pure chitosan and PLGA, respectively.

Reports in the literature have noted the potential application of PLGA/CH nanofibrous scaffolds for wound healing and skin tissue reconstruction [37-39]. However, no study has been carried out thus far to investigate the interaction of MSC with this hybrid scaffold. It should be noted that several researchers have focused upon the

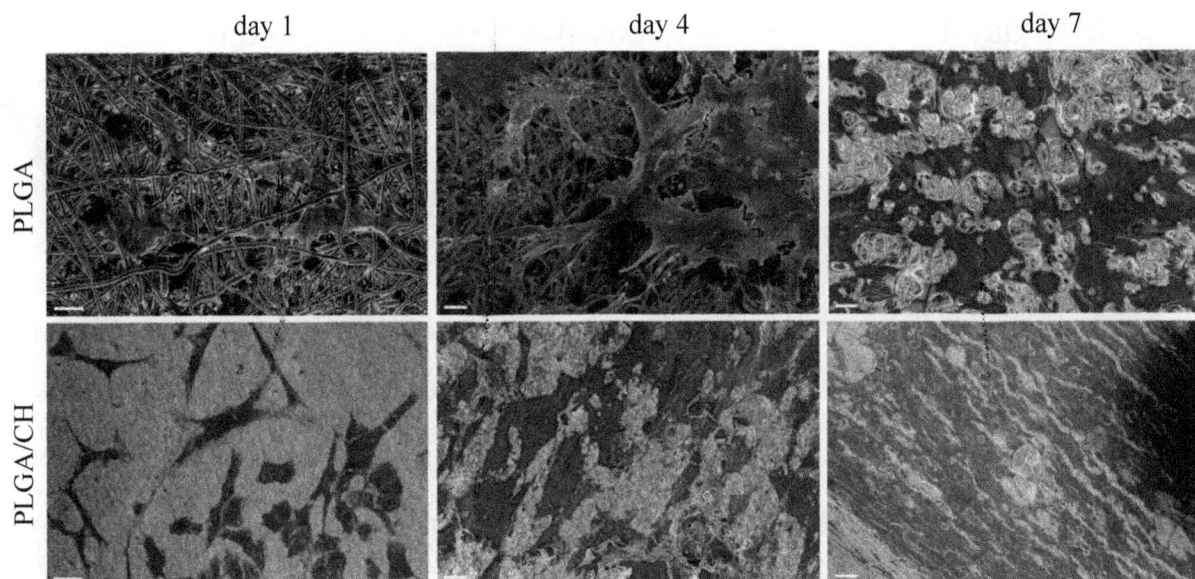

Figure 2: Adhesion of MSC onto electrospun scaffolds under *in vitro* culturing conditions. MSC were seeded onto PLGA and PLGA/CH scaffolds, and visualized by SEM after 1, 4, and 7 days in culture. MSC adhered, proliferated and stretched out on both of the PLGA and PLGA/CH scaffolds. Scale bars represent 20 µm.

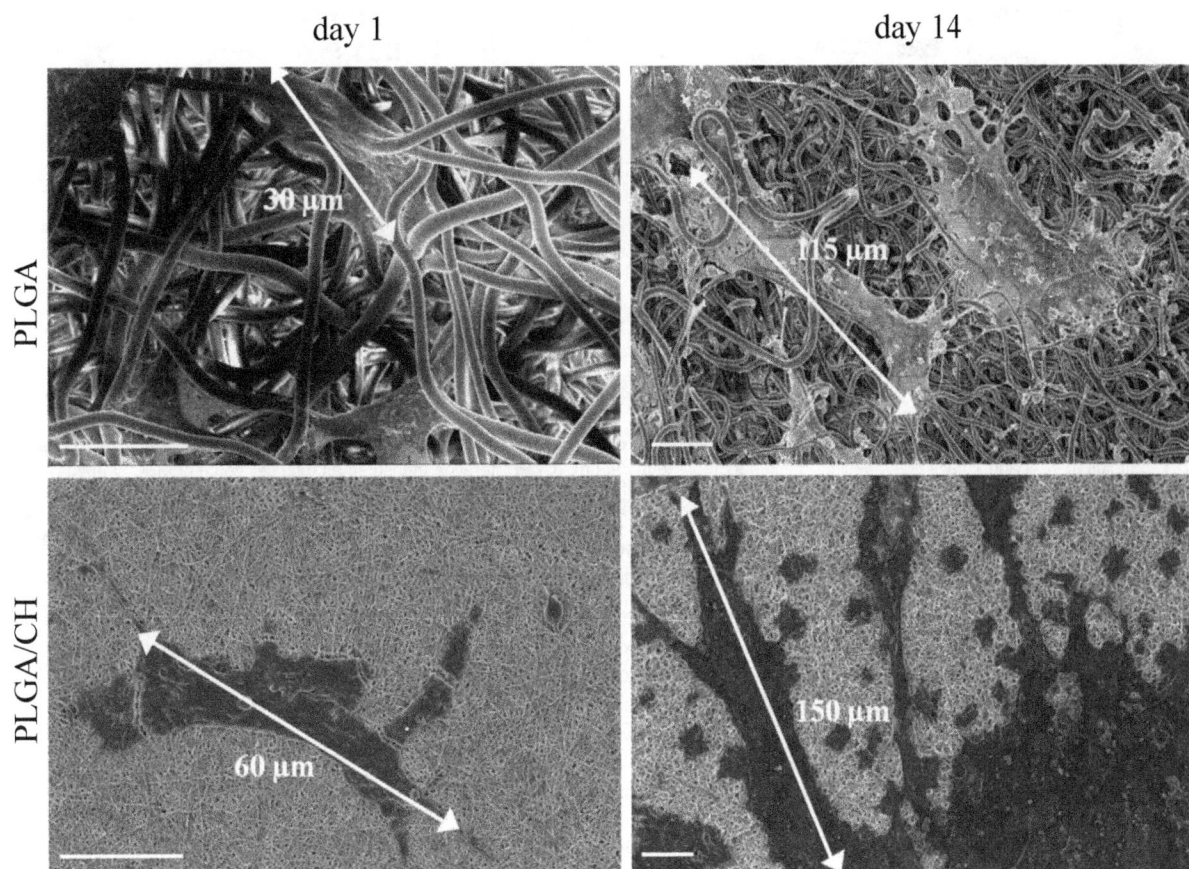

Figure 3: Morphology of MSC seeded onto PLGA and PLGA/CH nanofibers after 1 and 14 days in culture visualized by SEM. 1 day post seeding the MSC have adhered to PLGA and PLGA/CH scaffolds and demonstrate stretched out shape to over 100 µm after 14 days in culture. Scale bars represent 20 µm.

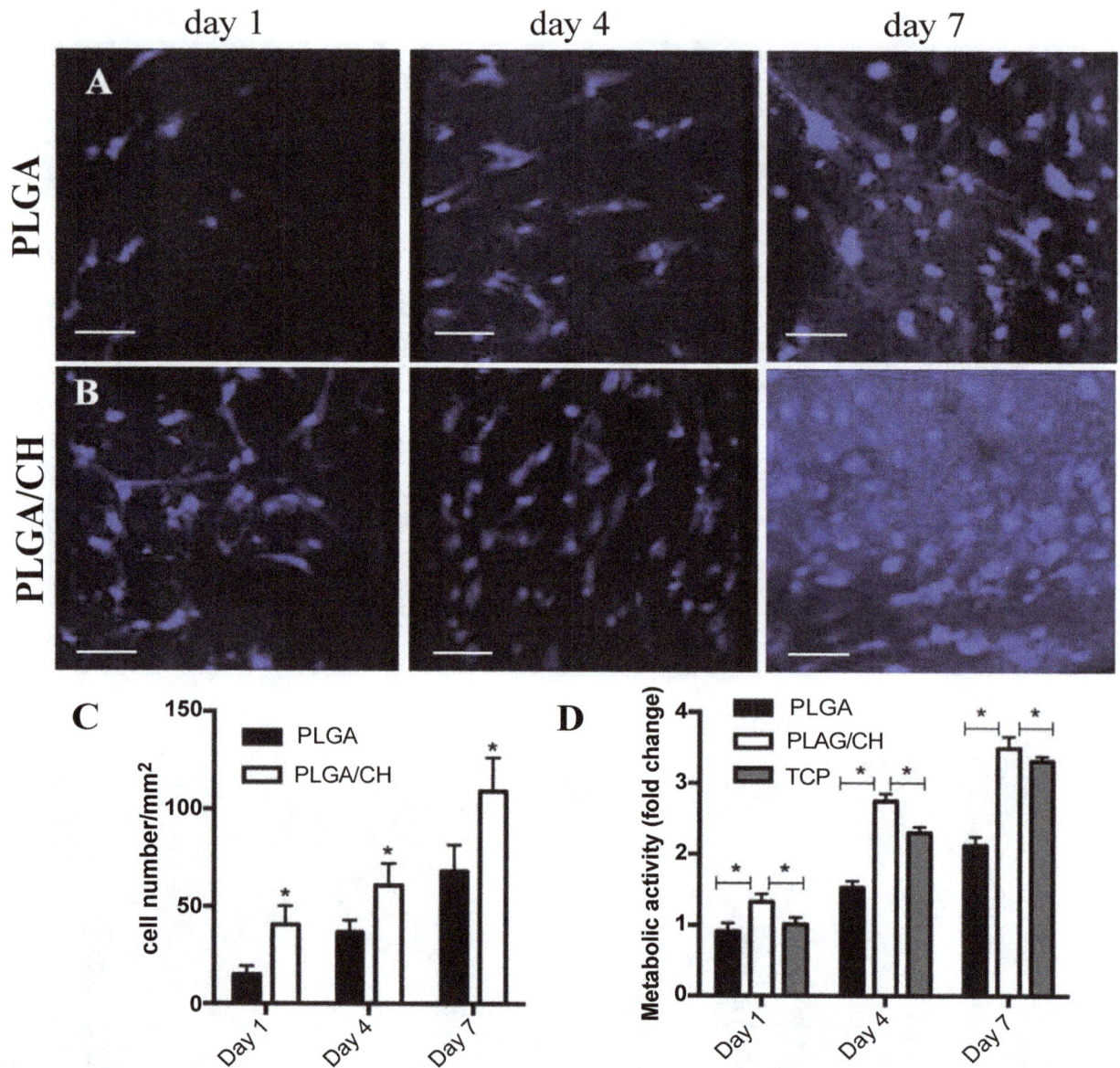

Figure 4: Proliferation and metabolic activity of MSC seeded onto PLGA and PLGA/CH nanofibers. MSC nuclei were stained with DAPI (blue) on **(A)** PLGA and **(B)** PLGA/CH scaffolds at 1, 4, and 7 days post seeding. Scale bars represent 100 μm. **(C)** Proliferation measured as total cell count of MSC seeded onto PLGA and PLGA/CH nanofibers at 1, 4, and 7 days in culture. Cell proliferation increased up to 7 days post seeding on both PLGA (black) and PLGA/CH (white) scaffolds. The total cell count of MSC on PLGA/CH was significantly higher at all of the investigated time points compared to PLGA (p<0.05). **(D)** Metabolic activity measured as a reflection of mitochondrial function using an MTS assay in MSC seeded onto PLGA (black) and PLGA/CH (white) nanofibers and tissue-culture plastic (TCP, gray). Metabolic activity of MSC was significantly increased when growing on PLGA/CH compared to PLGA (p<0.05) and TCP (p<0.05).

interaction between polyblend materials and adult stem cells. They have demonstrated improved attachment and growth of stem cells, enhanced mechanical performance, and increased wettability as compared to scaffolds prepared from either of the individual polymers making up the blend [40]. The interaction of MSC with PLGA/CH scaffolds has also been previously studied [31,33], but the scaffolds applied were not nanofibrous. For instance, Kuo et al. showed that MSC seeded onto PLGA/chitosan scaffolds fabricated by crosslinking and lyophilization were prone to differentiate toward osteoblasts without guidance, and toward neurons after introduction of nerve growth factor (NGF) [31]. In another work, Xue et al. incorporated autologous MSC into a neural scaffold consisting of a chitosan conduit with inserted PLGA filaments

[33]. They showed satisfactory repair and rehabilitation of large gaps after peripheral nerve injury in dogs, suggesting that a combination of autologous MSC with PLGA/CH could be used for neural tissue engineering.

Here, our comparison of PLGA/CH and PLGA nanofibers has demonstrated that PLGA/CH is more hydrophilic than PLGA, that it supports higher rates of cell proliferation, and that it has a mechanical strength similar to that of PLGA. We also demonstrated that both the scaffolds showed high porosities (77.3% for PLGA and 81.7% for PLGA/CH), which are characteristic of electrospun scaffolds. The difference between the porosity of two scaffold types was not statistically

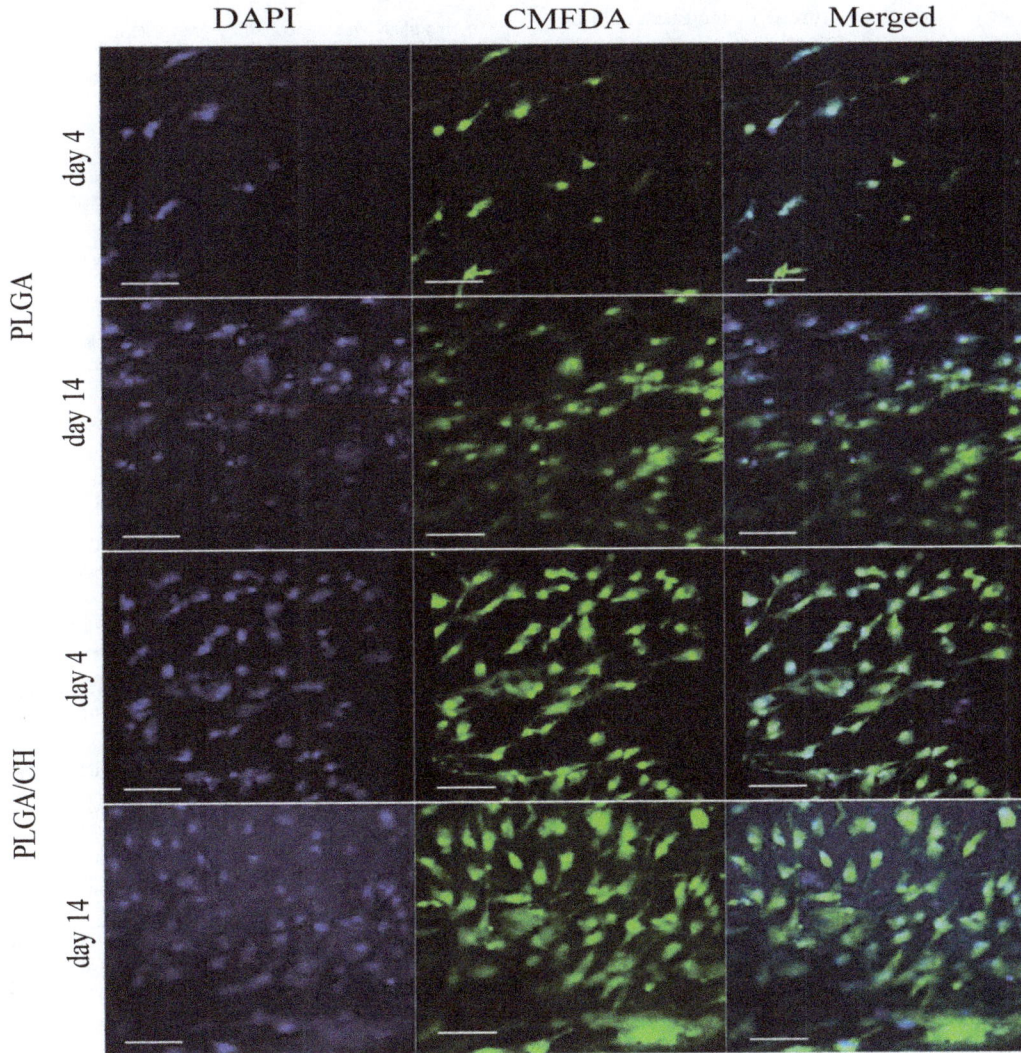

Figure 5: Morphology of MSC seeded onto PLGA and PLGA/CH nanofibers after 4 and 14 days in culture visualized by confocal microscopy. MSC are labeled with a cell tracking dye (CMFDA) here seen in green and nuclei stained with DAPI (blue). At 14 days post seeding the MSC stretched out to over 100 μm and the total number of cells covered the surface at a higher density on both nanofibers. Scale bars represent 100 μm.

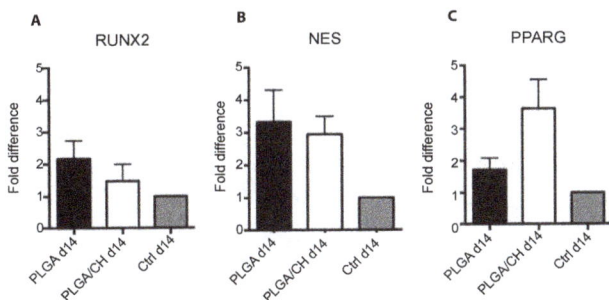

Figure 6: Genetic profile of MSC grown on PLGA and PLGA/CH on day14 (n=3). MSC grown on PLGA and PLGA/CH were analyzed by quantitative PCR for genetic markers in **(A)** bone (RUNX2): mean-fold difference, PLGA, 2.17; PLGA/CH, 1.46; **(B)** neurological marker (NES): mean-fold difference, PLGA, 3.33; PLGA/CH, 2.94; and **(C)** fat (PPARG): mean-fold difference, PLGA, 1.72; PLGA/CH, 3,63, when each was compared to MSC grown on tissue-culture plastic. .

significant. It should be noted that our final construct contained 80% PLGA and 20% chitosan. This ratio was selected because of the role of PLGA as a mechanical support (the main part of the construct) and chitosan as the agent to improve the cell-scaffold interaction (used in a lower ratio). As mentioned earlier, such hybrid scaffolds are expected to demonstrate improved mechanical properties over those made from the natural polymer only. According to previous studies, chitosan nanofibers undergo a sharp decrease in mechanical strength and Young's modulus when hydrated [41]. This property limits their biomedical application, given that specimens are kept in culture and/ or implanted in the body. However, the trends seen in the variation in mechanical properties of chitosan nanofibers under dry and wet conditions were not seen for our hybrid scaffold. As shown in Table 1, only a small decrease in strength and stiffness was seen for the PLGA/CH scaffold when dry and wet conditions were compared. Moreover, the difference in tensile strength between PLGA/CH and PLGA was almost insignificant.

It should also be noted that PVA extraction resulted in an increase in the extensibility of the hybrid structure. This increase could be a result of the longitudinal shrinkage of ~30% that was seen after PVA extraction; the increase provides an advantage for the PLGA/CH scaffold over chitosan nanofibrous scaffolds, which show only very little elongation at the break [41,42]. Taken together, our findings suggest that, given its improved mechanical properties when compared to chitosan nanofibers and strength similar to that of our PLGA scaffold, the hybrid PLGA/CH scaffold can be considered an appropriate candidate for biomedical and tissue engineering applications.

Our biocompatibility studies showed that both the PLGA and PLGA/CH scaffolds supported cell proliferation after 1, 4, and 7 days of culture. However, the increased metabolic activity and proliferation of MSC with PLGA/CH suggested that the polyblend construct provided more favorable growth and survival conditions for MSC than did PLGA alone. The improved attachment and growth of cells on substrates made from chitosan might be due to the cationic nature of the amine groups in chitosan, which enhances the attraction for negatively charged cells as well [28]. As we grow our MSC in media containing serum it is more likely that cells attach to adsorbed serum proteins. It should be taken into account that having chitosan incorporated into PLGA as a biopolymer to make it more bio-receptive also led to a decrease in the average diameter (and hence increased surface area) of the PLGA/CH nanofibers. Others have also reported a decrease in fiber diameter in polyblends of PLGA/gelatin and PLGA/collagen when compared to PLGA, as a result of adding a biopolymer to PLGA [43,44]. In our study, we performed electrospinning under optimal conditions that led to the formation of nanofibers of PLGA and PLGA/CH with an average diameter of 680 nm and 166 nm, respectively. The improved viability and attachment of MSC to PLGA/CH as compared to PLGA could be a function of the higher bioactivity, smaller fiber diameter, and larger surface area [45] of the hybrid construct. There are several studies investigating various nanofibres and their capacity to affect proliferation and function of stem cells [46-48]. Here we have investigated the effect of adding the natural polymer chitosan to PLGA nanofibers, showing increased proliferation and metabolic activity of human MSC most likely to an increase in biofunctionality of the scaffold. Our morphological studies using SEM and confocal microscopy not only confirmed that an increased number of cells reached confluency in 14 days but also showed extension of individual cells on both PLGA and PLGA/CH scaffolds that formed a connected network through interacting filopodia.

Quantitative PCR analysis demonstrated that both PLGA and PLGA/CH nanofibrous scaffolds supported the gene regulation of MSC into bone, fat and neuronal cells. The quantitative PCR results confirmed that the blending and electrospinning of chitosan with PLGA did not hamper the ability of PLGA [49] to promote the differentiation of bone, fat, and neuronal cells from MSC.

Conclusions

Herein, we have shown that the nanofibrous scaffold composed of PLGA/CH created a favorable substrate supporting the adherence, survival and proliferation of MSC with potential use in tissue engineering. Not only were the tensile strength and porosity similar to those of the electrospun scaffold of PLGA, but we saw increased hydrophilicty and higher viability and proliferation of MSC with the hybrid scaffolds than with PLGA alone. Our quantitative PCR analyses confirmed that both PLGA/CH and PLGA scaffolds could potentially have the ability to direct the MSC phenotype.

Acknowledgments

This study was supported by grants from EXODIAB (UFV-PA 2012/2330), Stem Therapy, the Swedish Research Council (A0290401, A0290402) and The Swedish Society of Medicine. We thank Dr. Deborah McClellan for editorial assistance.

References

1. Murphy MB, Moncivais K and Caplan AI (2013) Mesenchymal stem cells: environmentally responsive therapeutics for regenerative medicine. Exp Mol Med 45: e54.

2. Gustafsson Y, Haag J, Jungebluth P, Lundin V, Lim ML, et al. (2012) Viability and proliferation of rat MSCs on adhesion protein-modified PET and PU scaffolds. Biomaterials 33: 8094-8103.

3. Ringe J, Kaps C, Burmester GR, Sittinger M (2002) Stem cells for regenerative medicine: advances in the engineering of tissues and organs. Naturwissenschaften 89: 338-351.

4. Hwang NS, Varghese S, Elisseeff J (2008) Controlled differentiation of stem cells. Adv Drug Deliv Rev 60: 199-214.

5. Macchiarini P, Jungebluth P, Go T, Asnaghi MA, Rees LE, et al. (2008) Clinical transplantation of a tissue-engineered airway. Lancet 372: 2023-2030.

6. Elliott MJ, De Coppi P, Speggiorin S, Roebuck D, Butler CR, et al. (2012) Stem-cell-based, tissue engineered tracheal replacement in a child: a 2-year follow-up study. Lancet 380: 994-1000.

7. Atala A, Bauer SB, Soker S, Yoo JJ, Retik AB (2006) Tissue-engineered autologous bladders for patients needing cystoplasty. Lancet 367: 1241-1246.

8. Levenberg S, Huang NF, Lavik E, Rogers AB, Itskovitz-Eldor J, et al. (2003) Differentiation of human embryonic stem cells on three-dimensional polymer scaffolds. Proc Natl Acad Sci U S A 100: 12741-12746.

9. Hwang NS, Kim MS, Sampattavanich S, Baek JH, Zhang Z, et al. (2006) Effects of three-dimensional culture and growth factors on the chondrogenic differentiation of murine embryonic stem cells. Stem Cells 24: 284-291.

10. Huang ZM, Zhang YZ, Kotaki M, Ramakrishna S (2003) A review on polymer nanofibers by electrospinning and their applications in nanocomposites. Compos Sci Technol 63: 2223-2253.

11. Milleret V, Simona B, Neuenschwander P, Hall H (2011) Tuning electrospinning parameters for production of 3D-fiber-fleeces with increased porosity for soft tissue engineering applications. Eur Cell Mater 21: 286-303.

12. Li M, Mondrinos MJ, Chen X, Gandhi MR, Ko FK, et al. (2006) Co-electrospun poly(lactide-co-glycolide), gelatin, and elastin blends for tissue engineering scaffolds. J Biomed Mater Res A 79: 963-973.

13. Ma X, Che N, Gu Z, Huang J, Wang D, et al. (2013) Allogenic mesenchymal stem cell transplantation ameliorates nephritis in lupus mice via inhibition of B-cell activation. Cell Transplant 22: 2279-2290.

14. Haugh MG, Thorpe SD, Vinardell T, Buckley CT, Kelly DJ (2012) The application of plastic compression to modulate fibrin hydrogel mechanical properties. J Mech Behav Biomed Mater 16: 66-72.

15. Ajalloueian F, Zeiai S, Fossum M, Hilborn JG (2014) Constructs of electrospun PLGA, compressed collagen and minced urothelium for minimally manipulated autologous bladder tissue expansion. Biomaterials 35: 5741-5748.

16. Ajalloueian F, Zeiai S, Rojas R, Fossum M, Hilborn J (2013) One-stage tissue engineering of bladder wall patches for an easy-to-use approach at the surgical table. Tissue Eng Part C Methods 19: 688-696.

17. Bhattarai N, Edmondson D, Veiseh O, Matsen FA, Zhang M (2005) Electrospun chitosan-based nanofibers and their cellular compatibility. Biomaterials 26: 6176-6184.

18. Bhattarai N, Li Z, Gunn J, Leung M, Cooper A, et al. (2009) Natural-synthetic polyblend nanofibers for biomedical applications. Adv Mater 21: 2792-2797.

19. Gunn J, Zhang M (2010) Polyblend nanofibers for biomedical applications: perspectives and challenges. Trends Biotechnol 28: 189-197.

20. Qi H, Hu P, Xu J, Wang A (2006) Encapsulation of drug reservoirs in fibers by emulsion electrospinning: morphology characterization and preliminary release assessment. Biomacromolecules 7: 2327-2330.

21. Ajalloueian F, Tavanai H, Hilborn J, Donzel-Gargand O, Leifer K, et al. (2014) Emulsion electrospinning as an approach to fabricate PLGA/chitosan nanofibers for biomedical applications. Biomed Res Int 2014: 475280.

22. Shin HJ, Lee CH, Cho IH, Kim YJ, Lee YJ, et al. (2006) Electrospun PLGA nanofiber scaffolds for articular cartilage reconstruction: mechanical stability, degradation and cellular responses under mechanical stimulation in vitro. J Biomater Sci Polym Ed 17: 103-119.

23. Inanç B, Arslan YE, Seker S, Elçin AE, Elçin YM (2009) Periodontal ligament cellular structures engineered with electrospun poly(DL-lactide-co-glycolide) nanofibrous membrane scaffolds. J Biomed Mater Res A 90: 186-195.

24. Nakanishi Y, Chen G, Komuro H, Ushida T, Kaneko S, et al. (2003) Tissue-engineered urinary bladder wall using PLGA mesh-collagen hybrid scaffolds: a comparison study of collagen sponge and gel as a scaffold. J Pediatr Surg 38: 1781-1784.

25. Prabhakaran MP, Kai D, Ghasemi-Mobarakeh L, Ramakrishna S (2011) Electrospun biocomposite nanofibrous patch for cardiac tissue engineering. Biomed Mater 6: 055001.

26. Wang Q, Jamal S, Detamore MS, Berkland C (2011) PLGA-chitosan/PLGA-alginate nanoparticle blends as biodegradable colloidal gels for seeding human umbilical cord mesenchymal stem cells. J Biomed Mater Res A 96: 520-527.

27. Peniche H, Peniche C (2011) Chitosan nanoparticles: a contribution to nanomedicine. Polym Int 60: 883-889.

28. Moutzouri AG, Athanassiou GM (2011) Attachment, spreading, and adhesion strength of human bone marrow cells on chitosan. Ann Biomed Eng 39: 730-741.

29. Verma D, Katti KS, Katti DR (2010) Osteoblast adhesion, proliferation and growth on polyelectrolyte complex-hydroxyapatite nanocomposites. Philos Trans A Math Phys Eng Sci 368: 2083-2097.

30. Venkatesan J, Kim SK (2010) Chitosan composites for bone tissue engineering-an overview. Mar Drugs 8: 2252-2266.

31. Kuo YC, Yeh CF, Yang JT (2009) Differentiation of bone marrow stromal cells in poly(lactide-co-glycolide)/chitosan scaffolds. Biomaterials 30: 6604-6613.

32. Hu N, Wu H, Xue C, Gong Y, Wu J, et al. (2013) Long-term outcome of the repair of 50 mm long median nerve defects in rhesus monkeys with marrow mesenchymal stem cells-containing, chitosan-based tissue engineered nerve grafts. Biomaterials 34: 100-111.

33. Xue C, Hu N, Gu Y, Yang Y, Liu Y, et al. (2012) Joint use of a chitosan/PLGA scaffold and MSCs to bridge an extra large gap in dog sciatic nerve. Neurorehabil Neural Repair 26: 96-106.

34. Le Blanc K, Tammik L, Sundberg B, Haynesworth SE, Ringdén O (2003) Mesenchymal stem cells inhibit and stimulate mixed lymphocyte cultures and mitogenic responses independently of the major histocompatibility complex. Scand J Immunol 57: 11-20.

35. Wang CY, Zhang KH, Fan CY, Mo XM, Ruan HJ, et al. (2011) Aligned natural-synthetic polyblend nanofibers for peripheral nerve regeneration. Acta Biomater 7: 634-643.

36. Meng ZX, Zheng W, Li L, Zheng YF (2011) Fabrication, characterization and in vitro drug release behavior of electrospun PLGA/chitosan nanofibrous scaffold. Mater Chem Phys 125: 606-611.

37. Xie D, Huang H, Blackwood K, MacNeil S (2010) A novel route for the production of chitosan/poly(lactide-co-glycolide) graft copolymers for electrospinning. Biomed Mater 5: 065016.

38. Duan B, Yuan X, Zhu Y, Zhang Y, Li X, et al. (2006) A nanofibrous composite membrane of PLGA–chitosan/PVA prepared by electrospinning. Eur Polym J 42: 2013-2022.

39. Duan B, Wu L, Yuan X, Hu Z, Li X, et al. (2007) Hybrid nanofibrous membranes of PLGA/chitosan fabricated via an electrospinning array. J Biomed Mater Res A 83: 868-878.

40. Zhang Y, Ouyang H, Lim CT, Ramakrishna S, Huang ZM (2005) Electrospinning of gelatin fibers and gelatin/PCL composite fibrous scaffolds. J Biomed Mater Res B Appl Biomater 72: 156-165.

41. Wan Y, Cao X, Zhang S, Wang S, Wu Q (2008) Fibrous poly(chitosan-g-DL-lactic acid) scaffolds prepared via electro-wet-spinning. Acta Biomater 4: 876-886.

42. Schiffman JD, Schauer CL (2007) Cross-linking chitosan nanofibers. Biomacromolecules 8: 594-601.

43. Prabhakaran MP, Kai D, Ghasemi-Mobarakeh L, Ramakrishna S (2011) Electrospun biocomposite nanofibrous patch for cardiac tissue engineering. Biomed Mater 6: 055001.

44. Jose MV, Thomas V, Dean DR, Nyairo E (2009) Fabrication and characterization of aligned nanofibrous PLGA/Collagen blends as bone tissue scaffolds. Polymer 50: 3778-3785.

45. Chen M, Patra PK, Lovett ML, Kaplan DL, Bhowmick S (2009) Role of electrospun fibre diameter and corresponding specific surface area (SSA) on cell attachment. J Tissue Eng Regen Med 3: 269-279.

46. Chua KN, Chai C, Lee PC, Tang YN, Ramakrishna S, et al. (2006) Surface-aminated electrospun nanofibers enhance adhesion and expansion of human umbilical cord blood hematopoietic stem/progenitor cells. Biomaterials 27: 6043-6051.

47. Martino S, D'Angelo F, Armentano I, Kenny JM, Orlacchio A (2012) Stem cell-biomaterial interactions for regenerative medicine. Biotechnol Adv 30: 338-351.

48. Du J, Tan E, Kim HJ, Zhang A, Bhattacharya R, et al. (2014) Comparative evaluation of chitosan, cellulose acetate, and polyethersulfone nanofiber scaffolds for neural differentiation. Carbohydr Polym 99: 483-490.

49. Massumi M, Abasi M, Babaloo H, Terraf P, Safi M, et al. (2012) The effect of topography on differentiation fates of matrigel-coated mouse embryonic stem cells cultured on PLGA nanofibrous scaffolds. Tissue Eng Part A 18: 609-620.

10

Application of Central Composite Design and Artificial Neural Network for the Optimization of Fermentation Conditions for Lipase Production by *Rhizopus arrhizus* MTCC 2233

Aravindan Rajendran* and Viruthagiri Thangavelu

Biochemical Engineering Laboratory, Department of Chemical Engineering, Annamalai University, Annamalai Nagar– 608 002, Tamil Nadu, India

Abstract

The response surface optimization strategy was used to enhance the lipase production by *Rhizopus arrhizus* MTCC 2233 in submerged fermentation. Various vegetable oils were experimented as an inducer using the optimized medium to study the influence on lipase production, and corn oil was found to be the best inducer for lipase production by *Rhizopus arrhizus*. The optimization of fermentation conditions, temperature, initial pH and agitation speed was carried out using corn oil as the inducer. Statistical analysis of the experimental data showed that the temperature, agitation speed, quadratic effects of temperature, initial pH and agitation speed and interactive effects of temperature and agitation speed are significant parameters that affect lipase production. The optimum fermentation conditions were achieved at 32°C; pH 6.0 and agitation speed of 107 rpm with the maximum lipase activity of 4.32 U/mL. Artificial neural network model was used to predict the lipase activity and cell mass production under various fermentation conditions. Unstructured kinetic models, Logistic model, Luedeking-Piret model and modified Luedeking-Piret model were used to describe the cell biomass, lipase production and glucose utilization kinetics respectively.

Keywords: Lipase fermentation; Response surface methodology; Cell cultures; Kinetics; Optimization

Introduction

Hydrolase enzymes play significant roles in biotechnology because of their extreme versatility with respect to substrate specificity and stereoselectivity [1-3]. Lipases [triacylglycerol acylhydrolases (EC 3.1.1.3)] are enzymes which catalyze the hydrolysis of fatty acid ester bonds in triacylglycerol to give free fatty acids, diacylglycerols, monoacylglycerols and glycerol at the lipid water interface [2]. Various microbial lipases have been purified, characterized and studied for their biotechnological applications in food, dairy, detergents, pharmaceuticals, textile, cosmetics & paper industry, bioremediation of oil containing effluents and preparation of various flavor and fragrances as well [1-5]. Most of lipases produced commercially are currently obtained from fungi or yeasts [1-3]. *Rhizopus, Aspergillus, Mucor, Geotrichum, Penicillium* and *Candida* are the potential sources of commercial lipase production. The lipase from *Rhizopus* has 1,3-regioselectivity, for selectively catalyzing the hydrolysis of triacylglycerol [6]. However, the lipase production with *Rhizopus sp.* is relatively low-yield and economically unfavorable [6]. Various methods have been investigated in the literature to optimize the fermentation process to enhance lipase production yields, and most of literatures have reported improvement in lipase yield after optimization. Elibol and Ozer [7] optimized lipase production by *Rhizopus arrhizus* using response surface methodology and obtained 0.37 U/mL lipase activity (tributyrin as substrate). D'Annibale et al. [8] investigated the utilization of olive-mill wastewater as a growth medium for the microbial production of extra-cellular lipase using *Rhizopus arrhizus* NRRL 2286, *Rhizopus sp.* ISRIM 383 and *Rhizopus oryzae* NRRL 6431 and obtained a maximum lipase activity of 0.30, 0.35 and 0.36 IU/mL (β-naftilmyristate as substrate) respectively. Repeated-batch-fermentation by immobilized *R. arrhizus* was conducted by Yang et al. [6] and observed that the lipase productivity increased from 3.1 U/mL h in batch fermentation to 17.6 U/mL h in repeated batch fermentation, which was 5.6 times as high as that in batch fermentation.

Screening and evaluation of nutritional and environmental requirements of microorganism are important steps for bioprocess development. Microbial lipase fermentations are significantly affected by the environmental conditions such as medium pH, temperature, aeration, agitation and inoculation size. Response surface methodology (RSM) is a collection of experimental strategies, mathematical methods and statistical inference for constructing and exploring an approximate functional relationship between a response variable and a set of design variables [9,10].

Due to the metabolic complexity of microorganisms, the development of rigorous models for a given biological reaction system based on physical and chemical parameters is still a challenge. This is mainly due to the non-linear nature of the biochemical networks and lack of complete knowledge [11,12]. Artificial neural network (ANN) methods have been utilized with great success for system design, modeling, prediction, optimization and control. This is mainly due to their capacity to learn, filter noisy signals and generalize information through training procedures [12]. Mathematical models of bioprocess kinetics facilitate data analysis and definitely provide a strategy for solving problems encountered in industrial fermentation processes [13,14]. The goal of the fermentation kinetic studies is to increase the productivity of batch process and to optimally design and operate continuous fermentation. Objectives of the present study are;

(i)   Optimize the fermentation conditions such as pH, tempera-

*Corresponding author: Aravindan Rajendran, Assistant Professor, Biochemical Engineering Laboratory, Department of Chemical Engineering, Annamalai University, Annamalai Nagar– 608 002, Tamil Nadu, India

ture and agitation speed for lipase production by *Rhizopus arrhizus* using RSM in an optimized medium

(ii)   Utilize the ANN model for the prediction of lipase activity and cell biomass at various conditions

(iii)   Predict lipase fermentation kinetics using unstructured kinetic models.

Materials and Methods

Materials

The medium components and other chemicals used were procured from Himedia Ltd, Mumbai, India. All chemicals used were of analytical grade unless otherwise specified. The orbital shaker is a make of Remi Laboratories India, RIS - 24 BL; centrifuge from Remi Laboratories India, C-24 BL; homogenizer from Global-Butterfly India, 230V AC, 550W heavy duty and the spectrophotometer is a make of Elico India Ltd., (Double Beam UV-VIS Spectrophotometer-SL 164).

Microorganisms and culture maintenance

Rhizopus arrhizus MTCC 2233 was obtained from Microbial Type Culture Collection of gene bank (MTCC), Institute of Microbial Technology, Chandigarh, India. The *Rhizopus arrhizus* stock culture was maintained on potato dextrose agar slants containing potato infusion (infusion from 200 g potatoes) 4.0 g/L; Dextrose 20.0 g/L. Spirit blue agar (Himedia-Mumbai, India) was used for the detection of lipolytic activity of the microorganisms.

Submerged batch fermentation

The *R. arrhizus* MTCC 2233 slant was kept at 30°C for 2 days. After growth and sporulation, 10 mL sterile distilled water was aseptically added to the agar slant which was then scraped to release the spores. The spore suspension was centrifuged at 2500g for 10 min and, supernatant was discarded and then the spores were resuspended in 1 mL of sterile distilled water. Spore suspension (1 mL) was used as inoculum for 250 mL shake flask containing 100 mL of fermentation medium (glucose- 15 g/L; oil inducer -10 mL/L; peptone- 6 g/L; yeast extract- 6 g/L; K_2HPO_4- 0.1 g/L; KH_2PO_4- 2 g/L; $CaCl_2.2H_2O$- 1 g/L; $MgSO_4.7H_2O$- 0.5 g/L; $ZnSO_4$- 0.01 g/L; $FeSO_4.7H_2O$- 0.05 g/L; $CuSO_4$- 0.02 g/L; $MnSO_4 .H_2O$- 0.02 g/L). After the inoculation, flasks were placed on a rotary shaker at 30°C and 120 rpm incubated for 2 days and were then used to inoculate the production medium at 5% (v/v) level. The lipase production by *R. arrhizus* was studied in 250 mL Erlenmeyer flask with 100 mL of the production medium. The pH of fermentation medium was adjusted using 2M NH_4OH and sterilized at 121°C (15 psi) for 20 min. The flasks were incubated in an orbital shaker at constant agitator speed and temperature for the fermentation period of 108 h. Identical flasks were used for the fermentation and the cells were separated from the medium by centrifugation at 5000g for 15 min. The clear supernatant was used for the analysis of lipase activity, protease activity, total soluble protein and glucose. All the submerged batch fermentations were conducted in triplicate and average results were given in this report. Lipase production was measured and compared at different conditions.

Lipase activity assay

Lipase activity was estimated with olive oil emulsion by the procedure of Ota and Yamada [15]. Olive oil emulsion was prepared by homogenizing 25 mL of olive oil and 75 mL of 2 % w/v polyvinyl alcohol solution in a homogenizer for 6 min at 20000 rpm. The reaction mixture composed of 2 mL olive oil emulsion, 2.5 mL 0.05 M phosphate buffer and 0.5 mL enzyme solution was incubated at 37°C for 15 min. The emulsion was disrupted by addition of 10 mL acetone immediately after 15 min incubation and the liberated fatty acids content was titrated against 0.05 N NaOH. One unit (U) of lipase activity was defined as 1 µmole of free fatty acids liberated per mL of enzyme per minute at 37°C.

Protease activity assay

The protease activity was assayed by modified Anson method [16] using casein as the substrate. 2 mL of 1% w/v casein solution is mixed with 0.5 mL of enzyme solution and incubated at 37°C for 30 min and reaction was quenched by addition of 2.5 mL of 0.4 M trichloroacetic acid to the reaction solution. The solution with precipitate was filtered and to the 1 mL of filtrate, 5 mL of 0.4 M Na_2CO_3 and 0.5 mL of folin reagent were added. After 10 min of incubation at 37°C, the colour intensity was measured at 660 nm. One unit (U) of protease activity was defined as 1 µgram of tyrosine liberated per minute by 1 mL of enzyme at 37°C.

Biomass, glucose and protein determination

The biomass concentration (dry cell weight) was determined by gravimetric method. The glucose concentration in the cultivation broth was determined as described by Miller [17] using dinitrosalicylic acid. The total soluble protein in the medium was determined as described by Lowry et al. [18].

Central Composite Experimental Design (CCD)

An orthogonal $(2)^3$ factorial central composite experimental design with six star points and six replicates at the centre with a total of 20 experiments were used to optimize the fermentation conditions such as temperature, medium pH and agitation speed [19-22]. These conditions were tested at five coded levels, –1.682, –1, 0, +1 and +1.682. The experimental range and levels of the three fermentation conditions used in RSM in terms of coded levels and actual values and the CCD experimental plan were given in Table 1. The CCD experiment was designed using the MINITAB software package, version 14.0, The Math Works Inc. The variables were coded according to the following equation,

$$x_i = \frac{X_i - X_c}{\Delta X_i} , i = 1, 2, 3 \dots k. \tag{1}$$

Where,   is the coded value of an independent variable, X_i is the real value of an independent variable, X_c the real value of an independent variable at the center point and ΔX_i is the step change value. Multiple regression analysis of the experimental data gives the second order polynomial equation for optimization of variables. The behaviour of the system was explained by the following second degree polynomial equation,

$$Y = \beta_0 + \Sigma \beta_i x_i + \Sigma \beta_{ii} x_i^2 + \Sigma \beta_{ij} x_i x_j \tag{2}$$

Where, Y is the predicted response, β_0 the offset term, β_i the coefficient linear effect, β_{ii} the coefficient squared effect and β_{ij} the coefficient of interaction effect. The MINITAB software statistical program package was used for regression analysis of the experimental data obtained and to estimate the coefficients of the regression equation. The response surface plots were used to describe the individual and cumulative effects of the variables as well as the interactions between the variables on the lipase activity. The second degree polynomial equation was maximized by a constraint search procedure to obtain the optimal levels of the independent variables and the predicted maximum lipase activity.

Artificial Neural Network Model (ANN)

The feed forward back-propogation algorithm with one hidden layer was used in the training of the neural network, based on varying input/output pair data sets. ANN was applied for the purpose of simulation on the same experimental data (Table 1) used for RSM. The accuracy of the neural network prediction is dependent on the training patterns as well as the structure of the neural network [23]. Neural network consist of many processing elements called neurons interconnected by information channels. The number of neurons in the input and output layers are given by the number of input and output variables in the fermentation process. The input signals are amplified or dampened by a weight associated with each information channel. The neuron then sums all weighted inputs and passes them through a threshold to determine the activation value (the fired output signal) of this neuron. The inter neuron activity can be modeled by an activation function [24]. Sigmoid activation function which is commonly used in back propagation algorithm is given below,

$$y_j = - \left[1 + \exp\left(\sum_{i=1}^{n} a_i w_{ij} + \theta_j \right) \right]^{-1} \tag{3}$$

Where, a_i - input signal, y_j - fired output signal, w_{ij} -weight associated with the input signal ai, θ_j- threshold value of neuron j.

In back propagation networks, the process is executed according to an error feedback method by which it will first update the values of all the neurons corresponding to input data based on current weights and then adjust the weights according to the error between fixed outputs and desired outputs to reduce the error. The summation of output errors, E is given by,

$$E = \frac{1}{2} \sum_j \left(y_j - d_j \right)^2 \tag{4}$$

Where, d_j is the desired output from neuron j. A common scheme used for neural network training is the maximum gradient scheme, in which each connection weight w_{ik} was changed by Δw_{ik} as given below,

$$\Delta w_{ik} = - \eta \; \frac{\partial E}{\partial w_{ik}} \tag{5}$$

Where, η is the positive constant controlling the speed of learning. The Neural network toolbox in MATLAB software was used to construct the ANN topology. This topology of neural network was used in this study for the prediction of the lipase activity and cell mass under various environmental conditions.

Unstructured Model Development for Fermentation Kinetics

Various unstructured models were proved to be sufficient for characterizing the fermentation kinetics. In an unstructured model the cellular representations are single component representations [13,25,26]. The exponential growth phase can be characterized by the following first order equation which states that the rate of increase of cell biomass is proportional to the quantity of viable cell biomass at any instant time,

$$\frac{dX}{dt} = \mu X \tag{6}$$

Where, dX/dt is the growth rate (g/L·h); X is the concentration of biomass (g/L); μ is the specific cell growth rate (1/h). The growth of cell is governed by a hyperbolic relationship and there is a limit to the maximum attainable cell biomass concentration. Such growth kinetics is described by logistic equation [25] as,

$$\frac{dX}{dt} = \mu_0 \left(1 - \frac{X}{X_{max}} \right) X \tag{7}$$

Where μ_0 is the initial specific growth rate (1/h) and X_{max} is the maximum cell mass concentration (g/L). Equation (7) on integration using $X_0 = X \, (t = 0)$ gives a sigmoidal variation X (t) that may empirically represent both an exponential and a stationary phase.

$$X(t) = \frac{X_0 e^{\mu_0 t}}{1 - \left(\dfrac{X_0}{X_{max}} \right)\left(1 - e^{\mu_0 t} \right)} \tag{8}$$

The kinetic parameter, μ_0 in this equation is determined by rear-

Run No.	CCD Experimental Design matrix (Coded and real values)			Experimental			Predicted Lipase activity		Predicted Cell mass	
	x_1	x_2	x_3	Cell massa (g l⁻¹)	Lipase activityb (U ml⁻¹)	Protease activityc (U ml⁻¹)	RSM (U ml⁻¹)	ANN (U ml⁻¹)	RSM (g l⁻¹)	ANN (g l⁻¹)
1	-1 (28)	-1 (5)	-1 (120)	4.20	2.15	0.95	1.969	2.15	4.432	4.2
2	1 (32)	-1 (5)	-1 (120)	6.56	3.00	1.65	3.36	3.0	6.807	6.56
3	-1 (28)	1 (6.5)	-1 (120)	3.15	2.20	1.05	2.052	2.2	3.159	3.15
4	1 (32)	1 (6.5)	-1 (120)	5.82	3.8	0.45	3.518	3.8	5.499	5.82
5	-1 (28)	-1 (5)	1 (160)	3.30	1.90	0.85	1.975	1.9	3.772	3.3
6	1 (32)	-1 (5)	1 (160)	3.65	2.25	1.95	2.191	2.25	3.792	3.65
7	-1 (28)	1 (6.5)	1 (160)	3.12	2.65	0.65	2.083	2.65	3.024	3.12
8	1 (32)	1 (6.5)	1 (160)	3.09	2.40	1.26	2.374	2.4	3.008	3.09
9	-1.682 (26.6)	0 (5.75)	0 (140)	2.60	1.20	0.55	1.588	1.2	2.305	2.6
10	1.682 (33.3)	0 (5.75)	0 (140)	4.21	3.10	2.15	3.003	3.1	4.29	4.21
11	0 (30)	-1.682 (4.46)	0 (140)	5.81	2.60	0.88	2.383	2.6	5.232	5.81
12	0 (30)	1.682 (7.01)	0 (140)	3.14	2.10	0.45	2.607	2.1	3.503	3.14
13	0 (30)	0 (5.75)	-1.682 (106.4)	6.41	3.20	1.65	3.249	3.2	6.382	6.41
14	0 (30)	0 (5.75)	1.682 (173.6)	3.92	2.05	1.65	2.292	2.05	3.733	3.92
15	0 (30)	0 (5.75)	0 (140)	5.20	4.10	2.15	3.841	3.925	5.139	5.133
16	0 (30)	0 (5.75)	0 (140)	4.92	3.70	1.95	3.841	3.925	5.139	5.133
17	0 (30)	0 (5.75)	0 (140)	4.56	4.20	1.89	3.841	3.925	5.139	5.133
18	0 (30)	0 (5.75)	0 (140)	5.95	3.85	2.30	3.841	3.925	5.139	5.133
19	0 (30)	0 (5.75)	0 (140)	4.85	3.90	1.65	3.841	3.925	5.139	5.133
20	0 (30)	0 (5.75)	0 (140)	5.32	3.80	2.01	3.841	3.925	5.139	5.133

I_1: Temperature; x_2: Initial pH; x_3: Agitation speed

a, b, cThe maximum values of cell mass, lipase activity and protease activity respectively and they are the mean values of triplicates. 250 ml Erlenmeyer flask containing 100 ml production medium (constant) was incubated in an orbital shaker (variable) for the fermentation period of 48 hrs.

Table 1: Central composite experimental design and comparison of experimental values with RSM and ANN predicted values of lipase activity using R. arrhizus in submerged fermentation methods.

ranging equation (8) as,

$$\mu_0 t = \ln\left[\frac{X_{max}}{X_0} - 1\right] + \ln\left[\frac{\overline{X}}{1 - \overline{X}}\right] \tag{9}$$

Where $\overline{X} = \dfrac{X}{X_{max}}$, if the logistic equation describes the data suit-

ably, then a plot of $\ln\left[\dfrac{\overline{X}}{1-\overline{X}}\right]$ vs. t should give a straight line of slope 'μ_0'

and intercept $-\ln\left[\left(\dfrac{X_{max}}{X_0} - 1\right)\right]$

The kinetics of lipase production was described by Luedeking-Piret equation [27] which states that the product formation rate depends upon both the instantaneous biomass concentration (X) and growth rate (dX/dt) in a linear fashion.

$$\frac{dP}{dt} = \alpha \frac{dX}{dt} + \beta X \tag{10}$$

where α(gP/gX) and β(gP/gX•h) are empirical constants that may vary with fermentation conditions. Integrating equation (10) using equation (7),

$$P_t = P_0 + \alpha\, A(t) + \beta\, B(t) \tag{11}$$

Where P_0 and P_t are the product concentrations at initial time and at any time (at time t) respectively and,

$$A(t) = X_0 \left[\frac{e^{\mu_0 t}}{1 - \left(\dfrac{X_0}{X_{max}}\right)\left(1 - e^{\mu_0 t}\right)} - 1\right] \tag{12}$$

$$B(t) = \frac{X_{max}}{\mu_0} \ln\left[1 - \frac{X_0}{X_{max}}\left(1 - e^{\mu_0 t}\right)\right] \tag{13}$$

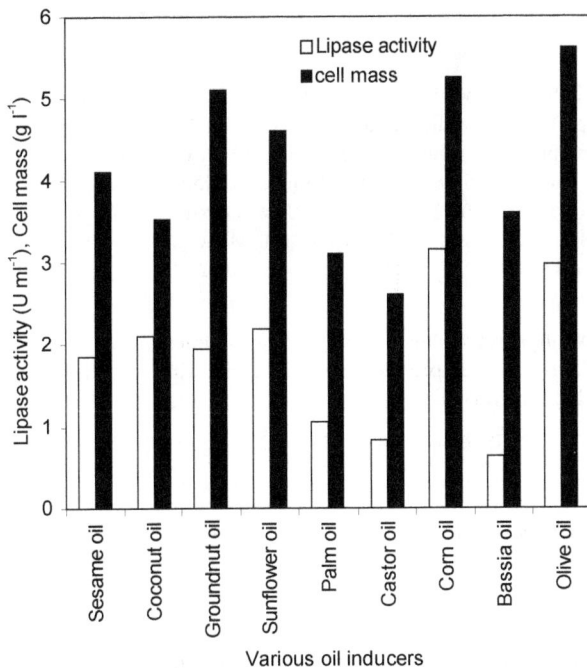

Figure 1: Effect of inducers on lipase production by *R. arrhizus* in submerged fermentation.

The parameters α and β in equation (11) are determined by plotting $- P_0/B(t)$ vs

$A(t)$ /B(t) which is a straight line with slope 'α' and intercept 'β'.

The substrate utilization kinetics is given by the following equation, which considers substrate conversion to cell mass, to product and substrate consumption for maintenance [25].

$$\frac{dS}{dt} = -\frac{1}{Y_{X/S}}\frac{dX}{dt} - \frac{1}{Y_{P/S}}\frac{dP}{dt} - k_e X \tag{14}$$

Where, $Y_{X/S}$ and $Y_{P/S}$ are yields of cell mass and product with respect to substrate and K_e is the maintenance coefficient for cells. Rearranging the substrate material balance equation (14),

$$\frac{dS}{dt} = -\gamma \frac{dX}{dt} - \eta X \tag{15}$$

Where,

$$\gamma(gS/gX) = \frac{1}{Y'_{X/S}} + \frac{\alpha}{Y'_{P/S}} \tag{16}$$

$$\eta(gS/gX.h) = \frac{\beta}{Y'_{P/S}} + k_e \tag{17}$$

Equation (15) is the modified Luedeking- Piret equation for substrate utilization kinetics.

Substituting for μ from equation (7) and integrating gives

$$S_t = S_0 - \gamma\, m(t) - \eta\, n(t) \tag{18}$$

Where, S_0 and S_t are the substrate concentrations at initial time and at anytime 't' respectively and,

$$m(t) = X_0 \left[\frac{e^{\mu_0 t}}{1 - \left(\dfrac{X_0}{X_m}\right)\left(1 - e^{\mu_0 t}\right)} - 1\right] \tag{19}$$

$$n(t) = \frac{X_m}{\mu_0} \ln\left[1 - \frac{X_0}{X_m}\left(1 - e^{\mu_0 t}\right)\right] \tag{20}$$

Kinetic parameters (γ,η) in equation (18) is determined by plotting $\dfrac{S_0 - S_t}{n(t)}$ vs $\dfrac{m(t)}{n(t)}$ which is a straight line with slope γ and intercept η.

Results and Discussion

Optimization of inducer

Lipase production by *Rhizopus arrhizus* was experimented in submerged batch culture in shake flasks. In the first step, the medium components that have the more influence on the lipase production were identified using Plackett-Burman (PB) statistical experimental design. The statistical analysis of PB design (data not presented) showed that, the medium components such as olive oil, peptone, $CaCl_2.2H_2O$ and $MgSO_4.7H_2O$ were found to have the most profound influence on the lipase production with a confidence level of more than 95%.

The composition of culture medium, in particular addition of different lipid substances can result in the production of different isoenzymes and oils play a vital role in the lipase synthesis [28]. The oils mainly contain various proportions of three main types of fatty acids such as saturated, monounsaturated and polyunsaturated fatty acids in addition to vitamins and growth factors. The basic constituents of vege-

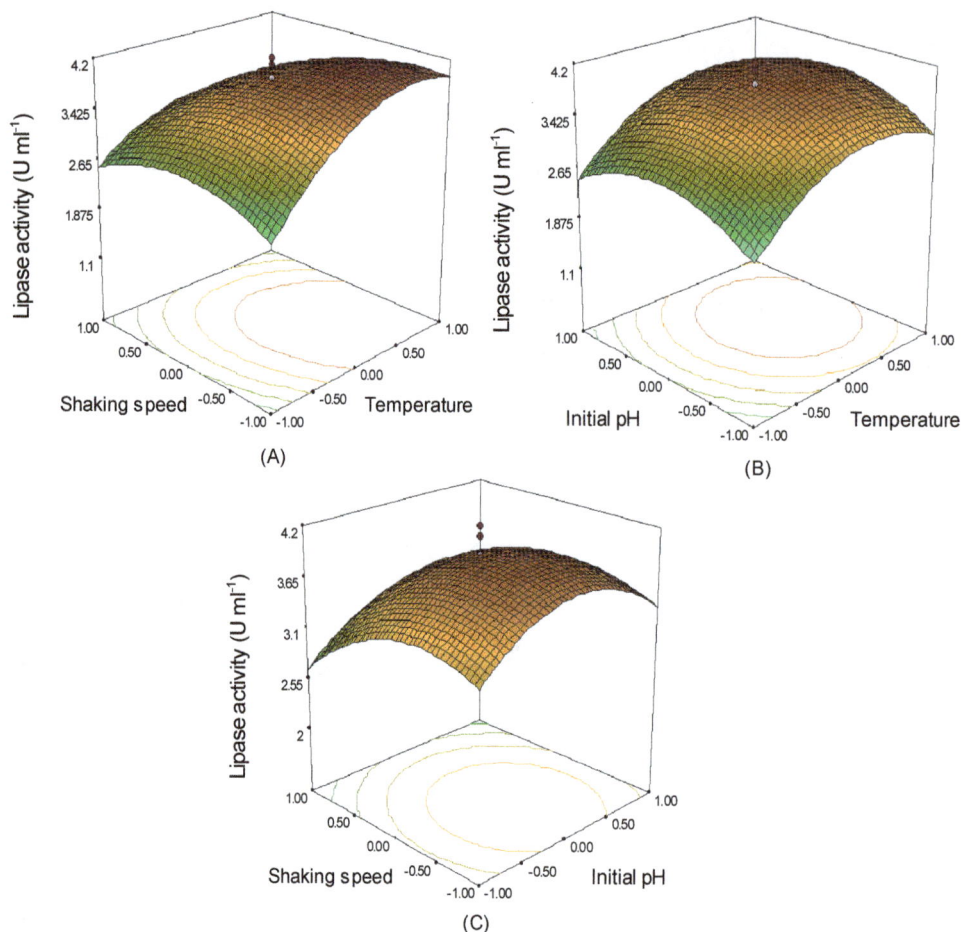

Figure 2: Response surface plot and contour plots showing the effects of (A) temperature and agitation speed, (B) temperature and initial pH, (C) initial pH and agitation speed, on lipase production by R. arrhizus with the remaining factors kept constant at the middle level of the central composite experimental design.

table oils are 16-carbon acids (palmitic and palmitoleic) and 18-carbon acids (stearic, oleic, linoleic and linolenic). The effect of vegetable oils on lipase production was studied by single factor optimization method keeping all other media constituents and process conditions constant. The optimized medium used for the inducer optimization studies which contains glucose- 15 g/L; peptone- 6 g/L; yeast extract- 6 g/L; K_2HPO_4- 0.1 g/L; KH_2PO_4- 2 g/L; $CaCl_2.2H_2O$- 1 g/L; $MgSO_4.7H_2O$- 0.5 g/L; $ZnSO_4$- 0.01 g/L; $FeSO_4.7H_2O$- 0.05 g/L; $CuSO_4$- 0.02 g/L; $MnSO_4$ $.H_2O$- 0.02 g/L and various vegetable oil inducers at the concentration of 10 mL/L. The vegetable oils used in this study were sesame oil, coconut oil, groundnut oil, castor oil, palm oil, sunflower oil, corn oil, bassia oil and olive oil. Among all oil inducers tested in this study, corn oil was found to be the best inducer for lipase production by *Rhizopus arrhizus* followed by sunflower oil, coconut oil, groundnut oil, olive oil, sesame oil, palm oil, castor oil, and bassia oil (Figure 1). Maximum lipase activity of 3.15 U/mL was observed at 84 h of cultivation using corn oil as an inducer. Elibol and Ozer have also reported similar results for the lipase production by *Rhizopus arrhizus* using corn oil [7].

Central Composite Experimental Design and Optimization of Fermentation Conditions by Response Surface Methodology

The experimental plan to determine the optimum combination of

fermentation conditions for enhancing the lipase production was done using CCD and results are presented in Table 1 along with experimental and predicted values using RSM and ANN. The production medium used for the optimum fermentation conditions contain: glucose- 15 g/L; corn oil-10 mL/L; peptone- 6 g/L; yeast extract- 6 g/L; K_2HPO_4- 0.1 g/L; KH2PO4- 2 g/L; $CaCl_2.2H_2O$- 1 g/L; $MgSO_4.7H_2O$- 0.5 g/L; $ZnSO_4$- 0.01 g/L; $FeSO_4.7H_2O$- 0.05 g/L; $CuSO_4$- 0.02 g/L; $MnSO_4$ $.H_2O$- 0.02 g/L.

Multiple regression analysis of the experimental data obtained using CCD for lipase production gave the following second order polynomial equation (21),

$$Y_1 = 3.9165 + 0.4207\ x_1 + 0.0666\ x_2 - 0.2844\ x_3 - 0.5721\ x_1^2 - 0.5014\ x_2^2 - 0.4042\ x_3^2 + 0.0187\ x_1x_2 - 0.2938\ x_1x_3 + 0.0062\ x_2x_3 \qquad (21)$$

where Y_1 is the lipase activity and x_1, x_2 and x_3 are the coded values of the independent variables, temperature, pH and agitation speed respectively.

The experimental results were analyzed using Minitab statistical software and the correlation coefficient (R), determination coefficient (R^2) and adjusted determination coefficient (Adj R^2) was determined to check the competence of polynomial model. The value of R is 0.956 which implies a high degree of correlation between the observed and

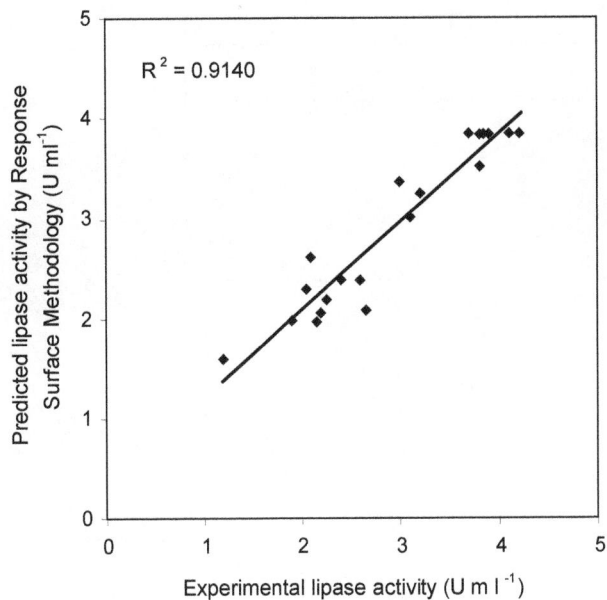

(A)

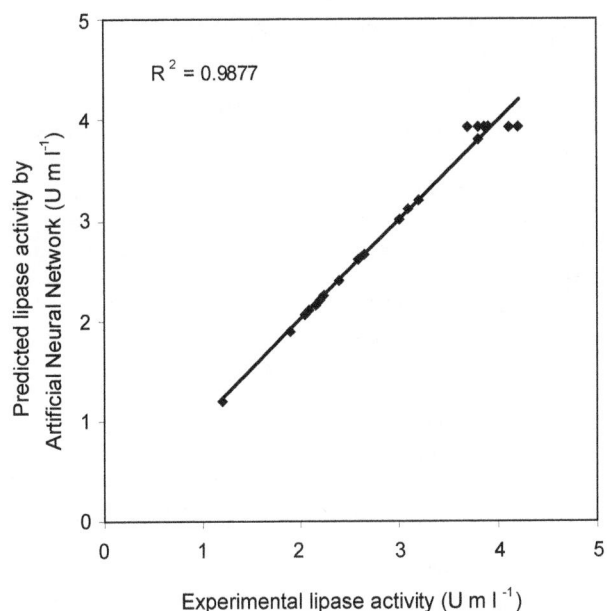

(B)

Figure 3: (A) Comparison of experimental and response surface methodology model predicted values of lipase activity by *R. arrhizus* (B) Comparison of experimental and artificial neural network model predicted values of lipase activity by *R. arrhizus*.

predicted values. The R^2 value is 0.914 suggests that only about 8.6% of the total variations are not explained by the model. The adj R^2 corrects the R^2 value for the sample size and the number of terms in the model. If there are many terms in the model and the sample size is not very large, the adjusted R^2 may be noticeably smaller than the R^2. The adjusted R^2 in this study was 0.837 which is close to R^2 value which indicates the better prediction of the model.

Statistical testing of the model was carried out using analysis of variance (ANOVA) technique, which is required to test the significance and adequacy of the model. The ANOVA of the regression model demonstrates that the model is highly significant as evident from the calculated F value (Explained variance /unexplained variance; F_{model} = 11.84) and a very low probability value (P_{model} > F =0). Moreover the computed F value is much greater than the tabulated F value ($F_{9,10}$ = 3.02 at 5% significance level) indicating that the treatment differences are highly significant. The Model F-value of 11.27 implies the model is significant. The students-t-distribution and the corresponding P values, along with parameter estimates for lipase activity was evaluated using MINITAB software. The P value signifies that the coefficient for the linear effect of temperature and agitation speed, the quadratic effects of temperature, pH and agitation speed and the interactive effects of temperature and agitation speed are significant. The regression equation was solved using MATLAB software and the optimal values of the test variables in the coded units were found to be x_1= 0.504, x_2= 0.078, x_3 = −0.533 and the corresponding uncoded values were x_1= 31°C, x_2 (pH) = 5.80, x_3 = 129.3 rpm with a predicted maximum lipase activity of 4.101 U/mL. The normal probability plot of the residuals is an important diagnostic tool to detect and explain the departures from the assumptions that errors are normally distributed, independent of each other and the error variances are homogenous. An excellent normal distribution confirmed the normality assumption and the independence of the residuals. The residual plot shows equal scatter of the residual data above and below the x- axis indicating that the variance was independent of the value of the lipase production and thus supporting the adequacy of the least square fit.

The response surfaces obtained using the MINITAB software were shown in Figure 2 for lipase production and the figures express the significance of various fermentation conditions. The shapes of the response surfaces may be circular or elliptical indicating the significance of the interactions between the variables. The elliptical nature of the contour plots between the conditions temperature and agitation speed indicates that the interaction between these set of variables has a significant effect on lipase yield. Comparison of experimental and RSM model predicted values of lipase activity and the comparison of experimental and ANN model predicted values of lipase activity by *R. arrhizus* is shown in Figures 3A & 3B respectively.

Figure 4 illustrates the main effects plot of mean lipase activity with the fermentation conditions temperature, pH and agitation speed. The mean lipase activity was high at the center of the experimental range of temperature, pH and agitation speed. Further increase in the variables causes a decrease in lipase activity. The mean lipase activity was very low at the lowest temperature setting (26.6°C) when compared to the other factor settings. The mean lipase activity was high (3.2 U/mL) at the lowest level of the agitation speed (106 rpm) when compared to the tested higher levels (173 rpm). Probably at lower agitation speed the microorganism may attain its complete morphological structure for high extracellular lipase secretion.

An experiment was conducted at the optimized fermentation conditions determined by RSM to confirm the predicted optimum response. The maximum lipase activity of 4.32 U/mL was obtained at 72 h of cultivation and illustrated in Figure 5. The experimental and the predicted values of enzyme activity showed good agreement with one another with high degree of accuracy of the model substantiating the model validation under the experimental conditions. The lipase production was found to increase gradually after the 24 h of the fermentation period when the growth of microorganisms reached the early

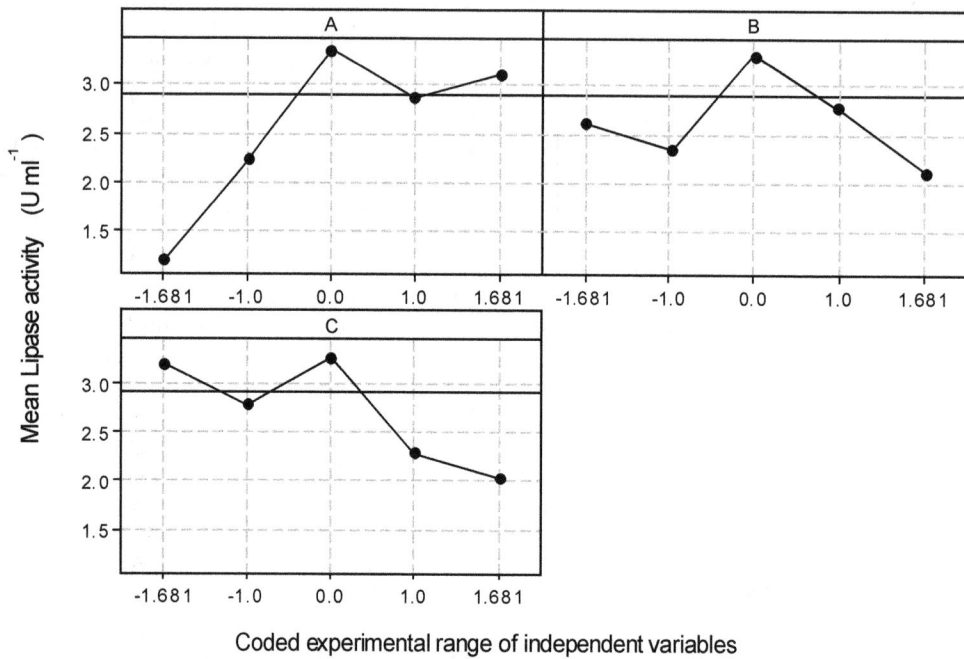

Figure 4: Main effects plot for lipase production by *R. arrhizus* with the fermentation conditions (A) Temperature, (B) Initial pH, (C) Agitation speed.

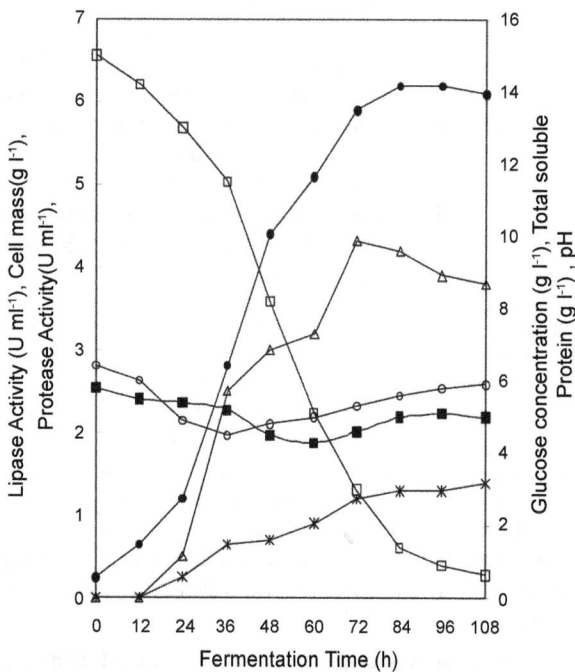

Figure 5: Fermentative production of lipase by R. arrhizus in submerged fermentation. Profile of lipase activity (Δ), protease activity (✳), pH (■), cell mass concentration (●), glucose concentration (□) and total soluble.

uct inhibition such as the formation of more protease enzyme at the post exponential growth phase of the microorganism or due to the non availability or less availability of inducer oil and the carbon source in the fermentation medium or may be due to the accumulation of fatty acids. Thermal deactivation and digestion of lipases by proteases were prevented due to the presence of vegetable oil inducers in the fermentation medium [29]. The cell mass concentration reached a maximum of 6.2 g/L at 84 h during the stationary phase and there was no further increase in the cell mass concentration after 84 h. The maximum protease activity of 1.4 U/mL at 108 h of the fermentation was obtained at the stationary phase of the microorganism. The rate of lipase production was found to be much greater than the rate of protease production. The total soluble protein was found to decrease gradually till 36 h of fermentation due to the consumption of medium components containing proteins by the microorganism for its growth and metabolic formation and was found to increase from 36 h to 108 h due to the secretion and accumulation of various proteins and enzymes in the medium. The pH of the medium was found to decrease gradually from the initial pH of 5.9 to pH 4.3 at 60 h during the exponential phase and the pH increases from pH 4.3 to pH 5 during the stationary phase of the microorganism. The decrease in the pH might be due to some organic acid production during the enzyme production and free fatty acid production by *R. arrhizus* and the increase in pH might be due to the production of free amino acid [30]. The rate of glucose utilization by the microorganism was found to increase rapidly after 12 h of the fermentation when the microorganism reaches the mid exponential phase. The rate of glucose utilization was directly proportional to the biomass production at any instant and 95.6% of the glucose was consumed at the end of the fermentation at 108 h, and maximum lipase production was obtained at 72 h of fermentation.

A well-trained artificial neural network was employed for the prediction of lipase production by *Rhizopus arrhizus* under various fer-

exponential phase being a growth associated product. The maximum lipase activity was found in the late exponential phase and early stationary growth phase of *R. arrhizus*. The lipase activity was found to decrease after 72 h of fermentation, which may be due to the prod-

mentation conditions. The developed neural network model using the experimental data obtained from CCD gave better predictions when compared to RSM model in predicting the lipase activity and cell mass. It was found that the neural network predictions were very close to the actual experimental values. A relatively good fit of the experimental data predicted by ANN was clearly evident from Figure 3B with R^2 values of 0.987 for lipase production.

Multiple regression analysis of the experimental data for cell mass production gives the following second order polynomial equation 22,

$$Y_2 = 5.139 + 0.59\, x_1 - 0.5141\, x_2 - 0.787\, x_3 - 0.651\, x_1^2 - 0.272\, x_2^2 -$$
$$0.0288\, x_3^2 - 0.008\, x_1 x_2 - 0.588\, x_1 x_3 + 0.131\, x_2 x_3 \tag{22}$$

where Y_2 is the response variable the cell mass and x_1, x_2 and x_3 are the coded values of the independent variables, temperature, pH and agitation speed respectively. Both the RSM and ANN models predicted the experimental lipase activity and cell mass extremely well with high R^2 value and the ANN predictions are more accurate than RSM. The experimental results agree closely with the results predicted by regression equation substantiating that the RSM using the statistical design of experiments can be effectively used to optimize the process conditions for lipase production.

Unstructured Kinetic Models for Prediction of Lipase Fermentation

Various unstructured kinetic models were used to predict the kinetics of lipase fermentation by *R. arrhizus*. Logistic model for cell growth, Logistic incorporated Luedeking Piret model for lipase production and Logistic incorporated modified Luedeking Piret model for substrate utilization provides an accurate approximation of the fermentation kinetics with high R^2 values of 0.997, 0.94 and 0.972 respectively. The value of μ_0 is 0.082 1/h for the cell growth model. The value of α and β is 0.65 and 0.002 respectively which shows that the lipase production by *R. arrhizus* is growth associated since the magnitude of the growth associated parameter 'α' is much greater than the magnitude of non-growth associated parameter 'β' in Luedeking Piret model. The product formation models were able to predict the kinetics of lipase production during the exponential phase of the microorganism accurately. But during the stationary phase, the lipase production models did not predict the lipase activity precisely since the unstructured models did not contain a term for inhibitory action of the protease enzyme during the later stages of the fermentation.

For lipase production by *Rhizopus arrhizus* from Figure 5 it was observed that the cell growth rate was found to be maximum at 36 h and the glucose utilization rate was found to be maximum at 48 h of fermentation time. The lipase production rate was found to be almost constant between 36 to 48 h of the fermentation and reduced with the reduction in the cell growth rate and reached to zero at 84 h. The cell growth rate declined after 36 h and the growth rate stopped at 84 h. The increase in cell growth rate is accompanied by the increase in substrate utilization rate and lipase production rate. There was a reduction in substrate utilization rate even when the growth rate of the microorganism reached to zero at 84 h. The substrate utilization rate declined after 48 h and stopped at the 108 h of fermentation.

Conclusion

Response surface methodology was employed to optimize the fermentation conditions temperature, pH of the fermentation medium and agitation speed. The present study identified the effect of various process parameters on the lipase yield and the lipase production was found to be significantly influenced by temperature, agitation speed, the quadratic effects of temperature, initial pH and agitation speed and the interactive effects of temperature and agitation speed are significant parameters that affect lipase production. The optimized fermentation conditions for lipase production by *R. arrhizus* using the statistical approach was found to be: Temperature, 32°C; initial pH 6 and agitation speed, 107 rpm. Maximum lipase activity of 4.32 U/mL and maximum cell mass concentration of 6.2 g/L was obtained at 72 h and 84 h respectively. The experimental results and RSM model predictions shows that the response surface methodology can be successfully used to optimize and to study the interaction effect among the fermentation conditions for lipase production.

Acknowledgments

The authors gratefully acknowledge the Chemical Engineering Department, Annamalai University for providing the facilities to carry out this research work.

x_i	Coded value of an independent variable
X_i	Real value of an independent variable
X_c	Real value of an independent variable at the center point
ΔX_i	Step change value
Y_1	U/mL Predicted response (lipase activity)
Y_2	g/L Predicted response (cell mass)
β_0	Offset term
β_i	Coefficient of linear effect
β_{ii}	Coefficient of squared effect
β_{ij}	Coefficient of interaction effect
a_i	Input signal
y_j	Fired output signal
w_{ij}	Weight associated with the input signal a_i
θ_j	Threshold value of neuron j
d_j	Desired output from neuron j
θ	Positive constant controlling the speed of learning
μ_0	1/h Initial specific growth rate
X_0	g/L Initial cell mass concentration
X_{max}	g/L Maximum cell mass concentration
$X(t)$	g/L Cell mass concentration at any time 't'
μ_{max}	1/h Maximum specific growth rate
S	g/L Concentration of the limiting substrate
α	gP/gX Growth associated parameter
β	gP/gX· h Non-growth associated parameter
γ, η	Constants in the modified Luedeking-Piret model
P_0, P_t	U/mL Product concentrations at initial time and at anytime 't' respectively
S_0, S_t	g/L Substrate concentrations at initial time and at anytime 't' respectively

$Y_{X/S}$	Yield coefficient of cell mass with respect to substrate
$Y_{P/S}$	Yield coefficient of product with respect to substrate
$Y_{P/X}$	Yield coefficient of product with respect to cell mass
dX/dt	g/L·h Cell growth rate
dP/dt	U/mL·h Product formation rate
dS/dt	g/L·h Substrate utilization rate
K_e	Maintenance coefficient for cells

References

1. Jaeger KE, Ransac S, Dijkstra BW, Colson C, van Henrel M, et al. (1994) Bacterial lipases. FEMS Microbiol Rev 15: 29-63.

2. Ghosh PK, Saxena RK, Gupta R, Yadav RP, Davidson S (1996) Microbial lipases: production and applications. Sci Prog 79: 119-157.

3. Pandey A, Benjamin S, Soccol CR, Nigam P, Krieger N, et al. (1999) The realm of microbial lipases in Biotechnology. Biotechnol Appl Biochem 29: 119-131.

4. Aravindan R, Anbumathi P, Viruthagiri T (2007) Lipase applications in the food industry. Indian J Biotechnol 6: 141-158.

5. Jaeger KE, Reetz MT (1998) Microbial lipases form versatile tools for biotechnology. Trends Biotechnol 16: 396-403.

6. Yang XH, Wang BW, Cui FN, Tan TW (2005) Production of lipase by repeated batch fermentation with immobilized Rhizopus arrhizus. Process Biochem 40: 2095-2103.

7. Elibol M, Ozer D (2002) Response surface analysis of lipase production by freely suspended Rhizopus arrhizus. Process Biochem 38: 367-372.

8. D'Annibale A, Sermanni GG, Federici F, Petruccioli M (2006) Olive-mill wastewaters: a promising substrate for microbial lipase production. Bioresour Technol 97: 1828-1833.

9. Kalil SJ, Maugeri F, Rodrigues MI (2000) Response surface analysis and simulation as a tool for bioprocess design and optimization. Process Biochem 35: 539-550.

10. Rojan PJ, Rajeev KS, Madhavan Nampoothiri K, Pandey A (2007) Statistical optimization of simultaneous saccharification and L (+)-lactic acid fermentation from cassava bagasse using mixed culture of Lactobacilli by response surface methodology. Biochem Eng J 36: 262-267.

11. Ezequiel FL, Hannes L, Dirk WB (2006) Evaluation of artificial neural networks for modelling and optimization of medium composition with a genetic algorithm. Process Biochem 41: 2200-2206.

12. Zorzetto LFM, Maciel Filho R, Wolf-Maciel FR (2000) Processing modelling development through artificial neural networks and hybrid models. Comput Chem Eng 24: 1355-1360.

13. Chandrasekhar K, Arthur Felse P, Panda T (1999) Optimization of temperature and initial pH and kinetic analysis of tartaric acid production by Gluconobacter suboxydans. Bioprocess Biosyst Eng 20: 203-207.

14. Ghaly AE, Kamal M, Correia LR (2005) Kinetic modeling of continuous submerged fermentation of cheese whey for single cell protein production. Bioresour Technol 96: 1143-1152.

15. Ota Y, Yamada K (1966) Lipase form Candida paralipolytica part I. Anionic surfactant as the essential activator in the systems emulsified by polyvinyl alcohol. Agric Biol Chem 30: 351-358.

16. Anson ML (1938) The estimation of pepsin, trypsin, papain and cathepsin with hemoglobin. J Gen Physiol 22: 79-89.

17. Miller GL (1959) Use of dinitrosalicylic acid reagent for determination of reducing sugar. Anal Chem 31: 426-428.

18. Lowry OH, Rosenbrough MJ, Farr AL, Randell RJ (1951) Protein measurement with folin phenol reagent. J Biol Chem 193: 265-275.

19. Chakravarti R, Sahai V (2002) Optimization of compactin production in chemically defined production medium by Penicillium citrinum using statistical methods. Process Biochem 38: 481-486.

20. Box GEP, Hunter WG, Hunter JS (1978) Statistics for Experimenters: An Introduction to Design, Data Analysis, and Model Building, 1st Ed. John Wiley and Sons, New York, pp 30-85.

21. Cochran WG, Cox GM (1957) Experimental Designs, 1st Ed. John Wiley and Sons, New York, pp. 85-100.

22. Gawande BN, Patkar AY (1999) Application of factorial designs for optimization of cyclodextrin glycosyltransferase production from Klebsiella pneumoniae pneumoniae AS-22. Biotechnol Bioeng 64: 168-173.

23. Dutta JR, Dutta PK, Banerjee R (2004) Optimization of culture parameters for extracellular protease production from a newly isolated Pseudomonas sp. using response surface and artificial neural network models. Process Biochem 39: 2193-2198.

24. Linko S, Luopa J, Zhu YH (1997) Neural networks as 'software sensors' in enzyme production. J Biotechnol 52: 257-266.

25. Weiss RM, Ollis DF (1980) Extracellular microbial polysaccharides I. Substrate, biomass and product kinetic equations for batch Xanthan gum fermentation. Biotechnol Bioeng 22: 859-873.

26. Young TB, Bruley DF, Bungay HR (1970) A dynamic mathematical model of the chemostat. Biotechnol Bioeng 12: 747-769.

27. Luedeking R, Piret EL (1959) A kinetic study of the lactic acid fermentation. Batch process at controlled pH. Biotechnol Bioeng 1: 393-412.

28. Dalmau E, Montesinos JL, Lotti M, Casas C (2000) Effect of different carbon sources on lipase production by Candida rugosa. Enzyme Microb Technol 26: 657-663.

29. Chen SJ, Cheng CY, Chen TL (1998) Production of an alkaline lipase by Acinetobacter radioresistens. J Ferment Bioeng 86: 308-312.

30. Singh J, Vohra RM, Sahoo DK (2004) Enhanced production of alkaline protease by Bacillus sphaericus using fed-batch culture. Process Biochem 39: 1093-1101.

11

Structure and Catalytic Mechanism of a Glycoside Hydrolase Family-127 β-L-Arabinofuranosidase (HypBA1)

Chun-Hsiang Huang[1], Zhen Zhu[1], Ya-Shan Cheng[2,3], Hsiu-Chien Chan[1], Tzu-Ping Ko[4], Chun-Chi Chen[1], Iren Wang[4], Meng-Ru Ho[4], Shang-Te Danny Hsu[4,6], Yi-Fang Zeng[7], Yu-Ning Huang[7], Je-Ruei Liu[5,7,8*] and Rey-Ting Guo[1]*

[1]Industrial Enzymes National Engineering Laboratory, Tianjin Institute of Industrial Biotechnology, Chinese Academy of Sciences, Tianjin 300308, China
[2]Genozyme Biotechnology Inc., Taipei 106, Taiwan
[3]AsiaPac Biotechnology Co., Ltd., Dongguan 523808, China
[4]Institute of Biological Chemistry, Taiwan
[5]Agricultural Biotechnology Research Center, Academia Sinica, Taipei 115, Taiwan
[6]Institute of Biochemical Sciences, Taiwan
[7]Institute of Biotechnology, Taiwan
[8]Department of Animal Science and Technology, National Taiwan University, Taipei 106, Taiwan

Abstract

The β-L-arabinofuranosidase from *Bifidobacterium longum* JCM 1217 (HypBA1), a DUF1680 family member, was recently characterized and classified to the glycoside hydrolase family 127 (GH127) by CAZy. The HypBA1 exerts exo-glycosidase activity to hydrolyze β-1,2-linked arabinofuranose disaccharides from non-reducing end into individual L-arabinoses. In this study, the crystal structures of HypBA1 and its complex with L-arabinose and Zn^{2+} ion were determined at 2.23-2.78 Å resolution. HypBA1 consists of three domains, denoted N-, S- and C-domain. The N-domain (residues 1-5 and 434-538) and C-domain (residues 539-658) adopt β-jellyroll architectures, and the S-domain (residues 6-433) adopts an $(\alpha/\alpha)_6$-barrel fold. HypBA1 utilizes the S- and C-domain to form a functional dimer. The complex structure suggests that the catalytic core lies in the S-domain where Cys^{417} and Glu^{322} serve as nucleophile and general acid/base, respectively, to cleave the glycosidic bonds via a retaining mechanism. The enzyme contains a restricted carbohydrate-binding cleft, which accommodates shorter arabino oligosaccharides exclusively. In addition to the complex crystal structures, we have one more interesting crystal which contains the apo HypBA1 structure without Zn^{2+} ion. In this structure, the Cys417-containing loop is shifted away due to the disappearance of all coordinate bonds in the absence of Zn^{2+} ion. Cys417 is thus diverted from the attack position, and probably is also protonated, disabling its role as the nucleophile. Therefore, Zn^{2+} ion is indeed involved in the catalytic reaction through maintaining the proper configuration of active site. Thus the unique catalytic mechanism of GH127 enzymes is now well elucidated.

Keywords: Glycoside hydrolase; Arabinofuranose; β-L-Arabinofuranosidase; Crystal structure; Synchrotron radiation

Abbreviations

Araf : Arabinofuranose; BSA: Bovine Serum Albumin; DUF: Domain of Unknown Function; EDTA: Ethylenediaminetetraacetic Acid; GH: Glycoside Hydrolase; HEPES: (4-(2-hydroxyethyl)-1-piperazineethanesulfonic acid); HRGPs: Hydroxyproline-Rich Glycoproteins; Hyp: Hydroxyproline; MIRAS: Multiple Isomorphous Replacement with Anomalous Scattering; MR: Molecular Replacement; RMSD: Root-Mean-Square Deviations; SEC/MALS: Size Exclusion Chromatography-Multi-Angle Light Scattering; SLS: Static Light Scattering

Introduction

Arabinofuranose (Araf) is broadly distributed in nature as an important component of glycoconjugates, most existing as α-L-forms. However, certain β-L-arabinofuranosyl linked residues are also found in bacteria and plants, including mycobacterial cell wall arabinans, the hydroxyproline (Hyp)-rich glycoproteins (HRGPs), glycopeptide hormones and biopolymers [1-9]. In spite of the abundance of β-L-Araf-containing sugars (L-Araf) in bacterial and plant cells, degradation and metabolism of these polysaccharides remain largely undiscovered due to the lack of knowledge of corresponding degradative enzymes.

In the previous studies, the β-L-arabinofuranosidase (HypBA1, belonging to GH127 family) and β-L-arabinobiosidase (HypBA2, belonging to GH121 family) from *Bifidobacterium longum* JCM 1217 that participate in the β-L-arabinooligosaccharides metabolisms have been characterized [10,11]. In the hydrolysis of Hyp-linked β-L-arabinooligosaccharides, HypBA2 releases the β-1,2-linked Araf disaccharide (β-Ara₂) from Araf-β-1,2-Araf-β1,2-Arafβ-Hyp (Ara₃-Hyp). HypBA1 subsequently liberates L-arabinoses by hydrolyzing β-Ara₂ [10,11]. In addition, Ara-, Ara₂- and Ara₃-Hyp as well as Ara₂- and Ara-Me (methanol) are also preferred substrates for HypBA1, but Ara₄-Hyp, extensin, potato lectin, pNP-arabino-, galacto-, gluco- and xylo-pyranosides are not. Based on the substrate specificity and catalytic products, HypBA1 was proposed to hydrolyze β-L-arabinooligosaccharides in an exo-acting manner [11]. Moreover, Fujita mutated three glutamate residues which are strictly conserved among the HypBA1 homologous (Glutamates are commonly observed

*Corresponding author: Rey-Ting Guo, Industrial Enzymes National Engineering Laboratory, Tianjin Institute of Industrial Biotechnology, Chinese Academy of Sciences, Tianjin 300308, China

Je-Ruei Liu, Department of Animal Science and Technology, National Taiwan University, Taipei 106, Taiwan

catalytic important residues among all GH families) and proposed that Glu[322], Glu[338] and Glu[366] might be potential catalytic residues [11].

Very recently, the crystal structures of the ligand-free and L-Ara*f* complex forms of HypBA1 (PDB ID: 3WKW and 3WKX) were determined [12]. Based on the structural analyses, biochemical experiments and quantum mechanical calculations, Ito showed that the nucleophile function is likely served by Cys[417] rather than a glutamate [12]. In the meantime, we have determined the crystal structures of HypBA1 in native form, in apo form and in complex with its product L-Ara*f*. Here, by analyzing these structures, the relationship between the glutamate residues, the catalytic cysteine, the Zn^{2+} ion, and the substrate in HypBA1 is elucidated, and implications on the retaining mechanism of GH127-family enzymes are discussed.

Materials and Methods

Protein expression, purification, crystallization and data collection

The expression and purification methods employed for the protein have been described before [13]. To obtain phase information by multiple isomorphous replacement (MIR), the apo HypBA1 crystals grown in 0.4 M ammonium acetate and 18% w/v polyethylene glycol 3350 [13] were used for preparing heavy atom derived crystals by using the Heavy Atom Screen Hg kit (Hampton Research). The apo crystals (isomorphous to native crystal) were soaked with various mercury-containing reagents (final concentration 2 mM) in cryoprotectant solution (0.5 M ammonium acetate, 25% w/v polyethylene glycol 3350 and 5% w/v glycerol) for at least 1 hr.

To remove any metal ion from protein, the purified enzyme (both Se-Met protein and native protein are used here) was dialyzed against 100 mM EDTA for two times (at least 12 hrs for each time) before crystallization. The crystals of Se-Met protein without Zn^{2+} ion (apo crystal) diffracted X-rays better than the native crystals. The HypBA1-Ara*f* complex crystal was obtained by soaking the native crystal switch a cryoprotectant solution containing 50 mM Ara*f*.

The X-ray diffraction datasets were collected at the beam line BL13B1 and BL13C1 of the National Synchrotron Radiation Research Center (NSRRC, Taiwan). The data was processed using the program of HKL2000 [14]. Prior to structural refinements, 5% randomly selected reflections were set aside for calculating R_{free} as a monitor [15].

Size exclusion chromatography-multi-angle light scattering (SEC/MALS)

Absolute molecular weight of the purified protein was determined by static light scattering (SLS) using a Wyatt Dawn Heleos II multiangle light scattering detector (Wyatt Technology) coupled to an AKTA Purifier UPC10 FPLC protein purification system using a Superdex 200 10/300 GL size-exclusion column (GE Healthcare). HypBA1 protein with 0.75, 1.4, 1.6, 10, and 20 mg/ml concentration were applied to the size-exclusion column with a buffer containing 20 mM HEPES (pH 7.0) and 0.02% NaN₃ by a flow rate of 0.5 ml/min. A 1.8 mg/ml concentration of BSA was used for the system calibration as a control. The absolute molecular weights of individual peaks in the size-exclusion chromatogram were determined by the SLS data in conjunction with the refractive index measurements (Wyatt Optilab rEX, connected downstream of the LS detector). A standard value of refractive index, d*n*/d*c*=0.185 ml/g, was used for proteins and the buffer viscosity h=1.0164 cP at 25°C was calculated using SEDNTERP. The value of reference refractive index, 1.3441 RIU, was taken directly from the measurement of the Wyatt Optilab rEX when buffer only passing through the reference cell.

Sedimentation velocity

Sedimentation velocity was performed using an XL-A analytical ultracentrifuge (Beckman Coulter, Fullerton, CA) using absorption optics. HypBA1 and control reference buffer (approximately 400 μL) were added to 12 mm thick epon double-sector centerpieces in an AN60-Ti rotor and spun at 20°C and 45,000 rpm after an initial 90-min temperature equilibration period. Detection of concentrations as a function of radial position and time was performed by optical density measurements at wavelength of 280 nm and absorbance profiles were recorded every 3 min. The protein samples were in the buffer containing 20 mM HEPES, pH 7 with a concentration of ca. 6 μM. The buffer density and viscosity were calculated by SEDNTERP [16] and the sorted data were analyzed with the standard c(s) model [17] in SEDFIT version 14.3. The plot of analytical ultracentrifugation concentration profiles and their fits were made with the software GUSSI, provided by Dr. Chad Brautigam in Department of Biophysics, University of Texas Southwestern Medical Center, Dallas, USA).

Structural determination and refinement

The HypBA1 structure was solved by MIR with anomalous scattering (MIRAS) using one native and two mercury datasets [$(C_2H_5HgO)HPO_2$ and $C(HgOOCCH_3)_4$]. In our previous studies, mercury can bind to free Cys residues very easily and the phase problem can be easily solved by using at least two mercury datasets by MIR [18-21]. The mercury atom binding sites were located using AutoSol wizard of PHENIX using 2.4 Å resolution cutoff [22]. Two mercury atom binding sites were identified and following density modification for phasing improvement was performed (figure of merit = 0.33). Automatic initial model building was carried out by AutoBuild of PHENIX. A 94% completeness model (620 residues with side chains) was built. Subsequent model building and structural refinement were carried out by using the programs COOT [23] and CNS [24].

The apo (treated with EDTA; no Zn^{2+} ion observed) and product-complex (HypBA1-L-Ara*f*) structures were determined by using the molecular replacement (MR) method with PHASER [25] using refined native structure as a search model. The $2F_o$-F_c difference Fourier map showed clear electron densities for most amino acid residues. Subsequent refinements by incorporating ligands and water molecules were according to 1.0 σ map level. The refined structures were validated by using RAMPAGE [26]. Data collection and refinement statistics of these crystals are summarized in Table 1 and Supplementary Table S1. All Figures were prepared by using PyMOL [27].

X-ray fluorescence scan analysis

To investigate the metal ion presented in the HypBA1 crystals, fluorescence scan analysis of the protein crystal was performed at beam line BL13B1 of the National Synchrotron Radiation Research Center (NSRRC, Taiwan). The wavelength from 12 eV to 25588 eV was used to scan the crystal to see if there is any metal ion existed.

Results and Discussion

The overall structure

In this study, the crystal structures of native (with a Zn^{2+} ion) and apo (no Zn^{2+} ion observed) and in complex with L-arabinose (HypBA1-L-Ara*f*) were determined. HypBA1 folds into three domains, designated N-domain (N-terminal domain), S-domain (all α-domain;

Data collection	Native	Apo (SeMet)	HypBA1-L-Araf
Resolution (Å)	25.00 - 2.78(2.88 - 2.78)	25.00 - 2.25(2.33 - 2.25)	25.00 - 2.23(2.31 - 2.23)
Space group	$P3_221$	$P3_221$	$P3_221$
Unit-cell a, b, c (Å)	75.9, 75.9, 254.0	75.9, 75.9, 254.6	77.6, 77.6, 254.2
No. of Unique reflections	21797 (2178)	41517 (4078)	43124 (4226)
Redundancy	7.6 (9.0)	7.1 (7.1)	5.6 (5.7)
Completeness (%)	97.9 (100.0)	100.0 (100.0)	97.9 (98.3)
Mean I/σ(I)	36.9 (6.1)	23.3 (2.6)	36.2 (4.4)
R_{merge} (%)	6.8 (48.8)	8.1 (49.2)	5.0 (44.7)
Refinement			
No. of reflections	20120 (1778)	40022 (3803)	40524 (3318)
R_{work} (95% of data)	22.3 (0.34)	21.5 (30.0)	20.4 (28.9)
R_{free} (5% of data)	27.2 (0.40)	25.5 (30.7)	24.4 (32.8)
Geometry deviations			
Bond lengths (Å)	0.005	0.004	0.008
Bond angles (°)	1.15	1.06	1.38
Ramachandran plot (%)			
Most favored	90.9	95.9	95.9
Allowed	8.6	3.9	3.7
Disallowed	0.5	0.2	0.4
No. of atoms / mean B-values (Å²)			
Protein atoms	4971 / 66.8	5018 / 44.2	5152 / 52.2
Water molecules	58 / 57.8	279 / 43.6	312 / 57.8
zinc ion	1 / 78.4	-	1 / 44.4
Ligand atoms	-	-	10 / 43.7
PDB ID	3WRE	3WRF	3WRG

Values in parentheses are for the outermost resolution shells

Table 1: Data collection and refinement statistics for HypBA1.

substrate-binding domain) and C-domain (C-terminal domain) (Figure 1A). Both adopting β-jellyroll folds, the N- (residues 1-5 and 434-538) and C-domain (residues 539-658) consist of 10 β-strands, and 7 β-strands and 3 small α-helices, respectively. The S-domain (residues 6-433) comprises an $(α/α)_6$-barrel and accommodates an L-Araf-binding cavity, so it is considered as the catalytic domain (Figure 1A).

In order to assess the unique structural feature of the N- and C-domains, a structural homology searches with DALI [28] were carried out (Figure S1). It reveals that N-domain (colored in green, broken square) of HypBA1 displays structural similarity to C-terminal domains of xyloglucanase (GH44, colored in cyan), human α-galactosidase (GH27, colored in magenta), β-xylosidase (GH39, colored in yellow) and α-L-rhamnosidase (GH78, colored in orange) with RMSD of 3.0 Å (512 Cα atoms with 14% sequence identity), 3.5 Å (390 Cα atoms with 10% sequence identity), 2.6 Å (501 Cα atoms with 10% sequence identity) and 2.8 Å (1029 Cα atoms with 9% sequence identity), respectively [29-32]. On the other hand, C-domain (colored in green, broken square) of HypBA1 shows structural similarity to N-terminal domain of ErbB4 kinase (colored in cyan) and LKB1 (colored in salmon) with RMSD of 2.6 Å (274 Cα atoms with 4% sequence identity) and 3.5 Å (311 Cα atoms with 15% sequence identity), respectively [33,34] (SI Figure S1). However, their functional roles remain to be investigated. A possibility is that these domains might associate with substrate to facilitate catalytic reaction, as a carbohydrate-binding motif does.

In order to investigate whether the L-Araf (product) could cause conformational change, the native, apo and complex structures were superimposed and no obvious conformational change was observed (data not shown). However, a few regions which are close to the L-Araf-

binding cavity were missing in both the native and apo structures (residues 31-50 and 247-250 for native, residues 35-50 and 413-415 for apo), but all these loop regions can be clearly seen in the complex structure. Interestingly, the longest loop consisting of residue 31-50 was found stretching across the surface of the S-domain to cover, or cap, the ligand-binding cavity (Figure 1B). From the complex structure, the capping loop undergoes induced fit and is then stabilized through interacting with the L-Araf. Accordingly, the capping loop might play an important role in the catalytic reaction by regulating the substrate binding because one of the ligand binding residues, Gln[45], is located on the capping loop and a similar case has been reported [35].

On the other hand, although not supplemented in the crystallization solution, a Zn²⁺ ion was observed adjacent to the active site as a strong

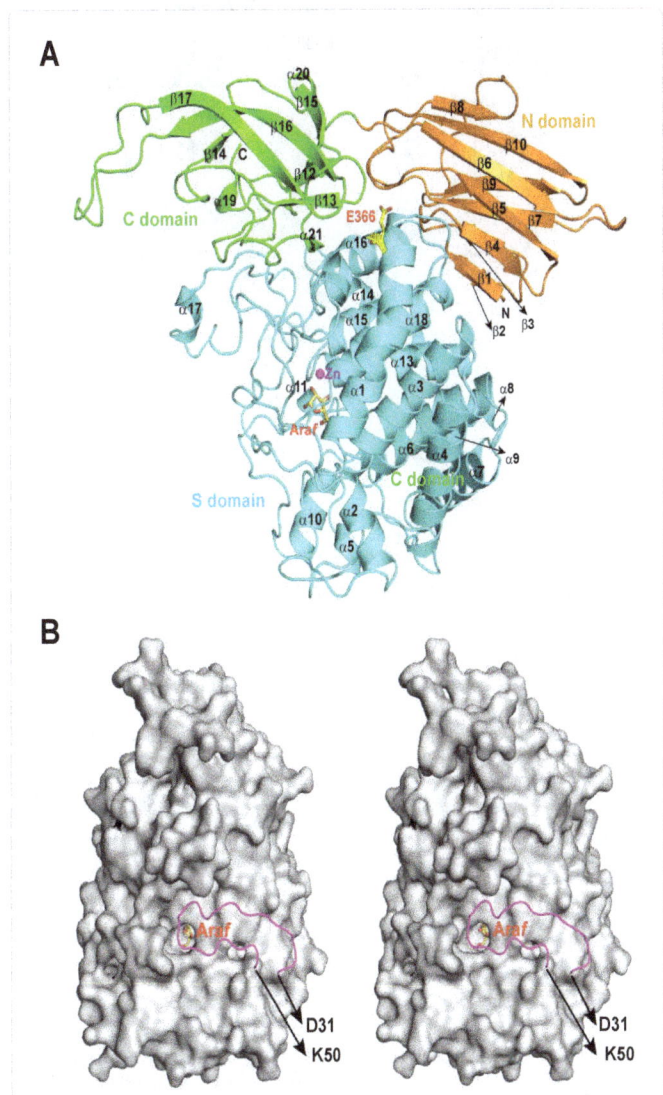

Figure 1: The overall structure of HypBA1-L-Araf. (A) A cartoon model of the HypBA1-L-Araf complex structure. Each domain is presented in different color: N-, S- and C-domain are in orange, cyan and green colors, respectively. The Zn²⁺ ion is shown as a magenta sphere, the bound L-Araf molecule and Glu³⁶⁶ is shown as yellow stick model. (B) The obvious difference between the native, apo and ligand-bound HypBA1 is the capping loop (colored in magenta). The surface representation of the complex structure by omitting the capping loop clearly shows that the loop covers the active site cleft upon ligand binding.

electron density. It was further validated by using X-ray fluorescence scan analysis and anomalous difference Fourier map (Figure 2A). Based on the $|F_o|$ - $|F_c|$ omit map, it is clear that the Zn^{2+} ion forms a typical tetrahedral complex with four Zn^{2+} ion-chelating residues, including Glu[338], Cys[340], Cys[417] and Cys[418] and all the metal-to-ligand distances are about 2.3 Å (Figure 2A). Interestingly, even though no significant conformational change was caused by Zn^{2+} binding, the two Zn^{2+} ion-chelating residues Cys[417] and Cys[418] in the apo structure were shifted away from the original positions (Figure 2B). Therefore, when the Zn^{2+} ion is absent only a broken loop can be observed because of the absence of stabilizing interactions (Figure 2C).

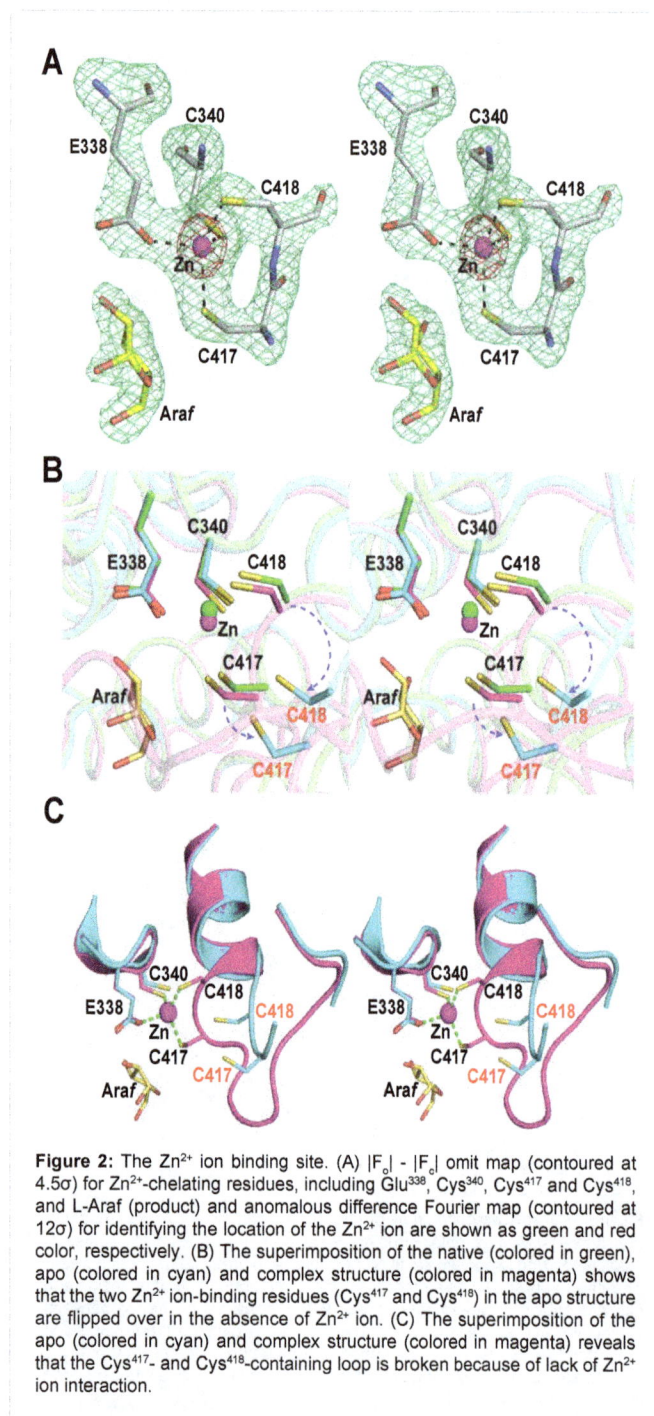

Figure 2: The Zn^{2+} ion binding site. (A) $|F_o|$ - $|F_c|$ omit map (contoured at 4.5σ) for Zn^{2+}-chelating residues, including Glu[338], Cys[340], Cys[417] and Cys[418], and L-Ara$f$ (product) and anomalous difference Fourier map (contoured at 12σ) for identifying the location of the Zn^{2+} ion are shown as green and red color, respectively. (B) The superimposition of the native (colored in green), apo (colored in cyan) and complex structure (colored in magenta) shows that the two Zn^{2+} ion-binding residues (Cys[417] and Cys[418]) in the apo structure are flipped over in the absence of Zn^{2+} ion. (C) The superimposition of the apo (colored in cyan) and complex structure (colored in magenta) reveals that the Cys[417]- and Cys[418]-containing loop is broken because of lack of Zn^{2+} ion interaction.

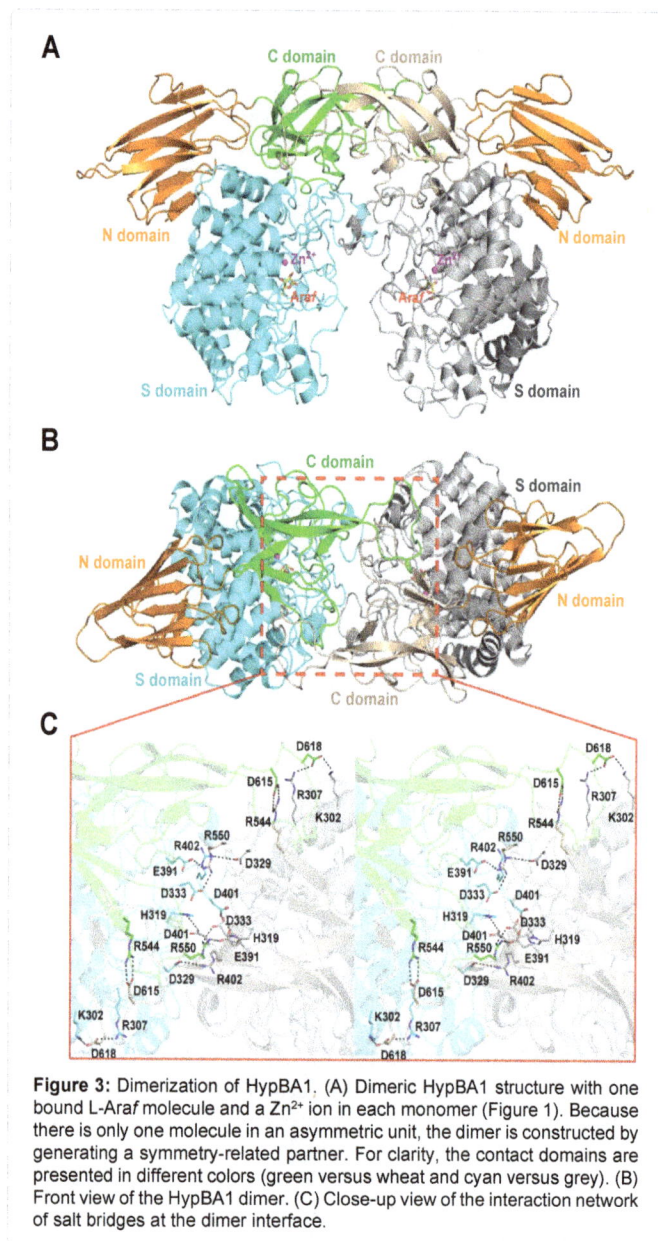

Figure 3: Dimerization of HypBA1. (A) Dimeric HypBA1 structure with one bound L-Ara$f$ molecule and a Zn^{2+} ion in each monomer (Figure 1). Because there is only one molecule in an asymmetric unit, the dimer is constructed by generating a symmetry-related partner. For clarity, the contact domains are presented in different colors (green versus wheat and cyan versus grey). (B) Front view of the HypBA1 dimer. (C) Close-up view of the interaction network of salt bridges at the dimer interface.

Dimerization

Interestingly, although there is only one HypBA1 monomer in an asymmetric unit, HypBA1 forms a dimer with a crystallographic symmetry-related molecule (Figure 3A and 3B). An analysis with PDBePISA shows that the contact interface encompasses 77 residues that bury a total surface area of about 2781 Å^2 on the S- and C-domain [36]. The intermolecular forces include hydrogen bonds (not shown) and salt bridges (Figure 3C). To confirm that HypBA1 also forms a dimer in solution, size exclusion chromatography coupled with multi-angle light scattering (SEC/MALS) was conducted. SEC/MALS offers an estimate of the absolute molecular weights in solution based on the angular dependence of scattered light intensity, which is less dependent on the molecular shapes. At protein concentrations of 0.75-1.60 mg/ml, the SEC/MALS analysis using a Superdex 200 10/300 GL column indicates that its molecular mass is 134.9-138.8 kDa, corresponding to a dimeric form of HypBA1 (Figure 4A). As the elution peak of the

SEC/MALS is relatively symmetric, the calculated molecular weight distribution indicates that the sample is monodispersed. Accordingly, our SEC/MALS data suggests that the HypBA1 protein exists as a very stable dimer in solution under low protein concentrations or at increased protein concentrations of 10 mg/ml or 20 mg/ml (data not shown).

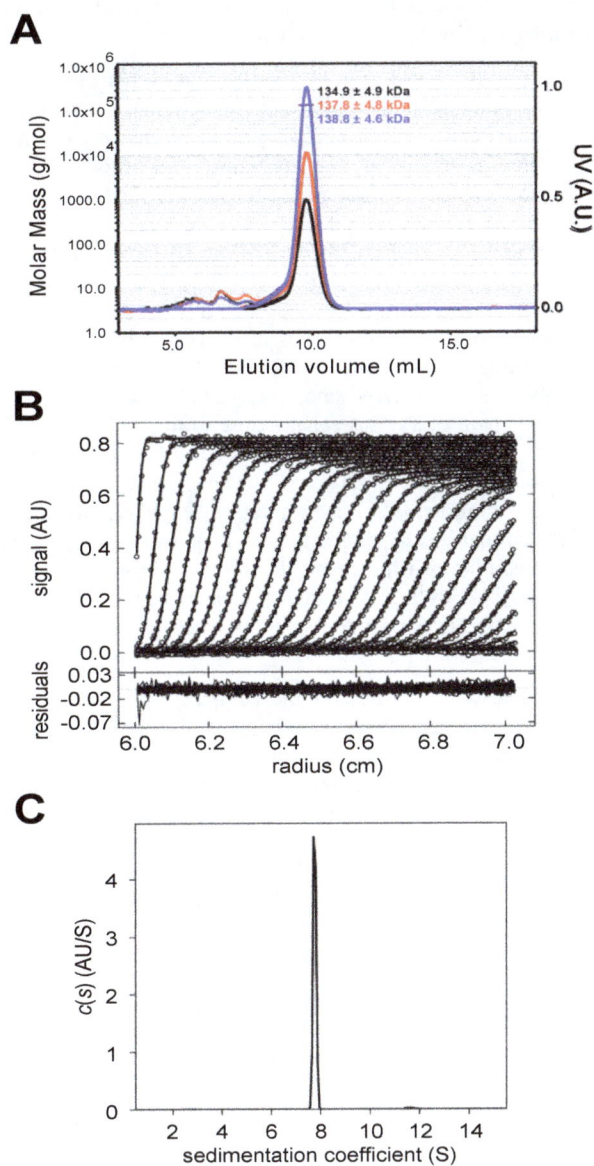

Figure 4: SEC/MALS and AUC-SV analysis. (A) SEC/MALS profiles of 0.75 (black), 1.20 (red) and 1.65 mg/ml (blue) of HypBA1 proteins (100 μl per injection). The thin line segments represent the calculated molar masses and the corresponding molecular weights of each peak (left ordinate axis) are given in number with the unit of kDa. The solid lines represent normalized UV absorbance (280 nm, right ordinate axis) and dashed lines are signals of light scattering. (B) A typical tracing of absorbance at 280 nm of the data of HypBA1 during the sedimentation velocity experiment. The symbols (open circles) are experimental data and the lines are computer-generated results by fitting the experimental data to the Lamm equation using the SEDFIT program. The residuals of the model fitting were shown in the lower panel. (C) Sedimentation coefficient distributions resulting from the fitting in (B). The major peak is compatible with a dimeric form of HypBA1 with its hydrodynamic properties.

Furthermore, attempting to determine the molecular weight of HypBA1 at an even lower concentration in solution, we applied sedimentation velocity experiment by using analytical ultracentrifugation (AUC-SV). At a concentration of 6 μM, the polypeptide was detected as a single species with a sedimentation coefficient of 7.77 S, which corresponds to a molecular mass of 143.0 kDa, the mass of a HypBA1 dimer (Figure 4B and 4C). Taken together, the results by employing two independent biophysical methods elucidate the oligomeric state of the recombinant HypBA1 in solution. At a much higher protein concentration (ca. 10 mg/ml), a symmetric elution peak was still observed and the corresponding molecular weight also coincided with the dimeric forms of HypBA1. The results are consistent with the AUC-SV data, which showed that HypBA1 was dimeric at a low protein concentration and only a small population of higher oligomers emerged when the protein concentration was increased (Figure 4). Both the results of the SEC/MALS and AUC-SV analyses are in a good agreement with the crystallographic findings that HypBA1 exhibits as a very stable dimerization propensity.

Ligand binding and substrate modeling

The $|F_o| - |F_c|$ omit map and anomalous difference Fourier map of the bound L-Araf and Zn^{2+} ion are both very clear, respectively (Figure 2A). Based on the binding mode of L-Araf, there are ten hydrogen bonds between the sugar unit and eight amino acid residues including Gln^{45}, His^{142}, His^{194}, His^{270}, Glu^{322}, Glu^{338}, Tyr^{386} and Cys^{415} (Figure 5A), but not Glu^{366}. According to the relative spatial positions of L-Araf and Zn^{2+} ion in the complex structure, Zn^{2+} ion might not directly involve in the catalytic reaction because the Zn^{2+} ion is distantly located to the C1 of product (5.0 Å) (data not shown). Moreover, the configuration of Zn^{2+} ion also makes it unlikely to activate a water molecule for catalysis (which would turn out an inverted α-sugar). This is because the Zn^{2+} ion has formed an almost perfect tetrahedral coordination with Glu^{338}, Cys^{340}, Cys^{417} and Cys^{418} (Figure 2A).

By analyzing the HypBA1-L-Araf complex structure, we believe that the bound L-Araf (product) corresponds to the -1 subsite. The space adjacent to the O2 atom is too small to accommodate a sugar residue. To further elucidate possible substrate binding mode, a two-sugar-units substrate (Araf-β1,2-Araf; β-Ara$_2$) was manually modeled into the potential subsites from -1 to +1 in the substrate-binding cavity (Figure 5B). The model was subsequently subjected to several cycles of energy minimization with CNS [24]. Interestingly, the location of the simulated model seems almost fit the size and shape of calculated cavity map (Figure 5B), which is generated by using the web server POCASA (POcket-CAvity Search Application) [37]. Therefore, the accuracy of the simulated model is justified. In this model, the side chain of Glu^{322} also binds to the O1 of the +1 sugar, which is in turn bound to Tyr^{386}, and two other residues Gln^{44} and Tyr^{250} can interact with the O5 of the same +1 sugar. Beyond the +1 sugar, there is no room to accommodate additional L-Araf and Hyp units unless the capping loop is opened. How the enzyme binds to Ara$_2$- and Ara$_3$-Hyp remains to be elucidated.

Proposed catalytic mechanism

As previously mentioned, three potential catalytic residues (Glu^{322}, Glu^{338} and Glu^{366}) have been proposed. Among them, Glu^{366} is too far away from the substrate-binding cavity and unlikely to participate directly in the catalytic reaction (Figure 1A). However, the mutant E366A had 16% activity left in a previous study [11]. Consequently, Glu^{366} might play a role in structural stability, although not participating directly in the catalytic reaction. By contrast, the residues Glu^{322} and Glu^{338} are more reasonable catalytic amino acids due to their proximal

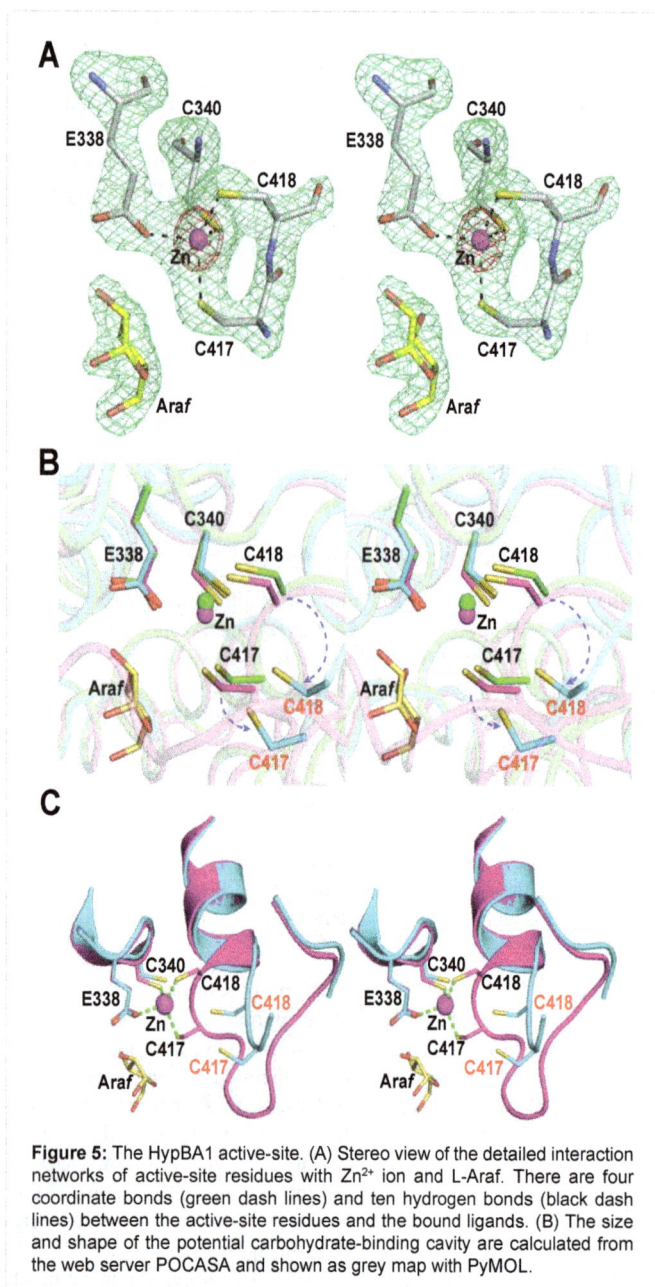

Figure 5: The HypBA1 active-site. (A) Stereo view of the detailed interaction networks of active-site residues with Zn²⁺ ion and L-Araf. There are four coordinate bonds (green dash lines) and ten hydrogen bonds (black dash lines) between the active-site residues and the bound ligands. (B) The size and shape of the potential carbohydrate-binding cavity are calculated from the web server POCASA and shown as grey map with PyMOL.

containing HypBA1 structure in ligand-free and complex forms [12]. The subsequent structure-based mutagenesis and biochemical analysis, in conjunction with quantum mechanical calculations, allowed Ito to make a clear proposal that the nucleophile should be Cys⁴¹⁷ rather than Glu³³⁸ [12]. In our apo structure, the Cys⁴¹⁷-containing loop is shifted away due to the disappearance of all coordinate bonds in the absence of Zn²⁺ ion (Figure 2C). Cys⁴¹⁷ is thus diverted from the attack position, and probably is also protonated, disabling its role as the nucleophile. Therefore, Zn²⁺ ion is involved in the catalytic reaction through maintaining the proper configuration of active site.

As said by Ito, however, we cannot rule out other possibility of catalytic reaction mechanism, such as the utilization of two carboxylate residues (Glu³²² and Glu³³⁸) separated by a suitable distance (5.4 Å) for retaining mechanism. In this case, the bound L-Araf should represent the +1 sugar and the -1 sugar would be severely skewed to fit into the limited space, which is not likely. On the other hand, a recent review suggests that some GH families employ novel mechanisms instead of typical carboxylate base/nucleophile, including substrate-assisted mechanisms, proton transferring network, utilization of non-carboxylate residues and utilization of an exogenous base/nucleophile [38]. Interestingly, apart from Glu³²², Glu³³⁸, Cys³⁴⁰, Cys⁴¹⁷and Cys⁴¹⁸, Tyr³⁸⁶ is also strictly conserved among several GH127 members (Figure 3). The side chain of Tyr³⁸⁶ is equally close to the C1 of L-Araf (3.2 Å) as is that of Cys⁴¹⁷, and it may correspond to the non-carboxylate residue in an alternative mechanism. However, the lack of a base to subtract its proton renders Tyr³⁸⁶ a weak nucleophile. Consequently, the most reasonable catalytic mechanism may involve a Cys⁴¹⁷-sugar intermediate, as shown in Scheme 1. Besides the use of a different nucleophile (Cys⁴¹⁷ rather than an Asp or Glu) in the first step, the remaining steps are almost the same as those of classic retaining mechanism.

In summary, the native, apo and complex crystal structures of HypBA1 give us a first glimpse of the GH127 family with respect to protein folding and catalytic mechanism. The results presented here shall provide a critical starting point and a firm basis for further studies of the GH127 family. In addition to the catalytic S-domain, HypBA1 also contains N-domain and C-domain, the latter participating in dimer formation. To investigate the functions of this novel multi-domain protein, further experiments with mutagenesis and truncation are required.

Acknowledgement

This work was supported by National High Technology Research and Development Program of China (2012AA022200), Academy of Sciences (2011B09030042), Tianjin Municipal Science and Technology Commission (12ZCZDSY12500) and National Science Council of Taiwan (NSC 102-2628-B-002-007-MY2). The synchrotron data collection was conducted at beam line BL13B1 and BL13C1 of NSRRC (National Synchrotron Radiation Research Center, Taiwan).

locations. In the crystal structure of HypBA1-L-Araf complex, Glu³²² is hydrogen bonded to the O1 atom of the sugar and it is in a good position for the general acid/base role. In the absence of L-Araf, Glu³²² is probably hydrogen bonded to His²⁷⁰, which is supposed to be protonated. When His²⁷⁰ turns to interact with the sugar, the proton may remain bound to Glu³²², making it a good general acid catalyst. On the other side Glu³²² is hydrogen bonded to Tyr³³⁶, which may also serve as a proton donor. Both His²⁷⁰ and Tyr³³⁶ are strictly conserved among the GH127 enzymes (Figure 3).

Regarding the nucleophile, Glu³³⁸ binds to the O2 atom but it is nearly 4 Å away from C1, too far and unlikely to undertake this role, although it can be fully ionized by binding to Zn²⁺. However, the finding that the mutant E338A only had 0.0013% activity left in a previous study [11] clearly shows that Glu³³⁸ plays an important role in ligand recognition and Zn²⁺ binding. Very recently, Ito solved the Zn²⁺ ion-

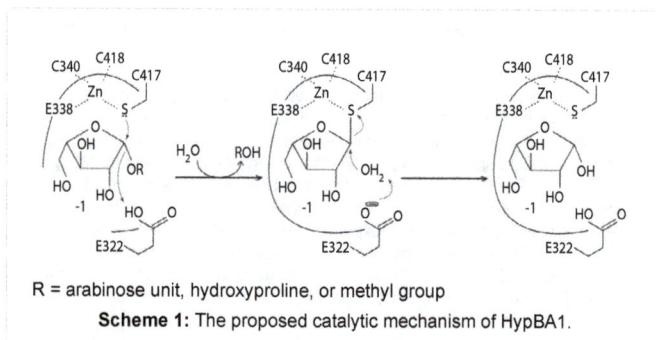

R = arabinose unit, hydroxyproline, or methyl group

Scheme 1: The proposed catalytic mechanism of HypBA1.

Chun-Hsiang Huang and Zhen Zhu are the co-authors & have contributed equally.

References

1. Besra GS, Khoo KH, McNeil MR, Dell A, Morris HR, et al. (1995) A new interpretation of the structure of the mycolyl-arabinogalactan complex of Mycobacterium tuberculosis as revealed through characterization of oligoglycosylalditol fragments by fast-atom bombardment mass spectrometry and 1H nuclear magnetic resonance spectroscopy. Biochemistry 34: 4257-4266.

2. Tam PM, Lowary TL (2009) Recent advances in mycobacterial cell wall glycan biosynthesis. Curr Opin Chem Biol 13: 618-625.

3. Amano Y, Tsubouchi H, Shinohara H, Ogawa M, Matsubayashi Y (2007) Tyrosine-sulfated glycopeptide involved in cellular proliferation and expansion in Arabidopsis. Proc Natl Acad Sci USA 104: 18333-18338.

4. Ohyama K, Shinohara H, Ogawa-Ohnishi M, Matsubayashi Y (2009) A glycopeptide regulating stem cell fate in Arabidopsis thaliana. Nat Chem Biol 5: 578-580.

5. Matsubayashi Y (2011) Post-translational modifications in secreted peptide hormones in plants. Plant Cell Physiol 52: 5-13.

6. Matsubayashi Y (2012) Recent progress in research on small post-translationally modified peptide signals in plants. Genes Cells 17: 1-10.

7. Mohnen D (2008) Pectin structure and biosynthesis. Curr Opin Plant Biol 11: 266-277.

8. Okamoto S, Shinohara H, Mori T, Matsubayashi Y, Kawaguchi M (2013) Root-derived CLE glycopeptides control nodulation by direct binding to HAR1 receptor kinase. Nat Commun 4: 2191.

9. Kieliszewski MJ, Lamport DTA, Tan L, Cannon MC (2011) Hydroxyproline-Rich Glycoproteins: Form and Function. Annual Plant Reviews: Plant Polysaccharides, Biosynthesis and Bioengineering 41: 321.

10. Fujita K, Sakamoto S, Ono Y, Wakao M, Suda Y, et al. (2011) Molecular cloning and characterization of a beta-L-Arabinobiosidase in Bifidobacterium longum that belongs to a novel glycoside hydrolase family. J Biol Chem 286: 5143-5150.

11. Fujita K, Takashi Y, Obuchi E, Kitahara K, Suganuma T (2011) Characterization of a novel beta-L-arabinofuranosidase in Bifidobacterium longum: functional elucidation of a DUF1680 protein family member. J Biol Chem 286: 38079-38085.

12. Ito K, Saikawa K, Kim S, Fujita K, Ishiwata A, et al. (2014) Crystal structure of glycoside hydrolase family 127 β-L-arabinofuranosidase from Bifidobacterium longum. Biochem Biophys Res Commun 447: 32-37.

13. Zhu Z, He M, Huang CH, Ko TP, Zeng YF, et al. (2014) Crystallization and preliminary X-ray diffraction analysis of a novel β-L-arabinofuranosidase (HypBA1) from Bifidobacterium longum. Biochem Biophys Res Commun 447: 32-37.

14. Otwinowski Z, Minor W (1997) Processing of X-ray diffraction data collected in oscillation mode. Methods in Enzymology 276: 307-326.

15. Brunger AT (1993) Assessment of phase accuracy by cross validation: the free R value. Methods and applications. Acta Crystallogr D Biol Crystallogr 49: 24-36.

16. Hayes D, Laue T, Philo J (1995) Program Sednterp: sedimentation interpretation program, Durham. NH: University of New Hampshire, UK.

17. Schuck P (2000) Size-distribution analysis of macromolecules by sedimentation velocity ultracentrifugation and lamm equation modeling. Biophys J 78: 1606-1619.

18. Guo RT, Kuo CJ, Chou CC, Ko TP, Shr HL, et al. (2004) Crystal structure of octaprenyl pyrophosphate synthase from hyperthermophilic Thermotoga maritima and mechanism of product chain length determination. J Biol Chem 279: 4903-4912.

19. Sun HY, Ko TP, Kuo CJ, Guo RT, Chou CC, et al. (2005) Homodimeric hexaprenyl pyrophosphate synthase from the thermoacidophilic crenarchaeon Sulfolobus solfataricus displays asymmetric subunit structures. J Bacteriol 187: 8137-8148.

20. Cheng YS, Ko TP, Wu TH, Ma Y, Huang CH, et al. (2011) Crystal structure and substrate-binding mode of cellulase 12A from Thermotoga maritime. Proteins 79: 1193-1204.

21. Ren F, Ko TP, Feng X, Huang CH, Chan HC, et al. (2012) Insights into the mechanism of the antibiotic-synthesizing enzyme MoeO5 from crystal structures of different complexes. Angew Chem Int Ed Engl 51: 4157-4160.

22. Adams PD, Afonine PV, Bunkoczi G, Chen VB, Davis IW, et al. (2010) PHENIX: a comprehensive Python-based system for macromolecular structure solution. Acta Crystallogr D Biol Crystallogr 66: 213-221.

23. Emsley P, Cowtan K (2004) Coot: model-building tools for molecular graphics. Acta Crystallogr D Biol Crystallogr 60: 2126-2132.

24. Brunger AT (2007) Version 1.2 of the Crystallography and NMR system. Nat Protoc 2: 2728-2733.

25. McCoy AJ, Grosse-Kunstleve RW, Adams PD, Winn MD, Storoni LC, et al. (2007) Phaser crystallographic software. J Appl Crystallogr 40: 658-674.

26. Winn MD, Ballard CC, Cowtan KD, Dodson EJ, Emsley P, et al. (2011) Overview of the CCP4 suite and current developments. Acta Crystallogr D Biol Crystallogr 67: 235-242.

27. DeLano WL (2001) The PyMOL Molecular Graphics System. DeLano Scientific, San Carlos, CA, USA.

28. Holm L, Rosenstrom P (2010) Dali server: conservation mapping in 3D. Nucleic Acids Res 38: W545-W549.

29. Fujimoto Z, Jackson A, Michikawa M, Maehara T, Momma M, et al. (2013) The structure of a Streptomyces avermitilis alpha-L-rhamnosidase reveals a novel carbohydrate-binding module CBM67 within the six-domain arrangement. J Biol Chem 288: 12376-12385.

30. Guce AI, Clark NE, Rogich JJ, Garman SC (2011) The molecular basis of pharmacological chaperoning in human alpha-galactosidase. Chem Biol 18: 1521-1526.

31. Ariza A, Eklof JM, Spadiut O, Offen WA, Roberts SM, et al. (2011) Structure and activity of Paenibacillus polymyxa xyloglucanase from glycoside hydrolase family 44. J Biol Chem 286: 33890-33900.

32. Czjzek M, Ben David A, Bravman T, Shoham G, Henrissat B, et al. (2005) Enzyme-substrate complex structures of a GH39 beta-xylosidase from Geobacillus stearothermophilus. J Mol Biol 353: 838-846.

33. Zeqiraj E, Filippi BM, Deak M, Alessi DR, van Aalten DM (2009) Structure of the LKB1-STRAD-MO25 complex reveals an allosteric mechanism of kinase activation. Science 326: 1707-1711.

34. Qiu C, Tarrant MK, Choi SH, Sathyamurthy A, Bose R, et al. (2008) Mechanism of activation and inhibition of the HER4/ErbB4 kinase. Structure 16: 460-467.

35. Chan HC, Huang YT, Lyu SY, Huang CJ, Li YS, et al. (2011) Regioselective deacetylation based on teicoplanin-complexed Orf2* crystal structures. Mol Biosyst 7: 1224-1231.

36. Krissinel E, Henrick K (2007) Inference of macromolecular assemblies from crystalline state. J Mol Biol 372: 774-797.

37. Yu J, Zhou Y, Tanaka I, Yao M (2010) Roll: a new algorithm for the detection of protein pockets and cavities with a rolling probe sphere. Bioinformatics 26: 46-52.

38. Vuong TV, Wilson DB (2010) Glycoside hydrolases: catalytic base/nucleophile diversity. Biotechnol Bioeng 107: 195-205.

12

Performance of Anaerobic Bioreactors under Diurnally Cyclic Air Temperatures: A Spectral Analysis Approach to Biogas Production

EA Echiegu[1], AE Ghaly[2]* and VV Ramakrishnan[2]

[1]Department of Agricultural and Bioresources Engineering, University of Nigeria, Nsukka, Enugu State, Nigeria
[2]Department of Process Engineering and Applied Sciences, Dalhousie University, Halifax, Nova Scotia, Canada

Abstract

The effects of two diurnally cyclic temperature ranges (20-40°C and 15-25°C) and four levels of hydraulic retention times (25, 20, 15, and 10 d) on the performance of anaerobic reactors operated on screened dairy manure were evaluated. The reactor temperature exhibited a lag relative to the chamber air temperature. For the 20-40°C temperature cycle, the average lags period at the maximum and minimum chamber temperatures 3.75 and 4.37 h, respectively. For the 15-25°C temperature cycle, the average lag periods at maximum and minimum chamber temperatures were 3.61 and 4.34 h, respectively. The effluent solids content were not adversely affected by the reactor diurnally cyclic temperature. The effluent total solids and methane content of the biogas diurnally cyclic patterns were out of phase with the diurnally cyclic pattern of the reactor temperature by about 12 hours under most operating conditions. The reductions in total solids and methane yield were all significantly affected by the diurnal temperature range and hydraulic retention time. Biogas production from a healthy digester operating under a diurnally cyclic temperature environment follow a sinusoidal pattern which can be described by a Fourier equation of the form: $\gamma_t = \gamma_o + \sum_{n=1}^{\infty}[a_n \cos(2\pi n\varphi t) + b_n \cos(2\pi n\varphi t)]$. However, where the operating conditions are not favourable, the production followed a sinusoidal pattern which may be embedded in some harmonic and noise.

Keywords: Anaerobic reactors; Solids; Volatile fatty acids; Diurnally cyclic Temperature; Biogas; Fourier function; Spectral analysis

Introduction

Anaerobic digestion is a biological process in which organic materials are decomposed in the absence of free oxygen to yield methane (CH_4), carbon dioxide (CO_2) and small quantities of other gases (H_2S, N_2 and H_2O). The process requires the concerted actions of a symbiotic population of three groups of facultative and obligate anaerobic bacteria. The first group degrade carbohydrates, lipids and proteins into alcohols and long chain fatty acids. The second group produce acetate, carbon dioxide and hydrogen from the fermentation of degraded organic materials. The third group produce methane and carbon dioxide from the fermentation of acetate.

Compared to other biological treatment processes, anaerobic digestion has several advantages including: (a) production of usable biogas that is about 60-80% methane with a fuel value of 17-23.9 MJ/m³, (b) the digested residue is almost odourless with partially stabilized solids content, (c) the inorganic nutrients are conserved resulting in an enhanced fertilizer value of the digested sludge and (d) pathogenic microorganisms such as Salmonella sp. and Brucella sp. as well as weed seeds are destroyed [1-5]. The latter implies that livestock can be grazed on pastures which have been spread with sludge from the anaerobic digester sooner than would be acceptable if raw manure was used [6]. However, despite these advantages, anaerobic digestion has not enjoyed widespread acceptance among many farmers due to the relative instability of the system resulting from the sensitivity of anaerobic bacteria to changes in environmental conditions [7]. Temperature is considered to be one of the most important environmental parameters affecting the growth and survival of anaerobic microorganisms [1,4,6,8-13]. Diurnally cyclic variation in slurry temperature is a common phenomenon in reactors operated under ambient conditions. Echiegu [14], Ghaly et al. [15] Ghaly and Hattab [6] showed that gas production as well as some other operating indices followed a diurnally cyclic pattern with some lag relative to the diurnally cyclic environmental temperature.

The aim of this study was to investigate the performance of a continuous mix anaerobic reactor operating under diurnally cyclic temperatures by evaluating reductions in the effluents solids and the profile of VFA and biogas. Also, the gas production data was subjected to spectral analysis to determine whether gas production under the above operating condition can be described by a perfectly sinusoid ally periodic function.

Modeling

Deterministic data is one which can be described by an explicit mathematical relationship; a non-deterministic data cannot. Deterministic data can be classified as periodic or non-periodic. Periodic data and for which the following relationship holds:

$$Y_t = f\left(t + T_p\right) \qquad (1)$$

Where:

Y_t is the value of the variable at time t in seconds

T_p is the period in seconds.

Many periodic data can be described in the form of a sine or cosine wave. Such a data is said to be sinusoidal. A sinusoidal data can be represented by a time dependent function of the following form:

*Corresponding author: Abdel E Ghaly, Process Engineering Department, Dalhousie University, Halifax, Nova Scotia, Canada

$$Y_t = A(cos\omega_o t + \phi) \tag{2}$$

Where:

A is the amplitude

ω_o is the angular frequency (rad/s)

φ is the phase angle (rad)

The angular frequency can be expressed as follows:

$$\omega_o = 2\pi f \tag{3}$$

Where:

f is the frequency in Hz = $1/T_p$

By invoking trigonometric relationships, equation (2) can be expressed as follows:

$$Y_t = a_1 \cos(\omega_o t) + b_1 \sin(\omega_o t)) \tag{4}$$

Where:

$$a_1 = A \cos\phi \tag{5}$$

$$b_1 = A \sin\phi \tag{6}$$

Many real life periodic data are not sinusoidal. They are rather complex in nature and can be described as follows:

$$Y_t = f\left(t + nT_p\right) \tag{7}$$

Where:

n is harmonic number

More generally, arbitrary periodic functions can be represented by an infinite series of sinusoids of harmonically related frequencies such as Fourier series. A Fourier series representation of a complex periodic data can be written as follows:

$$Y_t = Y_0 + \sum_{n=1}^{\infty}[a_n \cos(n\omega_0 t) + b_n \sin(n\omega_0 t)] \tag{8}$$

Where:

Y_o is the mean value of the periodic function

The terms Y_o, a_n and b_n are defined as follows:

$$Y_0 = \frac{1}{T_p}\int_0^{T_p} Y_t dt \tag{9}$$

$$a_n = \frac{2}{T_p}\int_0^{T_p} Y_t \cos(n\omega_0 t)dt \quad n = 0,1,2,..... \tag{10}$$

$$b_n = \frac{2}{T_p}\int_0^{T_p} Y_t \sin(n\omega_0 t)dt \quad n = 0,1,2,..... \tag{11}$$

For equally spaced data with N data points, equation (9) to (11) can be written respectively as follows:

$$Y_0 = \sum_{i=1}^{N} \frac{Y_t(t)}{N} \tag{12}$$

$$a_n = \frac{2}{N}\sum_{i=1}^{N}[Y_t \cos(n\omega_0 t)dt] \quad n = 0,1,2,..... \tag{13}$$

$$b_n\tau = \frac{2}{N}\sum_{i=1}^{N}[Y_t \sin(n\omega_0 t)dt] \quad n = 0,1,2,..... \tag{14}$$

Equation (8) indicates that a periodic data can be represented by a static component Y_o and an infinite number of sinusoidal components with a fundamental frequency ω_o and harmonic whose frequencies are constant multiples of the fundamental frequency. In periodic data, the

sequence repeats, although the repetition may not be exact. To measure the degree of similarity between two successive positions, the sequence would have to be compared with itself at two successive positions. This is achieved by calculating autocorrelation function. Autocorrelation is defined as the linear correlation between a time series at time (t) and the same series at a later time ($t + \tau$) [16].

Mathematically:

$$R_{xx}(t, t+\tau) = \lim_{n\to\infty}\left\{\frac{1}{n}\sum_{i=1}^{n}[Y_t(t)Y_t(t+\tau)]\right\} \quad n = 0,1,2.... \tag{15}$$

Where:

R_{xx} (t, t + τ) is the autocorrelation coefficient

τ is the time interval in seconds.

Lag is the amount of offset between two successive series being compared. If each observation in a segment of time series is numbered from 1 to i, the autocorrelation has zero lag when element 1, 2, 3, ...i is compared to elements 1, 2, 3, ...i of another segment of the series. When the correlation coefficient of elements 1, 2, 3 ...i-1 of one segment of the series are compared to elements 2, 3, 4, ...I of another segment of the same series, the autocorrelation has a lag of 1. For a perfectly periodic data such as sine wave, the autocorrelation at zero lag equals 1. A typical autocorrelation will start from +1 at zero lag, decrease to a negative value less than or equal to -1 and then rise again. For a perfectly periodic data, the autocorrelation coefficient will vary from +1 to -1.

Materials and Methods

Experimental apparatus

The experimental apparatus (Figure 1) consisted of four bioreactors, a feeding system, a temperature control system, a gas collection system, a data acquisition and control system and computer.

Bioreactors: Each bioreactor (Figure 2) was constructed of a cylindrical PVC vessel, (458 mm length, 300 mm diameter and a 1 0 mm wall thickness). It provided a liquid volume of 25 L and a gas head space of 7.4 L. The cover of the bioreactor was made of a transparent Plexiglas plate (370 mm diameter and 1 0 mm thickness). It was secured to the bioreactor, through PVC lugs glued onto the top outer periphery of the vessel by means of six bolts. A Vaseline-coated rubber gasket was fitted between the cover and the vessel to provide a gas-tight seal. The

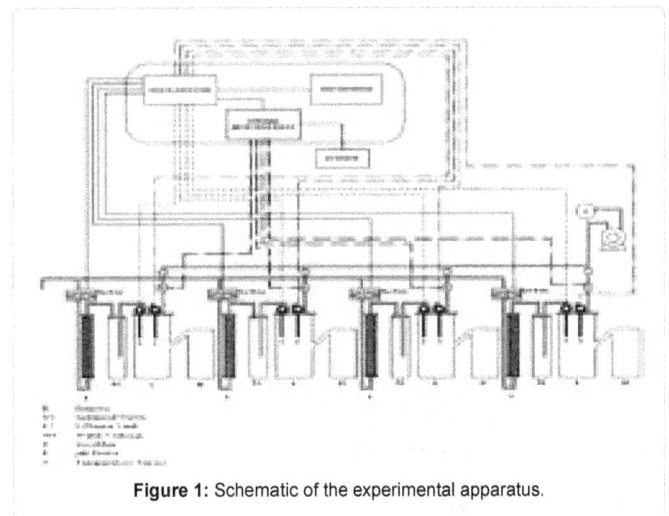

Figure 1: Schematic of the experimental apparatus.

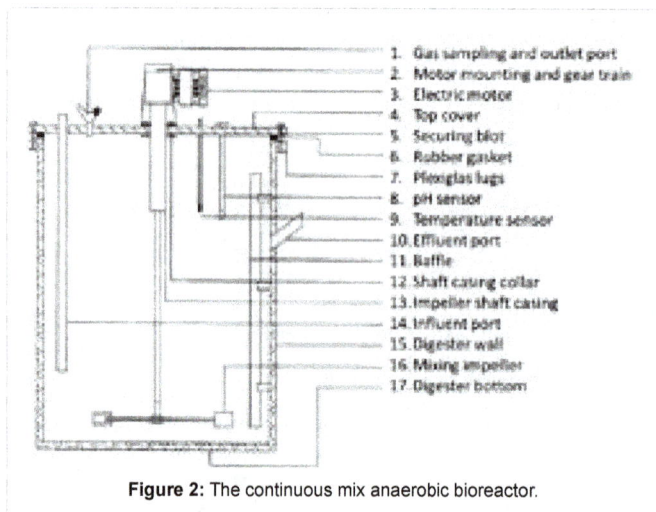

Figure 2: The continuous mix anaerobic bioreactor.

1. Gas sampling and outlet port
2. Motor mounting and gear train
3. Electric motor
4. Top cover
5. Securing blot
6. Rubber gasket
7. Plexiglas lugs
8. pH sensor
9. Temperature sensor
10. Effluent port
11. Baffle
12. Shaft casing collar
13. Impeller shaft casing
14. Influent port
15. Digester wall
16. Mixing impeller
17. Digester bottom

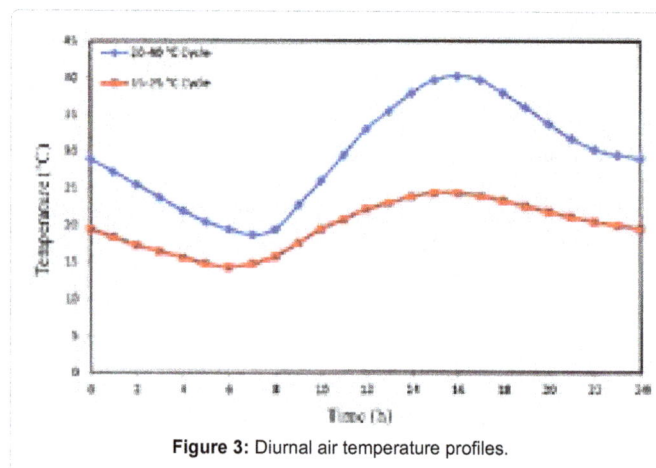

Figure 3: Diurnal air temperature profiles.

temperature and pH sensors were mounted on special holders threaded through the reactor cover and immersed to a depth of 1 00 mm below the liquid surface. The contents of the bioreactor were stirred by means of a flat-bladed impeller driven by a 1/12 HP electric motor (Model No6105061401, Franklin, Bluffton, Indiana, USA) at a speed of 125 rpm. The manure feed inlet consisted of a 25 mm diameter PVC tube inserted through the bioreactor cover to a depth of 246. mm below the liquid surface. The effluent port was located opposite to the feed inlet port at a distance of 300 mm above the bottom of the bioreactor.

Feeding system: The manure feeding system consisted of a manure storage tank, a feed pump, a set of tubing, a distribution manifold and a set of solenoid valves. A 60 L plastic container was used to store the manure and was fitted with a stirring paddle driven by a 1/12 HP electric motor (Model 55CP1OFG17AX, General Electric, Missisauga, Ontario, Canada) mounted on the tank cover. A feeding and ventilation ports were provided on the tank cover. The outlet port of the tank was located at the lower portion of the tank side. A peristaltic pump (Model 110-030, TAT Engineering Co., Logan, Ohio, USA) was used to feed the manure. At a rotational speed of 2 rpm and an output of 0.138 L of manure per revolution, the pump delivered the manure at a capacity of 16.56 L/h from the storage tank through the valve, tygon tubing, the distribution manifold and a set of automatically controlled solenoid valves to the individual reactors. A digital timer controlled and synchronized the operation of the feed tank stirrer (Model 5935932, Type NSI-10R93, Bodine Electric Co, North York, Ontario, Canada), the feed pump and the solenoid valves to ensure the delivery of a well mixed manure to the bioreactors at a predetermined sequence and loading rate. Eight feeding cycles were carried out daily at an interval of 3 hours.

Temperature control: The objective of the temperature control system was to maintain the profiles of the two temperature cycles (Figure 3) in the enclosed chamber in which the digesters were located. It consisted of an insulted chamber, a chiller, two heaters, two cross flow heat exchangers, two fans, temperature transducers and ducts for directing the circulation of air (Figure 4). The chamber was made of a 20 mm thick wooden box (2440 mm long, 710 mm wide and 525 mm deep) and divided into two compartments each was padded with 40 mm of Styrofoam insulation (R-value of 0.88 m2 KIW). Each compartment was maintained at one of the two temperature profiles and contained two bioreactors. On the floor of each compartment was a U-shaped galvanized steel duct of 1 00 mm inside diameter. A fan, a heat exchanger and a heater were located

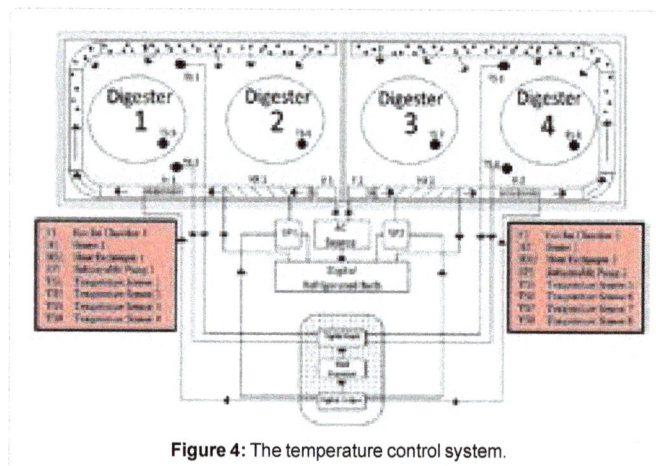

Figure 4: The temperature control system.

(in that order) inside one arm of each duct which circulated air in a closed loop. The fan drew air from the chamber into the duct and the air circulated over the heat exchanger and the heater and then left the duct through perforations on the other two sections of the U-shaped duct to bathe the bioreactor. The air was circulated using a thermally protected electric fan (Model 4C02A, Dayton Electric Manufacturing. Co., Nues, Illinois, USA). The capacities of the heaters were 750 and 500 W for the compartments operated at the higher and lower cyclic temperature ranges, respectively. A digital rapid cool refrigerated bath (Series 900, Polysciènce, Nues, Illinois, USA) with an operating temperature range of -35 to 150 °C) and an accuracy of ±0.02°C was use. The chiller had a maximum cooling capacity of 425 W at 20°C (or 225 W at -10°C), a bath capacity of 13 L and can deliver a maximum flow of 15 L/min at zero head. Two submersible pumps (Model 1 -MA, Little Giant, Bluffton, Indiana, USA), each delivered chilled water to one of the two heat exchangers, were immersed inside the chiller bath. With a power consumption of 3.73 W, each pump had a rated capacity of 10.72 L/min at 0.3 m and a maximum pumping height of 1.85 m. A Styrofoam cover of the same thickness as the wall insulation was provided for each of the chamber compartments in such a way that visual examination of the content of the reactors was possible without removing the covers. Four temperature sensors (TS 1 & TS2 and TS3 & TS4) were mounted on the chamber covers (two for each compartment). During operation, the temperature transducers continuously sampled the air temperature in each of the chamber compartments and transmitted the signals through the digital input node of the data acquisition system to the host processor. Every 30 seconds, the host processor computed the average temperature of each compartment and compared these with set point values. Depending on the error signal, the host processor generated a correction signal which either turned the heaters and the chilled water submersible pumps on and off (or vice versa), respectively.

Gas collection: Gas collection was effected through a Y-shaped nipple fitted on the cover. One end of the nipple carried a rubber septum through which gas samples were drawn with a syringe. The other arm of the nipples was fitted with flexible tubing which conveyed the evolved gases through a water column, gas scrubber and gas meter to the gas collection and storage system. The water column was used to provide back pressure in the reactor to maintain a constant liquid level. The scrubber was made of a gas-tight Plexiglas column (80 mm diameter and 450 mm long) filled with steel wool to strip the biogas of hydrogen sulfide. The gas meter was a tipping balance meter which indicated when 50 ± 2 mL of gas had been collected. The collected gas was subsequently pumped into a gas cylinder by means of a compressor.

Manure collection, preparation and storage: Dairy manure was collected from a commercial dairy farm located in Stewiacke, Nova Scotia, Canada, The manure was scrapped off the floor of the dairy barn and screened using a modified fish de-boner (Model SDX16, Bibun, Japan). During the sieving process, tap water was added to dilute the manure to 6.5% solid content. The manure was placed in buckets of 30 L and the buckets were sealed and transported to a commercial freezing plant (Associated Freezers, Dartmouth, Nova Scotia, Canada) where they were stored at -25°C until needed. When needed, buckets of manure were removed from the freezer and kept at room temperature in the Waste Management Laboratory to thaw for 48 hours and the manure was then loaded into the feeding tank. Some characteristics of raw manure are shown in Table 1.

Start-up procedure

The bioreactors were started by adding 18.0 L (to each reactor) of actively digesting sewage sludge obtained from Mill Cove Municipal Wastewater Treatment Plant (Bedford, Nova Scotia, Canada). This

was followed by the addition of 7.0 L of screened dairy manure. The temperature control and the data acquisition systems were activated and the digesters were operated in batch mode for 48 hours. The feeding system was used to feed the reactors. Reactors R1 and R2, were operated at the diurnal temperature cycle of 20-40°C and received feed at rates of 0.33 and 0.42 L/d, respectively. Reactors R3 and R1 were operated at the diurnal temperature cycle of 15-25°C and received feed at rates of 0.33 and 0.42 L/d, respectively. Feeding rates were equivalent to hydraulic retention times (HRTs) of 75 days for the feeding rate of 0.33 L/d and 60 days for the feeding rate of 0.42 L/d. The feeding rates were then adjusted to 0.5 L/d for reactors R1 and R3 and 0.63 L/d for R2 and R1 and were held constant for 72 hours. These rates corresponded to HRTs of 50 and 40 days, respectively. These rates were maintained for another 48 hours.

Operating procedure

After 7 days from start-up, R1 and R3 were operated at an HRT of 25 days while R2 and R4 were operated at an HRT of 20 days. The start-up period was concluded after a period of 32 days. Following the initial start-up period, monitoring of the biogas production was started on day 33 (from the start). Steady state was construed to have been achieved when a uniform gas production and/or uniform effluent quality were achieved. Once the steady-state was achieved at a given retention time, the system parameters were measured monitored for a period of at least five days. The feeding rates of the reactors were adjusted for the next set of retention times (15 days for reactors R1 and R3 and 10 days for reactors R2 and R4). When the steady state was attained, sampling monitoring was continued for five days.

Analyses: The liquid samples were analyzed for total solids (TS), total volatile solids (TVS), total fixed solids (TFS) and volatile fatty acids (VFA) whereas the biogas samples were analyzed for gas composition.

Solids: The total solids analyses were performed according to procedures described in the Standard Method for the Examination of Water and Wastewater [17].

Volatile fatty acids: The individual volatile acids (C2-C7) contained in a sample were determined using a Hewlett-Packard gas chromatograph (Model 5890 series II, Mississauga, Ontario, Canada) equipped with an HP 76734A automatic injector. Extraction of the VFA was carried out by acidifying 3.0 mL of each of the manure samples using 0.1 mL 30% sulphuric acid. The acidified samples were well mixed and centrifuged at 7000 rpm for 20 minutes. 2.0 mL of the supernatants were decanted and an equal amount of diethyl ether was added. The mixtures were well shaken and then centrifuged at 5000 rpm for 5 minutes to break down the emulsion layer. The upper layer which consisted of di-ethyl ether was removed for analysis. Volatile acids were, also, extracted from a volatile acid standard mixture (No4-6975, SupelCo, Oakville, Ontario, Canada) using di-ethyl ether. The chromatograph was calibrated by injecting 1 .0 mL of the extracted standard VFA mixture into the 25 mm x 0.2 mm capillary column of the liquid chromatography whose film thickness is 0.33 mm. 1.0 mL of the extracted samples were injected into the column. A split ratio of 1:5 was applied. The column temperature was first maintained at 60°C for 3 minutes and then increased at a rate of 10°C per minute until a temperature of 150 °C was attained. The column temperature was maintained at 150°C for 2 minutes. The injector was set to 180°C while the flame ionization detector was set at 250°C. The carrier gas was

Parameter	Mean Value*
√ Total Solids (g/L)	64.25
• Total Volatiles Solids (g/L)	50.26
• (% of Total Solids)	78.23
• Total Fixed Solids	13.99
√ Total Suspended Solids (g/L)	42.25
• Volatile Suspended Solids (g/L)	31.00
• Fixed Suspended Solids (g/L)	11.25
√ Total COD (g/L)	98.80
√ Soluble COD (g/L)	27.90
√ Total Kjeldahl Nitrogen (g/L)	3.80
• Ammonium Nitrogen (g/L)	2.45
• NH4-N/TKN Ratio	0.64
√ Volatile Fatty Acid (VFA)	
• Acetic (g/L)	1.55
• Propionic (g/L)	0.28
• i-Butyric (g/L)	0.04
• Butyric (g/L)	0.06
• i-Valeric (g/L)	0.04
• Valeric (g/L)	0.02
• i-Caprioc (g/L)	0.01
• Caprioc (g/L)	0.01
• Heptanoic (g/L)	0.04
Total VFA as acetic acid (g/L)	1.91

*Each value represents the average of 5 samples

Table 1: Characteristics of raw manure.

helium at a flow rate of 1.2 mL/mm.

Biogas composition: The composition of biogas was determined using a gas chromatograph (Model HP 5980A, Hewlett Packard, Mississauga, Ontario, Canada). Samples of 0.1 mL were taken from the gas collected in the sampling tube using a gas tight locked syringe. The samples were injected into 152.4 mm x 3.2 mm (6 in x 1/8 in) OD porapak Q stainless steel column of the gas chromatograph which is connected in a series bypass arrangement with a 152.4 mm x 3.2 mm (6 in X 1/8 in) OD molecular sieve 5A 60180 stainless steel column. The switch valve of the gas chromatograph was adjusted to permit the molecular sieve column to store nitrogen, methane and carbon monoxide until the elution of the CO_2, C_2H_2 and C_6H_6 through the porapak Q stainless steel column. The column was maintained at 45°C with helium as the carrier gas at 30 ml/min. The injector was set at 150°C while the thermal conductivity detector was set at 250°C.

Results and Discussion

Temperature

Typical variations of reactor temperatures with those of the chambers for the 20-40°C and 15-25°C diurnal temperature cycles are shown in Figure 5. A summary of the reactor temperature profiles at the various operating conditions is presented in Table 2. The mean minimum and maximum temperatures for the reactors operating under the 20-40°C cycle were 25.5°C and 33.8°C, respectively. The mean amplitude for the 20-40°C cycle was ± 8.3°C. The mean minimum and maximum temperatures for the 15-25°C cycle were 18.6°C and 22.8°C, respectively. The mean amplitude for the 15-25°C cycle was ± 4.2°C.

Relative to air temperature, there was a lag in the reactor temperature at both the maximum and minimum temperatures of both temperature cycles. The mean lag at the maximum and minimum temperatures for the 15-25°C cycle and the 20-40°C cycle were 3.37 h and 3.95 h and 3.38 h and 4.35 h, respectively. This was due to the significant differences between the density of air in the chamber and the liquid medium in the reactor and the thermal properties of the reactors' walls which affected the rate of heat transfer to and from the reactors.

Solids

The average values of total, fixed and volatile solids concentrations of the effluent samples collected during the steady state and their reductions are shown in Table 3. The diurnal variations in the total

solids of the effluent collected over a 24 h period during the steady state are shown in Figure 6.

Total solids

Diurnally cyclic variations in the effluent total solids were observed under the 20-40°C cycle. However, the diurnally cyclic variations in total solids were not distinct at the 15 and 10 d HRTs under the 15-25°C cycle. Similar diurnally cyclic variations were observed for the effluent total volatile solids and total fixed solids.

For the 20-40°C cycle, the TS reductions ranged from 30.3 to 36.8% (19.45 to 23.65 g/L), the TVS reductions ranged from 34.9 to 40.3% (17.56 to 20.26 g/L) while the TFS reductions ranged from 13.5 to 24.2% (1.89 to 3.39 g/L), depending on the hydraulic retention time. For the 15-25°C cycle, the TS reductions ranged from 24.4 to 38.3% (15.65 to 24.55 g/L), the TVS reductions ranged from 25.2 to 39.7% (12.66 to 19.96 g/L) while the TFS reductions ranged from 21.4 to 32.8% (2.99 to 4.59 g/L), depending on the hydraulic retention time. Higher TS reductions were achieved with the high diurnal temperature cycle (20-40°C) than those with the lower diurnal temperature cycle (15-25°C) at all HRTs.

Wilkie et al. [18] obtained the highest TS reduction of 46.7% during the treatment of dairy manure at a low temperature range of 11.7-32.4°C. Elango et al. [19] achieved a TS reduction of 87.6 % during the treatment of solid wastes over the temperature range of 26 to 36°C. The TS reduction achieved in the present study (25.2-39.7) is within the values reported in the literature. The hydraulic retention time (HRT) had a significant effect on the TS reduction in the present study. Generally, the shorter the HRT, the lower the solids reductions. Ribas and Barana [20] reported that a decrease in the HRT from 16.6 to 9.7 d increased the TS of anaerobically treated wastewater from 2.3 to 4.8%. Umana et al. [21] reported an increase in the TS removal efficiency from 14.6 to 68.8% when the HRT was increased from 1 to 5.5 d during the treatment of dairy manure. Rico et al. [22] noted that the concentration of TS increased from 39.2 to 45.3 g/L as the HRT was decreased from 20 to 10 d during the treatment of dairy manure.

Volatile solids

The absolute values of TVS reductions (g/L) were generally lower than those of the TS. This may be explained by the fact that volatile solids are converted to microbial biomass, volatile fatty acids (VFAs)

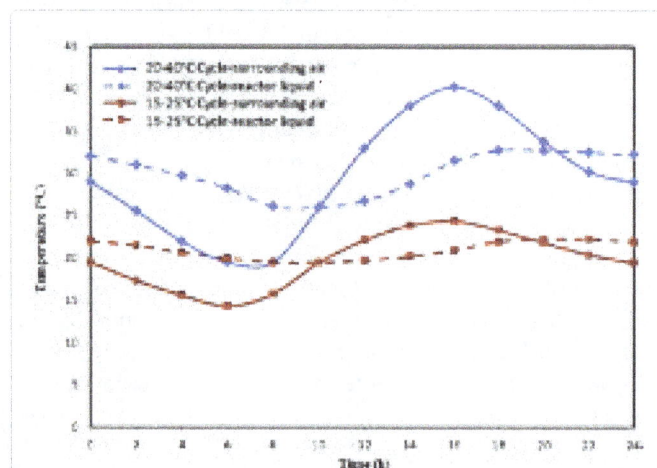

Figure 5: Diurnal variations in bioreactor and ambient temperatures.

Diurnal Temperature Cycle (°C)	HRT (d)	Reactor Temperature (°C)			Lag (h)	
		T_{min}	T_{max}	A	L_{min}	L_{max}
20-40	25	25.5	33.7	8.2	3.38	4.35
	20	25.1	33.6	8.4	3.30	4.69
	15	26.0	33.9	7.9	3.37	3.19
	10	25.2	33.8	8.6	3.27	3.49
	Mean	25.5±0.4	33.8±0.1	8.3±0.31	3.33±0.1	3.93±0.7
15-25	25	18.6	22.9	4.3	3.51	4.37
	20	18.3	22.4	4.1	3.47	4.47
	15	18.5	22.9	4.4	3.30	3.37
	10	18.8	22.8	4.0	3.19	3.60
	Mean	18.6±0.2	22.8±0.2	4.2±0.2	3.37±0.2	3.95±0.5

T_{max}	=	Maximum temperature
T_{min}	=	Minimum temperature
A	=	Amplitude
L_{min}	=	Lag at minimum temperature
L_{max}	=	Lag at maximum temperature

Table 2: Summary of reactor temperature profiles.

Diurnal Temperature Cycle (°C)	HRT (d)	TS			TFS			TVS		
		Mean	Reduction		Mean	Reduction		Mean	Reduction	
		(g/L)	(g/L)	(%)	(g/L)	(g/L)	(%)	(g/L)	(g/L)	(%)
20-40	25	40.60	23.65	36.8	10.60	3.39	24.2	30.0	20.26	40.3
	20	41.30	22.95	35.7	10.90	3.09	26.4	30.4	19.86	39.5
	15	44.10	20.45	31.4	11.80	2.19	15.7	32.3	17.96	35.7
	10	44.80	19.45	30.3	12.10	1.89	13.5	32.7	17.56	34.9
15-25	25	39.70	24.55	38.2	9.40	4.59	32.8	30.3	19.96	39.7
	20	44.90	19.35	30.1	10.20	3.79	27.1	34.7	15.56	30.9
	15	47.30	16.95	26.4	10.60	3.39	25.7	36.7	13.56	26.9
	10	48.60	15.65	24.4	11.00	2.99	21.4	37.4	12.66	25.2

TS	=	Total solids
TFS	=	Total fixed solids
TVS	=	Total volatile solids
Initial TS	=	64.25 g/L
Initial VS	=	50.26 g/L
Initial FS	=	13.99 g/L

Table 3: Total solids.

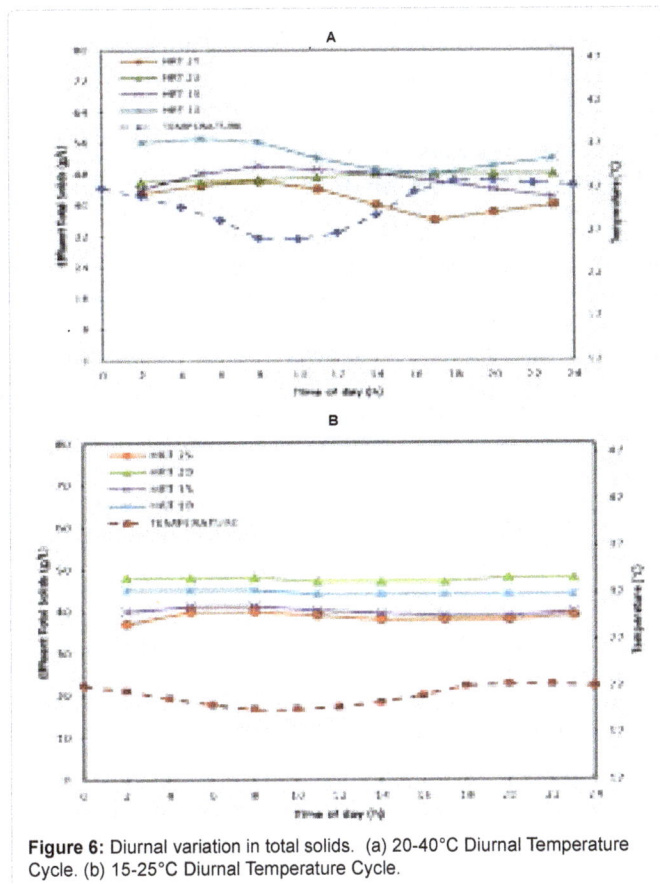

Figure 6: Diurnal variation in total solids. (a) 20-40°C Diurnal Temperature Cycle. (b) 15-25°C Diurnal Temperature Cycle.

and biogas. While microbial biomass contributes to the total solids content of the effluent, the VFAs are evaporated at 105 °C and therefore, are not detected in TS measurements [23].

The HRT had a significant effect on the effluent TVS. Generally, the shorter the HRT the lower the TVS reduction. De la Rubia et al. [24] reported an increase in the TVS reduction (from 53 to 73 %) as the HRT increased from 27 to 75 d. Ribas and Barana [20] reported an increase in the TVS from 1.6 to 4.3% in a wastewater treatment system as the HRT was decreased from 16.6 to 9.7 d. Umana et al. [21] reported an increase in the TVS reduction (from 26.3 to 78.8 %) during

the treatment of dairy manure as the HRT was increased from 1 to 5.5 d. Rico et al. [22] noted that the concentration of TVS increased from 23.7 to 27.8 g/L during anaerobic treatment of dairy manure as the HRT decreased from 20 to 10 d. The reduction in the TVS obtained in this study (25.2-39.7 %) was within the range reported in the literature.

The diurnal cycling temperature also affected the TVS reductions, the higher the temperature the higher the TVS reduction. Wilkie et al. [18] achieved the highest TVS reduction of 60.4% at a temperature of 32.4°C during the treatment of dairy manure wastewaters. Elango et al. [19] reported 88.1% reduction in TVS over the temperature range of 26 to 36°C. Trazcinski and Stuckey [25] reported TVS reductions in the range of 40-75% at the temperatures of 21-35°C during anaerobic digestion of solid waste. The TVS reductions achieved in this study (25.2-39.7%) are within the range reported in the literature.

Fixed solids

Fixed solids represent the inorganic components of the waste and it consists of potassium, sodium, calcium, magnesium, iron, copper and other minerals [23]. Unlike carbon and nitrogen, little or no significant losses in these inorganic compounds are generally expected to occur since they do not play a very significant role in the anaerobic digestion process. In this system, a fixed solid recovery ranging from 64.3 to 112.0 % was recorded.

Converse et al. [26] reported a fixed solid recovery ranging from 90 to 104 %. In this study, the losses in fixed solids may have been due to inadequate mixing of the bioreactor contents which may have resulted in the settling out of the inorganic mineral resulting in the lower concentration of these compounds in the effluent. They may, also, have been due to the fact that changes in the nature of these inorganic compounds may have occurred at the higher temperature (550°C) used in the determination of total ash content.

Volatile fatty acids

The concentrations of volatile fatty acids (VFAs) in the effluent samples taken during the steady state conditions are shown in Table 4. The VFAs included: acetic, propionic, isobutyric, iso-valeric, valeric, iso-caprioc and heptanoic acids.

The VFAs concentration in the effluent ranged from 45 to 1200 mg/L, depending on the operating conditions. The effluent VFAs concentration generally increased with decreases in the HRT and higher VFAs concentrations were recorded at the lower temperature cycle. At

Temp (°C)	HRT (d)	Volatile Acid Concentration (mg/L)									
		Acetic Acid	Propionic Acid	i-Butyric Acid	n-Butyric Acid	i-Valeric Acid	n-Valeric Acid	i-Caproic Acid	n-Caproic Acid	Heptanoic Acid	Total as Acetic Acid
20-40	25	5.5	26.6	15.6	8.4	9.8	<0.01	11.7	3.9	5.7	60.0
	20	44.0	<0.01	<0.01	<0.01	<0.01	<0.01	<0.01	<0.01	<0.01	44.7
	15	3.0	26.0	10.7	11.3	9.7	1.3	3.3	2.4	2.5	50.0
	10	53.8	56.9	35.7	8.2	11.5	5.	9.0	4.7	16.9	154.8
15-25	25	54.1	73.6	13.7	18.3	3.9	<0.01	2.2	5.4	20.0	151.0
	20	70.8	102.1	19.1	26.0	19.8	4.7	6.1	6.1	58.4	231.8
	15	207.4	92.0	18.5	27.6	19.7	4.1	6.6	4.3	59.4	362.5
	10	645.0	431.4	139.7	38.6	36.0	25.0	22.7	17.8	30.1	1187.0
Raw Manure		1548.4	283.5	44.5	60.5	40.2	21.0	7.0	11.3	37.1	1913.0

Table 4: Volatile fatty acids.

the 20-40°C diurnal temperature cycle, the VFAs (measured as acetic acid) first decreased from 60 to 44.7 mg/L as the HRT was reduced from 25 to 20 d. Further reductions in the HRT resulted in increasingly higher concentrations of VFAs, up to 154.8 mg/L was recorded at the 10 day HRT. Among the individual volatile acids, propionic acid had the highest concentration at all but 20 d HRT. At the 20 d HRT, only acetic acid was detected in measurable levels. At the 15-25°C diurnal temperature cycle, the total volatile acids concentration increased from 151.0 mg/L at the 25 d HRT to 1187.0 mg/L at the 10 d HRT. Propionic acid concentration was higher than that of the rest at the 25 and 20 d HRTs while acetic acid concentration was the highest at the 15 and 10 d HRTs.

The HRT has a significant influence on the production of VFAs, the shorter the HRT the higher the VFAs concentration. Colmenarejo et al. [27] reported that the highest VFAs production was achieved when the HRT was decreased from 5.2 to 1.7 h. El-Mashad et al. [28] reported that the VFA concentration increased when the HRT was decreased from 24 to 3 h during the treatment of domestic wastewater. Elefsiniotis and Oldham [29] reported an increase in VFA production when the HRT was increased from 6 to 12 h which then decreased with further increase in the HRT to 15 h. Maharaj and Elefsiniotis [30] noted highest VFA concentration at an HRT of 30 h as appose to 48 and 60 h while anaerobically treating wastewaters. De la Rubia et al. [24] found that the VFA production increased from 5532 to 6432 mg/L as the HRT decreased from 75 to 20 days.

The results also showed the importance of temperature on the production of VFAs. Higher VFA production was observed with lower diurnal cyclic temperature. Maharaj and Elefsiniotis [30] noted increased VFA concentrations as the temperature was increased from 8 to 25°C, which then decreased upon further increase in temperature (from 25 to 35°C). Choorit and Wisarnwan [31] noted that the VFA concentration slightly increased from 160.71 to 166.64 mg/L as the temperature decreased from 43 to 37°C and from 245.07 to 338.71 mg/L as the temperature decreased from 55 to 49°C. De la Rubia et al. [24] observed an increase in VFA production from 830 to 5688 mg/L during digestion anaerobic municipal sludge as the temperature increased from 35 to 55°C.

However, values of effluent VFAs concentrations reported in the literature vary depending on the type of waste. Converse et al. [27] reported effluent VFAs concentrations of 423-1113 mg/L (as acetic acid) using CSTR operated on dairy manure of 12.8-13.4 % TS content and a temperature of 35°C. Jeyanayagam and Collins [32] reported effluent VFAs concentrations of 754-1515 mg/L using laboratory digesters operated on dairy manure of 4.0-6.5 % TS content at hydraulic

retention times of 15-20 d and a temperature of 35°C. Lo and Liao [33] reported a VFAs concentration of 584 mg/L using a fixed film reactor operated on a screened dairy manure of 3.7 % TS content at an HRT of 20 d and a temperature of 12°C. Peck et al. [34] reported effluent VFAs concentrations of 64-98 mg/L using manure of 3458-4109 mg/L TS at 10-25 d HRTs and a temperature of 35°C. The VFA concentrations observed in this study are within the range of values reported in the literature.

Organic wastes are first converted to long chain fatty acids in the acidogenic stage of anaerobic digestion processes. Volatile acids are further converted to acetic acid which together with hydrogen and formate serve as substrates for methane bacteria. Different species of anaerobic bacteria are involved in the conversion of specific long chain fatty acids to acetic acid. The increased proportion of propionic acid relative to acetic acid indicated that the different species of anaerobic bacteria did not respond in a similar manner to the effect of diurnally cyclic temperature. The fact that individual VFAs are removed at different rates under adverse conditions such as a drop in temperature has been observed by other researchers [34] A lower methane yield was recorded under the conditions in which the propionic acid concentration was high indicating that propionic acid was not available for conversion to methane.

Biogas

Figure 7 shows the daily biogas production under various operating conditions. The diurnal variations in the mean rate of biogas production for the 25 h HRT are shown in Figure 8. A summary of the average composition of the biogas collected under the various treatment combinations is shown in Table 5. Typical results of the autocorrelation and Fourier analysis are shown in Figures 9-12.

Biogas production

There appears to be a definite relationship between the bioreactor temperature and the rate of biogas production. Biogas production followed a diurnally cyclic pattern similar to that of the bioreactor temperature under most operating conditions. The highest reactor temperature and the maximum rate of biogas production occurred at the 18 hour while the lowest reactor temperature and the minimum biogas production rate occurred at the 9 hour. This trend was observed at all HRTs with 20-40°C cycle and at the 20 d HRT with 15-25°C cycle. The daily amplitude of the gas production cycle was higher under the 20-40°C cycle than that observed under the 15-25°C cycle. Smaller amplitudes in the gas production cycle were recorded in the lower temperature cycle at all HRTs because of the smaller diurnal amplitude of the reactor temperature. With digesters operated under stressful

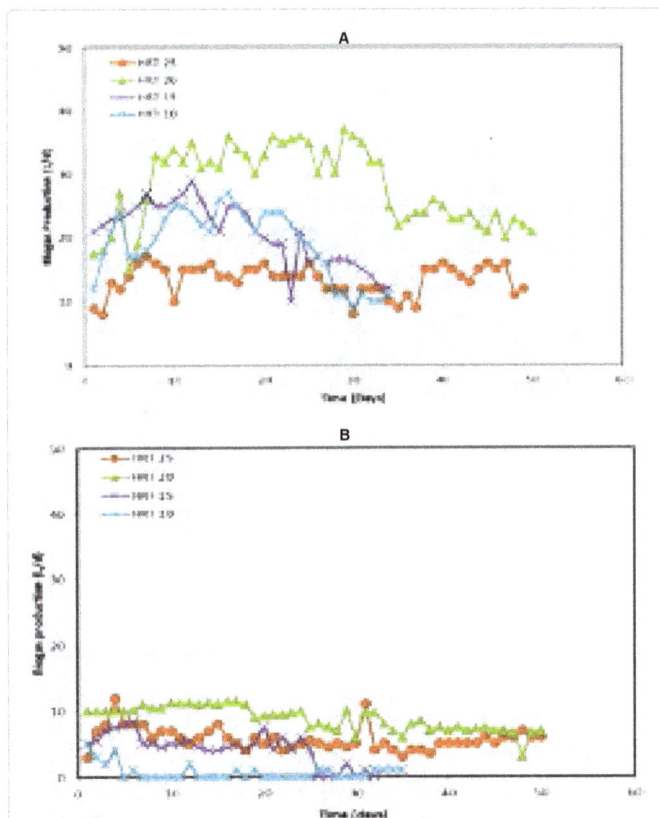

Figure 7: Daily biogas production. (a) 20-40°C Diurnal Temperature Cycle. (b) 15-25°C Diurnal Temperature Cycle

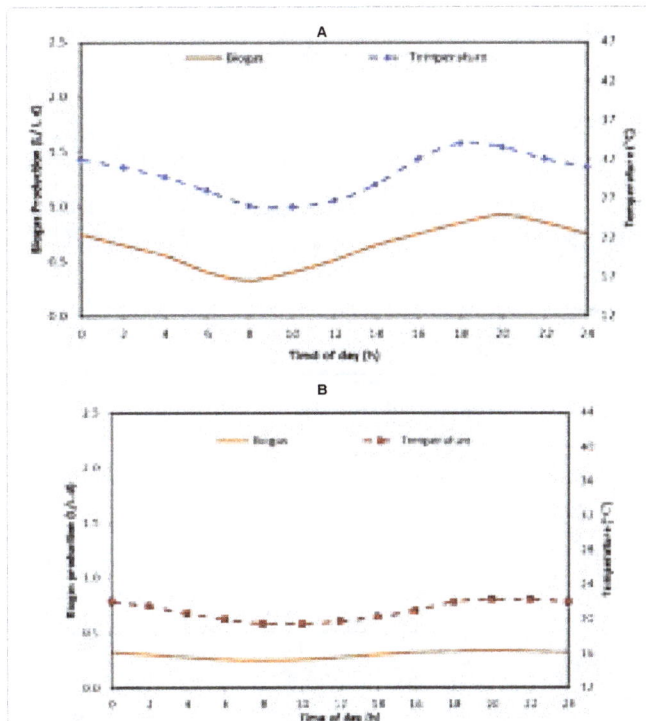

Figure 8: Diurnal variations in biogas production at the 25 d HRT. (a) 20-40°C Diurnal Temperature Cycle. (b) 15-25°C Diurnal Temperature Cycle.

conditions such as a combination of low temperature (15-25°C) and high loading rate (short HRT), the recorded diurnal variations in the rate of biogas production were either less pronounced or non-existent. The direct relationship between the rate of biogas production and the reactor temperature is supported by the known fact that microbial activity increases with increases in the reaction temperature.

Under the operating temperature cycle of 15-25°C, a steady gas production was achieved at the 25 d HRT only after a few days whereas at the 20 d HRT, a steady, but not rapid, decline in biogas production was observed after the 15th day, following the initial rise in biogas production. At the 15 d HRT, the gas production declined unsteadily after the first 8 days until it completely stopped at the 28th day. At the 10 d HRT, the gas production declined rapidly from the start, ceasing completely after the 5th day. These results indicated that the bioreactors did not operate satisfactorily at HRTs shorter than 25 d under the 15-25°C diurnal temperature cycle. The failure of the digesters at 15 and 10 d HRTs is due to the combined effects of the low cyclic diurnal temperature and high loading rate.

The biogas production observed at the 20-40°C diurnal temperature cycle was characterized by rapid rise in production followed by a rapid decline, the shorter the retention time the more pronounced was the decline. This rapid rise in biogas production experienced at the beginning of each retention time can be attributed to the response of anaerobic bacteria to increased food supply as a result of the higher loading rate. Apparently, the acid forming bacteria responded faster to the increased loading rate. However, the accumulation of VFAs resulted in inhibition of the methane producing bacteria and hence reduction in

Diurnal Temperature Cycle (°C)	HRT (d)	Gas Composition		
		CH$_4$ (%)	CO$_2$ (%)	Others** (%)
20-40	25	69.0±1.7	25.8±1.5	5.2
	20	67.6±1.3	28.6±2.3	3.8
	15	62.6±1.0	37.3±1.0	0.1
	10	64.9±1.0	34.9±1.0	0.2
15-25	25	71.8±2.1	25.3±1.0	3.0
	20	69.6±1.8	27.3±1.0	3.1
	15	47.8±0.5	44.0±1.2	8.2
	10	44.6±0.5	43.6±1.8	11.8

*Each mean represents the average of 12 samples
**Nitrogen, hydrogen sulphide, carbon monoxide, etc
Table 5: Biogas composition*.

biogas production. The higher rate of biogas production attained at a 20 d retention time indicated that at HRT longer than 20 d, the reactors may have been underfed whereas at shorter retention times the reactors may have been overloaded.

In this study, higher biogas production was obtained with the 20 h HRT. However, literature reports on effect of HRT on biogas production vary depending on the type of waste and system used. Hossain et al. [35] noted that the biogas production increased in wastewaters as the HRT increased from 2 to 8 h, but a further increase in the HRT from 8 to 14 h lead to a decrease in biogas production. Huang et al. [36] reported that the higher biogas yields were achieved at lower HRTs (8-12 h) during wastewater treatment. Rico et al. [22] noted that as the HRT decreased from 20 to 12.5 d, the biogas production increased from 0.66 to 1.21 m^3 during the anaerobic digestion of dairy manure.

In this study, higher biogas production was observed with higher temperatures (within the temperature range studied). However the

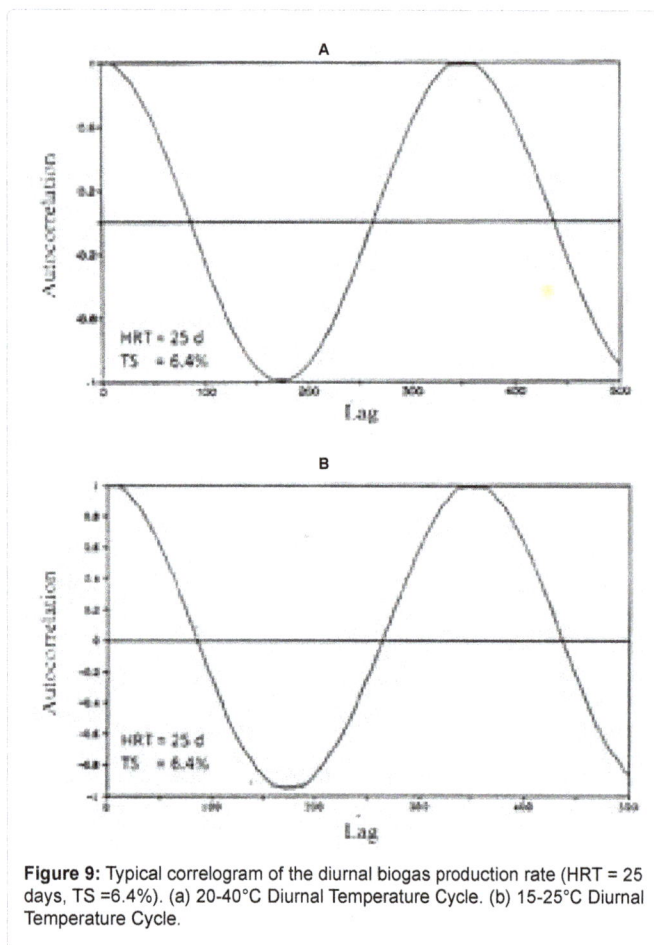

Figure 9: Typical correlogram of the diurnal biogas production rate (HRT = 25 days, TS =6.4%). (a) 20-40°C Diurnal Temperature Cycle. (b) 15-25°C Diurnal Temperature Cycle.

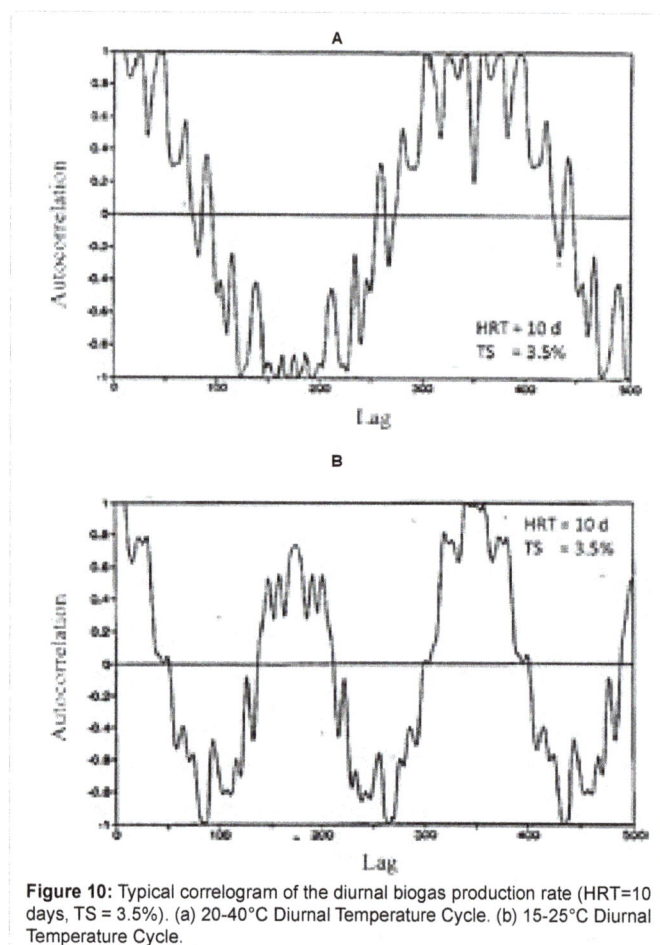

Figure 10: Typical correlogram of the diurnal biogas production rate (HRT=10 days, TS = 3.5%). (a) 20-40°C Diurnal Temperature Cycle. (b) 15-25°C Diurnal Temperature Cycle.

reports on the effect of temperature on biogas production vary. Hossain et al. [35] reported that biogas production increased as the temperature increased from 30 to 40°C during wastewater treatment, but further increase in temperature from 40 to 45°C resulted in a decrease in biogas production. Wu et al. [37] noted that the biogas production increased as the temperature increased from 20 to 40°C during anaerobic digestion of solid wastes and decreased with further increase in temperature to 45°C while a further increase to 55°C lead to a rapid increase in biogas production. Yu et al. [38] reported that more biogas production was achieved at temperatures of 55°C during anaerobic digestion of wastewaters than those obtained at 37°C. Banerjee and Biswas [28] illustrated that during anaerobic digestion of wastes over the temperature range of 35 to 55°C maximum biogas production was achieved at 50°C.

Gas composition

Diurnal variations in gas composition were observed at the all HRTs under the 20-40°C diurnal temperature cycle. Similar variations were observed at the 20 d HRT under the 15-25°C diurnal temperature cycle. A better gas quality (higher methane content) was obtained at all HRTs under the 20-40°C diurnal temperature cycle. Higher methane content was also obtained at the 25 and 20 HRTs under 15-25°C diurnal temperature cycle. Rises in the methane content of biogas as a result of a decrease in the bioreactor temperature have been reported by other researchers [34]. The increase in the quality of biogas is attributed to the raised solubility of carbon dioxide at the lower temperature cycle. Exceptions were observed with failed digesters. In stressed

methanogenic systems, hydrogen production and volatile fatty acids (more reduced than acetic acid such as propionate, butyrate etc.) have been reported to be produced in greater proportions [39].

The mean percentage of methane in the biogas ranged from 44.6 to 71.8 %, the carbon dioxide content ranged from 25.3 to 44.0 % while the composition of the other gases (nitrogen, hydrogen sulphide etc.) ranged from 0.1 to 11.8 %. Methane values reported in the literature ranged from 40 to 75 % [40,26,32-34]. The observed values are, thus, within the range reported in the literature.

Autocorrelation analysis

A typical correlogram of diurnal biogas production is shown in Figure 9. Figure 9a was obtained at 25 days HRT using manure of 6.4% TS content and the temperature cycle of 20–40°C, while Figure 9b was obtained using similar HRT and manure TS content but a temperature cycle of 15–25°C. From the correlogram, it is seen that the diurnal biogas production rate predominantly followed a pattern similar to a sine wave. This pattern was observed with the operating temperature cycle of 20–40°C for all HRTs using the manure of 6.4% TS content and for the 25, 20 and 15 days HRTs using the manure of 3.5% TS content. However, at the 15–25°C temperature cycle, the sinusoidal pattern was only observed at the 25 day HRT using the manure of 6.4% TS content.

The autocorrelation analysis shown in Figure 10 indicated that under low HRT (10 days) and low influent TS Concentration (3.5%), the biogas production rate follows a sine wave sandwiched in some

Temperature Cycle (°C)	Solids Content (%)	HRT (days)	Sinusiodally Periodic
20 – 40	6.4	25	Yes
		20	Yes
		15	Yes
		10	Yes
	3.5	25	Yes
		20	Yes
		15	Yes
		10	No
15 – 25	6.4	25	Yes
		20	No
		15	No
		10	No
	3.5	25	No
		20	No
		15	No
		10	No

*Each mean represents the average of 12 samples
**Nitrogen, hydrogen sulphide, carbon monoxide, etc

Table 6: Summary of Autocorrelation Analysis.

harmonics and noise. Various variants of Figure 10 were observed under the operating temperature cycle of 15–25°C for all retention times using the manure of 3.5% TS content and under the operating temperature cycle of 20–40°C with retention times of 10 to 20 days using the manure of 6.4% TS content. Furthermore, a similar correlogram was obtained under the retention time of 10 days using the manure of 3.5% TS concentration at the operating temperature cycle 20–40°C. These results are summarized in Table 6.

The sinusoidal periodic variations in biogas production rates were obtained at almost all loading rates under the 20–40°C temperature cycle because the temperature of the reactor used in the study was designed to vary in a diurnally cyclic (sinusoidal) manner and microbial activity and gas production increases with increase in temperature. However, sinusoidal variations were not obtained under most of the loading rates at the temperature cycle of 15–25°C due to fact that the amplitude of diurnal fluctuation in reactor temperature was relatively so small and it was impossible to detect a true sine wave. At the higher loading rate, pure sinusoidal variation was not obtained probably because of the higher frequency of the feeding cycle (i.e. shorter feeding interval) as most of the evolved gases are scavenged from the headspace of the reactors and measured during the feeding process.

The result of the fast Fourier analysis on the diurnal biogas production data shown in Figures 11 and 12 showed that, where the diurnal biogas production cycle followed true sinusoidal relationships, the highest Fourier amplitude appeared to occur at a frequency of about 1.22×10^{-5} Hz. Thus, the dominant frequency of such gas production cycle lies at about 1.22×10^{-5} Hz. This frequency roughly corresponds to a period of 24 hours. However, when the biogas production cycle did not follow a true sinusoidal cycle, the dominant frequency (indicated by the highest Fourier amplitude) appeared to lie between 0.0013 and 0.0014.

For the dominant frequency, the following regression equation, based on Fourier series, can be written to describe the diurnal variation in biogas production under diurnally cyclic temperature environment:

$$\gamma_t = \gamma_o + \sum_{n=1}^{\infty}[a_n \cos(2\pi n\varphi t) + b_n \cos(2\pi n\varphi t)] \qquad (16)$$

Where:

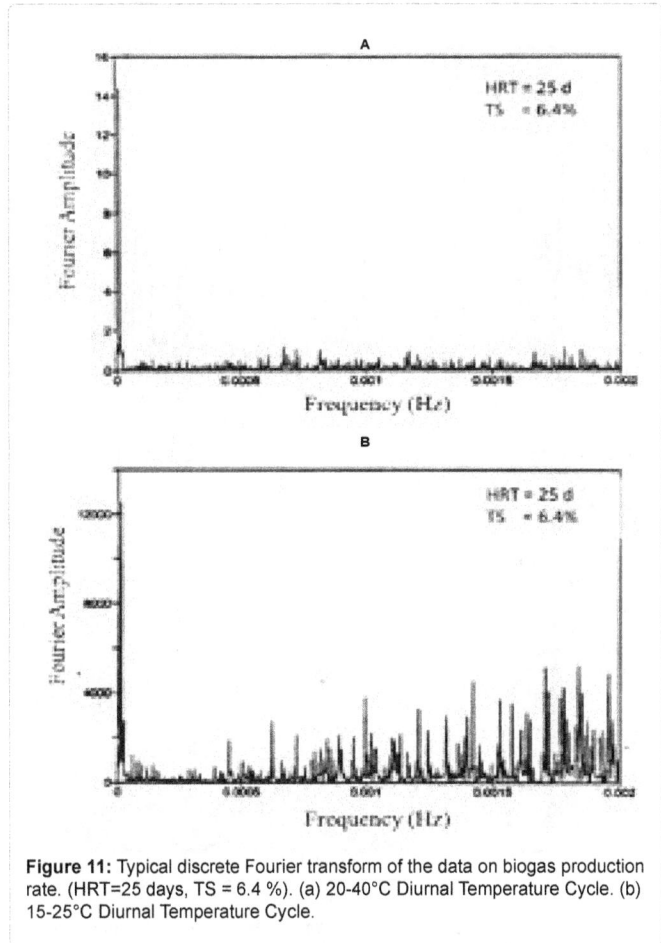

Figure 11: Typical discrete Fourier transform of the data on biogas production rate. (HRT=25 days, TS = 6.4 %). (a) 20-40°C Diurnal Temperature Cycle. (b) 15-25°C Diurnal Temperature Cycle.

γ_t is the biogas production at a time t (L/L.d)

γ_o is the average biogas production per unit volume of reactor (L/L.d)

φ is the dominant frequency of the cycle = 1.22×10^{-5} (Hz)

a_n, b_n are Fourier coefficients.

The Fourier coefficients and mean daily biogas production rate can be obtained by regression analysis. These values will vary with operating conditions (cyclic temperature ranges and loading rates).

Conclusion

The effects of two diurnally cyclic temperature ranges (20-40°C and 15-25°C) and four levels of hydraulic retention times (25, 20, 15, and 10 d) on the performance of anaerobic reactors operated on screened dairy manure were evaluated. The reactor temperature exhibited a lag relative to the chamber air temperature. For the 20-40°C temperature cycle, the average lag period at the maximum and minimum chamber temperatures 3.75 and 4.37 h, respectively. For the 15-25°C temperature cycle, the average lag periods at maximum and minimum chamber temperatures were 3.61 and 4.34 h, respectively. The effluent solids content were not adversely affected by the reactor diurnally cyclic temperature. The effluent total solids and methane content of the biogas diurnally cyclic patterns were out of phase with the diurnally cyclic pattern of the reactor temperature by about 12 hours under most operating conditions. The reductions in total solids and methane yield

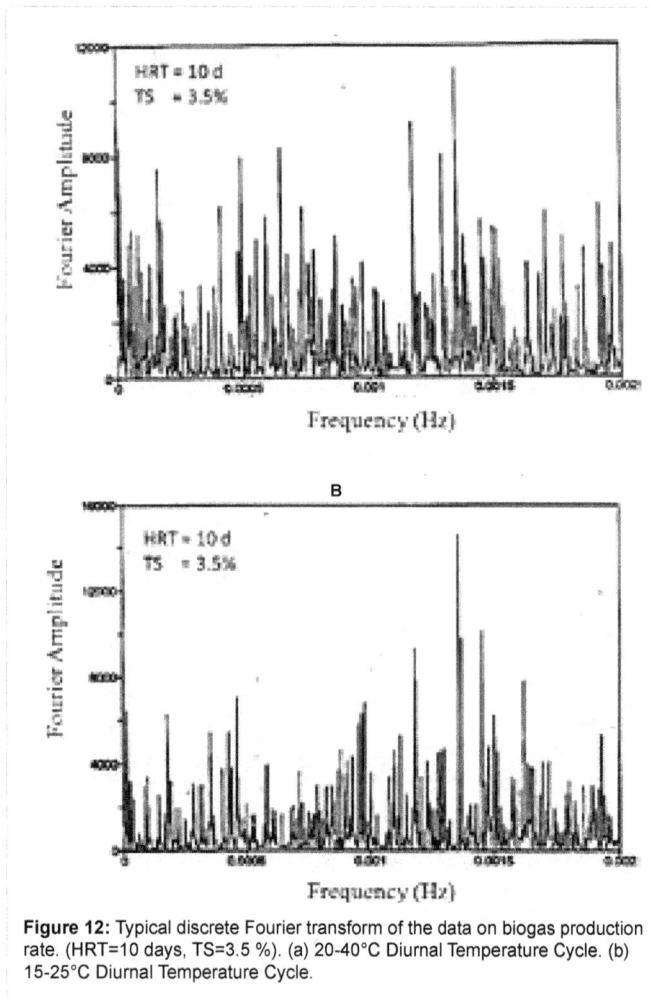

Figure 12: Typical discrete Fourier transform of the data on biogas production rate. (HRT=10 days, TS=3.5 %). (a) 20-40°C Diurnal Temperature Cycle. (b) 15-25°C Diurnal Temperature Cycle.

were all significantly affected by the diurnal temperature range and hydraulic retention time. Biogas production from a healthy digester operating under a diurnally cyclic temperature environment follow a sinusoidal pattern which can be described by a Fourier equation of the form: $\gamma_t = \gamma_o + \sum_{n=1}^{\infty}[a_n \cos(2\pi n\varphi t) + b_n \cos(2\pi n\varphi t)]$. However, where the operating conditions are not favourable, the production followed a sinusoidal pattern which may be embedded in some harmonic and noise.

References

1. El-Mashad HM, Zeeman G, Van Loon WKP, Bot GPA, Lettinga G (2004) Effect of Temperature and Temperature Fluctuation on Thermophilic Anaerobic Digestion of Cattle Manure. Bioresource Technol 95: 191-201.

2. Masse D, Gilbert Y, Topp E (2011) Pathogen Removal in a Farm-Scale Psychrophilic Anaerobic Digesters Processing Swine Manure. Bioresources Technol 102: 641-646.

3. Porpatham E, Ramesh A, Nagalingam B (2008) Investigation on the Effect of Concentration of Methane in Biogas when used as a Fuel for AaSpark Ignition Engine. Fuel 87: 1651-1659.

4. Burke DA (2001) Dairy Waste Anaerobic Digestion Handbook: Options for Recovering Beneficial Products From Dairy Manure.

5. Dieu TTM (2009) Food Processing and Food Waste. In C.J. Baldwin (Ed.), Sustainability in the Food Industry, Wiley-Blackwell, Ames, Iowa.

6. Ghaly AE, Al-Hattab MT (2011) Effect of Diurnally Cyclic Temperature on the Performance of a Continuous Mix Anaerobic Digester. A J Biochem Biotechnol 7: 146-162.

7. Ghaly AE, Sadaka SS, Huzzaa A (2000) Kinetics of an Intermittent Flow Continuous Mix Anaerobic Reactor. Energ Sources 22: 525.

8. Angelidaki I, Ahring BK (1994) Anaerobic Thermophilic Digestion of Manure at Different Ammonia Loads: Effect of Temperature. Water Res 28: 727-731.

9. Kim JK, Oh BR, Chun YN, Kim SW (2006) Effects of Temperature and Hydraulic Retention Time on Anaerobic Digestion of Food Waste. J Biosci Bioeng 102: 328-332

10. Chae KJ, Jang A, Yim SK, Kim IS (2008) The Effects of Digestion Temperature and Temperature Shock on the Biogas Yields from the Mesophilic Anaerobic Digestion of Swine Manure. Bioresource Technol 99: 1-6.

11. Appels L, Degreve J, Van der Bruggen B, Van Impe J, Dewil R (2010) Influence of Low Temperature Thermal Pre-Treatment on Sludge Solubilisation Heavy Metal Release and Anaerobic Digestion. Bioresource Technol 101: 5743-5748.

12. Lianhua L, Dong L, Yongming S, Longlong M, Zhenhong Y, et. al., (2010) Effect of Temperature and Solid Concentration on Anaerobic Digestion of Rice Straw in South China. Int J Hydrogen Energ 35: 7261-7266.

13. Gustin S, Marinsek-Logar R (2011) Effect of Ph, Temperature and Air Flow Rate on the Continuous Ammonia Stripping of the Anaerobic Digestion Effluent. Process Saf Environ 89: 61-66.

14. Echiegu EA (1992) Performance of a Continuous-Mix Anaerobic Reactor Operating on Dairy Manure Under Two Diurnal Temperature Ranges. Unpl. PhD thesis. Technical University of Nova Scotia.

15. Ghaly AE, Echiegu EA, Ben-Hassan RM (1992) Performance Evaluation of a Continuous Mix Anaerobic Reactor Operating of Dairy Manure under Diurnally Cyclic Temperature. ASAE paper No 92-6025. Presented at the Summer meeting of ASAE, Charlotte, North Carolina, USA.

16. Davis JC (1973) Statistics And Data Analysis In Geology. John Willey and Sons, NY.

17. American Public Health Association (APHA). 1985. Standard Methods for the Examination of Water and Wastewater, 16th ed. APHA, Washington, D.C.

18. Wilkie AC, Castro HF, Cubinski KR, Owens JM, Yan SC (2004) Fixed-Film Anaerobic Digestion of Slushed Dairy Manure after Primary Treatment: Wastewater Production and Characterisation. Biosystems Eng 89: 457-471.

19. Elango D, Pulikesi M, Baskaralingam P, Ramamurthi V, Sivanesan S (2007) Production of Biogas from Municipal Solid Waste with Domestic Sewage. J Hazard Mater 141: 301-304.

20. Ribas MM, Barana AC (2003) Start-Up Adjustment of a Plug-Flow Digester for Cassava Wastewater (Manipueira) Treatment. Scientia Agricola 60: 223-229.

21. Umana O, Nikolaeva S, Sanchez E, Borja R, Raposo F (2008) Treatment of Screened Dairy Manure by Upflow Anaerobic Fixed Bed Reactors Packed with Waste Tyre Rubber and a Combination of Waste Tyre Rubber and Zeolite: Effect of the Hydraulic Retention Time. Bioresources Technol 99: 7412-7417.

22. Rico C, Rico JL, Tejero I, Munoz N, Gomez B (2011) Anaerobic Digestion of the Liquid Fraction of Dairy Manure in Pilot Plant for Biogas Production: Residual Methane Yield of Digestate. Waste Manage 31: 2167-2173.

23. Ghaly AE, Echiegu E (1993) Kinetics of a Continuous Flow No-Mix Anaerobic Reactor. Energ Source 15: 433-449.

24. De la Rubia MA, Perez M, Romero LI, Sales D (2002) Anaerobic Mesophilic and Thermophilic Municipal Sludge Digestion. Chem Biochem Eng Q 16: 119-124.

25. Trazcinski AP, Stuckey DC (2011) Parameters Affecting the Stability of the Digestate from a Two-Stage Anaerobic Process Treating the Organic Fraction of Municipal Solid Waste. Waste Manage 31: 1480-1487.

26. Converse JC, Zeikus JG, Graves RC, Evans GW (1977) Anaerobic Degradation of Dairy Manure Under Mesophilic And Thermophilic Temperatures. Transactions of the ASAE 20: 336-340.

27. Colmenarejo MF, Sanchez E, Bustos A, Garcia G, Borja R (2004) A Pilot-Scale Study of Total Volatile Fatty Acids Production by Anaerobic Fermentation of Sewage in Fixed-Bed and Suspended Biomass Reactors. Process Biochem 39: 1257-1267.

28. El-Mashad HM, Van Loon WKP, Zeeman G (2003) A Model of Solar Energy Utilisation in the Anaerobic Digestion of Cattle Manure. Biosys Eng 84: 231-238.

29. Elefkiniotis P, Oldham WK (1994) Effect of HRT on Acidogenic Digestion of Primary Sludge. J Environ Eng 120: 645-660.

30. Maharaj I, Elefsiniotis P (2001) The Role of HRT and Low Temperature on the Acid-Phase Anaerobic Digestion of Municipal and Industrial Wastewaters. Bioresource Technol 76: 191-197.

31. Choorit W, Wisarnwan P (2007) Effect of Temperature on the Anaerobic Digestion of Palm Oil Mill Effluent. Electron J Biotechnol 10: 376-385.

32. Jeyanayagam SS, Collins ER (1984) Weed Seed Survival in a Dairy Manure Anaerobic Digester. Transactions of the ASAE 27: 1518-1523.

33. Lo KV, Liao PH (1986) Thermophilic Anaerobic Digestion of Screened Dairy Manure Using A Two-Phase Process. Energ Agr 5: 249-255.

34. Peck MW, Skillon JM, Hawkes FR, Hawkes DL (1986) Effect of Temperature Shock Treatments on the Stability of Anaerobic Digesters Operated on Separated Cattle Slurry. Water Res 20: 453-462.

35. Hossain M, Anantharaman N, Das M (2009) Anaerobic Biogas Generation from Sugar Industry Wastewaters in Three-Phase Fluidized-Bed Bioreactor. Indian J Chem Technol 16: 58-64.

36. Huang Z, Ong SL, Ng HY (2011) Submerged Anaerobic Membrane Bioreactor for Low-Strength Wastewater Treatment: Effect of HRT and SRT on Treatment Performance and Membrane Fouling. Water Res 45: 705-713.

37. Wu M, Sun K, Zhang Y (2006) Influence of Temperature Fluctuation on Thermophilic Anaerobic Digestion of Municipal Organic Solid Waste. J Zhejiang Univ Sci B 7: 180-185.

38. Yu HQ, Fang HHP, Gu GW (2002) Comparative Performance of Mesophilic and Thermophilic Acidogenic Upflow Reactors. Process Biochem 38: 447-454.

39. McInerney JM, Bryant MP (1981) Basic Principles of Bioconversion in Anaerobic Digestion and Methanogenesis. Biomass Conversion Process for Energy and Fuel 277-296.

40. Hills DJ, Stephens JR (1980) Solar Energy Heating of Dairy-Manure Anaerobic Digesters. Agr Wastes 2: 103-118.

13

Optimisation of Rhamnolipid: A New Age Biosurfactant from *Pseudomonas aeruginosa* MTCC 1688 and its Application in Oil Recovery, Heavy and Toxic Metals Recovery

Pathaka AN[1] and Pranav H Nakhate[2]*

[1]*Research Dean and Head of Institution, Amity Institute of Biotechnology, Amity University Rajasthan, India*
[2]*Biochemical Engineering Scholar Jaipur, Amity Institute of Biotechnology, Amity University Rajasthan, India*

Abstract

Rhamnolipid is a new age biosurfactant, commonly produced biotechnologically with *Pseudomonas aeruginosa* in batch cultivations whereas novel substrates like karanja oil and soybean oil cake were employed, giving yield of 3.609 gm/lit of rhamnolipid, which shows effective and enhanced production of rhamnolipid, compared to other vegetable oil as a carbon sources, mentioned in the literature, at optimised pH of 7.0 and optimised substrate concentration at 3.0%. The optimum yield in terms of substrate was observed as 3.609 gm/lit of rhamnolipid produced per 5.255 ml of oil consumed, while yield in terms of biomass was observed as 3.609 gm/lit of rhamnolipid produced per 2.5 gm of dry biomass. The chloroform:methanol (2:1) extraction system was found to be the best solvent extraction system, where 83% of the rhamnolipid was recovered. The Rhamnolipid was successfully applied for the heavy and toxic metals recovery, where rhamnolipid reduces heavy metal concentration to 73%, 65% and 71% for $FeCl_3$, $ZnSO_4$ and $Pb(NO_3)_2$ respectively, while 43% in the case of toxic metal i.e. NaF. The produced rhamnolipid was found efficient in recovering 31% no-edible oil from oil sludge.

Keywords: Rhamnolipid; Karanja oil; Soybean oil cake; Heavy metal; Toxic metal; Oil recovery

Introduction

Biosurfactants

"Biosurfactants" are microbially produced, structurally diverse group of surface active biochemical molecules. The huge diversity of biosurfactants makes them an interesting moiety for application in many areas including agriculture, health care, public health, food, waste utilization, and environmental pollution control, like degradation of hydrocarbons present [1].

Microbially produced biosurfactants are adequate to fulfil many of the roles for which petrochemical and oleochemical surfactants are currently used [2]. The global market for surfactants is approximately 15 million tonnes per annum with a global average annual growth of approximately 3 per cent [3]. As demand on "crude oil" supplies increases the use of "sustainable biosurfactants" instead of petrochemically derived surfactants becomes more fascinating.

Biosurfactants often have interesting characteristics not possessed by petrochemical or oleochemical surfactants, like their application in "pharmaceuticals", "bioremediation", "food processing", nevertheless, for this to happen actually in an industrial scale, there needs to be the improved development of fermentation processes and downstream separation techniques for effective biosurfactant production [1,4,5]. Downstream separation techniques require deep knowledge, as they contribute approximately 60% to the total cost of biosurfactant production [6]. Current research and industrial interest rests in many biosurfactants, including "surfactin" obtained from *Bacillus subtilis*, hydrophobin proteins from various filamentous fungi such as *Schizophyllum commune* and *Trichoderma reesei*, and rhamnolipids from *Pseudomonas aeruginosa* [7,8].

Rhamnolipid

Rhamnolipids are "anionic glycolipids" consisting of "L-rhamnose" and "β-hydroxy fatty acids" produced by *P. aeruginosa* strain. The hydrophilic rhamnose moiety is attached by a glycosidic linkage to the lipid fatty acid tail. Rhamnolipids are generated as a mixture of different rhamnolipids [2,6,9].

It has been explained that, the rhamnolipid can be produced from various microorganism, but the *Pseudomonas aeruginosa* is found to be the most effective and prominent strain responsible for the production of rhamnolipid [9,10]. Rhamnolipid is one of the most important biosurfactant, as it can used for the various purposes. The problems, that hinder the bulk production of rhamnolipid are, the use of cheap raw materials, and effective downstreaming. Various journals reported the use of vegetable oils, waste oil etc. as a carbon source. Some other journals reported use of cheap whey, or molasses as a carbon source [11].

For the production of rhamnolipid, the search of low cost source ended at the plant derived oil, i.e. karanja. Nowhere in the literature, is it reported so far, that uses karanja oil as a source for the production of rhamnolipid. Karanja oil contains more amounts of fatty acids, which can be made available to the microorganism easily, and high production of rhamnolipid can be expected. Waste soybean oil cake is the complete waste, and generally used as an animal feed. Hence, use of such oil cake will definitely decrease the bulk production cost [12,13]

*****Corresponding author:** Pranav H Nakhate, Biochemical Engineering Scholar, Amity University Rajasthan, 14-Gopal Bari, Ajmer Road, Jaipur 302001, India

(Figure 1).

Many researchers reported that, the rhamnolipid is late-growth phase product, and its production increases in the stationary phase. The results obtained from our research supports the findings of such researchers like SD Wadekar et al. [14]. From the graphs, explained later in the report, it was crystal clear that, rhamnolipid production starts at the late growth phase, indicating that, microorganisms utilises most of the carbon sources available first, and then produces product. The sudden increase in production was observed during the stationary phase. The media used contains less amount of phosphorus source in order to stringent the growth condition, hence ultimately microorganism growth ceases, and the culture enters into the stationary phase and high production can be achieved [15].

The production costs of biosurfactants, compared to synthetic compounds, are at least 50 times higher, on the surfactant market. It is estimated that the production costs of rhamnolipids produced in 25-200 m^3 scale is at about 6-25 US$/kg. Compared with petrochemical bulk surfactants like ethoxylates or alkyl polyglycosides, ranging at 1-3 US$/kg, rhamnolipids are not competitive in this field [3].

Material and Methodology

Inoculum preparation procedure

First step in the inoculum preparation is the culture revival. The *Pseudomonas aeruginosa* MTCC 1688 mother culture was revived according to the standard protocol given in the various microbiological journals. The mother culture was carefully inoculated into the freshly prepared nutrient broth, strictly under aseptic conditions in vertical laminar air flow.

After the successful microbial transfer, the 100 ml shake flask, containing nutrient broth was kept in Incubator for 24 hrs to grow the microorganisms. After 24 hrs, a loop full of inoculation from nutrient broth was taken and spread over the nutrient agar plate using quadrant technique. Four nutrient agar plates were kept at 4°C for further use.

Gram staining procedure

In order to check the Gram's nature of the microorganism, Gram staining was performed. The drop wise addition of crystal violet stains the bacterial in blue colour, while Gram's iodine is used to fix the crystal violet stain. Decolourisation occurs with addition of 70% ethanol, followed by addition of counter stain Safranin, which stains bacteria in pink colour.

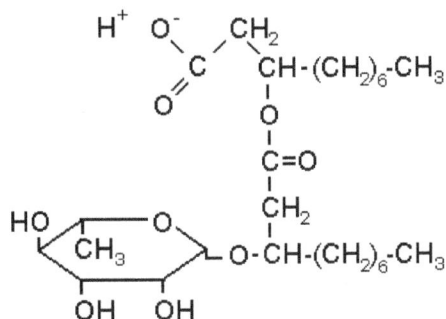

Figure 1: Rhamnolipid structure.

Shake flask cultivation

Cultivation experiment was carried out in 500 ml shake flask. As discussed in literature review, cheap carbon source had to use, hence, plant oil karanja was used, as it is a cheap hydrocarbon source. Soybean oil cake is also a waste, which is used as another hydrocarbon source. Glucose is used as a standard carbon source. The phosphate regulation was carried out by adapting minimal salt media (MSM) composition from Cameotra et al. [16]. Carbon to the nitrogen ratio i.e. C:N was maintained 20, by using $NaNO_3$ as a nitrogen source.

The optimization strategy was carried out by varying the percentage of carbon source used. 1%, 3% and 5% of carbon sources mentioned above were examined for the highest production of rhamnolipid. At the same time, pH of the media was varied using 0.1N HCl and 0.1N NaOH. Shake flask was operated at pH 5, pH 7 and pH 9 respectively after the Carbon source and its percentage optimisation.

The optimum carbon source, its concentration and optimum pH was calculated for highest rhamnolipid production. The other operating conditions, like temperature, agitation speed were properly maintained. The operating temperature was set 30°C and agitation speed was maintained 150 rpm in remi orbital shaker-incubator. During the cultivation period, biomass concentration, substrate concentration and rhamnolipid concentration were checked every after 24 hrs. After obtaining higher productivity, the cultivation days were calculated and the scale up was done (Figure 2).

Determination of biomass concentration

There are various methods which can be used for the determination of biomass. The cell dry mass was measured and calculated. The technique was carried out through following steps:

10 ml culture broth was taken

↓

Centrifugation carried out at 8000 rpm for 10 min.

↓

Pellet was collected and suspended in 0.9% saline tube

↓

Proper vortexed in order to re-suspend solid particles

↓

Allowed to settle the solid particles

↓

Filtered and dried the sample in order to measure dry mass

Determination of glucose concentration

Glucose is the simplest carbon source, utilised by almost all the microbes for their growth. The glucose decreases with the course of time, hence its concentration over the course of time is required for the calculations of yield of the product i.e. rhamnolipid product formed per unit substrate consumed. Product $Y_{p/s}$. The glucose concentration was measured by using standard DNSA curve.

DNSA method was invented by Miller in 1959. DNSA method is basically used to detect free aldehyde groups present in the sugar, i.e. to detect reducing sugar. Oxidation of aldehyde group results into the formation of carboxyl group, which then reduce to 3-amino, 5-nitro

Figure 2: Shake flask study.

Tube No.	Amount of stock added	Amount of D/W added	Final volume	Dilution	Concentration
1	0 ml	10 ml	10 ml	0	0 mg/ml
2	0.25 ml	9.75 ml	10 ml	1:40	2.5 mg/ml
3	0.5 ml	9.5 ml	10 ml	1:20	5 mg/ml
4	1 ml	9 ml	10 ml	1:10	10 mg/ml
5	2 ml	8 ml	10 ml	1:5	20 mg/ml
6	5 ml	5 ml	10 ml	1:2	50 mg/ml
7	10 ml	0 ml	10 ml	1:1	100 mg/ml

Table 1: Preparation of dilutions for glucose standard curve using DNSA method.

salicylic acid in presence of DNSA, which is calculated.

The stock solution of 100 mg/ml was prepared and further dilutions were made as mentioned below. 3 ml of DNSA solution was added to 3 ml of test sample and mixture was heated at 90°C for 15 min. Then, 1 ml of potassium-sodium tartarate i.e. Rochelle salt was added to solubilise the colour. The mixture was then allowed to cool down and OD was measure on colorimeter at 540 nm.

Determination of free fatty acid content

Another important substrates used for the rhamnolipid production, were karanja oil and soybean oil cake. Both the substrates are plant originated and contains high amount of unsaturated fatty acid, i.e. oleic acid. Hence free fatty acid content in terms of oleic acid was calculated. The hexane was added in order to extract oil residues from broth culture (Table 1).

The procedure was followed as given below:

10 ml culture broth was taken

↓

Centrifugation carried out at 8000 rpm for 10 min

↓

Pellet was collected and suspended in 0.9% saline tube

↓

Proper vortexed in order to re-suspend solid particles

↓

Allowed to settle the solid particles

↓

Filtered and dried the sample in order to measure dry mass

The titration readings were noted down and acid value was calculated from the equation,

$$Acid\ Value = \frac{(56.1 \times Burette\ reading \times Normality\ of\ KOH)}{Weight\ in\ grams}$$

Rhamnolipid detection [14]

Rhamnolipid detection was carried out using orcinol reagent. The culture broth was centrifuged at 8000 rpm for 15 min and the supernatant as collected. The supernatant was then treated with orcinol reagent, containing concentrated H_2SO_4 which causes L-rhamnose moiety to separate it out from lipid moiety and the L-rhamnose concentration can be calculated. This is the indirect way of detection of rhamnolipid, as pure rhamnolipid was not available.

Orcinol Reagent was produced by adding 53% H_2SO_4 solution to 0.19% orcinol solution. For each 1 ml sample, 9 ml orcinol reagent was summed. The formed mixture was then heated at 70°C for 30 min. and then allowed to cool down for 30 min. The optical density was

Tube No	Amount of stock added	Amount of D/W added	Final volume	Dilution	Concentration
1	0 ml	10 ml	10 ml	0	0 mg/ml
2	0.25 ml	9.75 ml	10 ml	1:40	0.25 mg/ml
3	0.5 ml	9.5 ml	10 ml	1:20	0.5 mg/ml
4	1 ml	9 ml	10 ml	1:10	1 mg/ml
5	2 ml	8 ml	10 ml	1:5	2 mg/ml
6	5 ml	5 ml	10 ml	1:2	50 mg/ml
7	10 ml	0 ml	10 ml	1:1	10 mg/ml

Table 2: Preparation of dilutions for L-rhamnose standard curve using orcinol reagent.

System used	Phase	O.D. at 540 nm
Ethyl acetate	Organic phase	0.41
	Aqueous phase	0.15
Chloroform: Methanol	Organic phase	0.48
	Aqueous phase	0.12

Table 3: Solvent extraction readings.

Process used	O.D. at 540 nm
Column chromatography	0.51
Rotary vacuum chromatography	0.48

Table 4: Different process readings.

S.No.	Stretching present	Wavelength (cm⁻¹)
1	C-H	2930.18
2	C-H	2856.55
3	C=O	1734.35
5	CH3	1401.17
6	C-H/O-H deform	1384.98
7	OH deform	1315.45
8	C=O stretching	1041.63
9	C-H deform	874.07

Table 5: Structural analysis of rhamnolipid using FTIR.

then measured at 520 nm. The standard L-rhamnose 10 mg/ml stock solution was prepared (Table 2).

CTAB agar method

CTAB is the cetyl trimethyl ammonium bromide method. This is also called methyl assay for rhamnolipid detection. Concentration of "anionic surfactant" from the mixture of solution can be dictated. CTAB, being "cationic surfactant", bind with the rhamnolipid, which is "anionic surfactant", forming insoluble ion precipitation. This "insoluble ion precipitation" was detected by appearance of 'dark blue' colour on 'light' background (Table 3).

The CTAB agar was prepared by using components mentioned in tables 4 and 5. After autoclaving, petri-plates were poured and allowed to get solidified. After solidification, agar gel was punctured at proper positions using borer, and culture broth supernatant was added into the well using micropipette. The plate was then kept for the 24 hrs incubation and the precipitation zone around the hole were observed.

Production scale up

Production scale up was done by moving from shake flask to fermenter. For the scale of the production, 2 litre fermenter was used. The important operating conditions which were chosen include, Agitation speed, which was maintained 150 rpm. The temperature was maintained 35°C throughout. Aeration rate was maintained 0.8 vvm and pH was kept 7 throughout the process.

Before sterilisation, all the probes and Heating jacket was removed properly. All the filters and a sampling port were taken out and washed properly with alcohol. All the reactor assembly was washed neatly through water and then wiped with alcohol. The reactor assembly was then filled with 1.4 litres of fermentation media, mentioned in table 4. Then, the reactor assembly along with the fermentation media was autoclaved for 45 min at 20 lbs pressure. After the successful sterilisation, the assembly was allowed to cool down (Figure 3).

After cooling down, the fermenter was then inoculated with seed media, prepared already. The 100 ml Seed media was firstly prepared to grow the 10% microbes, to give them desired environment. In the middle of the exponential phase, this seed media was transferred aseptically to the fermenter. For the preparation of Seed media, minimal salt media was used.

After the successful transfer of seed media, all the probes and filters were assembled to the reactor. For regulation of pH during the fermentation course, 0.1M H_2SO_4 and 0.1N NaOH were prepared, and attached to the reactor. The reactor was then started and operated for continuous 7 day. From starting the fermenter, after every 24 hrs, sampling was done.

Down streaming of product

Down streaming is very important aspect in the fermentation process. The down streaming process was followed as me mentioned below.

Solvent extraction

First important step in the down streaming of rhamnolipid fermentation was solvent extraction. For the solvent extraction method, two different solvent systems were used, including ethyl acetate and chloroform:methanol system.

The solvent extraction scheme is shown below:

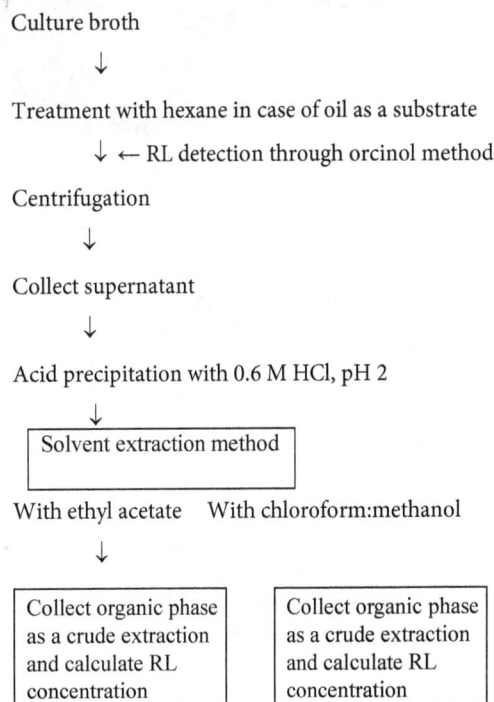

Culture broth

↓

Treatment with hexane in case of oil as a substrate

↓ ← RL detection through orcinol method

Centrifugation

↓

Collect supernatant

↓

Acid precipitation with 0.6 M HCl, pH 2

↓

Solvent extraction method

With ethyl acetate   With chloroform:methanol

↓

Collect organic phase as a crude extraction and calculate RL concentration

Collect organic phase as a crude extraction and calculate RL concentration

For ethyl acetate solvent system, ethyl acetate:extract ratio was used 1:1, while using chloroform:methanol:extract, 2:1:1 ratio was used. The extraction efficiency and partition coefficient was also calculated for both the systems by calculating the rhamnolipid concentration present in the organic phase.

Chromatography

Chromatography is one of the major down streaming technique, which purifies rhamnolipid. Paper chromatography was used as an analytical technique, to calculate the retention factor, i.e. R_f value, while another chromatography, i.e. column chromatography was performed to purify the rhamnolipid.

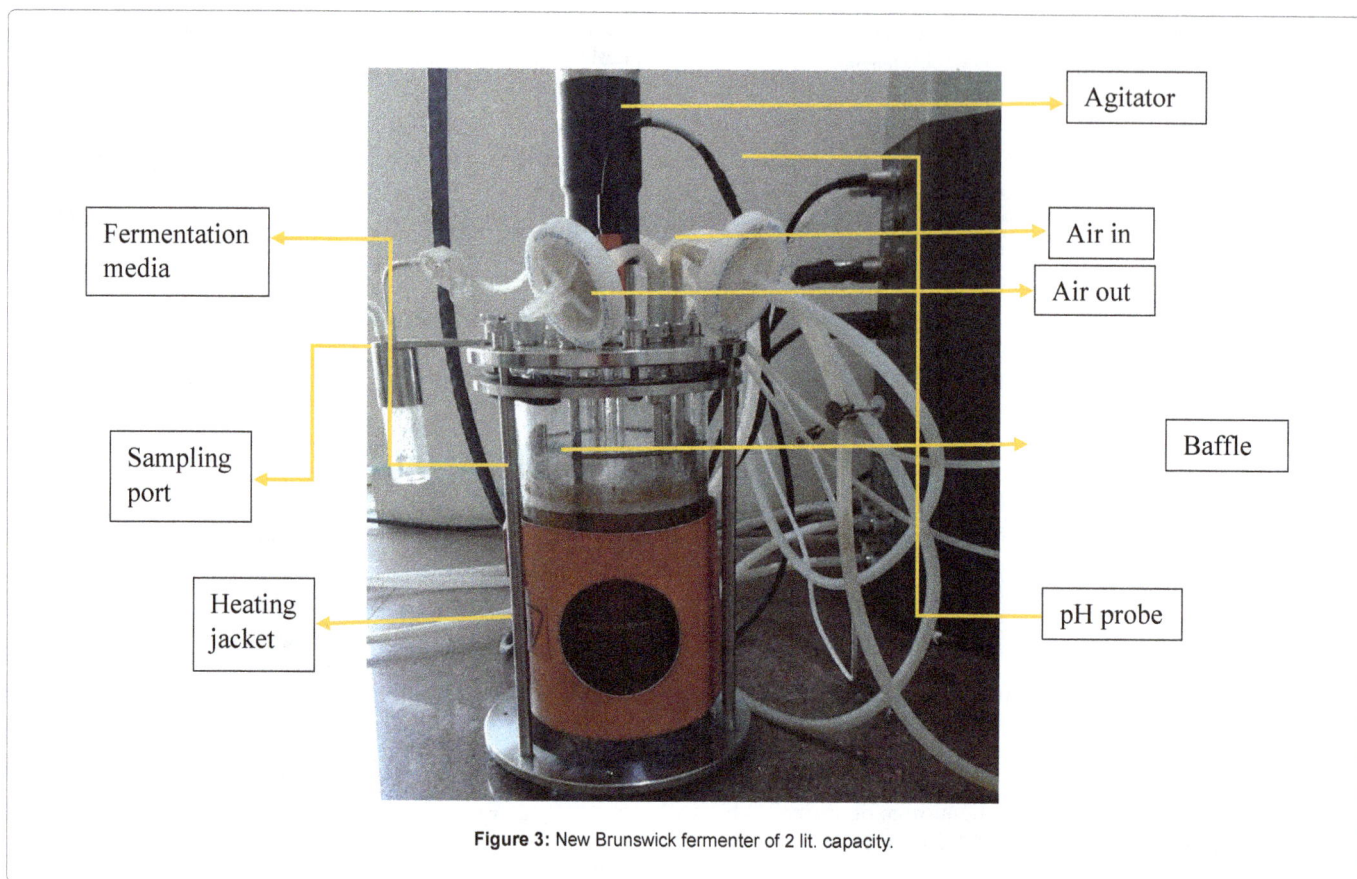

Figure 3: New Brunswick fermenter of 2 lit. capacity.

Paper chromatography: For paper chromatography, the mobile phase used was chloroform:methanol in ratio 2:1. Chromatography paper was marked initially for the sample loading. The sample was loaded at the marking position. The paper was then put into the chromatography jar. After some time, the sample runs on the chromatograph paper in up-ward position, leaving a mark. Hence, the distance travelled by sample to the mobile solvent was calculated.

Column chromatography: Column chromatography was performed with using silica-60 as a packing material or stationary phase. The silica gel-60 was packed tightly under the column. Primary cleaning was done by eluting the methanol from the column. As soon as the methanol was completely eluted, the crude sample was run in a column along with mobile phase. The mobile phase used was chloroform:methanol in 3:2 ratios. The methanol in the mobile phase was used to remove the hydrophobic impurities. Hence, the light brown coloured sample, which was eluted after the dark brown coloured sample, was collected and rhamnolipid concentration was measured.

Rotary vacuum evaporator

The concentrated product was collected by using the rotary vacuum evaporator (RVE). The collected crude sample was loaded in the round bottom flask of the rotary vacuum evaporator, integrated with vacuum system. The flask was rotated at the speed of 5 rpm, and the temperature was set 70°C, in order to evaporate the water content from the sample, using integrated vacuum. The concentrated sample was then analysed, using FTIR technique.

Emulsification index [9,10]

The emulsion index was calculated in order to check the capacity of formation of emulsion by rhamnolipid. The emulsion index was calculated after 24 hours; hence it is also called E_{24}. Emulsion index was measured in percentage by diving the height of the emulsion to the total height of the mixture. The followed procedure is described below.

6 ml D/W+2 ml immersion oil

↓

The sample was vortexed for 10 min

↓

The mixture was allowed to settle down

↓

2 ml crude rhamnolipid sample was added

↓

Again the sample was vortexed for 10 min and settled for 24 hrs

Emulsion was observed

Emulsion index stability [10,15]

The same procedure as described for the emulsion index was carried out for emulsion index stability test. The only difference with the previous technique is that, the crude rhamnolipid was kept at various temperatures and various pH. Rhamnolipid was kept at 4°C, 30°C and 80°C for around 30 min and the above procedure was carried out. Similarly, rhamnolipid was kept at pH 3, pH 7 and pH 9 for 30 minutes, and its emulsification index was calculated.

Oil spray assay [16,17]

Another important, but simple technique for the detection of emulsion capacity of rhamnolipid is oil spray assay. The following procedure was carried for the same.

20 ml D/W was poured in petri plate

↓

Immersion oil drop was added on the same D/W

↓

Rhamnolipid of 200 μl was added on the immersion drop

↓

Emulsion formation was observed

Heavy and toxic metal recovery [18]

The three major heavy metals along with the one toxic metal were recovered from the solution prepared. Three important heavy metals used were, $Pb(NO_3)_2$, $ZnCl$, $FeCl_3$. The toxic metal recovered was NaF. The standard concentration of heavy metal produced was 100 mg/ ml. The recovery procedure is given below for the 10 ml of the sample used each (Figure 4).

Standard metal conc. of 10mg/ml was taken 9 ml in test tube

↓

1 ml crude rhamnolipid was added in the sample

↓

The mixture was vortexed and allowed to settle for 2 hours

↓

The mixture was filtered through Whatman filter paper

↓

The weight of Whatman filter paper before and after the filtration was measured

The percentage of efficiency of removal of metals i.e. η were calculated as,

$$\eta = \frac{(\text{Initial heavy metal} - \text{Final heavy metal})}{\text{Initial metal concentration}} \times 100$$

Oil recovery [17,19]

Figure 4: Standard concentration of metal.

Rhamnolipid is very efficient in the oil recovery from oil sludge. The efficiency rhamnolipid for oil recovery was observed from the following protocol.

Components were mixed properly

↓

10 ml of immersion oil mixed with hot dry sand

↓

10 ml crude rhamnolipid was added

↓

The whole mixture was vortexed for 10 min and allowed to settle for 5 min

↓

Whole mixture was centrifuged at 6000 rpm for 20 min

↓

Different layers were observed, including oil emulsion

Results

Pseudomonas aeruginosa MTCC 1688 strain was successfully revived under the aseptic conditions. The green coloured pigment formation was observed, after 48 hrs of microbial inoculation, indicating presence of *Pseudomonas aeruginosa*. The strain was then identified using Gram staining procedure (Figures 5 and 6).

Gram staining

The pink coloured and rod shaped microorganisms were observed under the oil-immersion 100× lens of the microscope. Hence, gram staining procedure shows Gram-negative nature of *P. aeruginosa* (Graphs 1 and 2).

Shake flask cultivation

Shake flask optimisation study was carried out using 3 different substrates, i.e. glucose, karanja oil and soybean oil cake. The rhamnolipid production during the course of fermentation was determined by L-rhamnose standard curve. The equation of the straight line was obtained as, y=0.1712x+0.0042 (Graphs 3-5).

After analysing above experimentation, It has been observed that, 3% concentration of a substrate is efficient in rhamnolipid production. Hence, taking 3% substrate concentration, the experimentation for pH optimisation was carried out (Graphs 6-8).

After the shake flask studies, the data was analysed for higher rhamnolipid production, at a different substrates, their concentration and pH. After the successful data analysis of shake flask, the production study was carried out in 2 litre New-Brunswick fermenter, applying optimum conditions from shake flask (Figures 7-10) (Graph 9).

Fermentation studies

Confirmation of rhamnolipid production: Retention factor, i.e. R_f value=5.3/10=0.53

Solvent extraction

Ethyl acetate:extract ratio was used 2:1 for the extraction, and chloroform:methanol:extract ratio used was 2:1:1 (Figures 11-13) (Tables 3-5).

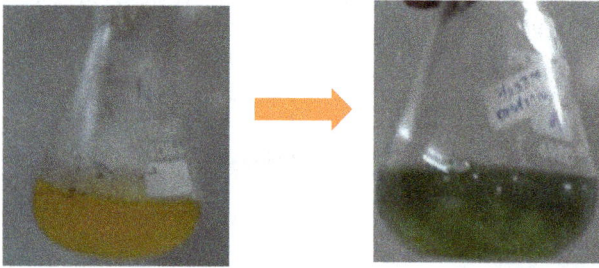

Figure 5: Change of color of *P. aeruginosa* inoculated nutrient broth after 48 hrs.

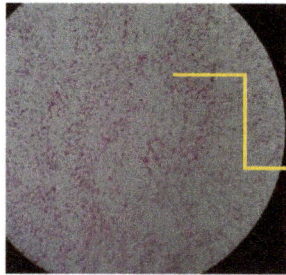

Figure 6: Gram staining.

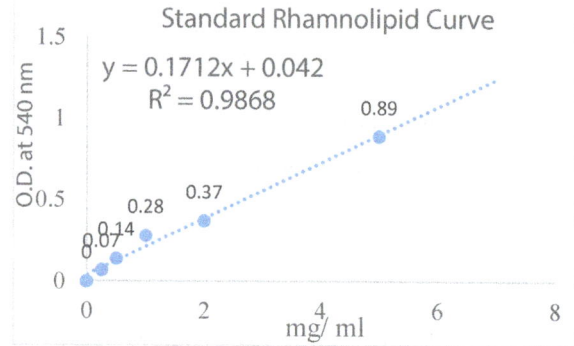

Graph 1: Standard glucose curve.

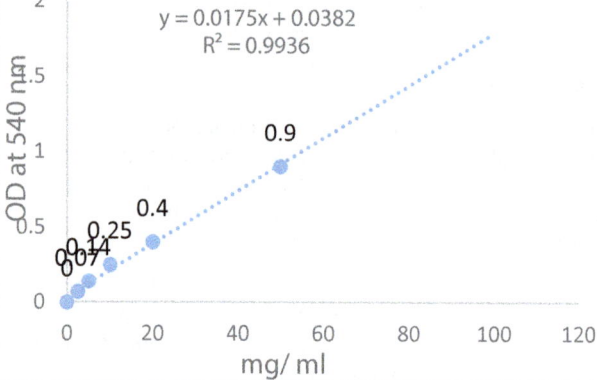

Structural analysis using FTIR

The FTIR spectrum analysis of the rhamnolipid, produced from *Pseudomonas aeruginosa* MTCC 1688 on 3% soybean oil cake as a substrate, is given below [9] (Table 6).

The data obtained was compared with the literature present, as well as standard spectrum data available, and found that, almost same stretching and deforms were observed.

Emulsification index

The emulsification index is determined (Figures 14-16), $E_{24} = 59.64$.

Graph 2: Standard rhamnolipid curve.

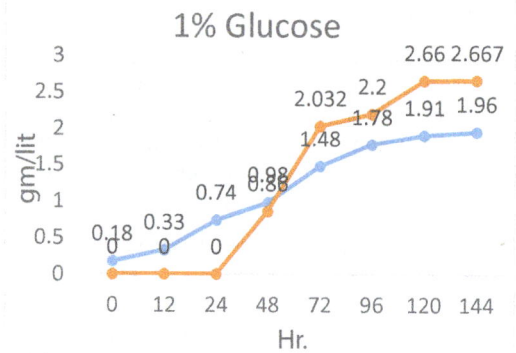

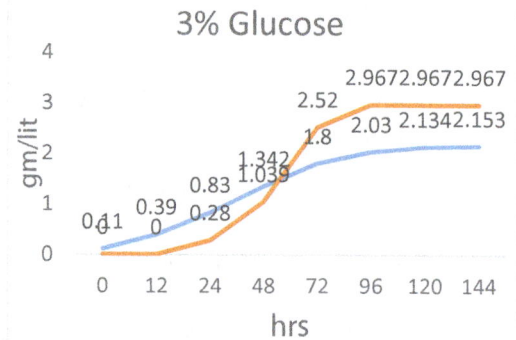

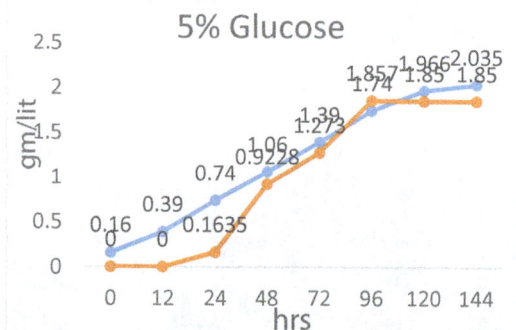

Graph 3.1- 3.3: Shows dry biomass concentration and rhamnolipid production at a glucose concentration of 1%, 3% and 5%, where line ⬤— shows dry biomass concentration and line ⬤— shows rhamnolipid concentration.

1% Oil Cake

3% Oil Cake

5 % Oil Cake

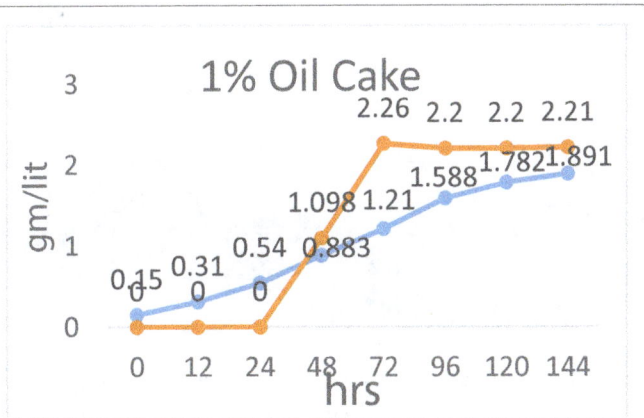

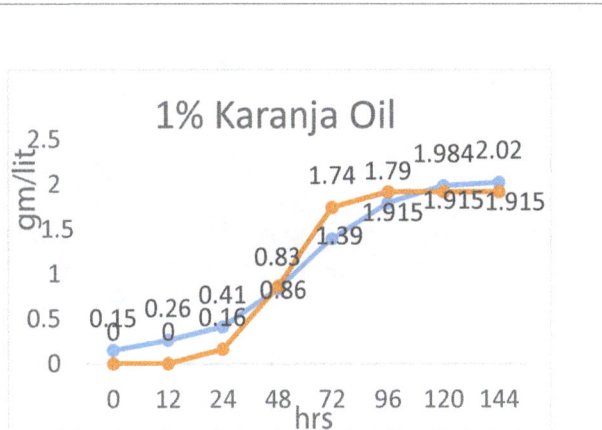

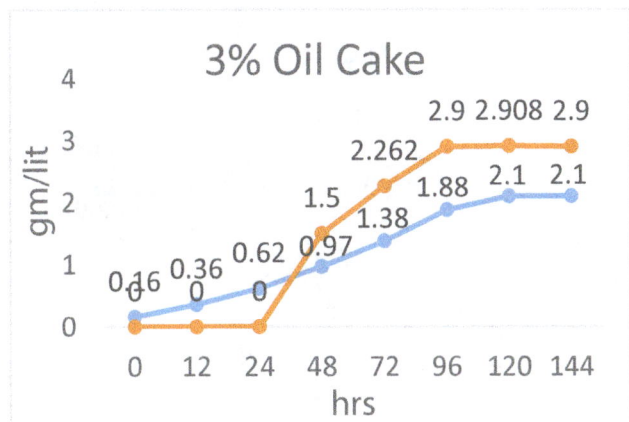

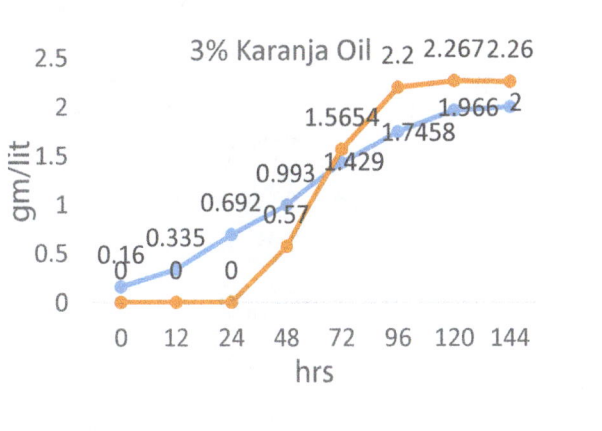

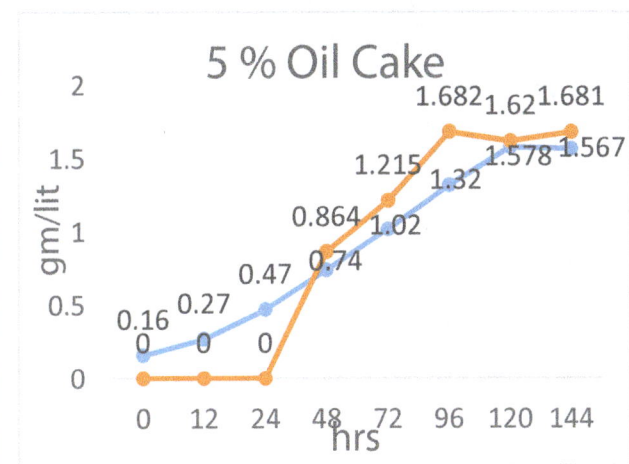

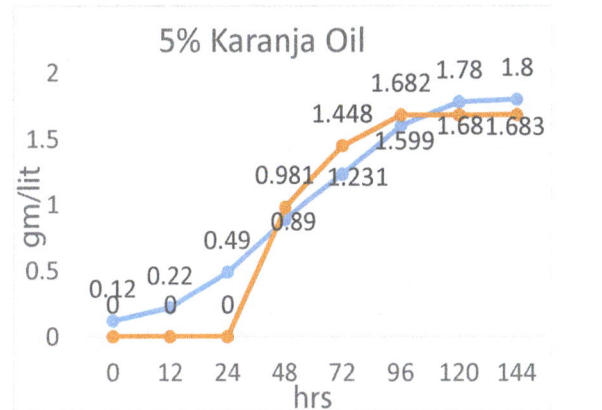

Graph 4.1- 4.3: Shows dry biomass concentration and rhamnolipid production at a soybean oil cake concentration of 1%, 3% and 5%, where line ———— shows dry biomass concentration and line ———— shows rhamnolipid concentration.

1% Karanja Oil

3% Karanja Oil

5% Karanja Oil

Graph 5.1- 5.3: Shows dry biomass concentration and rhamnolipid production at a karanja oil concentration of 1%, 3% and 5%, where line ———— shows dry biomass concentration and line ———— shows rhamnolipid concentration.

Emulsification index stability: Emulsification index stability at various pH and temperatures is plotted in Graphs 10 and 11.

Heavy and toxic metal recovery

The recovery of heavy $FeCl_3$, $ZnSO_4$, $Pb(NO_3)_2$ was done at concentration 73%, 65%, 71% respectively and NF at concentration 43% (Tables 6 and 7, Figures 17 and 18).

Conclusion

The results show that, the rhamnolipid was successfully produced from *Pseudomonas aeruginosa* MTCC 1688 strain. The strategic search for cheap and effective substrate ends in karanja oil and soybean oil cake, which shows effective and enhanced production of rhamnolipid,

3% Glucose at pH 5

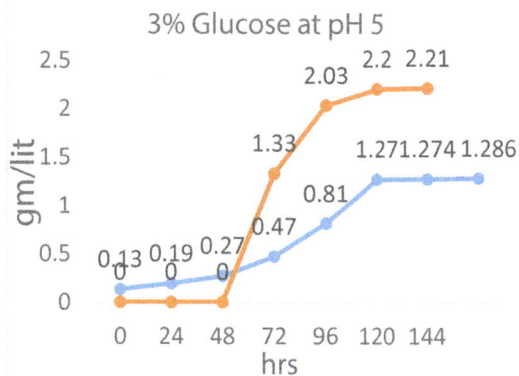

3% Oil Cake at pH 5

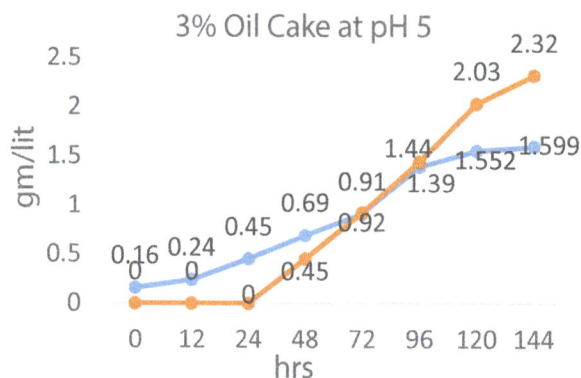

3% glucose at pH 7

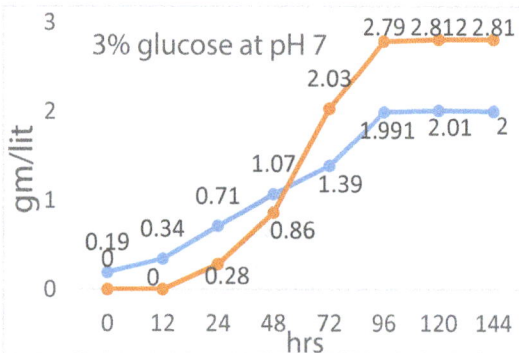

3% Oil Cake at pH 7

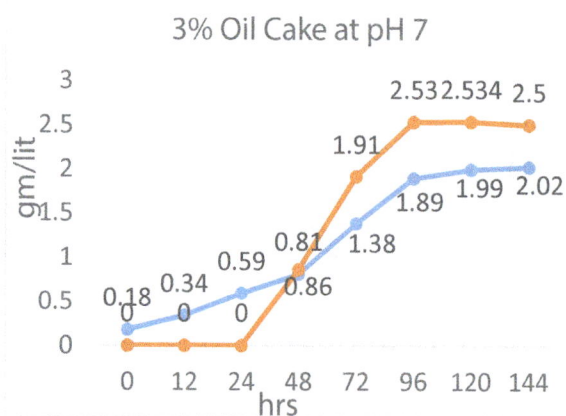

3% Glucose at pH 9

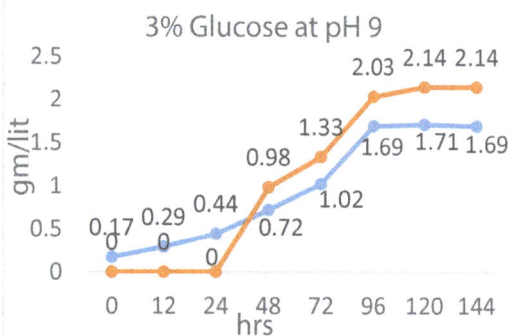

Graph 6.1- 6.3: Shows dry biomass concentration and rhamnolipid production at a 3% glucose concentration, at pH 5, 7 and 9 respectively, where line ———— shows dry biomass concentration and line ———— shows rhamnolipid concentration.

3% Oil Cake at pH 9

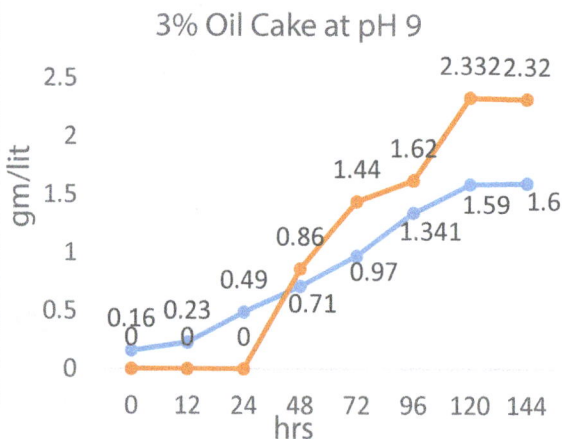

Graph 7.1-7.3: Shows dry biomass concentration and rhamnolipid production at a 3% soybean oil cake concentration, at pH 5, 7 and 9 respectively, where line ———— shows dry biomass concentration and line ———— shows rhamnolipid concentration.

at optimised pH of 7 and optimised concentration of 3%, compared to the various past researches. The optimum yield in terms of substrate was observed as 3.609 gm/lit of rhamnolipid produced per 5.255 ml of oil consumed, while yield in terms of biomass observed as 3.609 gm/lit of rhamnolipid produced per 2.5 gm of dry biomass. The chloroform:methanol extraction system was found to be the best solvent extraction system, where 83% of the rhamnolipid as recovered. The rhamnolipid was successfully applied for the heavy metals and toxic metals recovery, where rhamnolipid reduces heavy metal concentration to 73%, 65%, 71% for $FeCl_3$, $ZnSO_4$, $Pb(NO_3)_2$ respectively, while 43% in the case of toxic metal i.e. NaF. The produced rhamnolipid was found efficient in recovering 31% oil from oil sludge. Although rhamnolipid production has been intensively studied since the 1990´s, rhamnolipids

have not widely succeeded in substituting synthetic surfactants; rather their use is restricted to specific applications where biocompatibility is required. The main reason for this situation can be found in the high costs for synthesis and downstream processing of rhamnolipids. The development of new production processes is the key issue in overcoming these economic obstacles.

3% Karanja oil at pH 5

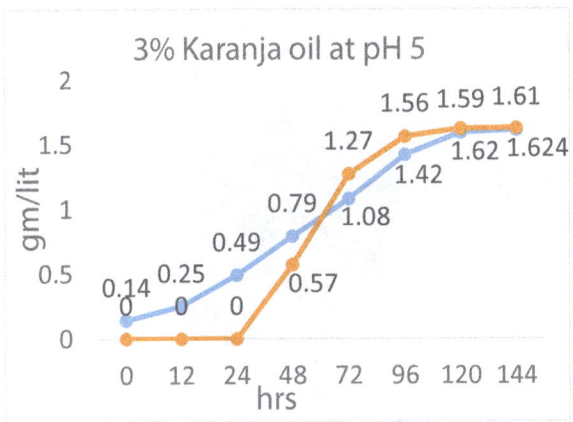

3% Karanja oil at pH 7

3% Karanja oil at pH 9

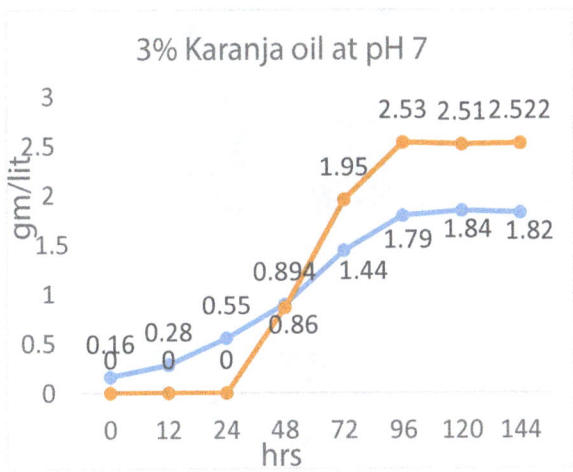

Graph 8.1- 8.3: Shows dry biomass concentration and rhamnolipid production at a 3% Karanja Oil concentration, at pH 5, 7 and 9 respectively where line ▬▬▬ shows dry biomass concentration and line ▬▬▬ shows rhamnolipid concentration.

Figure 7: Foam formation during the fermentation.

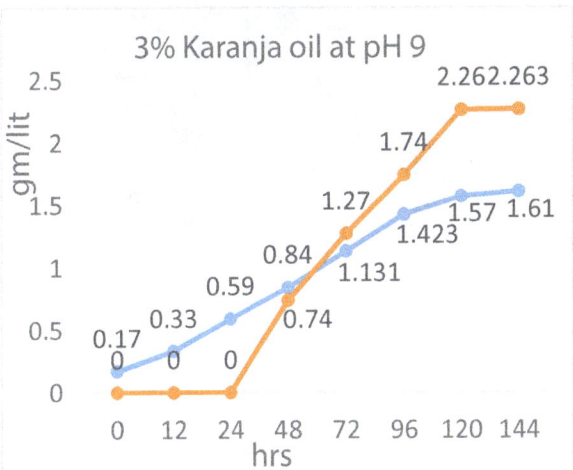

Figure 8: Foam formation using magnetic stirrer after the fermentation in order to enhance bio surfactant production.

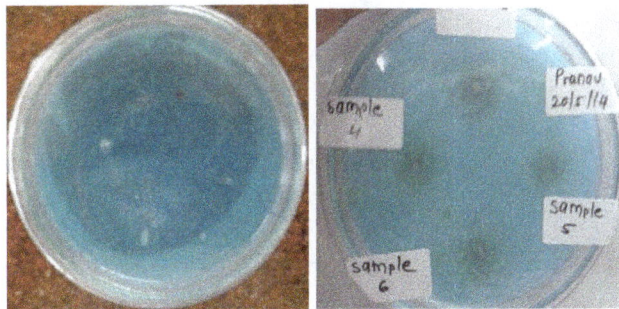

| Agar Plate before Incubation | Agar Plate after 24 hrs Incubation |

Figure 9: CTAB- methylene blue agar plate.

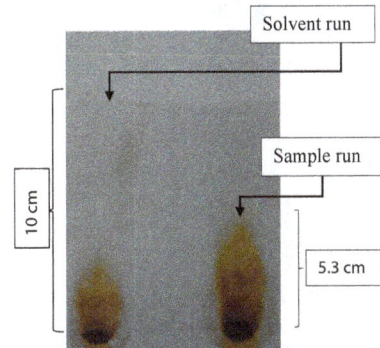

Figure 10: Paper chromatography result.

Biomass Concentration

Dry Biomass in gm/lit vs Time in hrs

Data points: 12, 0.31; 24, 0.67; 48, 0.989; 72, 1.56; 96, 1.99; 120, 2; 144, 2.38

Rhamnolipid Concentration

O.D. at 540 nm vs Time in hrs

Data points: 0, 0; 24, 0.16; 48, 0.31; 72, 0.48; 96, 0.56; 120, 0; 144, 0.67

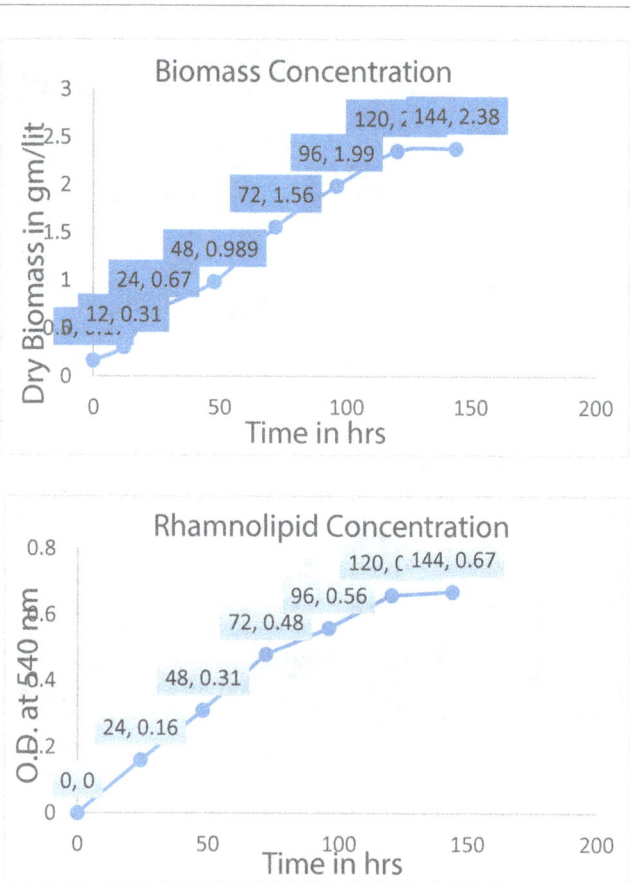

Graph 9.1- 9.3: Shows dry biomass concentration and rhamnolipid production at a 3% soybean oil cake concentration, at pH 5 in fermenter.

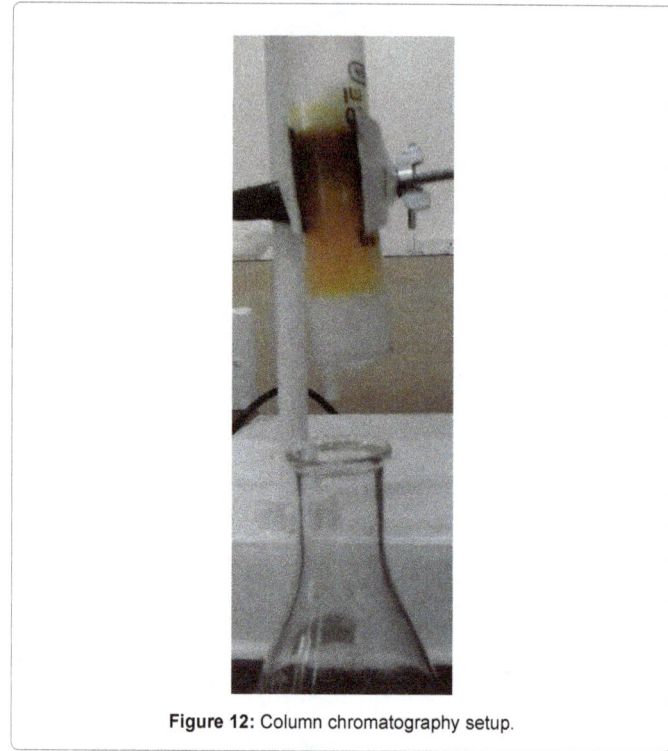

Figure 11: Solvent extraction using separating funnel with different extraction schemes

Figure 12: Column chromatography setup.

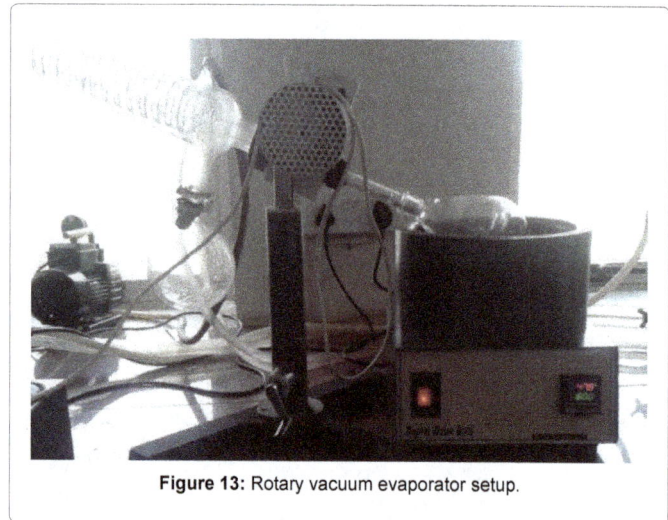

Figure 13: Rotary vacuum evaporator setup.

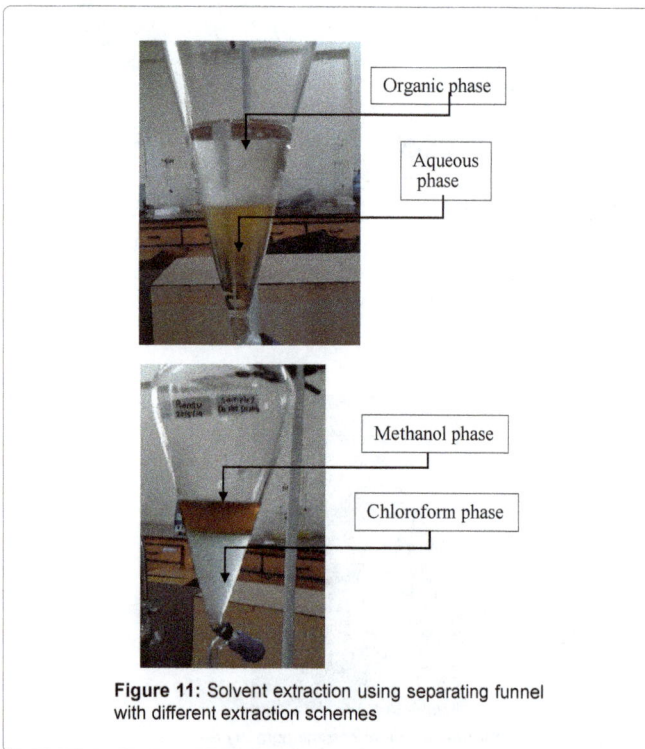

S. No	Metal	Wt. of Filter paper before (gm)	Wt. of Filter paper after filtration and drying (gm)	Difference (gm)	Efficiency of metal recovery
1	FeCl$_3$	1.244	1.271	0.027	73%
2	ZnSO4	1.273	1.308	0.035	65%
3	Pb(NO$_3$)$_2$	1.302	1.331	0.029	71%
4	NaF	1.242	1.285	0.043	43%

Table 6: Calculation table for heavy and toxic metal recovery.

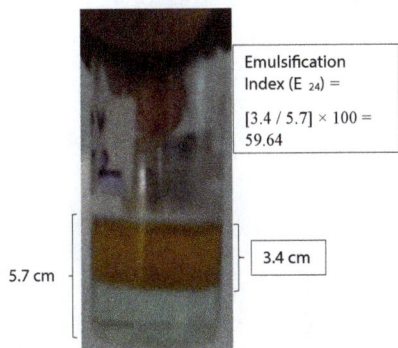

Emulsification Index (E 24) =

[3.4 / 5.7] × 100 = 59.64

5.7 cm

3.4 cm

Figure14: Formation of emulsion.

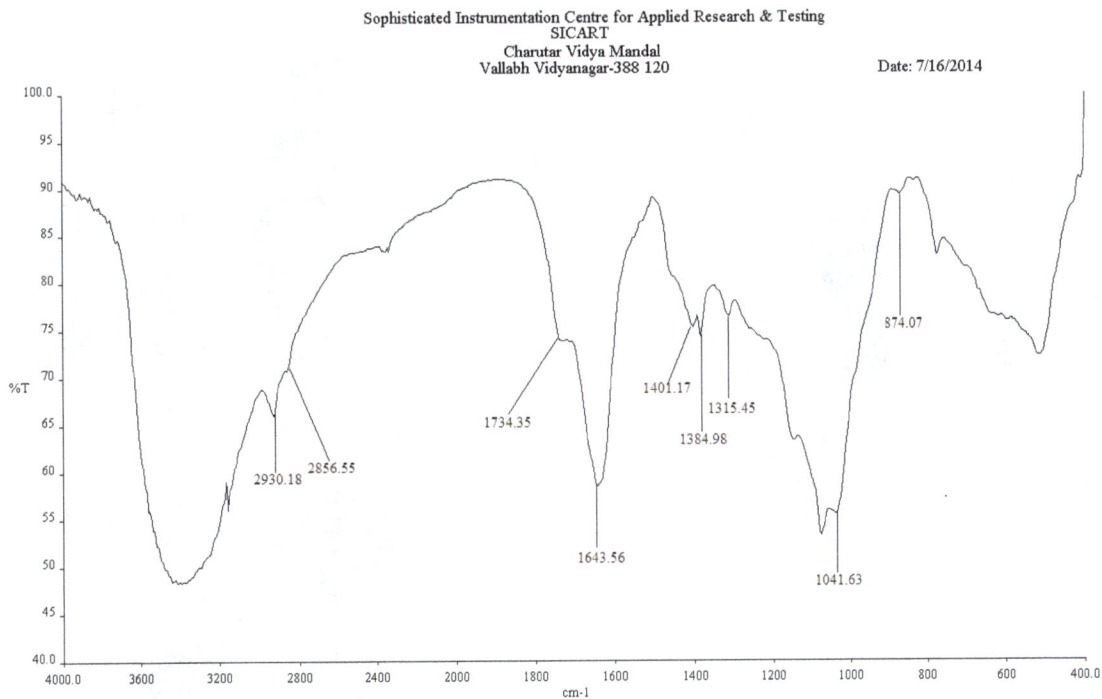

Sophisticated Instrumentation Centre for Applied Research & Testing
SICART
Charutar Vidya Mandal
Vallabh Vidyanagar-388 120   Date: 7/16/2014

Figure15: FTIR graph.

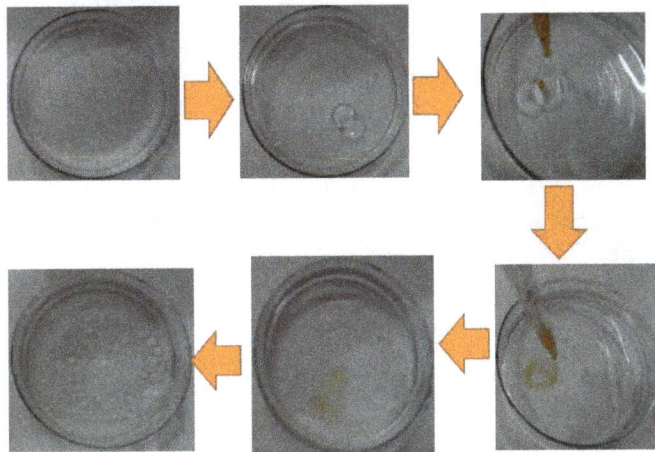

Figure 16: Formation of emulsion via oil spray technique.

Emuision Index Stability at Various pH

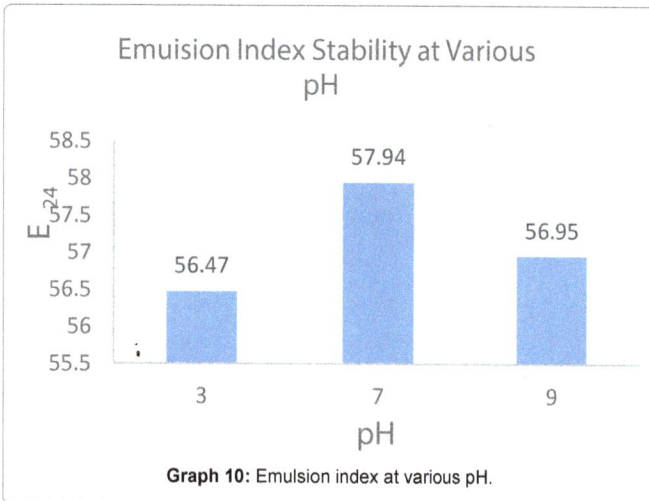

Graph 10: Emulsion index at various pH.

Emuision Index Stability at Various Temp.

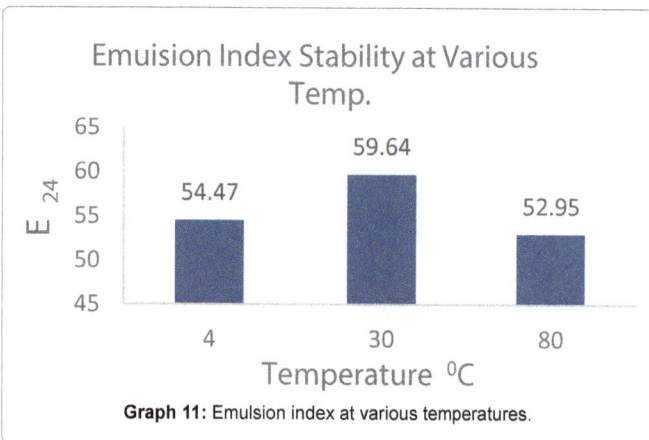

Graph 11: Emulsion index at various temperatures.

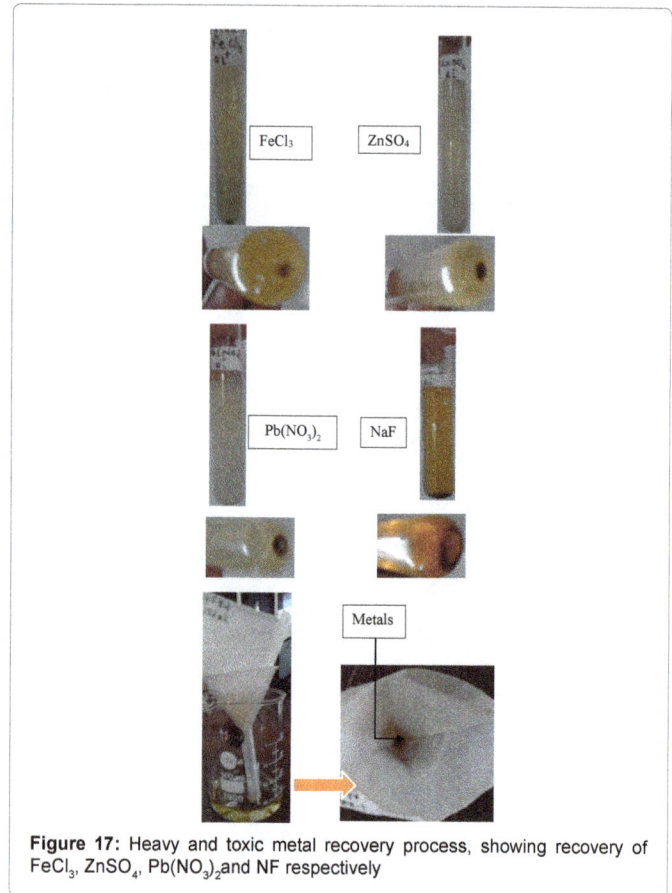

Figure 17: Heavy and toxic metal recovery process, showing recovery of $FeCl_3$, $ZnSO_4$, $Pb(NO_3)_2$ and NF respectively

S. No.	Experimental parameter	Value
1	Amount of sand taken	10 gm
2	Volume of oil added	10ml
3	Volume of RL added	10 ml
4	Vortexing time	5 min
5	Centrifugation speed	6000 rpm
6	Hexane used	10 ml
7	Oil recovered	3.1 ml
8	Percentage of oil recovered	31%

Table 7: Percentage recovery of oil from sludge.

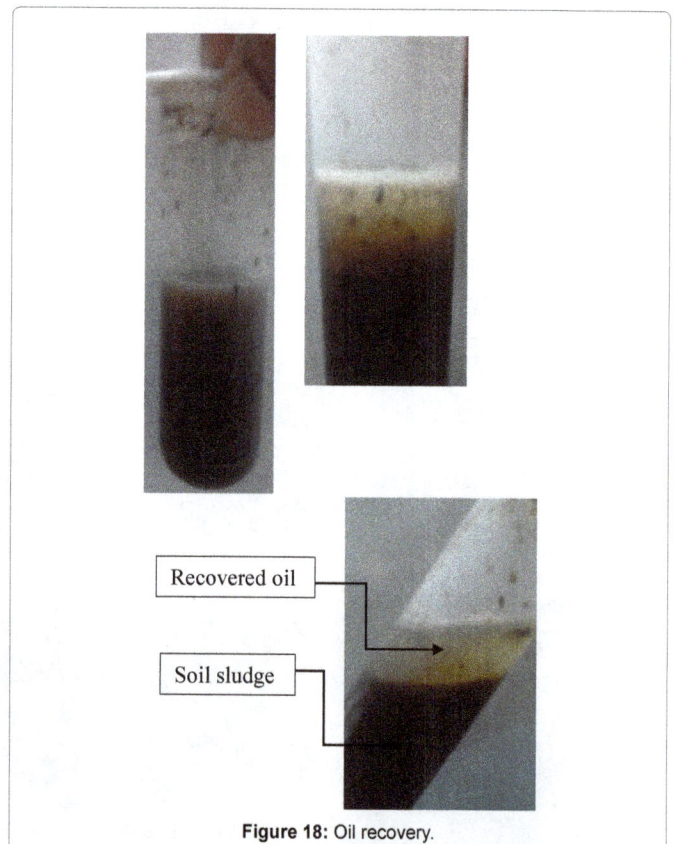

Figure 18: Oil recovery.

References

1. Al-Araji L, Abdul Rahman RNZR, Basri M, Salleh AB (2007) Microbial Surfactant - Miniriview. Asia Pac J Mol Biol Biotechnol 15: 99-105.

2. Benincasa M, Abalos A, Oliveira I, Manresa A (2004) Chemical structure, surface properties and biological activities of the biosurfactant produced by *Pseudomonas aeruginosa* LBI from soapstock. Antonie Van Leeuwenhoek 85: 1-8.

3. Edser C (2008) Status of global surfactant markets. Focus on Surfactants 11: 1-2.

4. Sharma D, Saharan BS, Chauhan N, Procha S, Lal S (2015) Isolation and functional characterization of novel biosurfactant produced by *Enterococcus faecium*. Springerplus 4: 4.

5. Banat IM, Franzetti A, Gandolfi I, Bestetti G, Martinotti MG, et al. (2010) Microbial biosurfactants production, applications and future potential. Appl Microbiol Biotechnol 87: 427-444.

6. Saharan BS, Sahu RK, Sharma D (2011) A Review on Biosurfactants: Fermentation, Current Developments and Perspectives. Genet Eng Biotechnol J 2011: GEBJ-29.

7. Šonc A, Grilc V (2004) Batch Foam Fractionation of Surfactants from Aqueous Solutions. Acta Chim Slov 5: 687-698.

8. Desai JD, Banat IM (1997) Microbial production of surfactants and their commercial potential. Microbiol Mol Biol Rev 61: 47-64.

9. Rikalovic MG, Gojgic-Cvijovic G, Vrvic MM, Karadžic I (2012) Production and characterization of rhamnolipids from *Pseudomonas aeruginosa* san-ai. J Serb Chem Soc 77: 27-42.

10. Mukherjee S, Das P, Sen R (2006) Towards commercial production of microbial surfactants. Trends Biotechnol 24: 509-515.

11. Henkel M, Müller MM, Kügler JH, Lovaglio RB, Contiero J, et al. (2012) Rhamnolipids as biosurfactants from renewable resources: Concepts for next-generation rhamnolipid production. Process Biochem 47: 1207-1219.

12. Reis RS, Pacheco GJ, Pereira AG, Freire DMG (2013) Biosurfactants: Production and Applications. In: Chamy R, Rosenkranz F (eds). Agricultural and Biological Sciences. Biodegradation - Life of Science. InTech Open.

13. Zhi L, Xingzhong Y, Hua Z, Guangming Z, Zhifeng L, et al. (2013) Optimizing rhamnolipid production by *Pseudomonas aeruginosa* ATCC 9027 grown on waste frying oil using response surface method and batchfed Fermentation. J Cent South Univ 20: 1015-1021.

14. Wadekar SD, Kale SB, Lali AM, Bhowmick DN, Pratap AP (2012) Utilization of sweetwater as a cost-effective carbon source for sophorolipids production by *Starmerella bombicola* (ATCC 22214). Prep Biochem Biotechnol 42: 125-142.

15. Cameotra SS, Makkar RS (1998) Synthesis of biosurfactants in extreme conditions. Appl Microbiol Biotechnol 50: 520-529.

16. Kannahi M, Sherley M (2012) Biosurfactant production by *Pseudomonas putida* and *Aspergillus niger* from oil contaminated site. IJCPS 3: 37-42.

17. Urum K, Pekdemir T, Copur M (2005) Screening of Biosurfactants for Crude Oil Contaminated Soil Washing. J Environ Engg Sci 4: 487-496

18. Pinto E, Aguiar AARM, Ferreira IMPLVO (2014) Influence of Soil Chemistry and Plant Physiology in the Phytoremediation of Cu, Mn, Zn. Crit Rev Plant Sci 33: 351-373.

19. Shafiei Z, Abdul Hamid A, Fooladi T, Yusoff WMW (2014) Surface Active Components: Review. Current Research Journal of Biological Sciences 6: 89-95.

14

C1-Esterase Inhibitor from Human Plasma-An Improved Process to Achieve Therapeutic Grade Purity

Nuvula Ashok Kumar*, Korla Lakshmana Rao, Archana Giri and Komath Uma Devi

Centre for Biotechnology, JNTUH, Hyderabad, India

Abstract

C1-Esterase Inhibitor (C1-INH) is a protein present in human plasma and is a member of serine protease inhibitors super family. C1-INH is a molecule with lot of therapeutic importance and is widely used in the treatment of Angioedema and in the current research an improved purification scheme is developed to fractionate C1-INH from human plasma. By a series of chromatography steps C1-INH is purified to homogeneity. A simple three step chromatography procedure has been described excluding the more inefficient or cumbersome methods like precipitation or affinity capture. The C1-INH purified by this scheme is characterized along with the reference molecule. C1-INH produced by this method is a single-chain glycoprotein with a molecular weight of 85-93 KDa with increased purity and yields of 100-120 mg/L of plasma. The chromatography steps are easily scalable and the process is economical and adaptable for therapeutic protein manufacture.

Keywords: C1-esterase inhibitor; Cellufine phenyl; DEAE Sepharose; Fractogel TMAE; IC_{50}

Abbreviations: C1-INH: C1 Esterase Inhibitor; PEG: Poly Ethylene Glycol; DEAE: Diethyl Amino Ethyl; TMAE: Tri Methyl Amino Ethyl; IC_{50}: Half Maximal Inhibitory Concentration

Introduction

Deficiency of C1-INH results in Hereditary Angioedema (HAE) which is characterized by localized swelling in the extremities, face, gut or upper airways; the most serious and potentially life-threatening manifestation of the disease is laryngeal edema [1,2]. Based on the combination of total protein and functional C1- INH levels, there are 3 types of HAE. Type-I is because of deficiency in the amount of C1-INH protein present in the plasma. In type-II, the circulating C1-INH concentration is normal or high but not fully functional where as type-III is acquired deficiency which is very rare. In type III HAE, there is no abnormality found with C1 inhibitor. Instead, it has several potential causes. This condition occurs mainly in women and can be made worse with pregnancy or use of birth control medications. It is also associated with a mutation in the Factor XII gene (F12 gene). In people with type III HAE, too much coagulation Factor XII is produced. C1-INH is administered to treat HAE but studies in animals and observations in patients indicated that administration of C1-INH may have a beneficial effect as well in other clinical conditions such as sepsis, cytokine-induced vascular leak syndrome, acute myocardial infarction, or other diseases [2]. C1-INH inhibits factor XIIa, factor XII fragment (XIIf), kallikrein, and plasmin. Thus, in its absence, there is marked activation of the bradykinin-forming cascade which is a vasodilator resulting in severe angioedema. [2].

C1-INH is a protein derived from human plasma. It is a member of serine proteinase inhibitor (serpin) super family. Other proteinases in this family are: Alpha-1-antitrypsin, Antithrombin, Alpha-2-macroglobulin, Heparin cofactor II, and Alpha-2-antiplasmin [1,2]. C1-INH is a major regulator of the classical complement pathway, contact activation pathway and fibrinolytic pathway and represents the only inhibitor of the classical pathway proteases-C1r and C1s [1-3]. C1-INH is a single polypeptide chain containing 478 amino acid residues with a molecular weight of about 90-105 KDa under non-reducing and 85-93 KDa under reducing conditions derived from human plasma.

C1-INH protein isoelectric point lies near 2.7-2.8 [1]. The high degree of glycosylation is unique to C1-INH among human blood plasma proteins, which is heavily glycosylated with approximately 50% of its total weight is contributed by carbohydrates; the molecular weight of peptide chain is ~53 KDa [1,3]. During the initial investigations that resulted in the isolation and characterization of C1, human serum was found to contain a heat sensitive substance that inhibited the enzymatic activity of C1. This inhibitory activity, termed CI esterase inhibitor, was further defined by Lepow [4] and was first isolated by Pensky et al. [5].

Several methods have been described for the purification of C1-esterase inhibitor involving Polyethylene Glycol (PEG) precipitation [6,7] or affinity chromatography or by using multiple chromatography steps [8-11]. The precipitation method suffers from low yields and less specific activity and the affinity method with lectin columns are not suitable for therapeutic protein manufacture due to the strong possibility of ligand leaching. The process described herein for human plasma derived C1-INH eliminates the need for PEG precipitation or affinity chromatography to purify this protein in high yields. A more efficient process helps to ensure optimal use of a scarce and valuable commodity like human plasma. It also ensures a more desirable quality for the end product by continuously upgrading the existing technology.

Materials and Methods

General equipment's in the laboratory

XK16/20 chromatography columns, AKTA Purifier chromatography system, DEAE Sepharose resin from GE Healthcare,

*Corresponding author: Ashok Kumar N, Centre for Biotechnology, JNTUH, Hyderabad, Telangana, India

Uppsala, Sweden. Cellufine phenyl resin from JNC Corporation and Fractogel TMAE resin from EMD-Millipore.

All equipments were from standard manufacturers as given below.

UV Spectrophotometer (GE Health care)

pH and Conductivity meter (Thermo Scientific)

Ultra filtration unit-Stirred cell (Amicon, Millipore)

Mini Pellicon TFF system (Millipore)

Electrophoresis unit and power supply (Tarsons)

HPLC system (SHIMADZU-LC2010CHT)

ELISA plate Reader-MULTISKAN EX (Thermo scientific).

Reagents

Substrate: Z-Lys-SBzl (N-Carbo benzyloxy-Lys-thiobenzyl ester) (Sigma), Primary antibody: Mouse monoclonal C1-Inhibitor IgG1 antibody (R&D systems), Secondary antibody: Rabbit polyclonal against mouse IgG1 (Sigma Aldrich), Coomassie brilliant blue R-250 (Merck), DMSO (Dimethyl sulphoxide) (Sigma), DTNB (5,5'-Dithio bis (2 nitro benzoic acid) (Calbiochem).

Chemicals

Citric acid anhydrous GR, Tri Sodium Citrate dihydrate GR, Sodium Dihydrogen phosphate monohydrate GR, Di Sodium hydrogen phosphate anhydrous GR, Ammonium Sulphate GR, Glycine GR, Tris buffer, Sodium acetate trihydrate, Sodium dodecyl sulphate. All reagents and chemicals used were from Merck.

C1-INH Reference standard

This was procured from R&D systems. A human plasma-derived protein which is 93 KDa under reducing conditions on SDS-PAGE (Catalog No 2488-PI-200, Lot No: NNH0111111, Make: R&D systems). In this study this was used as a reference standard. Purity of reference standard is >95% by SDS-PAGE under reducing conditions. Storage temperature is -20°C to -70°C for the lyophilized form which is stable for 6 months and after reconstitution, it is stable for 3 months at -20°C to -70°C.

Recombinant C1s

This was procured from R&D systems. It was a homogeneous protein on SDS-polyacrylamide gel electrophoresis which could form a stable complex with C1 esterase inhibitor, if this was added in excess (Catalog No: 2060-SE-010, Lot No: MTX0512051, Make: R&D systems). Purity of recombinant C1S is >95% by SDS-PAGE under reducing conditions. Storage temperature is -20°C to -70°C for the lyophilized form which is stable for 6 months and after reconstitution, it is stable for 3 months at -20°C to -70°C.

Human plasma for developing the process

Recovered plasma was procured from licensed and audited blood banks through cold chain shipment. Plasma was collected from the blood banks in and around Hyderabad, Visakhapatnam, and Bangalore which are part of the southern region of India. The serology test reports for each of the donor units were verified as negative for HIV 1 and 2 antibodies, HBS Ag, HCV antibody, VDRL and Malaria parasite. Plasma bags were also tested negative by Nucleic acid testing (NAT for HIV 1 and 2, HBV and HCV). These tested units were taken for the process.

Methodology adapted to purify C1-INH

All steps were carried out at room temperature (25-27°C). Prior to the purification process, the bags were removed from the freezer (-40°C) and left at room temperature i.e. 25°C for thawing for about 2 hr and then kept for further thawing in water bath set at 25°C -30°C for 30 min. After thawing, the bags were cut open under laminar air flow and the plasma from the bags was pooled in a collection vessel. The pooled plasma was filtered through 40 μm, Polypropylene filter (Domnick hunter) followed by 15 μm, Polypropylene filter (Domnick hunter) to remove the clots or any particulates and the filtrate was loaded on to chromatography columns to purify C1-INH. Filtration was carried out using a peristaltic pump. Below given was the sequence adapted to purify C1-INH to homogeneity.

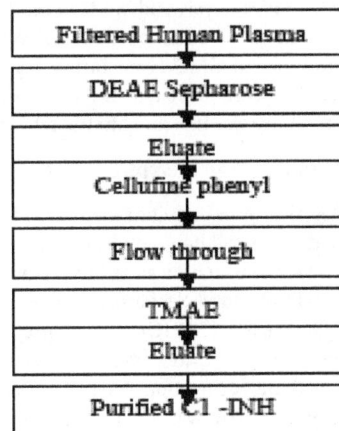

Capture step of C1-INH

Various chromatography resins were screened to check the efficiency to bind C1-INH from human plasma. To capture C1-INH from plasma, Cation exchange resins like CM Sepharose, SP Sepharose, Eshmuno S; Anion exchange resins-DEAE Sepharose,

Q Sepharose, Eshmuno Q and Mixed mode resin like Capto MMC were tried. Among all the resins, anion exchangers have the ability to bind C1-INH and in the group of anion exchangers, particularly DEAE Sepharose resin has the highest ability to bind C1-INH at pH 7.0 and 5 mS/cm conductivity. So, the column packed with DEAE Sepharose resin was optimized as the step to capture C1-INH.

The column was packed by Flow-packing method at a constant flow rate. Packing buffer was mixed with the resin to form 50-70% slurry. The slurry was poured into the column in a single continuous motion against the column wall to minimize the introduction of air bubbles. After pouring the resin, the adaptor was mounted on the top of the column and connected to a pump. The bottom outlet of the column was opened and the pump was set at flow rate 30% greater than that used during operation. Mobile phase was passed through the column in order to compress the resin. After the bed had compressed, the pump was paused and the adaptor was lowered onto the compressed bed, to complete the packing of the column. The efficiency of the packed column was monitored by determining HETP and Asymmetry factor. 1% Acetone solution was prepared and applied to the column. The column was washed with water at a flow rate of 30 cm/hr and the washing was continued till the UV peak at 280 nm reached the baseline. HETP and Asymmetry values were calculated for this peak to determine the efficiency of the packed column.

The filtered plasma without any particulate matter was loaded on to DEAE-Sepharose resin packed in XK 16/20 column to a height of 10 cm and this column was equilibrated thoroughly before loading the sample with 20 mM Sodium citrate pH 7.0 buffers which is having a conductivity of 5 mS/cm. The column was run at 90 cm/hr-120 cm/hr for loading, washing and elution. The column was washed with the same equilibration buffer to remove the other unbound plasma proteins. The fraction containing the C1-INH protein was eluted with the same buffer having increased conductivity of 20 mS/cm. The column was washed with 0.5 M NaCl to remove other tightly bound proteins and 0.5 M NaOH for regeneration and sanitization.

Other proteins collected in the flow through, wash fraction and regeneration step were not individually characterized in the current research but SDS-PAGE and Western blot analysis indicated that these were certain major proteins in plasma like Albumin, IgG, clotting factors etc. SDS-PAGE analysis and western blotting was performed for the column fractions to track for C1-INH .Western blot analysis confirmed the presence of C1-INH in elution-1 peak and SDS-PAGE picture indicated C1-INH protein band with less intensity and its electrophoretic mobility was matching to the reference standard.

Intermediate purification step of C1-INH

C1-INH after the initial capture step with lot of other plasma proteins was loaded on to the second chromatography column for purification. This step was mainly aimed to separate the other plasma proteins from C1-INH. Butyl Sepharose (GEHC), Phenyl Sepharose (GEHC), Octyl Sepharose (GEHC), Macro prep methyl (Bio-Rad), Cellufine phenyl (JNC) were the different hydrophobic resins tried out for the intermediate purification step of C1-INH but among all the resins, Cellufine phenyl resin was found to give good yield, purity and had better removal capacity for impurities at higher flow rates when compared to other resins which was confirmed by SDS-PAGE analysis. Column was packed with Cellufine phenyl resin in XK 16/20 column to a height of 10 cm by flow-packing method as detailed in the above section. 25 mM sodium phosphate buffer containing 0.8 M $(NH_4)_2SO_4$ was used for equilibrating the column and the same concentration of $(NH_4)_2SO_4$ was used for preparing the load and at this concentration of $(NH_4)_2SO_4$, it was observed that C1-INH did not bind to the column and was collected in the flow through fraction whereas the other plasma proteins bound to the column and eluted in the column water wash and regeneration steps. Other concentrations tried out were 0.4 M and 0.6 M $(NH_4)_2SO_4$ but under these conditions, C1-INH has bound to the column along with other proteins and at 1.0 M and 1.2 M $(NH_4)_2SO_4$ concentrations, precipitation was observed in the load sample and the sample became hazy. So, finally 0.8 M $(NH_4)_2SO_4$ concentration was found to be optimal to get purified C1-INH in the flow through fraction and to separate it from other plasma proteins. All the fractions from this column were analyzed on SDS-PAGE to check for the purity.

Polishing step of C1-INH

After the intermediate purification step, C1-INH was found to have two more extra bands which are carried over from plasma but not degradants of C1-INH which was confirmed by carrying out Western blot analysis for the purified C1-INH where only C1-INH band reacted with the antibodies and the two extra bands were not picked up in Western blot. Hence, those extra bands were presumed to belong to other proteins in plasma with lower molecular weights than C1-INH. In order to purify this further to homogeneity, C1-INH protein collected in flow through fraction in above step was diafiltered into 25 mM sodium citrate pH 7.0 buffer to remove ammonium sulfate. This diafiltered sample was loaded on to various cation exchanger resins like CM Sepharose, SP Sepharose, Eshmuno S and TMAE to remove the two extra bands and among all the resins, TMAE resin was able to selectively bind C1-INH protein. The TMAE resin was packed in XK 16/20 column to a height of 10 cm and equilibrated with 25 mM sodium citrate buffer containing 30 mM NaCl which is having a pH of 6.0 and conductivity of 8 mS/cm. The diafiltered sample was loaded onto TMAE column at a flow rate of 90 cm/hr and after loading the column was washed with equilibration buffer to remove any unbound proteins. C1-INH bound to the column was eluted by changing the pH and ionic strength of the buffer. Elution was carried out with 25 mM Sodium citrate containing 170 mM NaCl which is having a pH of 7.0 and Conductivity of 20 mS/cm. The impurities which did not bind to column were observed in the flow through fraction. The bound C1-INH which was eluted by the buffer of higher ionic strength was found to be free of impurities and was looking pure on SDS-PAGE.

Characterization of the purified c1-esterase inhibitor

The purified C1-INH preparation was analyzed for homogeneity on SDS-PAGE under reducing and non-reducing conditions, on Size Exclusion HPLC (SE-HPLC) and Reverse phase–HPLC (RP-HPLC) columns. All the analytical studies were carried out alongside a reference standard (obtained commercially from R&D systems).

a) SDS-PAGE analysis was performed to analyze the fractions and also the purified sample in comparison with the reference standard.

SDS-PAGE methodology

10% SDS-PAGE gel was casted where resolving gel was prepared by adding 3.1 ml of 30% Bis-Acrylamide solution, 1.9 ml of 1.5 M Tris buffer pH 8.8 to 2.5 ml of water. To this solution 37.5 μl of 10% APS and 10 μl TEMED were added and poured into the assembled gel plates and allowed to polymerise. Stacking gel was prepared by adding 0.4 ml of 30% Bis-Acrylamide solution, 0.75 ml of 0.5 M Tris buffer pH 6.8 to 1.8 ml of water. To this solution 20 μl of 10% APS and 5 μl TEMED were added and poured on top of the resolving gel. Immediately a clean comb was inserted into the stacking gel carefully avoiding trapping of air bubbles. The gel was allowed to polymerize. After the gel has polymerized the comb was removed carefully. Meanwhile, 40 μl of the samples to be loaded on the gel were aliquoted into micro centrifuge tubes and to that 10 μl of sample buffer (containing Tris, SDS, bromophenol blue, β-mercapto ethanol and glycerol) was added and kept in boiling water bath for 5 min. After boiling, the samples were removed and centrifuged at 10,000 rpm for 2 min. Once the 10% gel is ready, the samples were loaded into the wells slowly using micropipette. Then filled the gel and the casting unit with tank buffer (containing Tris, Glycine and SDS) and carried out electrophoresis at 300 V for 45 min till the tracking dye (bromophenol blue) has run out of the gel. Then the gel was carefully removed and stained with Coomassie Brilliant Blue R-250 dye prepared in methanol and acetic acid. Once the protein bands were developed, the gel was destained using solution containing methanol and acetic acid. Then this SDS-PAGE gel was scanned and documented.

Western blot methodology

The protein bands were separated on 10% SDS-PAGE as detailed above and the gel was trans-blotted onto a nitrocellulose (NC) membrane. Transfer was carried out for 2 hrs at 100 V in a buffer containing Tris–HCl, Glycine and methanol. After transfer, the NC membrane was incubated in 50 ml TBST (Tris buffered saline with Tween-20) containing 5% skimmed milk powder for 60 min. This step was to block additional protein binding sites. Following the incubation, the NC membrane was washed with TBST twice for 10-15 min. After washing, the membrane was incubated with 25 mL anti-Human serpin G1/C1 inhibitor Antibody (monoclonal mouse IgG, 0.5 mg/mL stock diluted 1:500 in TBST buffer containing 0.25% Bovine serum albumin) for 1 hr at room temperature with gentle shaking. The membrane was washed thrice for 10 min in 50 ml TBST and then incubated the membrane in 25 mL Rabbit anti-mouse-IgG polyclonal antibody conjugated to horseradish Peroxidase (1:1000 dilution with TBST) at room temperature for 1 hr with gentle shaking. After 1hr, the blot was washed with TBST. 1X TMB-H_2O_2 which is a chromogenic substrate was added to develop the specific protein bands which reacted with the antibodies.

b)   SE-HPLC and RP-HPLC methods were developed to characterize the purified molecule. RP-HPLC analysis was carried out in gradient mode on Inertsil c4 column using Acetonitrile/TFA solvent system. SE-HPLC analysis was carried on Shodex Protein KW-803 shodex column using Sodium citrate pH 6.8 buffers.

c)   The specific activity of the purified C1-INH protein was estimated in comparison with the reference standard by measuring the IC_{50} using a colorimetric method. The method [12,13] is based on the ability of C1-INH to inhibit the recombinant human complement C1s and estimation of the residual C1s esterase activity by measuring the increase in absorbance at 405 nm in a kinetic test using colorimetric peptide substrate Z-Lys-SBZl. The OD values obtained for C1-INH reference standard and in-house purified sample are plotted on Graph Pad prism. The activity of C1-INH may also be measured by checking its ability to prevent the release of H^+ ions which is monitored by the use of pH meter or a suitable acid base indicator [14]. The commercially available methods depends on measuring the C1 INH-C1S complex formation which is an immune assay method or inhibition of C1 esterase cleavage of artificial substrates like N-acetyl L-tyrosine-ethyl ester which is a colorimetric method [15,16] and is a non-competitive mechanism for which the Michaelis-Menten constant, Km, is 0.017 mol/l at 37°C, in the optimum pH range 7.2-7.4 [16]. This kinetic constant for the substrate is dependent on the pH of the solution which indicates that a terminal amino group may be involved in catalysis [17].

The rhC1s diluted to 10 μg/ml in assay buffer was used to prepare a curve of C1-INH reference standard in assay buffer with concentrations of 0.01 to 20 nM. Equal volumes of diluted rhC1s and reference standard or purified C1-INH at different concentrations were mixed. Two rhC1s blanks containing assay buffer were included. The mixtures were incubated at room temperature for 30 min. The reaction was started by adding 50 μl of 200 μM substrate/DTNB mixture. The absorbance was read at 405 nm in kinetic mode for every 5-10 min on MULTISKAN EX (Thermo scientific).

Results and Discussion

DEAE Sepharose step to capture C1-INH

Filtered plasma was loaded on to the packed DEAE Sepharose column which was equilibrated with Sodium citrate pH 7.0 buffer. The column was run at 90 cm/hr-120 cm/hr for loading, washing and elution. The column was washed with the same equilibration buffer to remove the other unbound plasma proteins. The fraction containing the C1-INH protein was eluted with the buffer of higher conductivity (DEAE Chromatogram (Figure 1a). The column was washed with 0.5 M NaCl to remove the tightly bound proteins and 0.5 M NaOH for regeneration and sanitization. SDS-PAGE analysis and western blotting was performed for the column fractions to track for C1-INH (Figures 1b and 1c). Western blot analysis confirmed the presence of C1-INH in elution-1 peak and SDS-PAGE picture indicated C1-INH protein band with less intensity in elution-1 peak and its electrophoretic mobility was matching to the reference standard.

Cellufine phenyl step

The DEAE-Sepharose elution-1 peak with other impurities was loaded on to Cellufine phenyl column after adjusting the conductivity of sample equal to 0.8 M $(NH_4)_2SO_4$. The column was equilibrated with pH 7.0 buffer containing 0.8 M $(NH_4)_2SO_4$. The sample after adjusting the conductivity was loaded on to the equilibrated column at 90 cm/hr flow rate and under the applied conditions C1-INH did not bind to the column and was collected in the Flow through. The remaining impurities were eluted in the column regeneration steps (Figure 2a). The flow through sample was concentrated and diafiltered into 25

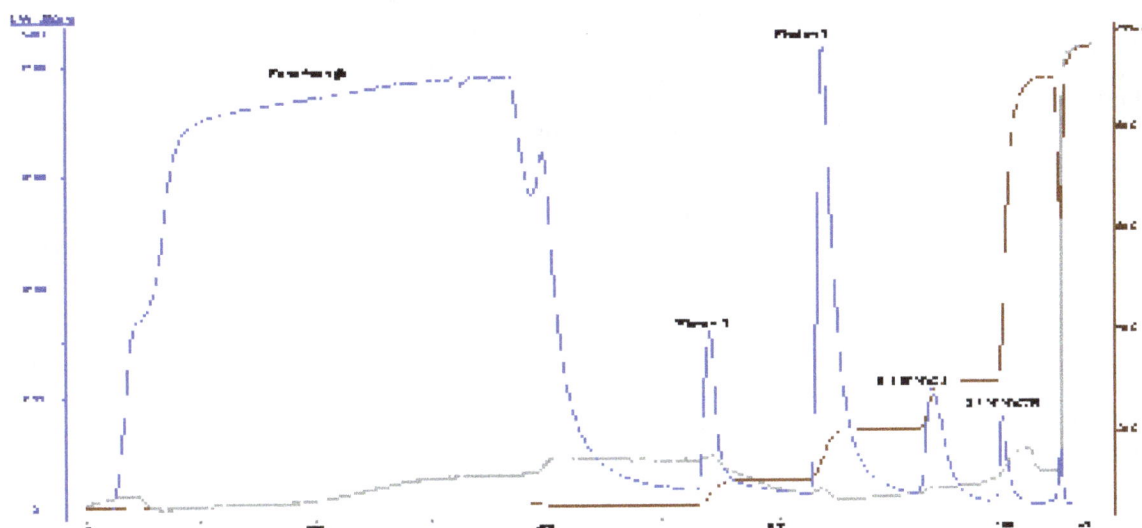

Figure 1a: DEAE Sepharose FF chromatography (1×10 cm, 20 ml) EQB: 20 mM sodium citrate pH: 7.0, Elution buffers: EQB containing NaCl. Eluate-1 contains C1-INH protein along with some impurities.Eluate-2 has other plasma proteins.

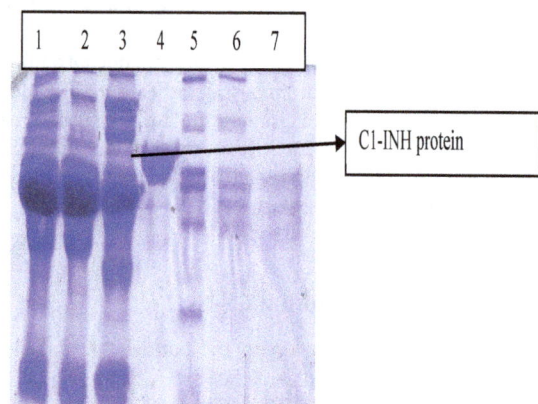

Figure 1b: SDS-PAGE analysis was performed to analyze the DEAE fractions. Samples were prepared in loading buffer containing 60 mM Tris-Cl pH 6.8, 2% SDS, 10% glycerol, 5% β-mercaptoethanol, 0.01% bromophenol blue and run on a 10% polyacrylamide gel. Lane 1: Load sample, 40 µg, + βME; Lane 2: Flow through, 40 µg, +βME; Lane 3: Elution-1 peak sample, 40 µg, +βME; Lane 4: Ref. standard, 20 µg, +βME; Lane 5: Elution-2 peak sample, 40 µg, +βME; Lane 6: 0.5M NaCl wash, 40µg, +βME; Lane 7: 0.5M NaOH peak, 40 µg, +βME; C1-INH protein can be seen in elution-1 fraction, a less intense band whose electrophoretic mobility matched the reference standard in lane 4.

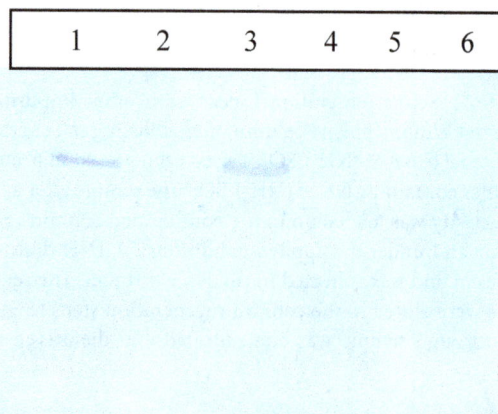

Figure 1c: Western blot analysis was performed to identify the fraction containing C1-INH protein. Samples were prepared in loading buffer containing 60 mM Tris-Cl pH 6.8, 2% SDS, 10% glycerol, 5% β-mercaptoethanol, 0.01% bromophenol blue and run on a 10% polyacrylamide gel and the gel is transferred on to a blot. Lane 1: DEAE load, 40 µg, +βME; Lane 2: Flow through, 40 µg, +βME; Lane 3: Elution-1 peak sample, 40 µg, +βME; Lane 4: Elution-2 peak sample, 40 µg, +βME; Lane 5: 0.5M NaCl wash, 40 µg, +βME; Lane 6: 0.5M NaOH peak, 40 µg, +βME; the protein of interest C1-INH was identified in load and elution-1 fraction.

mM sodium citrate pH 7.0 buffer using Amicon stirred cell loaded with 10 KDa molecular weight cut off membrane. This diafiltered sample was analyzed on SDS-PAGE. The C1-INH band was enriched in intensity but it was also found to have two more extra bands which are low molecular weight impurities (Figure 2b). In order to purify this further to homogeneity, the diafiltered sample was processed on TMAE column.

TMAE step to purify C1-INH to homogeneity

The diafiltered sample was loaded onto TMAE column at a flow rate

of 90 cm/hr and after loading the column was washed with equilibration buffer to remove any unbound proteins. C1-INH bound to the column was eluted by changing the pH and ionic strength of the buffer. The bound impurities were removed by washing the column with 0.5 M NaCl and 0.5 M NaOH solutions (Figure 3a). All the fractions from the column were concentrated on a stirred cell in order to analyze on SDS-PAGE and the analysis indicated that the low molecular weight impurities did not bind to the column under the applied conditions where as C1-INH had bound to the column. The sample from the elution peak showed a single band on SDS-PAGE (Figure 3b).

The purified C1-esterase inhibitor was pooled, concentrated and subsequently exchanged into 10 mM sodium citrate buffer pH 6.8, containing 0.13 M Glycine and 0.14 M sodium chloride. This buffer exchange was carried out using 10 KDa TFF cassettes on a mini Pellicon TFF system. The observed yield was 100-120 mg of C1-INH per liter of human plasma by adapting the developed process and the purified molecule was characterized by the methods described in the following section.

SDS-PAGE analysis of the purified C1-INH

The product obtained after the TMAE step is typically a single chain protein which was found to be more than 98% pure as shown by SDS-polyacrylamide gel electrophoresis (Figure 4). On comparison with reference standard its molecular weight is 85-93 KDa under reducing conditions and 90-105 KDa under non-reducing conditions and the mobility matched that of the reference standard.

SE-HPLC analysis of the purified C1-INH

The identity and purity check by SE-HPLC analysis in comparison to standard indicated that the purified C1-INH protein is more than 98% pure (Figure 5). The analysis was carried out on Protein KW-803 Shodex column (8.0 mm I.D×300 mm length) with particle size of 5 µm. The retention time of the purified C1-INH peak (15.827 min) closely matched the retention time of the reference standard (15.820 min) and also showed a better purity profile (98.9%) with a major single peak and a very minor peak at 11.779 min, when compared to the reference standard (92.4% purity) with at least two other contaminants on SE-HPLC chromatogram. Monomer along with the aggregate peak would be >95% purity for C1-INH reference standard. Rapid reactivation of C1INH-sample with polymers and dimers to functional monomers is possible by denaturation, on-column refolding method developed by Mathew Gauthier and Philip A. Patston [18].

RP-HPLC analysis of the purified C1-INH

RP-HPLC analysis indicated that the purified C1-INH protein is more than 99% pure (Figure 6). The analysis was carried out on Inertsil WP 300 C4 column (4.6 mm I.D×250 mm length) with particle size of 5 µm and the applied flow rate was 1 ml/min. Mobile phase A: 0.1% TFA in 10% Acetonitrile, Mobile phase B: 0.1% TFA in 90% Acetonitrile. The retention time of the purified C1-INH peak (28.50 min) also showed a good purity profile (99.43%) with a major single peak and two very minor peaks before the main peak on RP-HPLC chromatogram.

Specific activity

Specific activity was estimated by the colorimetric method and the IC_{50} value was determined by plotting the OD values obtained for C1-INH reference standard and the purified C1-INH sample using Graph Pad prism software (Figure 7). The IC_{50} of the purified C1-INH protein (1.367) was comparable in value to that of the reference standard (1.392).

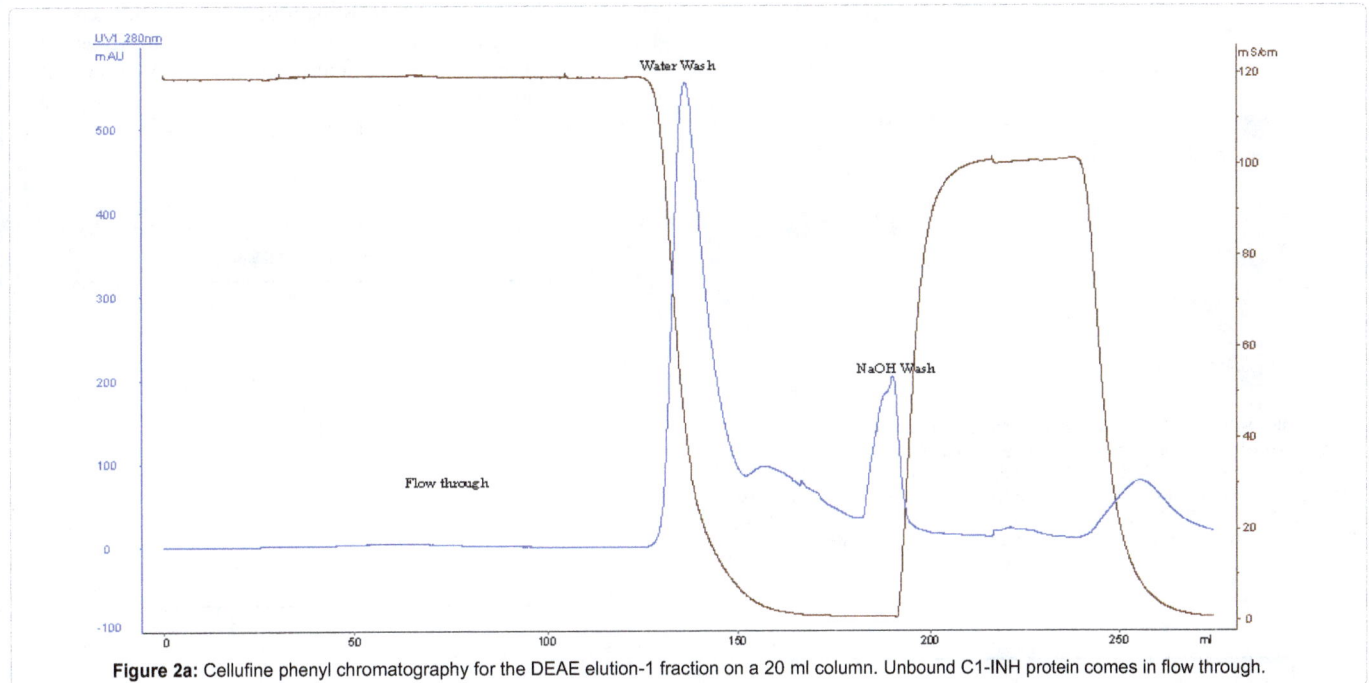

Figure 2a: Cellufine phenyl chromatography for the DEAE elution-1 fraction on a 20 ml column. Unbound C1-INH protein comes in flow through.

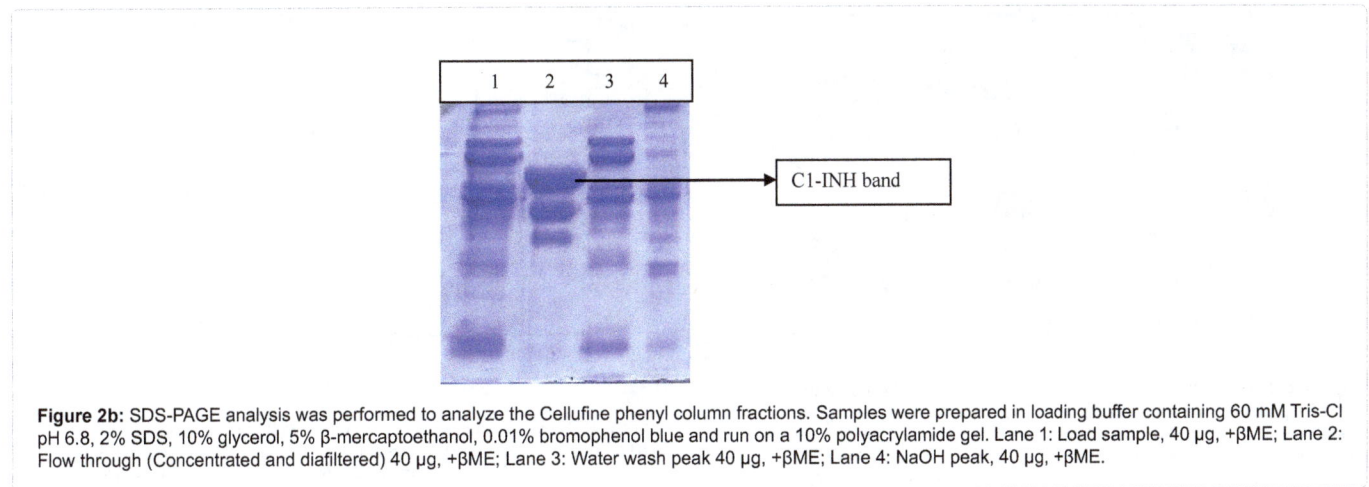

Figure 2b: SDS-PAGE analysis was performed to analyze the Cellufine phenyl column fractions. Samples were prepared in loading buffer containing 60 mM Tris-Cl pH 6.8, 2% SDS, 10% glycerol, 5% β-mercaptoethanol, 0.01% bromophenol blue and run on a 10% polyacrylamide gel. Lane 1: Load sample, 40 µg, +βME; Lane 2: Flow through (Concentrated and diafiltered) 40 µg, +βME; Lane 3: Water wash peak 40 µg, +βME; Lane 4: NaOH peak, 40 µg, +βME.

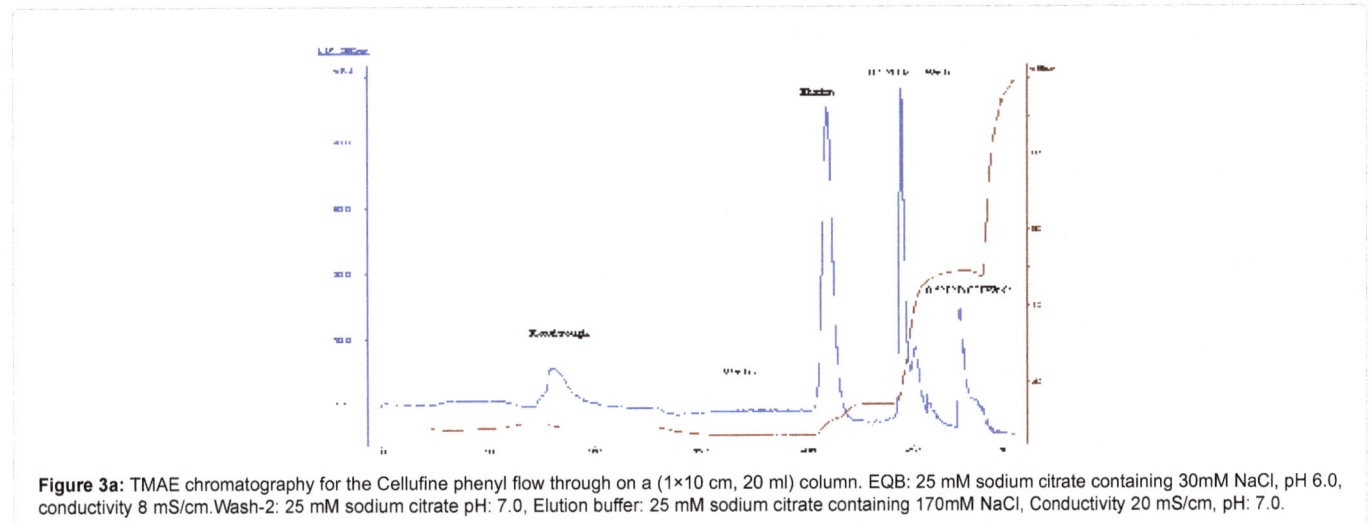

Figure 3a: TMAE chromatography for the Cellufine phenyl flow through on a (1×10 cm, 20 ml) column. EQB: 25 mM sodium citrate containing 30mM NaCl, pH 6.0, conductivity 8 mS/cm.Wash-2: 25 mM sodium citrate pH: 7.0, Elution buffer: 25 mM sodium citrate containing 170mM NaCl, Conductivity 20 mS/cm, pH: 7.0.

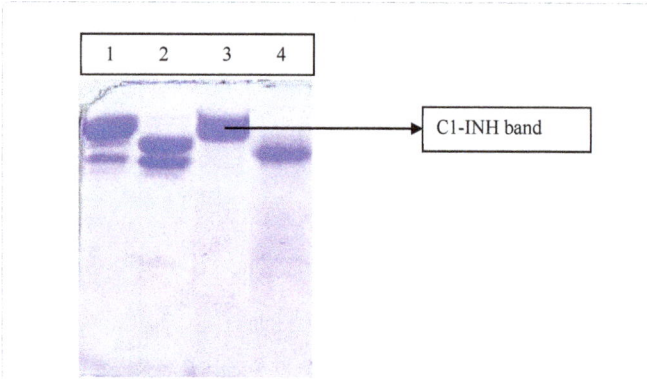

Figure 3b: SDS-PAGE analysis was performed to analyze the fractions from TMAE column. Samples were prepared in loading buffer containing 60 mM Tris-Cl pH 6.8, 2% SDS, 10% glycerol, 5% β-mercaptoethanol, 0.01% bromophenol blue and run on a 10% polyacrylamide gel. Lane 1: Column load sample, 40 μg, +βME; Lane 2: Column flow through 40 μg, +βME; Lane 3: Elution peak sample; Lane 4: 0.5M NaCl elution peak 40 μg, +βME.

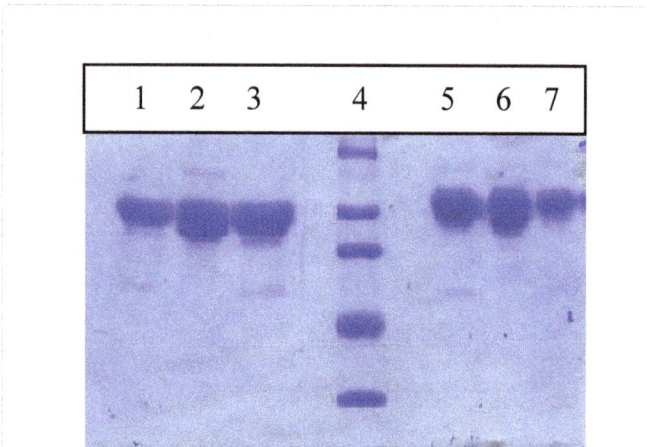

Figure 4: SDS-PAGE analysis was performed to check the purity of purified C1-INH protein in reference to C1-INH standard (R&D systems). Samples were prepared in loading buffer containing 60 mM Tris-Cl pH 6.8, 2% SDS, 10% glycerol, 5% β-mercaptoethanol, 0.01% bromophenol blue and run on a 10% polyacrylamide gel. Lane 1: Purified C1-INH, 10 μg, +βME; Lane 2: Reference standard 15 μg, +βME; Lane 3: Purified C1-INH, 15 μg, +βME; Lane 4: Molecular weight marker, 29-205 KDa; Lane 5: Purified C1-INH, 15 μg,-βME; Lane 6: Reference standard 15 μg,-βME; Lane 7: Purified C1-INH, 10 μg,-βME.

Conclusion

There are many reports of purification of C1-INH protein [4-6,8,10,12,13] and these previously published reports detail the usage of precipitation and affinity chromatography methods for purification of C1-INH and some of them have problem with low yields, time consuming process and does not result in pure protein of therapeutic grade. In the current research, an attempt was made to look for a purification scheme that is both economical and scalable for the purification of C1-esterase inhibitor, without compromising on yields, purity and its other therapeutic characteristics. The process developed was economical because it was devoid of multiple steps to purify C1-INH unlike some of the previous published reports which had-PEG precipitation, DEAE, Zinc chelate, Immuno adsorption steps and lysine Sepharose, DEAE Sephadex, Sephadex G-150, Hydroxy apatite

chromatography . Some of the existing processes use affinity columns which are generally expensive. Besides an additional purification step will be required to remove the ligands as there is always a possibility of ligand leaching from the affinity column. But the current process does not require any affinity column and thereby eliminates the need

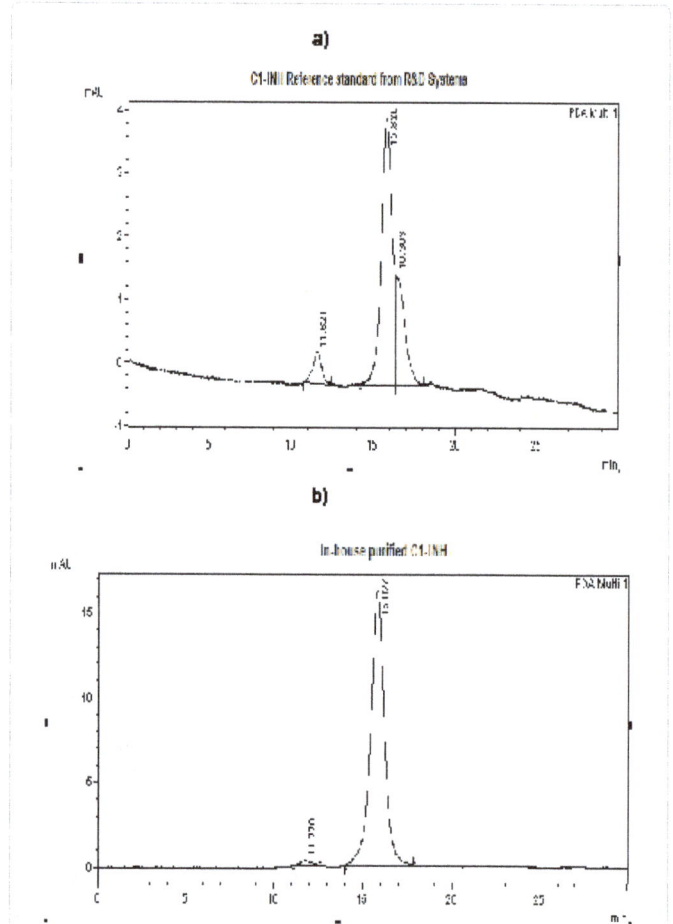

Figure 5: SE-HPLC analysis was performed to check the purity of purified C1-INH protein in reference to C1-INH standard shows that (a) Reference standard is 92.40% pure (b) In-house purified C1-INH molecule is 98.90% pure.

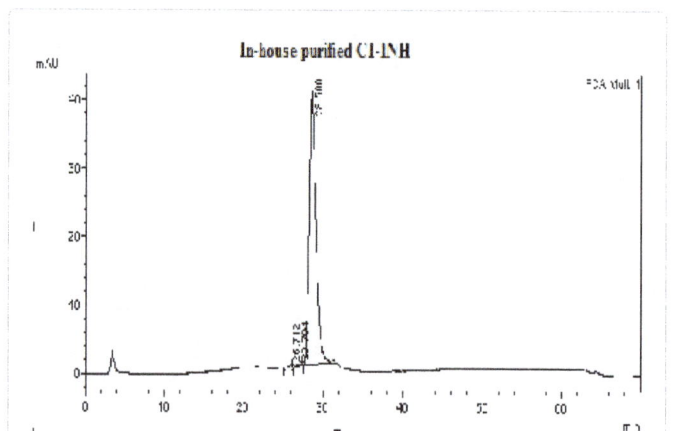

Figure 6: The above RP-HPLC chromatogram shows that the In-house purified C1-INH protein is 99.43% pure.

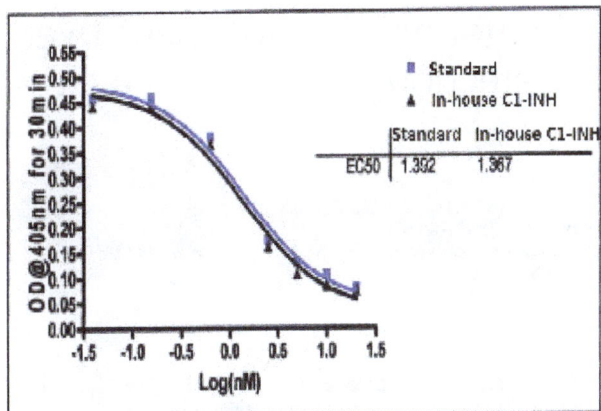

Figure 7: IC_{50} activity of the purified C1-INH molecule was estimated in reference to standard and the results indicated that it had similar activity to standard. O.D at 405nm was plotted versus log concentration using Graph Pad prism software.

for additional steps to remove leached ligands. Polyethylene glycol (PEG) was not used in the current process which brings down the cost of procuring PEG and also eliminates the usage of additional filters or equipments like centrifuge to separate the PEG precipitated protein from the solution. The developed process was optimized at 1 L scale and then scaled up to 10 L scale without any major challenges. This process was easily scalable as this was totally end to end chromatography based without any precipitation steps. Chromatography systems and columns are easily scalable from R&D to Production and the chromatography resins used in the process are also available in bulk quantities.

After an initial capture on DEAE resin, two different new generation resins were chosen for their ability to selectively bind and purify C1-INH from among the other plasma proteins. Figures 2a and 2b indicate how the use of Cellufine phenyl resin was able to purify C1-INH to a significant extent with the SDS-PAGE (Figure 2b) analysis revealing only a few protein impurities. As a polishing step for purity, the C1-INH sample after Cellufine phenyl step was loaded on Fractogel TMAE resin, which was successful in increasing the purity to a significant extent as shown in lane 3 of Figure 3b. The yield of the final purified C1-INH by this process was found to be in the range 100 to 120 mg/litre of human plasma. This increased recovery and purity helped us nail this as an ideal process scheme for therapeutic manufacturing, subject to the other biochemical characteristics being within the required specifications. Results of the characterization studies on purified C1-INH indicated the comparability of the protein with an existing reference standard with respect to its purity on SDS-PAGE, RP-HPLC analysis and specific activity (Figures 4-7). The overlap of reference protein standard curve with the purified C1-INH curve and their comparable IC_{50} values show that the functionality of the protein has not been compromised by this purification scheme. A scheme that uses only chromatography steps is easily scalable unlike the earlier reported procedures of precipitation and affinity capture. The increased yields and purity show that further studies can be done to adapt the method for industrial manufacturing of this very important therapeutic protein from human plasma.

References

1. Davis AE 3rd (1988) C1 inhibitor and hereditary angioneurotic edema. Annu Rev Immunol 6: 595-628.

2. Caliezi C, Wuillemin WA, Zeerleder S, Redondo M, Eisele B, et al. (2000) C1-Esterase inhibitor: an anti-inflammatory agent and its potential use in the treatment of diseases other than hereditary angioedema. Pharmacol Rev 52: 91-112.

3. Gower RG, Busse PJ, Aygören-Pürsün E, Barakat AJ, Caballero T, et al. (2011) Hereditary angioedema caused by c1-esterase inhibitor deficiency: a literature-based analysis and clinical commentary on prophylaxis treatment strategies. World Allergy Organ J 4: 9-S21.

4. Levy LR, Lepow IH (1959) Assay and properties of serum inhibitor of C'l-esterase. Proc Soc Exp Biol Med 101: 608-611.

5. Pensky J, Levy LR, Lepow IH (1961) Partial purification of a serum inhibitor of C'1-esterase. J Biol Chem 236: 1674-1679.

6. Teh LC, Froger M (1992) Evaluation of the chromatographic procedure for the preparation of a high-purity C1-esterase inhibitor concentrate from cryosupernatant plasma. J Chromatogr 582: 65-70.

7. Nilsson T, Wiman B (1982) Purification and characterization of human C1-esterase inhibitor. Biochim Biophys Acta 705: 271-276.

8. Prograis LJ Jr, Hammer CH, Katusha K, Frank MM (1987) Purification of C1 inhibitor. A new approach for the isolation of this biologically important plasma protease inhibitor. J Immunol Methods 99: 113-122.

9. Yendra L, Wolfgang S, Hans-Peter S, Oliver Z (2001) Method for production of a c1 esterase inhibitor (c1-inh)-containing composition. US patent 2001/0019839.

10. Pilatte YM, Hammer CH, Frank MM, Fries LF (1991) Process for the purification of C1-Inhibitor. US patent 5030578.

11. Pelzer H, Heber H, Heimburger N, Preis HM, Naumann H (1987) Process for the preparation of the c1 inactivator and its use. US patent 1223201 A1.

12. Procedure of Enzymatic assay for C1 Inhibitor to estimate IC 50. R&D Systems 2488-PI.

13. Nilsson T, Sjöholm I, Wiman B (1983) Structural and circular-dichroism studies on the interaction between human C1-esterase inhibitor and C1s. Biochem J 213: 617-624.

14. Smith AM, Thompson RA (1980) A method for the estimation of the activity of the inhibitor of the first component of complement. J Clin Pathol 33: 167-170.

15. Gompels MM, Lock RJ (2007) Laboratory testing for C1 inhibitor deficiency: a comparison of two approaches to C1 inhibitor function. Ann Clin Biochem 44: 75-78.

16. Schena FP, Manno C, D'Agostino R, Bruno G, Cramarossa F, et al. (1980) A kinetic test for the assay of the C1 esterase-inhibitor. J Clin Chem Clin Biochem 18: 17-21.

17. Canady WJ, Westfall S, Wirtz GH, Robinson DA (1976) The nature of human C1-esterase: the hydrophobic nature of its binding site and pH dependence of the kinetic constants. Immunochemistry 13: 229-233.

18. Gauthier M, Patston PA (1997) Reactivation of C1-inhibitor polymers by denaturation and gel-filtration chromatography. Anal Biochem 248: 228-233.

15

Current Advances and Prospects on Implementation of Highly Sensitive Aptamer-based Dual System for Melamine Detection: New Promising Tool of Great Affinity

Mukama Omar[1,2], Ndikubwimana Jean de Dieu[1], Muhammad Shamoon[3] and Byong H Lee[4*]

[1]Key Laboratory of Carbohydrate Chemistry and Biotechnology, School of Biotechnology, Jiangnan University, Wuxi 214122, P.R. China
[2]Department of Applied Biology, College of Science and Technology, Avenue de l'armée, P.O. Box: 3900 Kigali, Rwanda
[3]The synergistic Innovation Center of Food Safety and Nutrition, State Key Laboratory of Food Science and Technology, Jiangnan University, Wuxi 214122, P.R. China
[4]Department of Food Science/Agricultural Chemistry, McGill University, Saint-Anne-de-Bellevue, Qc H9X3V9, Canada

Abstract

Much attention has been devoted to melamine (MA) analysis in food products in accordance with the food safety standards. Aptamer-based analytical techniques thrived with the improvements in latest tools, analytical reagents and methods and most importantly iterative *in vitro* selection process known as systematic evolution of ligands by exponential enrichment (SELEX). Aptamer-based techniques possess very high affinity and specificity towards contaminants and play a pivotal role in MA detection. However, success depends on the starting aptamer, selection of appropriate nanoparticle, target molecule(s) and characterization of advances in aptamer selection, strategies of preparing, treating the nanoparticles analytical system methods. Current review has focused to elaborate the key recent innovation in aptamer-based dual system construction for MA detection. We have also highlighted the promising types of aptamer-conjugated nanomaterial for the specific recognition of some other potential adulterants and food hazards. Finally, we proposed future directives in developing novel aptamer and further condition optimizations that could give high-throughput food-safety analysis method and melamine "zero tolerance" towards food safety incidents.

Keywords: Aptamer; SELEX; Nanoparticles; Melamine; Detection

Abbreviations: CDC: Centers For Disease Control Prevention; Agnps: Silver Nanoparticles; TSE: Transmissible Spongiform Encephalopathy; Aunps: Gold Nanoparticles; HPLC: High-Performance Liquid Chromatography; CE: Capillary Electrophoresis; AFM: Atomic Force Microscopy; CTA: Chromotropic Acid; PDDA: Polydiallyldimethylammonium Chloride; GC: Gas Chromatography; ELISA: Enzyme-Linked Immunosorbent Assay; Haucl4: Chloroauric Acid; SELEX: Systematic Evolution of Ligands By Exponential Enrichment; LOD: Limit of Detection; STM: Scanning Tunneling Microscope; MA: Melamine; SPM: Scanning Probe Microscopy; MS: Mass Spectrometry; SPR: Surface Plasmon Resonance Assay; Prp: Prion Protein; SERS: Surface Enhanced Raman Spectroscopy; SAA: Salfanilic Acid; ELAAS: Enzyme-Linked Aptamer Assays; UTR: Untranslated Region

Introduction

Recently, melamine (MA) has become one of the major global food safety concerns being notorious for its effects on human and animal health. Owing to its water-repellent, shrink-resistant, stain-repellent and fire-retardant properties, the role of MA has gained a significant consideration in manufacturing of tableware, industrial coatings, paints, adhesives, glues paper, floor tiles, paperboard and plastic packages. It has common structure with triazine family where some of the members serve as herbicides. MA is a non-registered fertilizer in the U.S. but it is still practiced as a fertilizer in some other parts of the world [1]. In 2007, Food and Drug Administration (FDA) strictly prohibited MA as a source of non-protein nitrogen in food or feed even if it cannot solely cause side effects on animals like cats and dogs. However, when combined with its derivative cyanuric acid through gut microbiota mediation (Figure 1) (e.g. *Klebsiella terrigena*), it forms insoluble crystals in pets leading to renal failure [2]. The hallmark of kidney stones is renal colic, hematuria, little or no urine, kidney infection, and/or high blood pressure. The cases of pediatric ailments like kidney stones caused by MA has been reported in nearly 300,000

infants and deaths of six by the Pediatric Environmental Health Specialty Units (PEHSU) [3]. In 2008, the Centers for Disease Control Prevention (CDC) has also found MA in Chinese-manufactured infant formula [4]. To date, MA detection has extended beyond the milk products to meat, poultry, eggs and various vegetables. MA deposit in eggs from poultry fed with MA contaminated feed [5]. MA nitrogen content is above 65% by weight, thus it is use as food adulterant and traditional nitrogen-based tests of Kjeldahl and Dumas methods lack accuracy in discriminating between the proteinous and MA nitrogen [6]. Safety limit of MA set by UN food standard commission in food and beverages is 2.5 and 1 ppm respectively with exception of infant formula [7,8]; and this has led to search for the fast and accurate detection methods. Several analytical techniques are currently available such as fluorescence-based techniques [9], scanning probe microscopy (SPM, AFM, STM), surface enhanced raman spectroscopy (SERS) [10], electrochemical [11], surface plasmon resonance assay (SPR) [12] and impedance spectroscopy for MA [13]. Development of analytical techniques such as liquid chromatography [14], liquid chromatography-tandem mass spectrometry (LC-MS/MS) have been used for MA screening and confirmation due to their good selectivity and detection limits [15-17]. HPLC with fluorescence detection of MA [18], LC-Raman spectroscopy [19], capillary zone electrophoresis (CE) have been also utilized [20]. Despite accuracy of instrumental techniques such as HPLC, GC, and CE; they are time-consuming and

*Corresponding author: Byong H Lee, Distinguished Professor, Department of Food Science/Agricultural Chemistry, McGill University, Saint-Anne-de-Bellevue, Qc H9X3V9, Canada

expensive [21]. Furthermore, in detection of melamine in pet food samples, an inhibition ELISA was developed based on the monoclonal antibody that give efficient standard linear curve ranging from 0.03 to 9 ng/l with a limit of detection (LOD) 0.01 ng/ml and assay sensitivity 0.35 ng/ml [22]. However, the above mentioned techniques require more time of development and cannot be compared to aptamers-based techniques. The latter possess the ability to target molecules for which instrumental techniques and antibodies are not well suited with equal high specificity and affinity. Currently, aptamer-coupled techniques such as immunological (nano-ELAAS) [23], fluorescent chemicals (perlene) [24], nanoparticles (NPs) [25,26], electrochemical [27,28] and colorimetric aptamers [29] are available in form of biosensors. The use of nucleic acids bio-detectors 'Aptamers' showed high ability to recognize specific targets with great affinity, stability, quick, cheap and easy to develop using SELEX [30] or automated *in vitro* selection that reduces the duration of a selection process from several weeks to three days [31]. Aptamer is derived from the Latin word "aptus" meaning "to fit" because of its strong binding to specific targets based on structural conformation. More interestingly, its combination to colorimetric nanomaterials, gives naked-eye visual detection. In this review, we have elucidated recent implementation of nanomaterial-nucleic acid (aptamers) dual system construction in bioassays to detect the infinitesimal amount of MA in foodstuffs added either non-intentionally (e.g. melamine-formaldehyde resin migration from tableware to food or beverages) or intentionally as an adulterant.

Medical Significance of Melamine

Analytical techniques for MA detection are highly needed in biomedical and clinical field due to frequent use of plastic tableware, water dispensers and other similar sources of MA. The HPLC analysis showed that 20.78% of breast milk samples contained MA levels between 10.09 and 76.43 ng/l [32]. Despite the LOD was below than that established by WHO, fast diagnosis for MA is still a significant concern to prevent the acute renal failure development (Figure 1). Continuous consumption of MA in low dose may also lead to urolithiasis and nephrolithiasis [33], hippocampal synaptic plasticity and spacial memory [34]. Currently, Yin et al. revealed the effects of MA on mice spleen lymphocytes by decreasing CD4+/CD8+ in presence or absence of MA-cyanuric acid complex formation, suggesting that MA may act as inducer and suppressor of expression of Bax and Bcl-2, respectively [35]. As a consequence of this, Aptamer and nanoparticles assay in detection of molecules such as MA, DNA and proteins may prove a new tool of detection.

Selection of Aptamer Probes

RNA has ability not only in coding for proteins but also in catalyzing reactions, binding other RNAs, proteins or specifically can bind to a target ligand(s) [30]. Aptamers are obtained from a large random sequence pool containing about 1013-1015 single-strand RNAs composed of a random sequence region flanked by a binding site. Incubation of these oligonucleotides with the single target or a large variety of targets molecule(s) results binding with high affinity and specificity in three-dimensional shape [36]. Few nucleic acids bind to the target are considered as aptamers. This step is followed by washing which help to filter out unbound nucleic acids. Aptamers are separated from the target by elution and PCR amplification of bound nucleic acids is done to create a new library. The selected sequences used in a new round of SELEX for further optimization (Figure 2). The variant SELEX exists to enhance oligonucleotides conformation stability or resistance to nucleases depending on target molecule. Some of them are genomic SELEX or cDNA-SELEX [37], photo SELEX

[38], covalent-SELEX or cross-linking SELEX [39-41], multistage SELEX [42] used for starting pool rationalization or improve aptamer selectivity using negative SELEX protocol or counter SELEX, and deconvolution SELEX or subtractive SELEX. These help to remove undesirable adsorbed oligonucleotides from the poolby matrixes that used for immobilizing the targets and distinction of similar structures [43]. In addition, in order to enhance targets applicability, other variants SELEX used are: Whole bacteria-SELEX [43], blended SELEX [44], TECS-SELEX [45], complex target SELEX [46], toggle SELEX target-switching [47], expressions cassette also known as SELEX-SAGE [48], the mirror-image SELEX or Spiegelmer technology [49], whole cell-SELEX [50]. Efficacy is the prerequisite for any aptamer-based design; several SELEX strategies such as non-SELEX or NECEEM-SELEX, CE-SELEX [51], microfluidic SELEX [52], HTS-SELEX or automated SELEX [53], FluMag-SELEX [54], *in silico* SELEX [55] are applicable. However, currently improvements in SELEX have been made. For example: SELEX process needs a PCR step, the randomized region of the oligonucleotide libraries must be flanked by two fixed primers binding sequences. It has been designed an *in vitro* selection novel type of dual RNA library for mirror-image peptides detection like Ghrelin. It carries fixed sequences which constrains the oligonucleotides into a partially double-stranded structure that allows primer less selection in order to minimize the primer binding sequences to be part of the target-binding motif [56]. Moreover, DNA aptamers with only natural bases can often lack the desired specificity and binding affinity to target proteins compared to aptamers containing recombinant bases. Thus, over 100-fold affinity DNA aptamers have been developed using expanded genetic alphabet of nucleic acid (use of unnatural libraries bases). The DNA aptamer possessing the four natural nucleotides and unnatural nucleotides against two human target proteins, interferon-γ (IFN-γ) and vascular endothelial cell growth factor-165 (VEGF-165) that bind with the dissociation constant of 0.038 nM and 0.65 pM, respectively has also been developed in the same fashion [57].

For MA detection, the most commonly used aptamers are in form of poly-Ts aptamer (ssDNA) with purified sequence 5'TTTT TTTTTTTTTTTTTTTTTTTTTTTTTTTT [58]. Yun et al. used the selected MA aptamer 5'-TTTTTTTTTTTTTTTTTTTTTT-3' to detect melamine in milk sample. This aptamer has potential to protect gold nanoparticles under high salt concentration conditions by averting the immediate colour change of AuNPs upon addition of MA because the latter combine competitively with aptamer [59]. Thus, the well selected aptamer shows efficiency in stabilizing nanomaterials and consequently gives accurate results (Figure 2).

Gold Nanoparticles for MA Detection

The gold, silver and quantum dots nanoparticles have emerged as new tools that ensure the food safety via providing fast and reliable detection approach for targets with simplicity, lower cost compared with immunoassay-based and chromatography based methods. Gold nanoparticles (AuNPs) are synthesized using sodium citrate reduction method [60]. This method yields particles with desired dimensions by changing the gold precursor salt and sodium citrate molar ratio. It works both as a reducing agent for nucleation of AuNPs and as stabilizing agent by coating the nanoparticles' surface and inhibiting their aggregation in solution. When MA is introduced in the system, hydrogen bonds formation between MA amine groups and citrate ions decrease the electrostatic repulsion between individual AuNPs and results in the AuNPs aggregation. The major advantages are based on the facts that no sophisticated apparatus is needed, rapid and naked-eye color change observation, giving a precise quantification of MA when analyzed on UV-Vis spectrometer. For example, well-dispersed AuNPs solution is red, whereas aggregated AuNPs appear in blue or purple colour [61]. However, other methods of AuNPs synthesis are used because sodium citrate method provides small size nanoparticles concentration. AuNPs are synthesized from concentrated HAuCl$_4$ via addition of sodium hydroxide, pH and temperature control [62]. MA was detected using AuNPs with a LOD 0.4 mg/L displayed in 12 min including sample preparation [63]. The developed gold nanoparticle-based kit for MA detection in milk products reached working range of 1-120 mg/l improving time of detection about 10 min including sample pretreatment [64]. However, Cao et al. found that hydrogen-bonding can induce colorimetric detection of MA by label free, and nonaggregation-based AuNPs probe prepared using 3,5-dihydroxybenzoic acid (DBA) reducer interact with MA through strong hydrogen-bonding interaction. Consequently, the melamine can hinder nanoparticles generation resulting in color change from purple to yellow green with gradual increase of melamine concentration. The plasmon absorbance of the formed AuNPs allows the quantitative detection of MA with MA concentration sensitivity ranging from 1×10^{-9} M to 1×10^{-5} M, a linear coefficient of 0.993 and high selectivity to melamine with a low detection limit of 8×10^{-10} M [65]. As Media pH and reaction time are major influencing factors of AuNPs aggregation, Naveen et.al recently improved citrate reduction method [63] by preparing AuNPs with 5ml of 38.8 Mm trisodium citrate, 8.0 NaOH and further optimization in sample extraction and detection of MA. The AuNPs changes from its wine red color to blue or purple with MA detection down to a concentration of 0.05 mg/l determined by monitoring with the naked eyes or a UV-Vis spectrophotometer [66].

Although silver nanoparticles AgNPs are often used for MA detection, gold nanoparticles are more stable with less stringent requirements for storage and handling. In addition, during melamine detection, it has also been seen that AuNPs treated with citrate buffer induce highly an enhanced melamine signals in the Raman spectrum due to the formation of SERS "hot spots" caused by the ability of melamine to form AuNPs aggregates. The concentration range of 0.31-5.0 mg/l in milk with a limit of detection of 0.17 mg/l has been obtained in lower than 30 minutes [67]. Even though it improved time of detection, its LOD is of concern.

Silver Nanoparticles for MA Detection

Silver nanoparticles (AgNPs) can be prepared using borohydride reduction method [68], but some modifications improved the efficiency. Its detection principle consists of the ability of AgNPs to be stabilized and coated in aqueous solution with negatively-charged citrate ions [69] able to interact via Van der Waals' force and with the positively-charged MA to its three exocyclic amine group and a three-nitrogen hybrid ring. This means, when MA is added to a solution of AgNPs, attachment occurs with result of AgNPs aggregation followed by color change. It is thought that either three exocyclic amine group or a three-nitrogen hybrid ring results the label free AgNPs aggregation [70]. Compared with gold nanoparticles, AgNPs have some advantages, such as lower cost of preparation, higher extinction coefficients relative to AuNPs of the same size. Therefore, they are used as calorimetric probes of MA detection and other hazards to assure food safety and protect public health. Despite some advantages of Ag NPs over AuNPs, several papers proposed AuNPs as preferred substrates for melamine detection since the color of the particles changes after the MA-AuNPs interaction in solution. The use of label-free AgNPs detection of MA concentration in raw milk pretreated with chloroform and trichloroacetic acid to remove the protein and fat resulted a naked eye

or a UV-Vis spectrophotometer yellow to red visible color change with detection limit of 2.32 μM~0.29 mg/l, which is below the safety limit of 2.5 ppm recommended in USA and EU. Whereas for infant formula in China, only 1 ppm of MA is tolerated, thus it has an advantage over some methods used for MA detection but not conducive as AuNPs [71]. However, in one way, this method can be interfered by some amino acids and other positively charged molecules that can give false positive results due to some instability of prepared AgNPs. In another way, no MA-cyanuric acid aggregation, that can thus show no color display. Therefore, chromotropic acid (CTA) capping of AgNPs makes them to remain in stable and dispersed form. Once MA is added in nanomaterials solution, it will bind with CTA found on the CTA-AgNPs complex through hydrogen bonding with high selectivity to its structural analogs like cyanulic acid. MA only can act as hydrogen donor to sulfur- groups of the CTA-capped AgNPs and thus induces aggregation accompanied by yellow to orange color change observed by naked eyes [72]. However, it is still require a method for detection of MA in food and beverages with high sensitivity. Thus, the same group of Fangying developed an AgNPs conjugated with sulfanilic acid (SAA), this combination of techniques significantly increased the sensitivity and selectivity of melamine in pretreated milk and only can melamine react with SAA-AgNPs followed by rapid aggregation due to the MA affinity towards the SAA functional groups. The selectivity of this method is below 50 nM and LOD of 10.6 nM which is not only low compared to the multiple existing methods used to detect MA, but also cheap with one sample treatment with analysis costs of 0.5$ [73]. However, further meticulous research with an easy, cost-effective method lies ahead to achieve ultrasensitive and selective methods for screening MA level in food products. In the next section, we have focused on the aptamer-nanomaterial techniques and different treatment strategies as a potential solution to nanoparticles drawbacks mentioned above.

Aptamer-Conjugated Nanoparticle Based Novel System for MA Detection

Aptamer-modified AuNPs show high stability and selectivity compared to unmodified AuNPs in MA and other small molecules sensing [74]. Either ssDNA or RNA aptamers are used to label nanoparticles in the synthesis of dual system aptamer-NP probe for MA detection. The probe is stabilized at pH 6.6 or 7.0 NaH$_2$PO$_4$-

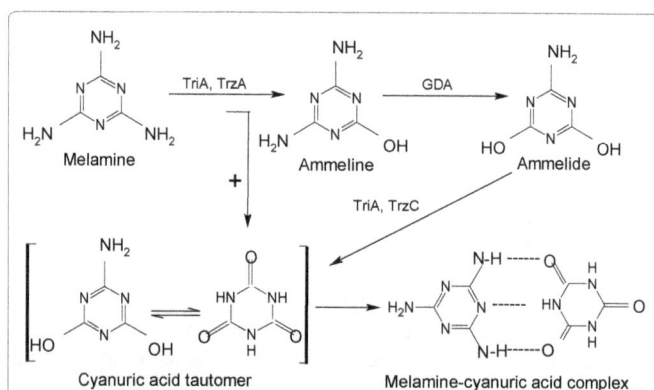

Figure 1: Melamine combined with its derivative product cyanuric acid to form melamine-cyanuric acid insoluble crystal complex that can cause renal failure (kidney stones, hematuria, high blood pressure etc.). This figure also depicts the gut microbiota mediation of enzymatic conversion of melamine to cyanuric acid. TriA stands for melamine deaminase, TrzA for S-trizine hydrolase, GD for Guanine deaminase, AtzC for n-isopropylammelide aminohydrolase and TrzC for Ammelide hydrolase.

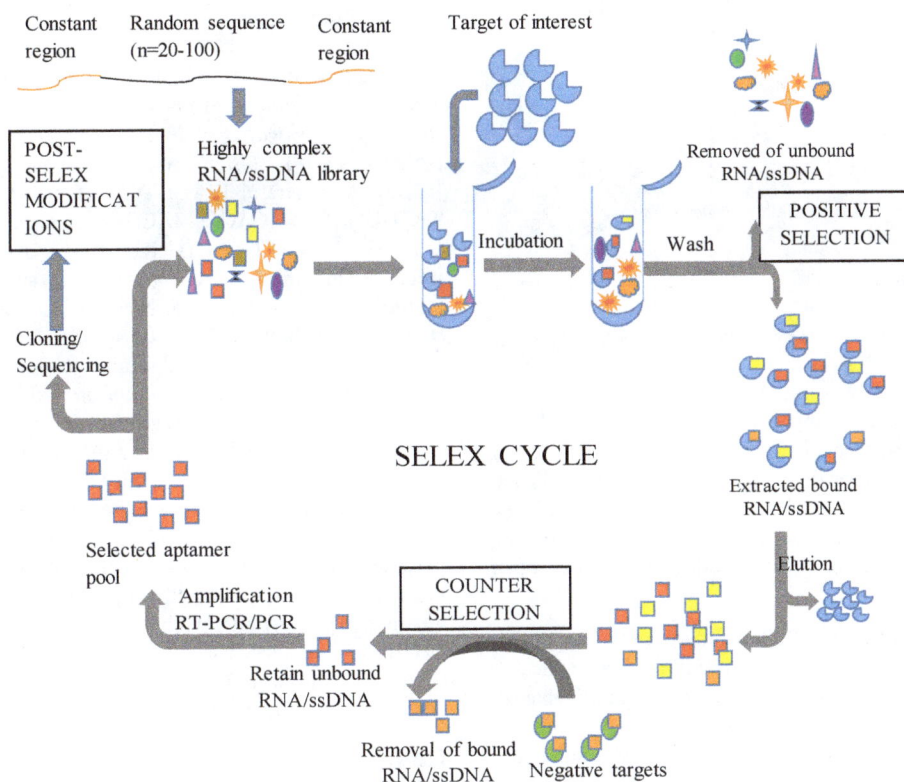

Figure 2: General principle of SELEX.

Na$_2$HPO$_4$ buffer solutions and in high concentrated electrolyte like 25-35 mmol/l of NaCl solution. High salt concentration neutralizes free negative charges of citrate when chloroauric acid (HAuCl$_4$) is used for AuNPs synthesis that leads to the AuNPs aggregation. Upon addition of MA, MA-aptamer-modified AgNP complex is formed that can result an increase of resonance scattering at 470 nm with a linear relation to added MA concentration [75]. The increase of MA decreases the unreacted NGssDNA and the RRS intensity decrease linearly accompanied by color changes from red to blue. While using this method in MA detection, the results ranged between 1.89-81.89 µg/l with LOD 0.98 µg/l determined by the NGssDNA probe [58]. For label free MA aptasensor, the AuNPs are incubated with anti-MA aptamer for 15 min followed by purification using centrifugation at 10,000 rpm for 5 min to obtain the aptamer-protected AuNPs. The salt concentration also influences results (Figure 3). The salt tolerance of aptamer-protected AuNPs should be tested using NaCl solutions at different concentrations. Initially, the higher concentration of salt solution results in the higher sensitivity. Conversely, it decreases the stability of intra-assay and the productivity of sensing system. Thus, the optimal concentration of 15 mM NaCl was adopted to induce the AuNPs aggregation and assure sensitivity. AuNPs should be treated because untreated ones show no color change after binding the target. However, this pretreatment doesn't intervene in detection optimization [76]. Intriguingly, while adding aptamer to AuNPs, there is no change in surface SPR absorption peak and stays 520 nm at UV-VIS spectra. By contrast, upon adding MA into AuNPs-aptamer solution the SPR absorbance 520 nm decreases up to 620 nm followed by color change from red wine to blue visible by naked eye or UV-VIS spectroscopy. Furthermore, it is important to take into account that

the reason of color change is still surmised that the addition of a target results to the aggregation of AuNPs due to competitive binding of MA to aptamer (Figure 3). Compared to bare nanoparticles used in MA detection illustrated above such as AuNPS [63] and AgNPs [70], this method promises the reliability, accuracy and detection time of MA in milk samples because aptamer has ability to stabilize NPs in salts. In addition, the LOD of this method possesses same range of that LOD reported for biosphenol detection in water samples [77]. However, its LOD is lower probably due to the complexity of milk composition which may affect the sensitivity of target induced-AuNPs aggregation (Figure 3).

Based on colorimetric detection, the aptamer modified AuNPs probe was reported as an easy, quick, stable and selective method for MA in milk with LOD 4.2 µg/l. Aptamers can adsorb on AuNPs surface via electrostatic forces and react with poly-diallyldimethylammonium chloride (PDDA) that prevents AuNPs aggregation. Principles of detection consist of forming MA-aptamer complex via hydrogen bonds upon addition of MA in solution and this results the change of color due to the formation of cationic polymer which can aggregate AuNPs. The produced color can be seen with naked eye without quantitative results or measured by UV-Vis spectrum [74]. However, the degree of errors can occur due to foreign substances interference or poor manipulation.

Merits of Aptamer-Based Techniques

Nanoparticles and aptamers have remarkable features of binding with analytes due to their strong localized surface plasmon resonance and high selectivity, respectively. In MA detection, these techniques and

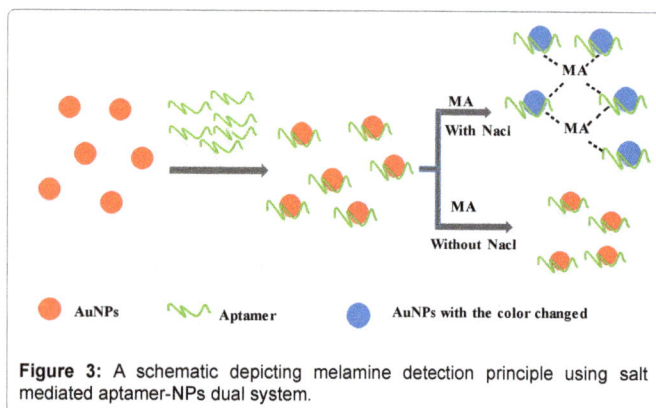

Figure 3: A schematic depicting melamine detection principle using salt mediated aptamer-NPs dual system.

their combination are efficient. Comparing current research on UPLC and HPLC, MA analysis reached a concentration of 2.59 µg/kg using the molecular imprinted solid-phase extraction-UPLC and applied as a specific sorbent for the selective solid phase extraction of MA and its metabolites [10]. However, sample preparation, instrumentation and validation of the experiment require FTIR, NMR and/or other sophisticated techniques.

Compared to newly developed ELISA and recombinant ELISA, their detection limit ranges from 0.5-7.0 µg/ml [78] which is higher than those of highest aptamer-based techniques (Table 1). However, this technique can be interfered by sample matrix and also related chemical compounds, thus possibility of false results may exist. Moreover, production of monoclonal antibody for target(s) generally has several inherent limitations *in vivo* such as: lack of MA immunogenicity due to non- recognition by the host (only proteins and few carbohydrates have the ability to induce immune response). However, Lin et al . have developed a disclose method of producing immunogens by polymerization of non-immunogens. These methods are specific for haptens with amine and carboxylic groups including MA. Crosslinking reagents have been used to polymerize those functional groups into macromolecules to obtain haptenic polymers which will be used to immunize animal for antibodies production. The non-immunogenic MA has been polymerized using glutaraldehyde to reach a preferred complete antigen size at least of 4 kDa [78,79]. However, *in vivo* produced antibodies still have experimental limitations in their applications, mainly non-specific binding that can lead to cross-reactivity and high background. Therefore, aptamers-recombinant antibody can be made using library methods instead of animal immunization technologies. Additional advantages of aptamers over antibodies include easy storage, transport, suitability in variable conditions, and stability in changeable kinetic parameters, reporter molecules and high discrimination of targets. Therefore, aptamer-NPs system is a fast and more versatile approach (Table 1).

Application of Aptamer in Food Safety

As a new platform technology, aptamer-based techniques bioassay has recently been adopted in food analysis, clinical, medical and environmental monitoring [80-85]. In food safety monitoring, aptamers are applied for detection of chemical and/or biological hazards. Illegal additives used in food adulteration such as 17-β estradiol and bisphenol A were detected using DNA aptamers5'-GGGCCGTTC-GAACACGAGCATGCCGGTGGGTGGTCAGGTGGGATA GCGT-3' and 5'-TCCGCGAATTACACGCAGAGGTAGCGGCTCTGCG-CATTCAATTGCTG-3' combined with 5'-CGCGCTGAAGCGCG-GAAGC-3', respectively [77,86,87]. This method is based on competi-

tive recognition of bisphenol with immobilized aptamer fixedon the surface of the electrode. MA detection in milk using ssDNA aptamer with sequence of 5'-TTTTTTTTTTTTTTTTTTTTTTTTTTTTTTTTTTTTTTT-3' [58] is highly effective. Aptamer detection and identification for various harmful microorganisms in clinical practice such as *Pseudomonas sp.* [88], *Campylobacter* sp. [89], *Escherichia coli* O157: H7 [90], *Staphylococcus* sp. [91], *Listeria* sp., and *Salmonella* sp. are also available [92,93]. Aptamers 5'-TTTGGTCCTTGTCTT ATGTCCAGAATGC-TATGGCGGCGTCACCCGACGGG-3' and 5'-GACTTGAATTATA-CA   GATTTCTCCTACTGGGATAGGTGGATTAT-3'   sequences were used in salmonella serovars detection and showed high detection affinity. The aptamer S8-7 has been used as a ligand for magnetic capture of serially diluted *Salmonella typhimurium* species by detecting the whole cell content in a 290 µl sample volume containing 10^2-10^3 CFU equivalents of *Salmonella typhimurium* which resulted in relatively high binding affinity to the target with an apparent dissociation of 1.73 ± 0.54 µM [94]. Traditionally, *Salmonella* genera detection in food requires standard culture methods which include sequential steps of pre-enrichment, selective enrichment and selective differential plating. All these steps are tedious and time consuming. For decades, toxins became problematic to food safety and aptamer based techniques have been developed for the detection of toxin like Ochratoxin A with DNA aptamer [95]. Currently, Ochratoxin A has also been detected using electrochemical aptasensor [96], aflatoxins [97-100], neurotoxins [101] such as botulinum [102,103], antibiotics [104-107] as well as pesticides such as nitenpyram, chlorpyrifos and imidacloprid [108].

Conclusion and Future Directives

Aptamer, nanoparticle or aptamer-modified NPs showed efficiency in detection of MA in food or beverages with a sensitivity of 1.89-81.89 µg/l and 0.98 µg/l in short time [58]. However, AuNPs by SERS showed the concentration range of 0.31-5.0 mg/l in milk with a limit of detection of 0.17 mg/l within 30 min by improving time of detection but its LOD is of anxiety. These achievements are due to three major aspects: a) development of new aptamer-based analytical methods, b) modification in nanoparticles treatment and c) combination of techniques. To date many workers use aptamers for protecting and stabilizing nanoparticles. So far, the improved SELEX protocol is the promising tool for the detection of various target analytes, but it is still under continuous evolution. Theoretically, SELEX protocol is used in target analytes screening, but the market is still dominated by antibody-based method. This is due to two major reasons: 1) development of sensitive and specific aptamer for specific categories of target analytes is still limited and 2) limitations in further SELEX technology in order to produce efficient and low cost aptamer. In addition, it should be kept in mind that many of the above mentioned aptamer-based techniques are conceptual in that much works need to be done before practical application, even though it showed more stability in salts compared to techniques that use bare nanoparticles. Although aptamer-conjugated nanomaterials are emerging as promising platform for MA detection, much works remain in development of other aptamer-based methods because natural and modified natural nucleotides used during aptamers production often lack specificity and desired binding affinity. Thus, we propose aptamers development using expanded genetic code whereby natural and unnatural bases can be mixed and optimized for MA detection. Moreover, other tasks are to be done in investigating aptamer-NPs array-based, discovering combinatorial techniques beyond two as mentioned in detection of *Salmonella typhimurium* using nano-ELAAS (gold nanoparticles, ELISA and aptamer) [23] and developing relevant kits could curtail the time, errors and elucidate

Technique	Type of Food	Sensitivity	LOD	Reference
HPLC	Milk	0.05-10.0 µg/ml	8.1 µg/l	[109]
	Edible plants	-	2.51 µg/kg	[17]
GC-MS	Milk	-	0.01 mg/kg	[110]
	Eggs		10 µg/kg	[102]
LC-MS	Eggs	-	8 µg /kg	[15,111]
UPLC-MS/MS	Eggs	-	5-10 µg/kg	[102]
ELISA	Milk, Milk products and animal feeds		5.210-5.460 µg /l	[79,112]
ELISA KITS	Milk, milk products and animal feeds	-	1.908,2.125,4.15 µg /l	[112]
Bare AuNPs	Milk	0.31-5.0 mg/l	0.17 mg/l	[66,113]
Bare AuNPs Kit	Milk products	1-120 mg/l	-	[64]
Bare AgNPs	Milk	2.32-0.29 µg/l	-	[70,73]
Bare Aptamer	Milk	0.2 to 24 µg/l	10.08 µg/l	[114]
AuNPs-Aptamer	Milk	1.89-81.89 µg/l	0.98 µg/l	[58,59]
AgNPs-Aptamer	Milk	6.3403.6 µg/l	1.2 µg/l	[115]
SAA-AgNPs	Milk	0.1-3.1 µM	10.6 nM	[73]
CTA-AgNPs	Milk	126 µg/l	4.5 µg/l	[72]

Table 1: Recent reports on MA detection in food samples.

some obstacles like interference of other compounds towards high throughput detection of MA and other targets in food stuff.

References

1. Osborne CA, Lulich JP, Ulrich LK, Koehler LA, Albasan H, et al. (2009) Melamine and cyanuric acid-induced crystalluria, uroliths, and nephrotoxicity in dogs and cats. Vet Clin North Am Small Anim Pract 39: 1-14.

2. Thompson M, Owen L, Wilkinson K, Wood R, Damant A (2004) Testing for bias between the Kjeldahl and Dumas methods for the determination of nitrogen in meat mixtures, by using data from a designed interlaboratory experiment. Meat Sci 68: 631-634.

3. Yasui T, Kobayashi T, Okada A, Hamamoto S, Hirose M, et al. (2014) Long-term follow-up of nephrotoxicity in rats administered both melamine and cyanuric acid. BMC Res Notes 7: 87.

4. Jooste PJ, Anelich L, Motarjemi Y (2014) Safety of Food and Beverages: Milk and Dairy Products. In: Motarjemi Y, Moy G, Todd E (eds.) Encyclopedia of Food Safety. Academic Press, Waltham, USA, pp. 285-296.

5. Chen Y, Yang W, Wang Z, Peng Y, Li B, et al. (2010) Deposition of melamine in eggs from laying hens exposed to melamine contaminated feed. J Agric Food Chem 58: 3512-3516.

6. Moore JC, DeVries JW, Lipp M, Griffiths JC, Abernethy DR (2010) Total protein methods and their potential utility to reduce the risk of food protein adulteration. Compreh Revi Food Sci Food Safety 9: 330-357.

7. Tsai IL, Sun SW, Liao HW, Lin SC, Kuo CH (2009) Rapid analysis of melamine in infant formula by sweeping-micellar electrokinetic chromatography. J Chromatogr A 1216: 8296-8303.

8. Dai H, Shi Y, Wang Y, Sun Y, Hu J, et al. (2014) Label-free turn-on fluorescent detection of melamine based on the anti-quenching ability of Hg^{2+} to gold nanoclusters. Biosens Bioelectron 53: 76-81.

9. Tang G, Du L, Su X (2013) Detection of melamine based on the fluorescence resonance energy transfer between CdTe QDs and Rhodamine B. Food Chem 141: 4060-4065.

10. Ge X, Wu X, Wang J, Liang S, Sun H (2015) Highly sensitive determination of cyromazine, melamine, and their metabolites in milk by molecularly imprinted solid-phase extraction combined with ultra-performance liquid chromatography. J Dairy Sci 98: 2161-2171.

11. Xue J, Lee PT, Compton RG (2014) Electrochemical detection of melamine. Electroanalysis 26: 1454-1460.

12. Liu S-S, Yi X-Y, Wang J-X (2014) Sensitive detection of Melamine at surface plasmon resonance chips pre-immobilized with bovine serum albumin-melamine conjugate. Chinese J Anal Chem 42: 695-700.

13. Wu B, Wang Z, Zhao D, Lu X (2012) A novel molecularly imprinted impedimetric sensor for melamine determination. Talanta 101: 374-381.

14. Tan J, Li R, Jiang Z-T (2011) Determination of melamine in liquid milk and milk powder by titania-based ligand-exchange hydrophilic interaction liquid chromatography. Food Anal Methods 5: 1062-1069.

15. Rodriguez Mondal AM, Desmarchelier A, Konings E, Acheson-Shalom R, Delatour T (2010) Liquid chromatography-tandem mass spectrometry (LC-MS/MS) method extension to quantify simultaneously melamine and cyanuric acid in egg powder and soy protein in addition to milk products. J Agric Food Chem 58: 11574-11579.

16. Beltran-Martinavarro B, Peris-Vicente J, Carda-Broch S, Esteve-Romero J (2014) Development and validation of a micellar liquid chromatography-based method to quantify melamine in swine kidney. Food Control 46: 168-173.

17. Ge X, Wu X, Liang S, Sun H (2014) A sensitive and validated HPLC method for the determination of cyromazine and melamine in herbal and edible plants using accelerated solvent extraction and cleanup with SPE. J Chromatogr Sci 52: 751-757.

18. Zhang Y, Lin S, Jiang P, Zhu X, Ling J, et al. (2014) Determination of melamine and cyromazine in milk by high performance liquid chromatography coupled with online solid-phase extraction using a novel cation-exchange restricted access material synthesized by surface initiated atom transfer radical polymerization. J Chromatogr A 1337: 17-21.

19. Liu F, Zou MQ, Zhang M, Zhang XF, Li M (2014) Study of fast pretreatment method in detection of melamine in liquid milk using liquid chromatography and Raman spectroscopy. Guang Pu Xue Yu Guang Pu Fen Xi 34: 685-688.

20. Kong Y, Wei C, Hou Z, Wang Z, Yuan J, et al. (2014) Stacking and analysis of melamine in milk products with acetonitrile-salt stacking technique in capillary electrophoresis. J Anal Methods Chem 2014: 212697.

21. Sun F, Ma W, Xu L, Zhu Y, Liu L, et al. (2010) Analytical methods and recent developments in the detection of melamine. TrAC Trends Anal Chem 29: 1239-1249.

22. Zhou Y, Li CY, Li YS, Ren HL, Lu SY, et al. (2012) Monoclonal antibody based inhibition ELISA as a new tool for the analysis of melamine in milk and pet food samples. Food Chem 135: 2681-2686.

23. Wu W, Li J, Pan D, Li J, Song S, et al. (2014) Gold nanoparticle-based enzyme-linked antibody-aptamer sandwich assay for detection of *Salmonella typhimurium*. ACS Appl Mater Interfaces 6: 16974-16981.

24. Lv Z, Liu J, Bai W, Yang S, Chen A (2015) A simple and sensitive label-free fluorescent approach for protein detection based on a Perylene probe and aptamer. Biosens Bioelectron 64: 530-534.

25. Xin JY, Zhang LX, Chen DD, Lin K, Fan HC, et al. (2015) Colorimetric detection of melamine based on methanobactin-mediated synthesis of gold nanoparticles. Food Chem 174: 473-479.

26. Wu Q, Long Q, Li H, Zhang Y, Yao S (2015) An upconversion fluorescence resonance energy transfer nanosensor for one step detection of melamine in raw milk. Talanta 136: 47-53.

27. Chen L, Chen ZN (2015) A multifunctional label-free electrochemical impedance biosensor for Hg(2+), adenosine triphosphate and thrombin. Talanta 132: 664-668.

28. Roushani M, Shahdost-fard F (2015) A novel ultrasensitive aptasensor based on silver nanoparticles measured via enhanced voltammetric response of electrochemical reduction of riboflavin as redox probe for cocaine detection. Sensors Actuators B-Chemical 207: 764-771.

29. Smith JE, Griffin DK, Leny JK, Hagen JA, Chávez JL, et al. (2014) Colorimetric detection with aptamer-gold nanoparticle conjugates coupled to an android-based color analysis application for use in the field. Talanta 121: 247-255.

30. Ellington AD, Szostak JW (1990) In vitro selection of RNA molecules that bind specific ligands. Nature 346: 818-822.

31. Bouvet P (2009) Identification of nucleic acid high-affinity binding sequences of proteins by SELEX. Methods Mol Biol 543: 139-150.

32. Yurdakok B, Filazi A, Ekici H, Celik TH, Sireli UT (2014) Melamine in breast milk. Toxicology Res 3: 242-246.

33. Li X, Lu J, Shang P, Bao J, Yue Z (2015) The selective NADPH oxidase inhibitor apocynin has potential prophylactic effects on melamine-related nephrolithiasis in vitro and in vivo. Mol Cell Biochem 399: 167-178.

34. Yang Y, Xiong GJ, Yu DF, Cao J, Wang LP, et al. (2012) Acute low-dose melamine affects hippocampal synaptic plasticity and behavior in rats. Toxicol Lett 214: 63-68.

35. Yin RH, Liu J, Li HS, Bai WL, Yin RL, et al. (2014) The toxic effects of melamine on spleen lymphocytes with or without cyanuric acid in mice. Res Vet Sci 97: 505-513.

36. Yang Y, Yang D, Schluesener HJ, Zhang Z (2007) Advances in SELEX and application of aptamers in the central nervous system. Biomol Eng 24: 583-592.

37. Shimada T, Fujita N, Maeda M, Ishihama A (2005) Systematic search for the Cra-binding promoters using genomic SELEX system. Genes Cells 10: 907-918.

38. Jensen AE, Hjeltnes N, Berstad J, Stanghelle JK (1995) Residual urine following intermittent catheterisation in patients with spinal cord injuries. Paraplegia 33: 693-696.

39. Kopylov AM, Spiridonova VA (2000) Combinatorial chemistry of nucleic acids: SELEX. Mol Biol 34: 940-954.

40. Spiridonova LN (2014) Introgression of nuclear and mitochondrial DNA markers of Mus musculus musculus to aboriginal populations of wild mice from central Asia (M. m. wagneri) and south Siberia (M. m. gansuensis). Mol Biol (Mosk) 48: 89-98.

41. Spiridonova VA, Levashov PA, Ovchinnikova ED, Afanasieva OI, Glinkina KA, et al. (2014) DNA Aptamer-Based Sorbents for Binding Human IgE. Russian J Bioorganic Chem 40: 151-154.

42. Wu L, Curran JF (1999) An allosteric synthetic DNA. Nucleic Acids Res 27: 1512-1516.

43. Torres-Chavolla E, Alocilja EC (2009) Aptasensors for detection of microbial and viral pathogens. Biosens Bioelectron 24: 3175-3182.

44. Kulbachinskiy AV (2007) Methods for selection of aptamers to protein targets. Biochemistry (Mosc) 72: 1505-1518.

45. Ohuchi SP, Ohtsu T, Nakamura Y (2006) Selection of RNA aptamers against recombinant transforming growth factor-beta type III receptor displayed on cell surface. Biochimie 88: 897-904.

46. Chen CK (2007) Complex SELEX against target mixture: stochastic computer model, simulation, and analysis. Comput Methods Programs Biomed 87: 189-200.

47. Hamula C, Guthrie J, Zhang H, Li X, Le X (2006) Selection and analytical applications of aptamers. TrAC Trends Anal Chem 25: 681-691.

48. Martell RE, Nevins JR, Sullenger BA (2002) Optimizing aptamer activity for gene therapy applications using expression cassette SELEX. Mol Ther 6: 30-34.

49. Klussmann S, Nolte A, Bald R, Erdmann VA, Fürste JP (1996) Mirror-image RNA that binds D-adenosine. Nat Biotechnol 14: 1112-1115.

50. Park HC, Baig IA, Lee SC, Moon JY, Yoon MY (2014) Development of ssDNA aptamers for the sensitive detection of Salmonella typhimurium and Salmonella enteritidis. Appl Biochem Biotechnol 174: 793-802.

51. Kasahara Y, Irisawa Y, Ozaki H, Obika S, Kuwahara M (2013) 2',4'-BNA/LNA aptamers: CE-SELEX using a DNA-based library of full-length 2'-O,4'-C-methylene-bridged/linked bicyclic ribonucleotides. Bioorg Med Chem Lett 23: 1288-1292.

52. Lin H, Zhang W, Jia S, Guan Z, Yang CJ, et al. (2014) Microfluidic approaches to rapid and efficient aptamer selection. Biomicrofluidics 8: 041501.

53. Hünniger T, Wessels H, Fischer C, Paschke-Kratzin A, Fischer M (2014) Just in time-selection: A rapid semiautomated SELEX of DNA aptamers using magnetic separation and BEAMing. Anal Chem 86: 10940-10947.

54. Stoltenburg R, Reinemann C, Strehlitz B (2005) FluMag-SELEX as an advantageous method for DNA aptamer selection. Anal Bioanal Chem 383: 83-91.

55. Savory N, Abe K, Yoshida W, Ikebukuro K (2014) In silico maturation: Processing sequences to improve biopolymer functions based on genetic algorithms. In: Valadi J, Siarry P (eds.) Applications of Metaheuristics in Process Engineering: Springer International Publishing, USA, pp. 271-288.

56. Jarosch F, Buchner K, Klussmann S (2006) In vitro selection using a dual RNA library that allows primerless selection. Nucleic Acids Res 34: e86.

57. Kimoto M, Yamashige R, Matsunaga K, Yokoyama S, Hirao I (2013) Generation of high-affinity DNA aptamers using an expanded genetic alphabet. Nat Biotechnol 31: 453-457.

58. Liang A, Zhou L, Qin H, Zhang Y, Ouyang H, et al. (2011) A highly sensitive aptamer-nanogold catalytic resonance scattering spectral assay for melamine. J Fluoresc 21: 1907-1912.

59. Yun W, Li H, Chen S, Tu D, Xie W, et al. (2014) Aptamer-based rapid visual biosensing of melamine in whole milk. Eur Food Res Technol 238: 989-995.

60. McFarland AD, Haynes CL, Mirkin CA, Van Duyne RP, Godwin HA (2004) Citrate synthesis of gold nanoparticles. J Chem Educ 81: 544A.

61. Ai K, Liu Y, Lu L (2009) Hydrogen-bonding recognition-induced color change of gold nanoparticles for visual detection of melamine in raw milk and infant formula. J Am Chem Soc 131: 9496-9497.

62. Li C, Li D, Wan G, Xu J, Hou W (2011) Facile synthesis of concentrated gold nanoparticles with low size-distribution in water: temperature and pH controls. Nanoscale Res Lett 6: 440.

63. Li L, Li B, Cheng D, Mao L (2010) Visual detection of melamine in raw milk using gold nanoparticles as colorimetric probe. Food Chem 122: 895-900.

64. Zhou Q, Liu N, Qie Z, Wang Y, Ning B, et al. (2011) Development of gold nanoparticle-based rapid detection kit for melamine in milk products. J Agric Food Chem 59: 12006-12011.

65. Cao Q, Zhao H, He Y, Li X, Zeng L, et al. (2010) Hydrogen-bonding-induced colorimetric detection of melamine by nonaggregation-based Au-NPs as a probe. Biosens Bioelectron 25: 2680-2685.

66. Kumar N, Seth R, Kumar H (2014) Colorimetric detection of melamine in milk by citrate-stabilized gold nanoparticles. Anal Biochem 456: 43-49.

67. Giovannozzi AM, Rolle F, Sega M, Abete MC, Marchis D, et al. (2014) Rapid and sensitive detection of melamine in milk with gold nanoparticles by Surface Enhanced Raman Scattering. Food Chem 159: 250-256.

68. Wagner J, Tshikhudo TR, Kohler JM (2008) Microfluidic generation of metal nanoparticles by borohydride reduction. Chem Eng J 135: S104-S109.

69. Pinto VV, Ferreira MJ, Silva R, Santos HA, Silva F, et al. (2010) Long time effect on the stability of silver nanoparticles in aqueous medium: Effect of the synthesis and storage conditions. Colloids and Surfaces A: Physicochem Eng Aspects 364: 19-25.

70. Ping H, Zhang M, Li H, Li S, Chen Q, et al. (2012) Visual detection of melamine in raw milk by label-free silver nanoparticles. Food Control 23: 191-197.

71. Buka I, Osornio-Vargas A, Karr C (2009) Melamine food contamination: Relevance to Canadian children. Paediatr Child Health 14: 222-224.

72. Song J, Wu F, Wan Y, Ma LH (2014) Visual test for melamine using silver nanoparticles modified with chromotropic acid. Microchimica Acta 181: 1267-1274.

73. Song J, Wu F, Wan Y, Ma L (2014) Colorimetric detection of melamine in pretreated milk using silver nanoparticles functionalized with sulfanilic acid. Food Control 50: 356-361.

74. Xing HB, Zhan SS, Wu YG, He L, Zhou P (2013) Sensitive colorimetric detection of melamine in milk with an aptamer-modified nanogold probe. RSC Adv 3: 17424-17430.

75. Liang A, Zhou L, Jiang Z (2011) A simple and sensitive resonance scattering spectral assay for detection of melamine using aptamer-modified nanosilver probe. Plasmonics 6: 387-392.

76. Mei Z, Chu H, Chen W, Xue F, Liu J, et al. (2013) Ultrasensitive one-step rapid visual detection of bisphenol A in water samples by label-free aptasensor. Biosens Bioelectron 39: 26-30.

77. Zhou L, Wang J, Li D, Li Y (2014) An electrochemical aptasensor based on gold nanoparticles dotted graphene modified glassy carbon electrode for label-free detection of bisphenol A in milk samples. Food Chem 162: 34-40.

78. Li W, Meng M, Lu X, Liu W, Yin W, et al. (2014) Preparation of anti-melamine antibody and development of an indirect chemiluminescent competitive ELISA for melamine detection in milk. Food Agric Immunol 25: 498-509.

79. Cao B, Yang H, Song J, Chang H, Li S, et al. (2013) Sensitivity and specificity enhanced enzyme-linked immunosorbent assay by rational hapten modification and heterogeneous antibody/coating antigen combinations for the detection of melamine in milk, milk powder and feed samples. Talanta 116: 173-180.

80. Tombelli S, Minunni M, Mascini M (2007) Aptamers-based assays for diagnostics, environmental and food analysis. Biomol Eng 24: 191-200.

81. Xu S, Yuan H, Chen S, Xu A, Wang J, et al. (2012) Selection of DNA aptamers against polychlorinated biphenyls as potential biorecognition elements for environmental analysis. Anal Biochem 423: 195-201.

82. Centi S, Silva E, Laschi S, Palchetti I, Mascini M (2007) Polychlorinated biphenyls (PCBs) detection in milk samples by an electrochemical magneto-immunosensor (EMI) coupled to solid-phase extraction (SPE) and disposable low-density arrays. Anal Chim Acta 594: 9-16.

83. Citartan M, Gopinath SC, Tominaga J, Tan SC, Tang TH (2012) Assays for aptamer-based platforms. Biosens Bioelectron 34: 1-11.

84. Kim SE, Su W, Cho M, Lee Y, Choe WS (2012) Harnessing aptamers for electrochemical detection of endotoxin. Anal Biochem 424: 12-20.

85. Xu S, Yuan H, Chen S, Xu A, Wang J, et al. (2012) Selection of DNA aptamers against polychlorinated biphenyls as potential biorecognition elements for environmental analysis. Anal Biochem 423: 195-201.

86. Marks HL, Pishko MV, Jackson GW, Coté GL (2014) Rational design of a bisphenol A aptamer selective surface-enhanced Raman scattering nanoprobe. Anal Chem 86: 11614-11619.

87. Yildirim N, Long F, He M, Shi HC, Gu AZ (2014) A portable optic fiber aptasensor for sensitive, specific and rapid detection of bisphenol-A in water samples. Environ Sci Process Impacts 16: 1379-1386.

88. Kim LH, Yu HW, Kim YH, Kim IS, Jang A (2013) Potential of fluorophore labeled aptamers for Pseudomonas aeruginosa detection in drinking water. J Korean Soc Appl Biol Chem 56: 165-171.

89. Suh SH, Dwivedi HP, Jaykus LA (2014) Development and evaluation of aptamer magnetic capture assay in conjunction with real-time PCR for detection of Campylobacter jejuni. Lwt-Food Sci Technol 56: 256-260.

90. Wu W, Zhao S, Mao Y, Fang Z, Lu X, et al. (2015) A sensitive lateral flow biosensor for Escherichia coli O157:H7 detection based on aptamer mediated strand displacement amplification. Anal Chim Acta 861: 62-68.

91. Chang YC, Yang CY, Sun RL, Cheng YF, Kao WC, et al. (2013) Rapid single cell detection of Staphylococcus aureus by aptamer-conjugated gold nanoparticles. Sci Rep 3: 1863.

92. Liu G, Lian Y, Gao C, Yu X, Zhu M, et al. (2014) In vitro selection of DNA aptamers and fluorescence-based recognition for rapid detection Listeria monocytogenes. J Integrative Agric 13: 1121-1129.

93. Suh SH, Dwivedi HP, Choi SJ, Jaykus LA (2014) Selection and characterization of DNA aptamers specific for Listeria species. Anal Biochem 459: 39-45.

94. Dwivedi HP, Smiley RD, Jaykus LA (2013) Selection of DNA aptamers for capture and detection of Salmonella typhimurium using a whole-cell SELEX approach in conjunction with cell sorting. Appl Microbiol Biotechnol 97: 3677-3686.

95. Cruz-Aguado JA, Penner G (2008) Determination of ochratoxin a with a DNA aptamer. J Agric Food Chem 56: 10456-10461.

96. Wu J, Chu H, Mei Z, Deng Y, Xue F, et al. (2012) Ultrasensitive one-step rapid detection of ochratoxin A by the folding-based electrochemical aptasensor. Anal Chim Acta 753: 27-31.

97. Ali WH, Pichon V (2014) Characterization of oligosorbents and application to the purification of ochratoxin A from wheat extracts. Anal Bioanal Chem 406: 1233-1240.

98. Luan Y, Chen Z, Xie G, Chen J, Lu A, et al. (2015) Rapid visual detection of aflatoxin B1 by label-free aptasensor using unmodified gold nanoparticles. J Nanosci Nanotechnol 15: 1357-1361.

99. Ma X, Wang W, Chen X, Xia Y, Duan N, et al. (2015) Selection, characterization and application of aptamers targeted to Aflatoxin B2. Food Control 47: 545-551.

100. Shim WB, Mun H, Joung HA, Ofori JA, Chung DH, et al. (2014) Chemiluminescence competitive aptamer assay for the detection of aflatoxin B1 in corn samples. Food Control 36: 30-35.

101. McConnell EM, Holahan MR, DeRosa MC (2014) Aptamers as promising molecular recognition elements for diagnostics and therapeutics in the central nervous system. Nucleic Acid Ther 24: 388-404.

102. Xia X, Ding S, Li X, Gong X, Zhang S, et al. (2009) Validation of a confirmatory method for the determination of melamine in egg by gas chromatography-mass spectrometry and ultra-performance liquid chromatography-tandem mass spectrometry. Anal Chim Acta 651: 196-200.

103. Chang TW, Blank M, Janardhanan P, Singh BR, Mello C, et al. (2010) In vitro selection of RNA aptamers that inhibit the activity of type A botulinum neurotoxin. Biochem Biophys Res Commun 396: 854-860.

104. Wang K, Tao ZH, Xu L, Liu YQ (2014) Research and development of functionalized aptamer based biosensor. Chinese J Anal Chem 42: 298-304.

105. Ma L, Wen GQ, Liu QY, Liang AH, Jiang ZL (2014) Resonance Rayleigh scattering determination of trace tobramycin using aptamer-modified nanogold as probe. Guang Pu Xue Yu Guang Pu Fen Xi 34: 2481-2484.

106. Lin Z, Ma Q, Fei X, Zhang H, Su X (2014) A novel aptamer functionalized CuInS2 quantum dots probe for daunorubicin sensing and near infrared imaging of prostate cancer cells. Anal Chim Acta 818: 54-60.

107. Yan L, Luo C, Cheng W, Mao W, Zhang D, et al. (2012) A simple and sensitive electrochemical aptasensor for determination of Chloramphenicol in honey based on target-induced strand release. J Electroanal Chem 687: 89-94.

108. He J, Liu Y, Fan M, Liu X (2011) Isolation and identification of the DNA aptamer target to acetamiprid. J Agric Food Chem 59: 1582-1586.

109. Finete Vde L, Gouvêa MM, Marques FF, Netto AD (2014) Characterization of newfound natural luminescent properties of melamine, and development and validation of a method of high performance liquid chromatography with fluorescence detection for its determination in kitchen plastic ware. Talanta 123: 128-134.

110. Pan XD, Wu P, Yang DJ, Wang LY, Shen XH, et al. (2013) Simultaneous determination of melamine and cyanuric acid in dairy products by mixed-mode solid phase extraction and GC-MS. Food Control 30: 545-548.

111. Wang PC, Lee RJ, Chen CY, Chou CC, Lee MR (2012) Determination of cyromazine and melamine in chicken eggs using quick, easy, cheap, effective, rugged and safe (QuEChERS) extraction coupled with liquid chromatography-tandem mass spectrometry. Anal Chim Acta 752: 78-86.

112. Gong Y, Zhang M, Wang M, Chen Z, Xi X (2014) Development of immuno-based methods for detection of melamine. Arabian J Sci Eng 39: 5315-5324.

113. Giovannozzi AM, Rolle F, Sega M, Abete MC, Marchis D, et al. (2014) Rapid and sensitive detection of melamine in milk with gold nanoparticles by Surface Enhanced Raman Scattering. Food Chem 159: 250-256.

114. Wang G, Zhu Y, He X, Chen L, Wang L, et al. (2013) Colorimetric and visual determination of melamine by exploiting the conformational change of hemin G-quadruplex-DNAzyme. Microchim Acta 181: 411-418.

115. Wen G, Zhou L, Li T, Liang A, Jiang Z (2012) A sensitive surface-enhanced Raman scattering method for determination of melamine with aptamer-modified nanosilver probe. Chinese J Chem 30: 869-874.

16

Solid State Cultivation of *Hericium erinaceus* Biomass and Erinacine: A Production

Blaz Gerbec[1], Eva Tavčar[2], Andrej Gregori[3], Samo Kreft[2] and Marin Berovic[1*]

[1]*Department of Chemical, Biochemical and Environmental Engineering, Faculty of Chemistry and Chemical Technology, University of Ljubljana, Askerceva 5, SI-1115 Ljubljana, Slovenia*
[2]*Department of Pharmaceutical Biotechnology, Faculty of Pharmacy, University of Ljubljana, Aškerčeva 7, SI-1115 Ljubljana, Slovenia*
[3]*Institute for Natural Science (Zavod za naravoslovje), Ulica Bratov Ucakar 108, SI-1000 Ljubljana; MycoMedica d.o.o. Podkoren 72, SI-4280 Kranjska Gora, Slovenia*

Abstract

Hericium erinaceus is a medicinal mushroom producing biologically active metabolite erinacin A. In this research we compared fungal growth and metabolite production in different reactor types and different substrates with added casein peptone and NaCl. Cultivation of fungal biomass was performed in glass jars and horizontal stirred tank reactor. The cultivation in reactor was carried out at 24°C and airflow 5 L/min. Periodical mixing of $N = 80$ rpm, was used. Solid state substrate based on mixture of husked millet and paddy millet was used. For simultaneous determination of erinacine A and ergosterol contents in a single sample new analytical method was developed. The amount of biomass obtained in glass jars averaged at 100 mg/g ± 12 mg/g, while in horizontal mixing bioreactor it easily raised up to 350 mg/g. The highest erinacin A concentration was detected in substrate with 0.56% NaCl and 3.4% casein peptone content.

Keywords: Solid state cultivation; Horizontal stirred tank reactor; *Hericium erinaceus*; LC-MS; Erinacine A; Ergosterol

Introduction

Medicinal properties of *Hericium erinaceus* (Bull.: Fr) Pers. (also known as lion's mane, monkey's head, hedgehog fungus, pom pom blanc and yamabushitake) have been well-known for hundreds of years in traditional Chinese and Japanese cuisine as well as herbal medicine to treat various human diseases [1]. Fungi of the genus *Hericium* contain various ingredients with antibacterial activity, cytotoxic effect on cancer cells and compounds that stimulate the synthesis of the Nerve Growth Factor (NGF). Although quite a few species of *Hericium* are known, the fruiting bodies cultivation on the waste sawdust was developed for *H. abietis* and *H. erinaceus* only [2].

In *H. erinaceus* various pharmaceutically active substances were found such as phytosterols (β-sitosterol and ergosterol), which lower the content of Low-Density Lipoproteins (LDL) and triglycerides that operate anticarcinogenic and as well reduce the metabolism of fats [3]. The fruit body is composed of numerous constituents such as polysaccharides, proteins, lectins, phenols, hericenones, erinacines and terpenoids. They strengthen the immune system, relieve gastritis and gastrointestinal infections, reflux and upset stomach due to stress [1,4].

H. erinaceus water-soluble polysaccharides increase activity of macrophages and other immune cells in the fight against cancer cells, but also demonstrate reduction of metastases formation. The most outstanding activity of *H. erinaceus* extract is that it strengthens the immune system and activates NGF synthesis [5]. Due to the increased proliferation of T and B-lymphocytes it strengthens the immune system and strengthens the body's natural defenses, thus expressing very positive effects on cancer patient's life quality [3].

Among the compounds isolated from *H. erinaceum* fruiting bodies and cultured mycelia, most interesting are the low-molecular-weight compounds belonging to a group of cyathin diterpenoids (erinacines A–K, P, and Q). Several of them, i.e. erinacines A–H, are known to have a potent stimulating effect on Nerve Growth Factor (NGF) synthesis *in vitro* [6-12].

NGF is an essential protein supporting growth and maintenance of peripheral sympathetic neurons. Due to degeneration of neurons NGF synthesis stimulators can be applied as potential drugs against central nervous system disorders, such as Alzheimer's disease [1].

Erinacine A, isolated from the cultured mycelia of *H. erinaceum*, the main representative of this compounds group, has a strong enhancing effect on NGF synthesis, much stronger than epinephrine [6]. Furthermore, this compound increases catecholamine and NGF content in the central nervous system of rats [13-15].

Erinacines and hericenons reduce anxiety as well as depression. For an accurate understanding of the overall mechanism of *H. erinaceus* diterpenoids action it requires additional clinical studies with physiological markers, such as hormones, or more profounded studies of autonomic nervous activity [16].

The structure of erinacine A was determined for the first time through the observation of chemical reactions, NMR and IR spectroscopy, and mass spectrometry data [6]. Krzyczkowski et al. [18] developed an isocratic HPLC method for quantitative determination of erinacine A [17]. Erinacine A was described in the form of white crystals [6] and yellow amorphous powder with the melting point of 72-74°C [18,19].

Besides reports on *H. erinaceus* fruiting bodies cultivation, reports on biomass cultivation in bioreactors, particularly in solid state reactors, are rare. In present research, cultivation of *H. erinaceus* was optimized in glass jars and transferred to Horizontal Stirred Tank Reactor (HSTR). Fast gradient HPLC analytical method for simultaneous determination of erinacine A and ergosterol was determined.

*Corresponding author: Marin Berovic, Department of Chemical, Biochemical and Environmental Engineering, Faculty of Chemistry and Chemical Technology, University of Ljubljana, Askerceva 5, SI-1115 Ljubljana, Slovenia

Materials and Methods

Microorganism

Hericium erinaceus strain He Erin (Institute for Natural Sciences, Slovenia) was used in all of the experiments. Strain was maintained on Potato Dextrose Agar (Biolife, Italy) at 24°C. It was maintained active by regular (every 14 day) transfers on fresh Potato Dextrose Agar plates.

Cultivation in glass jars

Substrate was based on mixture of husked and paddy millet in ratio=7: 2. The impact of NaCl and casein peptone, as basic additives of the medium was examined. Additives of 0, 0.1, 0.2 and 0.5 g of NaCl and 0, 1, 3 and 7 g of casein peptone to 88.8 g of dry media were used. The addition of NaCl to media corresponds to 0, 0.11, 0.23 and 0.56% and casein peptone addition corresponds 1.1, 3.4 and 7.9% of dry media weight respectively. The total amount of the substrate was in all cases 180 g, with 50% water content. Substrate were filled into jars and sterilized for 1 hour at T=121°C and pressure of P=1.2 × 10^5 Pa.

Cultivation proceeded for 5, 6 and 7 weeks at 24°C. All of the experiments were managed in triplicates. In total, 144 jars were used (4 different NaCl contents × 4 different casein peptone content × 3 cultivation times × 3 replicates). After cultivation the glass content was dried for 100 h at 60°C to achieve constant moisture content.

Cultivation in horizontal stirred tank reactor

Experiments were carried out in a 15 L Horizontal Stirred Tank Reactor (HSTR) of our own construction and design [20]. The cultivation was carried out at 24°C and 5 L/min airflow. Periodical mixing of N = 80 rpm, for 2 minutes (in the first 7 days every second day, and every day in the last part of cultivation) was used. A moistening air device with sterile distilled water was connected to the bioreactor to prevent the drying of the substrate (Figure 1).

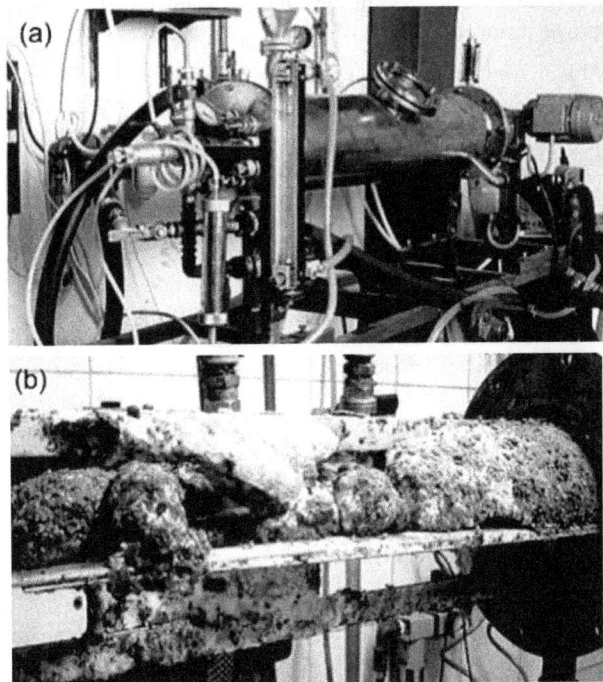

Figure 1: (a) Horizontal Stirred Tank Reactor, (b) The growth of fungal biomass in horizontal stirred tank reactor after 8 week of cultivation.

The cultivation substrate consisted of beech sawdust, paddy millet and hulled millet. The ratio paddy millet: husked millet in the substrate was 1:1 and substrate water content 50%. In solid state reactor in contrary with glass jars aeration of the solid matrix was applied. Sawdust (10%) was added to the substrate to increase the porosity and oxygen permeability. In accordance with the results of substrate optimization in the glass jar experiments, the addition of NaCl was 0.56% and casein peptone 3.4% (dry weight/dry weight).

Inoculum for the cultivation was prepared from a mixture of beech sawdust and paddy millet in the ratio 4: 1 incubated and overgrown with *H. erinaceus* mycelia. It represented 7% of the total substrate weight. Cultivation of fungal biomass in HSTR proceeded for 8 weeks.

Analytical methods

Biomass: For determination of biomass on solid particles, ergosterol assay was applied in all the experiments [21]. Ergosterol in fungi is included as an integral part of the cell membrane and therefore a relevant indicator of the fungal biomass content in solid matrix [22]. For determination of ergosterol, HPLC method was developed and described in Chromatographic analysis at the Results and discussion section.

Electron microscopy: For monitoring of microbial growth, Field Emission Scanning Microscopy (FE-SEM, Supra 35, Carl Zeiss, Germany) equipped with energy dispersive spectroscopy Oxford INCA 400, UK was used.

Extraction of ergosterol and erinacin A

2 g of dry overgrown substrate sample was milled and extracted with 20 ml of solvent. 30% methanol and ethanol were tested as solvents for the extraction of erinacine A and ergosterol. Ethanol proved as a better solvent and was further tested in concentrations of 20, 50, 70, 85 and 96%. 70% concentration was confirmed to be optimal. The extracts were sampled after 1, 4 and 24 h of the extraction and then re-extracted with fresh solvent for 2 and 120 h. The extraction conditions were further optimized in a test, comparing three procedures: 1) the maceration on a shaker (150 rpm, room temperature), 2) maceration on a shaker (150 rpm, room temperature) with additional 10 min sonification, and 3) maceration in a water bath (50°C) with three additional 10 min shakings (10 min, 150 rpm on a shaker). The 24 h extraction in a water bath (50°C) was selected as an optimal procedure. Prior to HPLC analysis, all samples were centrifuged (3 min, 4192.5 g) and filtered through a 0.22 μm filters (Supelco, PTFE).

Chromatographic analysis

The HPLC system (Shimadzu Prominence, Kyoto, Japan) consisted of a system controller (CBM-20A), column oven (CPO-20AC) and a solvent delivery pump with a degasser (DGU-20A5) connected to a refrigerated auto sampler (SIL-20AC) with a Photo Diode Array detector (SPD-M20A) that monitored the wavelengths 190-800 nm. The responses of the detectors were recorded using LC Solution software version 1.24 SP1. The chromatography was performed at 40°C and a flow rate of 1.5 mL/min using a Phenomenex Kinetex® C18 column (10 cm × 4.6 mm I.D., 2.7 μm particle size). The following gradient method using water (solvent A) and acetonitrile (solvent B), both containing 0.1% formic acid, was utilized: 0.01-3.00 (30-50% B), 3.00-9.00 (50-60% B), 9.01-18.00 (60-100% B), 18.01-23.00 (30% B). Since the standard od erinacine A was not available to make a calibration curve, all quantitative results on erinacin A content are expressed as the area under the curve (AUC) of erinacin A chromatogram peak.

Mass spectrometric analysis

The chromatographic separation was performed on a Waters Acquity ultra-performance liquid chromatograph® (Waters Corp., Milford, MA, USA) with a column identical to that used for the quantitative HPLC analysis. The flow rate was adjusted according to the capacity of the mass detector to 0.5 mL min⁻¹. The following gradient method was utilized with MilliQ water containing 0.1% formic acid (solvent A) and acetonitrile (solvent B): 0.01-9.00 (30-50% B), 9.00-27.00 (50-60% B), 27.01-54.00 (60-100% B), 54.01-69.00 (30% B). The injection volume was 10 μL, and the column temperature was maintained at 40°C. The LC system was interfaced with a hybrid quadrupole orthogonal acceleration time-of-flight mass spectrometer (Q-ToF Premier, Waters, Milford, MA, USA). The compounds were analyzed under positive (ESI (+)) and negative (ESI(-)) ion conditions. The capillary voltage was set at 3.0 kV, while the sampling cone voltage was 20 V. The source and desolvation temperatures were 120 and 200°C, respectively. The nitrogen desolvation gas flow rate was 500 L h⁻¹. The acquisition range was between m/z 50 and 1000 with argon serving as the collision gas at a pressure of $4.5 \cdot 10^{-3}$ mbar in the T-wave collision cell. The data were collected in centroid mode, with a scan accumulation time of 0.2 s and an interscan delay of 0.025 s. The data station utilized the Mass Lynx v4.1 operating software. Accurate mass measurements were obtained with an electrospray dual sprayer using the reference compound leucine enkephalin ([M+H]⁺=556.2271) at a high mass resolution of 10,000.

Statistics

All of the cultivation experiments were performed at least in three runs and the average values were calculated from the sum of the data. Data were expressed as mean ± S.D. The results were analyzed for statistical significance by one-way analysis of variance (ANOVA) test using the Statistical Package of the Social Science (SPSS) version 11.0 (SPSS Inc., Chicago, IL, USA). Statistical significance of the test effects was evaluated at $p < 0.05$.

Results and Discussion

Growth optimization in glass jars

First step were substrate optimization experiments. In glass jar cultivation, the influence of NaCl and casein peptone addition was studied. Mycelial growth in the glass jars was very intensive. After extended adaptive phase of 2 weeks, first mycelium appears, at 5 weeks mycelia overgrew half of the substrate height and after 8 weeks the whole substrate volume was overgrown.

The best production of erinacine A 225, 4 × 10³ AUC was indicated in 8 weeks of cultivation in the substrate containing a combination of 0.56% NaCl and 3.4% casein peptone (Figure 2). In the control substrate without additions, the biomass growth as well as the biosynthesis of erinacine A completely stopped after 2 weeks of cultivation. The amount of obtained biomass in glass jars was in average about 100 mg/g ± 12 mg/g.

Cultivation in horizontal stirred tank reactor

In first HSTR experiments the same substrate mixture composition optimized already in glass jar experiments were used. In HSTR a problem of water condensate on the bottom of reactor tube occurred. Therefore 10% of beech sawdust was added to paddy millet and hulled millet substrate to absorb the water condensate. Sawdust was added to the substrate also to increase solid bed porosity and oxygen

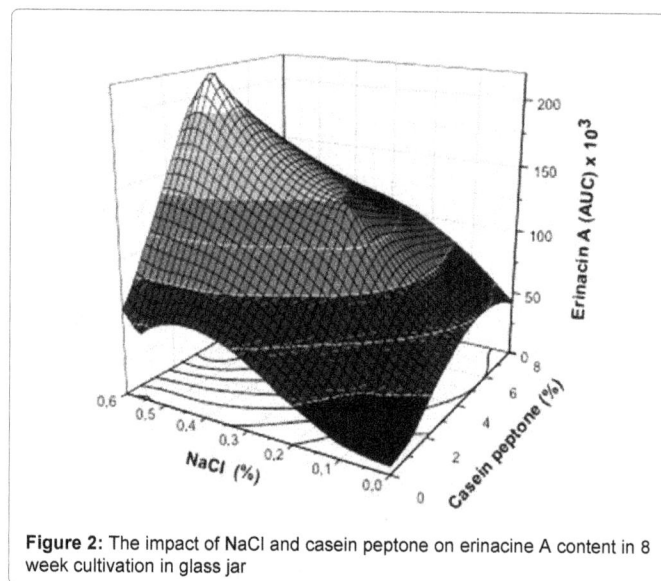

Figure 2: The impact of NaCl and casein peptone on erinacine A content in 8 week cultivation in glass jar

permeability. Permanent mixing could eliminate this problem, but it would disable the filamentous growth and biomass cluster formation. In contrary with glass jars cultivation the growth of fungal mycelia in HSTR proceeded at different condition. In solid state reactor aeration of solid matrix was applied during whole course of cultivation. Cultivation in HSTR proceeded for 8 weeks. At the end of cultivation substrate was completely overgrown in thick woody form (Figure 1b).

The SEM electron microscopy images had shown the growth situation and the porosity of overgrown solid matrix in HSTR. Although mixing was periodically applied 2 minutes per day in the first 2 weeks of cultivation, SEM images (Figures 3a and 3b) shown that it damaged the filaments and inhibited the growth. At 5 weeks the substrate was already fixed to the impeller, so further disintegration did not occur. After 8 weeks of cultivation solid matrix was completely overgrown by mycelia (Figures 3c and 3d).

Although out the experiments in glass jar cultivation the best results of erinacine A production were obtained at the combination of 0.56% NaCl and 3.4% casein peptone in HSTR, the highest concentration of ergosterol was obtained at 8 weeks of cultivation with addition of 0.56% NaCl and 2% casein peptone. Compared with glass jar cultivation in HSTR cultivation higher yield of ergosterol content was obtained while no erinacine A was detected (Figure 4).

Chromatographic analysis

For simultaneous determination of erinacine A and ergosterol contents in a single sample, new method was developed. Erinacine A was eluted at 6.7 min and ergosterol at 15 min on the HPLC chromatogram (Figure 5a). The chromatograms were observed at 340 nm for erinacine A and a 281 nm for ergosterol. Erinacine A can also be detected at 281 nm when only one wavelength detection is possible, but this yields a lower sensitivity.

Identification of the compounds

The absorption spectrum of erinacine A (Figure 5b) exhibited maxima at 345 and 295 nm and a minimum at 253 nm. A corresponding absorption maximum at 340 nm was previously reported [17]. Ergosterol was used for fungal biomass measurement. It was identified with the use of a standard compound and a characteristic absorption

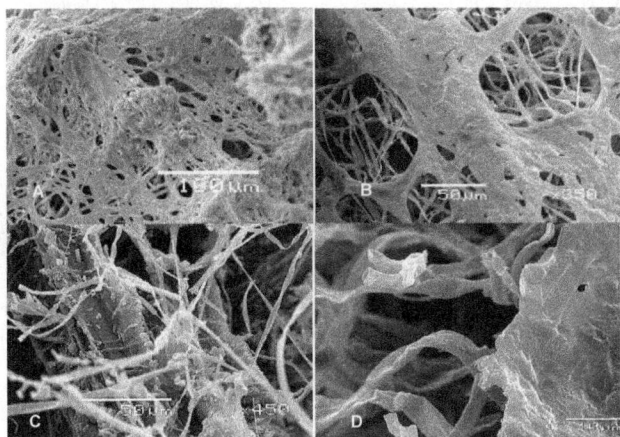

Figure 3: *Hericium erinaceus* growth mycelia in horizontal stirred tank reactor. SEM electron microscope images of mycelial growth. A and B at 6 weeks, C at 7 weeks and D at 8 weeks - thick mycelia cover at the end of cultivation.

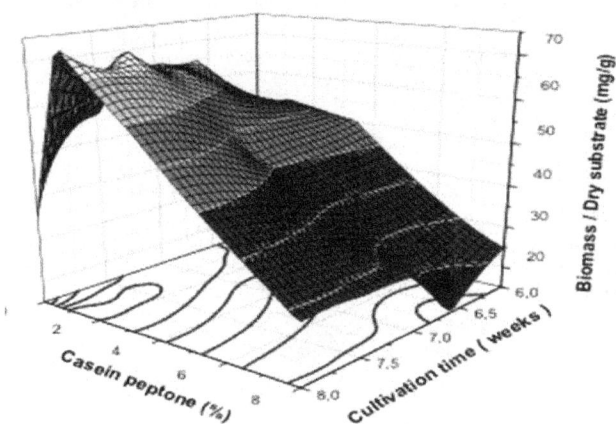

Figure 4: Production of biomass in HSTR on substrate with various addition of casein peptone and 0.56% NaCl

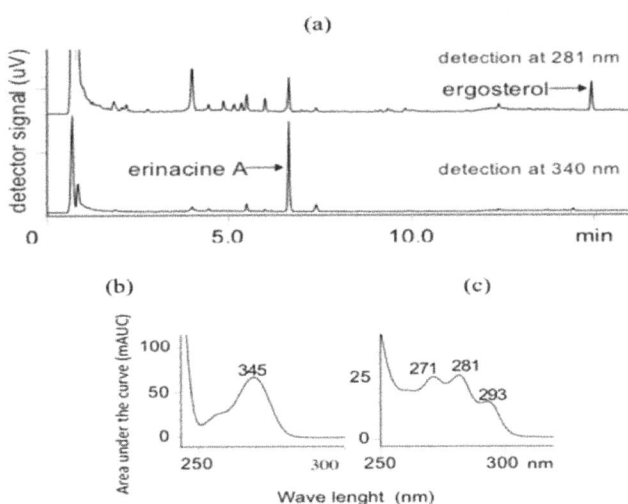

Figure 5: (a) HPLC chromatograms of a typical *Hericium erinaceum* extract including detection of ergosterol and erinacine A, (b) absorption of erinacine A and (c) ergosterol spectra with three characteristic maxima

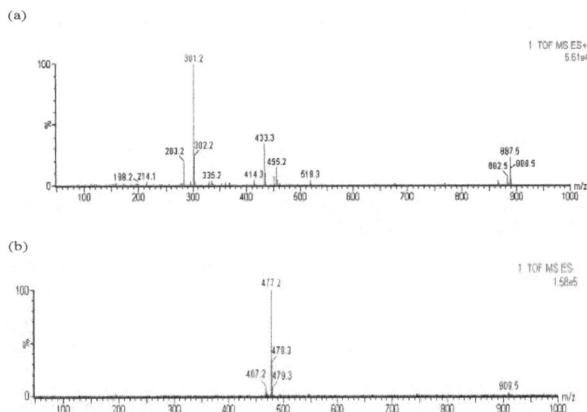

Figure 6: (a) LC-MS spectrum of erinacine A in TOF ESI+ mode (top), (b) ESI- mode (bottom), respectively.

spectrum with three maxima at wavelengths 271, 281 and 293 nm and minima at 276 and 290 nm (Figure 5c).

The LC/MS measurements provided mass spectra with the basic information of molecular masses ($[M+H]^+$ or $[M-H]^-$ for the chromatographically separated compounds. The mass measurements were simultaneously obtained using a Q-TOF high resolution mass analyzer. The proposed elemental compositions of the ions were obtained from the accurate mass measurement data. A molecular ion $[M+H]^+$ at m/z 433 with predicted elemental composition of $C_{25}H_{36}O_6$ could be seen on the positive spectrum. Signals with m/z 455 and 887 represented molecular ion adducts, $[M+Na]^+$ and $[2M+H]^+$. An acetic acid adduct $[M+HCOOH]^-$ formation with m/z 477 was observed on the negative spectrum. Our findings correspond and complement the literature data [6,17]. Using LC-MS, Erinacine A was detected at 340 nm at 6.7th min (bottom) and ergosterol was detected at 281 nm at 15th min (top) (Figures 6a and 6b).

Conclusions

The extraction procedure and HPLC analytical method for simultaneous determination of erinacine A and ergosterol from *Hericium* erinaceum have been developed, enabling simultaneous quantification of active metabolite (erinacine A) content and determination of biomass content in the medium. These analytical methods are a very useful tool in determination of optimal growth conditions for *H. erinaceus* erinacine A and fungal biomass production on solid-state substrates.

Although the experiments in glass jars enable large number of experiments in the small-scale, they can be used as a compass for *H. erinaceus* fruiting bodies cultivation. Great advantage of growing in solid-state bioreactor is aeration, temperature and mixing control, representing a more suitable method for large-scale fungal biomass production. In present case, the amount of biomass obtained in glass jars was 100 mg/g ± 12 mg/g on average, while in HSTR on sawdust containing substrates it easily raised up to 350 mg/g.

In the glass jars up to 225, 4×10^3 AUC of erinacine A was produced, while in a HSTR on sawdust containing substrates there was no erinacine A production detected. The main reason for this could be the lack of precursors needed for erinacine A biosynthesis. Substrate used in HSTR also did not include enough starch to support the fungal metabolism in sense of erinacine A synthesis. In contrary with glass jars in HSTR cultivation carbon dioxide was ventilated out of solid matrix and therefore eliminating its stimulatory effect and affecting erinacine A biosynthesis as well.

References

1. Mizuno T (1999) Bioactive Substances in *Hericium erinaceus* (Bull.:Fr.) Pers. (Yamabushitake), and its Medicinal Utilization. Int J Med Mushrooms 1: 105-119.

2. Stamets P (2006) Can mushrooms help save the world? Interview by Bonnie J. Horrigan. Explore (NY) 2: 152-161.

3. Mizuno T (1995) Bioactive biomolecules of mushrooms: food function and medicinal effect of mushroom fungi. Food Reviews International 11: 7-21.

4. Wong KH, Sabaratnam V, Abdullah N, Kuppusamy UR, Naidu M (2009) Effects of Cultivation Techniques and Processing on Antimicrobial and Antioxidant Activities of *H. erinaceus*. Food Technol Biotechnol 47: 47-55.

5. Kolotushkina EV, Moldavan MG, Voronin KY, Skibo GG (2003) The influence of *Hericium erinaceus* extract on myelination process in vitro. Fiziol Zh 49: 38-45.

6. Kawagishi H, Shimada A, Shirai R, Okamoto K, Ojima F, et al. (1994) Erinacines A, B and C, strong stimulators of nerve growth factor (NGF)-synthesis, from the mycelia of *Hericium erinaceum*. Tetrahedron Letters 35: 1569-1572.

7. Kawagishi H, Shimada A, Shizuki K, Mori H (1996) Erinacine D, a stimulator of NGF-synthesis, from the mycelia of *Hericium erinaceum*. Heterocyclic Communications 2: 51-54.

8. Kawagishi H, Shimada A, Hosokawa S, Mori H, et al. (1996) Erinacines E, F, and G, stimulators of nerve growth factor (NGF)-synthesis, from the mycelia of *Hericium erinaceum*. Tetrahedron Letters 37: 7399-7402.

9. Kawagishi H, Masui A, Tokuyama S, Nakamura T (2006) Erinacines J and K from the mycelia of *Hericium erinaceum*. Tetrahedron 62: 8463-8466.

10. Lee EW, Shizuki K, Hosokawa S, Suzuki M, Suganuma H, et al. (2000) Two novel diterpenoids, erinacines H and I from the mycelia of *Hericium erinaceum*. Biosci Biotechnol Biochem 64: 2402-2405.

11. Kenmoku H, Sassa T, Kato N (2000) Isolation of erinacine P, a new parental metabolite of cyathane-xylosides, from *Hericium erinaceum* and its biomimetic conversion into erinacines A and B. Tetrahedron Letters 41: 4389-4393.

12. Xing XW, Hawthorne WJ, Yi S, Simond DM, Dong Q, et al. (2009) Investigation of porcine endogenous retrovirus in the conservation population of Ningxiang pig. Transplant Proc 41: 4389-4393.

13. Kenmoku H, Shimai T, Toyomasu T, Kato N, Sassa T (2002) Erinacine Q, a new erinacine from *Hericium erinaceum*, and its biosynthetic route to erinacine C in the basidiomycete. Biosci Biotechnol Biochem 66: 571-575.

14. Riaz SS, Tomlinson DR (1996) Neurotrophic factors in peripheral neuropathies: pharmacological strategies. Prog Neurobiol 49: 125-143.

15. Yamada K, Nitta A, Hasegawa T, Fuji K, Hiramatsu M, et al. (1997) Orally active NGF synthesis stimulators: potential therapeutic agents in Alzheimer's disease. Behav Brain Res 83: 117-122.

16. Shimbo M, Kawagishi H, Yokogoshi H (2005) Erinacine A increases catecholamine and nerve growth factor content in the central nervous system of rats. Nutrition Research 25: 617-623.

17. Nagano M, Shimizu K, Kondo R, Hayashi C, Sato D, et al. (2010) Reduction of depression and anxiety by 4 weeks *Hericium erinaceus* intake. Biomed Res 31: 231-237.

18. Krzyczkowski W, Malinowska E, Herold F (2010) Erinacine A biosynthesis in submerged cultivation of *Hericium erinaceum*: Quantification and improved cultivation. Engineering in Life Sciences 10: 446-457.

19. Krzyczkowski W, Malinowska E, Suchocki P, Kleps J, Olejnik M, et al. (2009) Isolation and quantitative determination of ergosterol peroxide in various edible mushroom species. Food Chemistry 113: 351-355.

20. Maruyama H, Yamazaki K, Murofushi S, Konda C, Ikekawa T (1989) Antitumor activity of *Sarcodon aspratus* (Berk.) S. Ito and *Ganoderma lucidum* (Fr.) Karst. J Pharmacobiodyn 12: 118-123.

21. Berovic M, Habijanic J, Boh B, Wraber B, Petravic-Tominac V (2012) Production of Lingzhi or Reishi medicinal mushroom, *Ganoderma lucidum* (W.Curt. :Fr.) P. Karst. (higher Basidiomycetes), biomass and polysaccharides by solid state cultivation. Int J Med Mushrooms 14: 513-520.

22. Klamer M, Baath E (2004) Estimation of conversion factors for fungal biomass determination in compost using ergosterol and PLFA 18:2ω6,9. Soil Biology & Biochemistry 36: 57-65.

23. Nyuland JE, Wallander H (1992) 5 Ergosterol analysis as a means of quantifying mycorrhizal biomass. Methods in Microbiology 24: 77-88.

17

Food and Agriculture Residue (FAR): A Potential Substrate for Tannase and Gallic Acid Production using Competent Microbes

Swaran Nandini[1], Nandini KE[2] and Krishna Sundari S[2]*

[1]*King Lab Burnett School of Biomedical Sciences, College of Medicine, UCF, USA*
[2]*Plant & Microbial Technology (PMT) Group, Biotechnology Department, Jaypee Institute of Information Technology (JIIT), Noida, U.P., India*

Abstract

The study establishes potential of carefully designed formulations of Food and Agricultural Residues (FAR) as most viable and natural substrates for the production of commercially important enzyme tannase and a byproduct gallic acid through Solid State Fermentation (SSF). Novelty of this study was formulation of FAR to achieve better production of tannase and gallic acid by optimizing choice of FAR and its combinations for SSF. Twenty bacterial cultures were successfully isolated and among them sixteen were found with tannase producing ability. Of the entire group of bacteria isolated, two bacterial isolates (B 2.2 and B 2.7) emerged as the best performing candidates in terms of both enzyme and gallic acid production. One Fungal isolate (F1) has also been included in this study from our laboratory collection. A total of 6 FARs (PP, STP, TSP, CH, CC and BP) that are rich in natural tannins were tested in six different combinations. PP with STP in the ratio of 1:1 was found to be most preferred FAR combination by all three isolates (B 2.2, B 2.7, & F 1) for the production of tannase and gallic acid. Maximum tannase (19.02 U/g) and gallic acid (5.32 mg/g) production were achieved by F1, closely followed by B 2.2 and B 2.7. Amongst bacterial isolates, B 2.2 was leading in production of tannase (13.21 U/g) and gallic acid (3.51 mg/g) whereas B 2.7 proved second best registering 9.15 U/g of tannase and 3.36 mg/g of gallic acid. The combination PP with BP was observed to be the second best preferred FAR formulation for the production. Further variations in the formulation of FAR and relative ratios of individual FAR were tested and arrived at a conclusion that PP with STP mixed in a ratio of 1:1 as the most suitable FAR combination for optimal yield of enzyme and its byproduct.

Keywords: FAR formulation; FAR; Tannase; Gallic acid; SSF

Introduction

Enzymes are involved in all aspects of biochemical conversions and are the focal points of biotechnology research. Enzymes catalysed reactions will be 100 million to 10 billion times faster than any normal chemical reaction which makes them highly efficient. In several industries enzymes are used as cost effective eco-friendly substitutes in place of synthetic chemicals for processing complex substrates. Today about 200 enzymes are used in industries for various commercial applications. The global market for industrial enzymes is estimated at 3.3 billion dollars in 2010. This market is expected to reach more than 4 billion dollars by 2015 [1,2]. Tannase (tannin acyl hydrolase) is one such commercially important enzyme. Tannase catalyses the hydrolysis of ester bond and depside bond present in hydrolysable tannins to form glucose and gallic acid [3]. Techniques for production of tannase have been extensively studied and commercial production of tannase was achieved using synthetic tannic acid as substrate. Industrial bioconversion of tannic acid is accomplished by enzyme tannase producing gallic acid (3, 4, 5-tri hydroxyl benzoic acid), a pharmaceutically important intermediate used for the preparation of trimethoprim (antibacterial drug), pyrogallol, propyl gallate. Certain industrially important products derived by the action of tannases are high grade leather, clarified beer and fruit juice, coffee flavoured soft drinks, premium branded tea, paper, cosmetics etc. other applications include detannification of industrial effluents [4-7]. Tannases are produced by fungal as well as bacterial isolates. Depending on the isolate/strain studied and the culture conditions provided, the enzyme was expressed at various quantities and exhibited different levels of activity, pH and temperature stability and half-life. Amongst a wide variety of microbes tested, fungi from genus *Aspergillus* [8,9] and bacteria belonging to the genus *Bacillus* [10,11] are mostly used by commercial industries. The industrial process makes use of chemical tannic acid for tannase production but this process involving synthetic substrates has adverse environmental consequences. Conventionally gallic acid is also produced chemically by acid hydrolysis of synthetic tannic acid and suffers from disadvantages like high cost to yield ratio and low purity. Alternatively, gallic acid can be produced by the microbial hydrolysis of tannic acid (synthetic or natural) by using the enzyme tannase [3,6,12]. Hence any advancement in making the production of tannase more economical and environmental friendly would have far reaching benefits. This has led to generating interest in identifying natural sources for tannic acid that can be effectively utilized by microbes and produce the enzyme tannase and gallic acid during the fermentation process. It takes a lot of research efforts to produce tannase with properties in sync for optimized industrial output and is produced through microbial fermentation using natural sources of tannic acid thus reducing the dependency on synthetic tannic acid.

The present manuscript is an effort to assess the employment of Fruit and Agricultural Residues (FAR) as a natural and inexpensive substrate for the production of tannase and gallic acid and strengthen the methods of microbial hydrolysis. Hence the study was focused on isolating native microbes from soil and fruit wastes having the ability to produce tannase and gallic acid using natural sources of tannins/tannic acid. FAR explored in the study were Pomegranate Peel (PP),

***Corresponding author:** Krishna Sundari S, Plant & Microbial Technology (PMT) Group, Biotechnology Department, Jaypee Institute of Information Technology (JIIT), A-10, Sector-62, NOIDA, 201307, U.P., India*

Banana Peel (BP), Spent Tea Powder (STP), Tamarind Seed Powder (TSP), Corn Husk (CH) and Coconut Coir (CC) which are reported in literature to have high tannin content. Till date none of the reported literature formulated SSF combinations using these above mentioned substrates together as an alternative source for synthetic tannic acid. So the uniqueness of this study lies in formulating various combinations of FAR for SSF (Solid State Fermentation) on the basis of available tannins and carbohydrate content and to incorporate both tested and unreported natural tannic acid sources originating from FAR. The objective was to support microbial growth on FAR while achieving enhanced production of tannase and gallic acid. The study establishes dual benefit/applicability by: a. producing economically important products utilizing a resource (FAR), whose accumulation is a burden on environment and b. offering a viable and economical alternative to synthetic tannic acid for the production of tannase and gallic acid economically.

Materials and Methods

Sample collection

Multiple soil samples were collected from selected areas of fruit and vegetable market's dump site near Paschim Vihar, New Delhi, India. Surface soil debris was removed and a 15 cm hollow metal borer was plunged into the soil to lift the soil lump from 5-10 cm depth. Randomly collected soil samples were pooled thoroughly mixed and air dried for five days. For all subsequent studies including physico-chemical analysis and microbial isolation representative soil sample was drawn from the air dried sample stock.

Physico-chemical analysis of soil samples

At the time of sampling, the color and texture of the soil were observed and recorded. pH of the soil sample was analyzed by mixing the soil with de-ionized water in the ratio of 1:2 (soil:distilled water) as suggested by Clesceri et al. [13]. All analysis were carried out in triplicates.

Isolation of tannase producing bacterial strains

1 g of air dried soil sample was seeded in 10 ml of nutrient broth for isolating the bacterial cultures and incubated at 37°C. 50 μl of soil soup was spread over two nutrient agar plates (Plate 1 & Plate 2) and kept for overnight incubation at 37°C. Post incubation, ten physically distinct colonies were selected from each plate (1 & 2) and analyzed for their tannase producing ability. These twenty colonies were inoculated in different test tubes containing 10 ml nutrient broth supplemented with 0.2% filter sterilized tannic acid and incubated overnight at 37°C (Figure 1). Second round of isolation was carried out using 50 μl of inoculums from enrichment culture and spreading them over nutrient agar plates supplemented with 2% (w/v) filter sterilized tannic acid. Addition of tannic acid to nutrient agar forms a tannin-protein complex. Bacteria producing tannase cleave this complex thus forming a dark brown colored zone around the colonies [14]. Sixteen distinct colonies that exhibited tannase producing ability were taken ahead for biochemical characterization.

Fungal culture for SSF studies

Fungal isolate (F1) from the laboratory culture collection with reports for high tannase producing ability [30] was used as the fungal counterpart in these SSF studies. The culture was maintained constantly on Czapek-Dox agar plate supplemented with filter sterilized tannic acid (2%, w/v), incubated at 30°C with periodic sub-culturing.

Biochemical characterization of bacterial isolates

All sixteen morphologically distinct bacterial isolates were characterized by microscopy and biochemical assays. Cell morphology and gram nature of the isolated bacterial colonies was identified using gram staining technique. To characterize the bacteria up to genus level following biochemical tests were performed as per the protocols mentioned in Bergey's manual [15]. All biochemical tests were performed in triplicate and for all such tests fresh overnight grown culture broth / single CFU were used as per the demands of the experiment.

Triple sugar iron test: Inoculum of bacterial cultures (pure) was stabbed into the TSI slant separately and incubated overnight at 37°C. Lactose fermenting nature of the bacteria will be indicated by yellow slant/yellow butt (acid/acid reaction), whereas non-lactose fermenters will result in pink/yellow or yellow/yellow (if sucrose is fermented). Blackening of the butt due to H_2S production may mask the acid reaction (yellow) in the butt.

Motility test: Motility media was prepared and inoculum of bacterial culture was stabbed through center of the medium (one-half of the depth) and test tubes were incubated at 37°C for 18-48 h. A visually diffuse growth spreading from the line of inoculation indicates motility.

Catalase test: A loopful of bacterial culture was smeared on a slide containing a drop of 3% hydrogen peroxide. Catalase positive isolates were expected to produce bubbles due to breakdown of hydrogen peroxide into oxygen and water.

Lactose and sucrose test: Lactose broth was inoculated with bacterial culture and incubated overnight at 37°C. The solubilization of lactose/sucrose by bacterial isolates result in production of acids which will be indicated by change in color from red-orange color of the indicator phenol red to yellow.

Starch hydrolysis test: Bacterial isolates were grown on starch agar medium. Post inoculation, the plates were flooded with iodine and appearance of a clear zone indicates starch hydrolyzing capability of the organism.

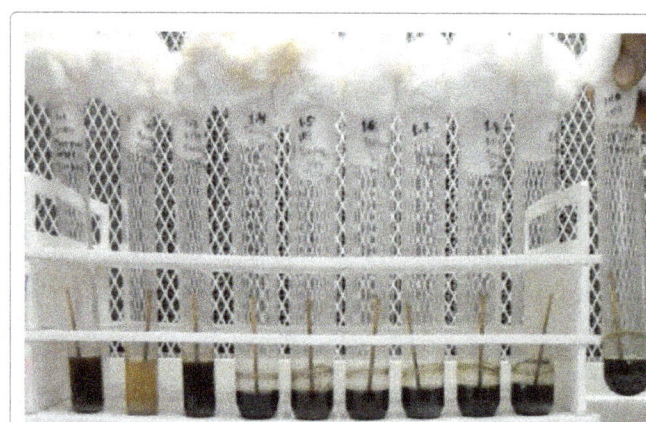

Enrichment culture of bacterial isolate

Tubes a) B 1.1; b) B 1.2; c) B 1.3; d) B 1.4; e) B 1.5; f) B 1.6; g) B 1.7; h) B 1.8; i) B 1.9; j) B 1.10

Figure 1: Different bacterial isolates grown on nutrient broth supplemented tannic acid

Citrate test: Simmon citrate slants were inoculated with bacterial isolates by stab culture method and incubated at 37°C for 48 h. A positive test for citrate utilization is represented by the development of deep blue color within 24 to 48 h.

Indole test: Peptone broth was inoculated with the bacterial culture. After overnight inoculation at 37°C, Kovac's reagent was added. Positive result was marked by the presence of red or red-violet color on the surface of alcohol layer in culture broth. Development of yellow color indicates negative result.

Bile esculin test: Bile esculin slant was inoculated by stab culture method and incubated overnight at 37°C for 48 h. Hydrolysis of esculin media will turn media into dark brown or black color registering a positive result.

Selection of bacterial isolates for tannase production on synthetic media

Preliminary screening of bacterial isolates was studied by estimating the gallic acid as an indicator for tannase producing ability where the isolates were screened for production of gallic acid- a byproduct of tannic acid hydrolysis. For the screening study all the sixteen isolates were grown on synthetic media (nutrient broth supplemented with 2% filter sterile tannic acid), incubated for 36 h. Post incubation, the culture broth was centrifuged at 8000 RPM to pellet out the bacterial biomass. The culture supernatant thus obtained was analyzed for gallic acid content and it was expressed in μg/ml. As a further step of confirmation, a quantitative measurement was made where diameter of the coloured zone was recorded for the bacterial isolates grown on tannic acid (2%, w/v) supplemented nutrient agar plates. Single point inoculations of a single CFU of each bacterial culture were made on separate plates in triplicates and incubated at 37°C. Zone diameter of top five bacterial isolates (isolates that showed very distinct and strong zone around colonies) was recorded.

Designing of FAR based substrate for SSF

FAR proposed in this study were Banana Peels (BP), Pomegranate Peels (PP), Spent Tea Leaves Powder (STP), Tamarind Seed Powder (TSP), Coconut Coir (CC) and Corn Husk (CH). These FARs were collected from the local market and dried separately in the hot air oven till the moisture was completely removed. The dried FAR were ground to form a fine powder and stored for further use. Selection and formulation of FAR combinations for SSF experiments was based on the composition of mineral content, natural tannins as well as the available essential nutrients present in these FARs as evidenced from the published literature (Table 1) [16-22]. Further combination of these FARs was designed so as to provide a balanced medium where all essential nutrients are present for obtaining optimal microbial growth and enzyme production.

A three stage approach was attempted for designing/formulating the FAR combinations. In the first stage, the microbial isolates were grown on individual FAR to check the efficiency in utilizing the complex substrate (FAR) by the organism for its growth and production of tannase and gallic acid. In the next stage, the combinations (2 FARs mixed in 1:1 ratio) were proposed so as to provide all the essential nutrients for enhanced/optimal production and also based on the results of preliminary studies on production of tannase and gallic acid on individual FAR. Further top two combinations of FAR on which maximum production was obtained were taken and combined in a different proportion (3:1) to know if the relative proportions of FAR influence the yield of tannase and/or gallic acid.

Formulation of FAR for solid state fermentation

Solid state fermentation by three microbial cultures i.e. two bacterial isolates (B 2.2 and B 2.7, native isolates) and one fungal isolate (F 1, from laboratory collection) were carried out on FAR to examine the tannase and gallic acid production by utilizing FAR as an alternate substrate. 2 g of FAR was individually taken in jam bottle of 250 ml capacity. Wetting media (containing the following salts (in g/l): K_2HPO_4, 0.76; KCl, 0.26; $MgSO_4$, 0.26; $NaNO_3$, 1.5; $FeSO_4$ and $ZnSO_4$, 0.005) was added on the basis of moisture retention capacity of the particular FAR. The contents were autoclaved at 15 lb, 120°C for 20 min and cooled. 1 ml of overnight grown bacterial isolates were added separately to the jam bottle and incubated for 48 h. CFU count of bacterial culture used for inoculation were 32×10^8 CFU/ml in case of B 2.2 and 48×10^8 CFU/ml for B 2.7 respectively. Media for fungal inoculation was prepared following similar procedure mentioned above. Two stubs of 5 mm diameter of the growing mycelium of 7 day old fungal culture were placed upside down (mycelia bearing side of the stub facing down i.e., touching the SSF substrate) on the SSF media and incubated at 35°C for 72 h. The results were analyzed and further combinations of FAR were designed to get maximum production of tannase and gallic acid. The tested combination included PP with BP/STP/TSP/CC in the ratio 1:1 and PP with STP and BP and STP in the ratio of (3:1). Extraction of extracellular tannase produced was carried out by mixing the contents of jam bottle after SSF with 30 ml of citrate buffer of pH 5.5 and kept for 2 h stirring. The content was centrifuged and the supernatant was used as the crude extract for the estimation of tannase and gallic acid. All the experiments were performed in triplicates and average values are reported.

Significant difference between means were determined by Anova two factor with replication and T-test (two samples assuming equal variance) using Microsoft Excel. The significance of differences was defined at P ≤ 0.05.

Estimation of tannase activity

Tannase activity was determined using spectrophotometric method based on the formation of chromogen between rhodanine and gallic acid (that is released by the action of enzyme-tannase on substrate i.e. methyl gallate). 100 μl of crude extract was used for the estimation of enzyme and absorbance was recorded against blank at 520 nm using Shimadzu UV-1800 spectrophotometer [23]. Tannase activity was calculated based on the standard graph of gallic acid and expressed in terms of U/g of substrate.

Gallic acid estimation

Gallic acid was estimated by spectrophotometric method using methanolic rhodanine at 520 nm using spectrophotometer (Shimadzu UV-1800) as per the procedure Sharma et al. [23]. Gallic acid content was calculated on the basis of standard curve of gallic acid. Gallic acid is expressed as mg/g of substrate.

Results and Discussion

Physicochemical analysis of soil samples

The collection of the soil samples was done during the early monsoon season (i.e. July-August). pH of the soil sample was slightly alkaline i.e. 7.37 and soil sample was brown, coarse, porous and rich in humus content. Soil sampling was done from the sub surface as such sub-surface soil usually possesses active microbial population in abundance.

Components	Pomegranate peel	Tamarind seed	Banana peel	Spent tea	Corn Husk	Coconut coir
Total solids (%)	94.50		95.66	91.80		
Moisture (%)	5.40	9.4-11.3	6.70		15	25.5
Total Sugars (%)	17.70	11.3-25.3				
Reducing Sugars (%)	4.34	7.40				
Protein (%)	4.90	13.3-26.9	4.77	18.60		
Crude Fiber (%)	16.30	7.4-8.8	31.70	23.60		
Fat content (%)	1.26	4.5-16.2				
Ash (%)	3.40	1.60-4.2	8.50	5.87	8.26	9.00
Carbohydrate (%)		73.68	59.00	-		
Total Phenol (%)	8.10					
Hydrolysable tannins (%)	6.11					
Condensed tannins (%)	1.20					
Tannins (%)	7.31	20.00	30-40	13-17		29
Calcium		36.60-786 (mg/100g)	19.20 mg/g	1.5	0.4	
Phosphorous		165-312 (mg/100g)	2.3 mg/g	0.53	0.5	
Magnesium		28.20-214 (mg/100g)				
Potassium		272.8-610 (mg/100g)	78.12 mg/g		0.9	
Sodium		19.2-28.8 (mg/100g)	24.30 mg/g			
Copper		1.6-19.0 (mg/100g)				
Iron		6.30-45.5 (mg/100g)	0.61 mg/g			
Zinc		7.0 (mg/100g)				
Manganese		0.68-12.1 (mg/100g)	76.20 mg/g			
Organic Matter		12.10 mg/g	91.50%			
Crude Lipid			1.70%			
Oxalate			0.51 mg/g			
Hydrogen Cyanide			1.33 mg/g			
Phytate			0.28 mg/g			
Saponins			24 mg/g			
Starch (%)		33.10%				

Table 1: Chemical composition of individual FAR included in SSF studies

Isolation of tannase producing bacterial strains

Direct soil culturing was carried out in nutrient broth. Consequent secondary culturing was by plating method via serial dilution yielded a total of twenty physically distinct bacterial colonies which were then analyzed for their tannase producing ability on tannic acid supplemented nutrient agar plates. This step eliminated four isolates as they were not able to grow on the tannic acid supplemented plates. The growth of the remaining sixteen isolates indicated that they are capable of degrading tannic acid into their simpler compounds such as gallic acid. Further these sixteen bacterial isolates were classified according to their gram nature and morphology.

Physiological typing (characterization) of bacterial strains

Gram staining and microscopy studies revealed that among sixteen isolates ten were found to be gram positive (B 1.1, B 1.3, B 1.4, B 1.5, B 1.7, B 1.8, B 1.10, B 2.1, B 2.5 and B 2.10) and six (B 1.2, B 1.9, B 2.2, B 2.6, B 2.7 and B 2.8) were gram negative. Based on cell morphology the cultures were classified as Bacilli and Cocci. Further the results of microscopic and biochemical test performed on different isolates were tabulated in Table 2. Results of biochemical assays of important isolates were represented in the figure (Figure 2). From the biochemical analysis we concluded that the gram positive isolates could belong to the genus *Bacillus, Lactobacillus, Enterococcus* or *Streptococcus* and that the gram negative isolates could belong to the genus *Klebsiella, Escherchia* or *Citrobacter*. All these species of bacteria have been reported to possess tannase producing ability and were reported in the literature [24-28].

Selection of bacterial isolates for tannase production on synthetic media

To select the bacterial isolates for tannase producing ability we used gallic acid as a marker. The isolate which produce higher amount of gallic acid as the end product was assumed to have higher tannase producing ability as gallic acid is a byproduct produced when tannase hydrolyses tannic acid. This study indicated among sixteen bacterial isolates, the strains B 1.1, B 1.7, B 2.2, B 2.7 and B 2.8 were the top five producers of gallic acid (Figure 3). These strains were better producers of gallic acid and in turn they are considered as better producers of tannase.

Further top five bacterial isolates thus selected were tested for their ability to produce zone of clearance on nutrient agar plate containing 2% tannic acid supplemented to confirm the gallic acid production results. This confirmation was needed as the gallic acid could not be the exact indicator of tannase producing ability. Gallic acid produced by the organisms sometimes further broken down to the simpler chemical compounds.

A visible dark green halo of gallic acid surrounding the bacterial colony was observed and measured as zone diameter. This was produced by the cleavage of tannin-protein complex; cleavage of this complex by bacteria producing tannase forms a zone around the colonies. Based on zone diameter 5 bacterial isolates produced statistically significant (ANOVA, t-test) result of which two bacterial isolates B 2.2 and B 2.7 that showed maximum zone diameter were selected for further studies on RAR based SSF (Table 3). This was confirmed by the production of zone of clearance around the bacterial growth and brown coloration after prolonged incubation [14]. It was evident from the literature that the strains having ability to utilize tannic acid to maximum extent will be the better producer of tannase which was also confirmed by observing high correlation between the zone of clearance with the

Bacterial isolates	Gram nature	Shape	TSI	Citrate	Bile Esculin	Sucrose	Indole	Motility	Starch	Catalase	Lactose
B 1.1	+ ve	Cocci	+ ve	- ve	+ ve	- ve	- ve	+ ve	- ve	- ve	+ ve
B 1.2	- ve	Cocci	+ ve	- ve	- ve	- ve	+ ve	- ve	- ve	- ve	+ ve
B 1.3	+ ve	Bacilli	+ ve	+ ve	+ ve	- ve	- ve	+ ve	+ ve	- ve	+ ve
B 1.4	+ ve	Bacilli	+ ve	+ ve	+ ve	- ve	- ve	+ ve	- ve	- ve	+ ve
B 1.5	+ ve	Cocci	+ ve	+ ve	+ ve	- ve	- ve	+ ve	- ve	- ve	+ ve
B 1.7	+ ve	Cocci	+ ve	+ ve	+ ve	- ve	+ ve	+ ve	- ve	- ve	+ ve
B 1.8	+ ve	Bacilli	+ ve	+ ve	+ ve	- ve	+ ve	+ ve	- ve	- ve	+ ve
B 1.9	- ve	Cocci	+ ve	+ ve	+ ve	- ve	- ve	+ ve	- ve	- ve	+ ve
B 1.10	+ ve	Bacilli	+ ve	+ ve	+ ve	- ve	- ve	+ ve	- ve	- ve	+ ve
B 2.1	+ ve	Bacilli	+ ve	- ve	+ ve	- ve	- ve	+ ve	- ve	- ve	+ ve
B 2.2	- ve	Cocobacilli	+ ve	+ ve	+ ve	- ve	- ve	+ ve	- ve	- ve	+ ve
B 2.5	+ ve	Cocci	+ ve	+ ve	+ ve	- ve	- ve	+ ve	- ve	- ve	+ ve
B 2.6	- ve	Cocci	+ ve	+ ve	+ ve	- ve	- ve	+ ve	- ve	- ve	+ ve
B 2.7	- ve	Bacilli	+ ve	+ ve	+ ve	- ve	+ ve	+ ve	- ve	- ve	+ ve
B 2.8	- ve	Cocci	+ ve	+ ve	+ ve	- ve	- ve	+ ve	- ve	- ve	+ ve
B 2.10	+ ve	Bacilli	+ ve	+ ve	+ ve	- ve	- ve	+ ve	- ve	- ve	+ ve

Table 2: Physiological typing (characterization) of sixteen tannase producing bacterial isolates

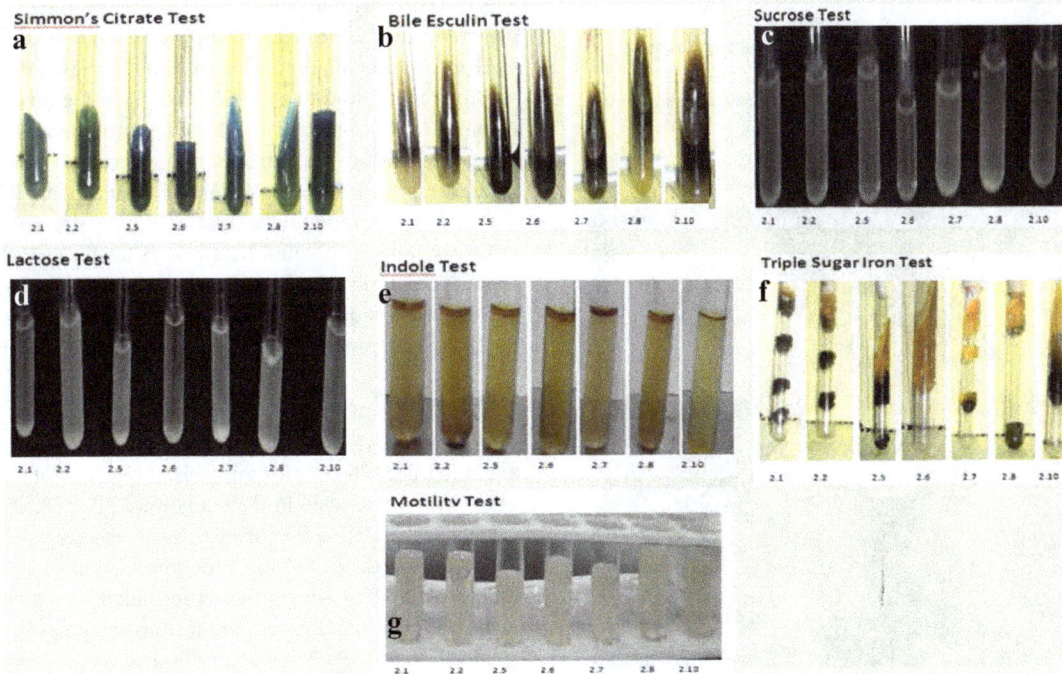

a) Simmons Citrate test; b)Bile Esculine test; c) Sucrose test; d) Lactose test; e) Indole test; f) Triple sugar iron test; g) Motility test and sequence of bacterial culture inoculation from left to right is: B 2.1;B 2.2; B 2.5;B 2.6; B 2.7; B 2.8; B 2.10 and was maintained same for all the tests conducted.

Figure 2: Biochemical assays of important bacterial isolates

quantitative enzyme production [4,29]. The growth of the two bacterial isolates (B 2.2 and B 2.7) and fungal isolate F1 on tannic acid agar plate were shown in the figure (Figure 4) and these isolates were selected for SSF studies.

Formulation of FAR for Solid state fermentation

The choice of substrates was made on the basis of tannin content and reducing sugar estimation previously carried out in our lab and reported [30]. The use of some of these FAR as alternative substrate to synthetic tannic acid was also evidenced by the reported literature [31-33]. A three stage approach was followed to design the combination of FAR for SSF study. The first stage of choosing FARs was based on

the published data of the available nutrients (Table 1) and estimated results of tannin and reducing sugar content from our previous studies [30] with FAR. This designing of FAR combination is the uniqueness of our study as the FARs were complex substrates and attaining balance of exact concentration of macro and micro nutrient is not feasible. Hence efforts were made to formulate such SSF substrate combination where required basic nutrients along with the principal substrate for enzyme i.e. tannase is present in moderate quantities along with readily metabolizable sugar to support the microbial growth in SSF.

In the first stage the organisms are grown on individual FAR. Amongst the individual FARs tested, PP and STP gave maximum yield (statistically significant at $P \leq 0.05$) (Table 4). Between bacterial

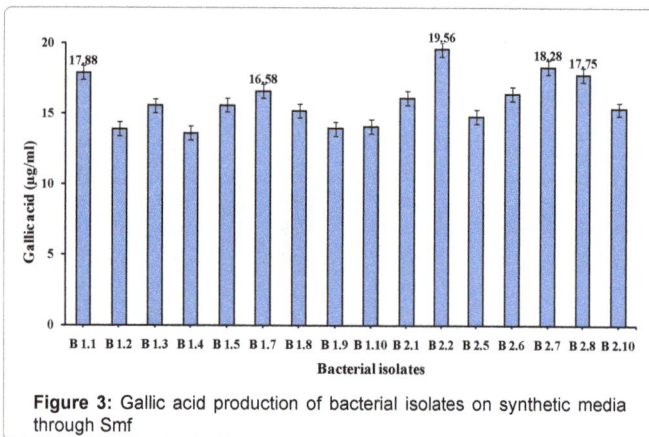

Figure 3: Gallic acid production of bacterial isolates on synthetic media through Smf

Bacterial isolates	Zone diameter (mm)
B 2.2	17 ± 0.37[a]
B 2.7	15 ± 0.38[b]
B 1.1	13 ± 0.45[c]
B 2.8	11 ± 0.43[d]
B 1.7	10 ± 0.57[e]

Data with different superscripts (a, b, c, d and e) differ significantly at the probability level p ≤ 0.05.

Table 3: Top five tannase producing bacterial isolates identified based on zone diameter

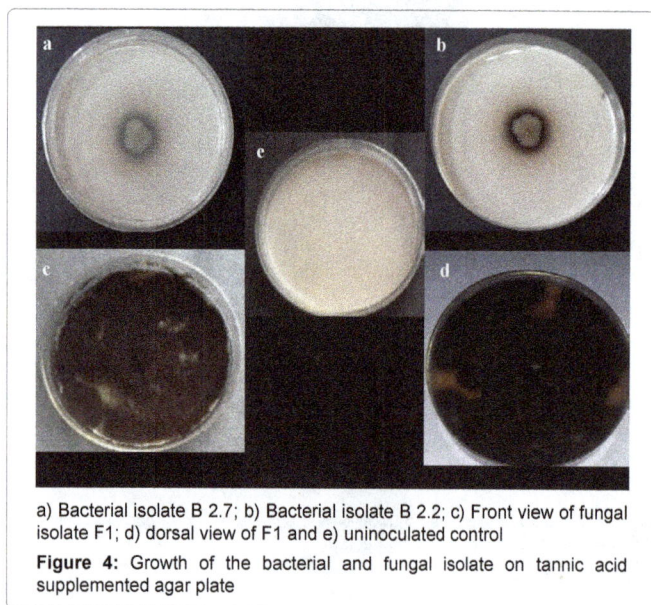

a) Bacterial isolate B 2.7; b) Bacterial isolate B 2.2; c) Front view of fungal isolate F1; d) dorsal view of F1 and e) uninoculated control

Figure 4: Growth of the bacterial and fungal isolate on tannic acid supplemented agar plate

and fungal isolates, F1 resulted in significantly higher production (ANOVA, T-test at P ≤ 0.05) of tannase and gallic acid. Maximum tannase production observed by F 1 was of 8.49 U/g on PP, followed by 7.62 U/g on STP, 5.75 U/g on TSP and 4.01 U/g on BP. Similar trend was observed for gallic acid production in case of PP and STP, 2.88 mg/g on PP and 2.52 mg/g on STP. Whereas the amount of gallic acid production on TSP and BP were almost similar i.e. TSP (1.07 mg/g) and BP (1.13 mg/g). Among bacterial cultures B 2.2 was better producer of tannase (6.88 U/g) and gallic acid (2.06 mg/g) on PP. Further the preference of FAR by all organisms showed similar pattern.

In the second stage FAR combinations were made in such a way

that the tannin content available should be less than 15% as the higher concentration of tannins will also hinder the tannase production. Also it was ensured that PP is included in all the combinations, as it was the found to be most favored SSF substrate (as compared to other FARs studied) for tannase and gallic acid production (Table 4). Significantly higher tannase production was observed in two FAR combinations (PP with STP and PP with BP) that involve three FARs i.e. PP, STP and BP. Highest production of tannase observed was 19.21 U/g and gallic acid was 5.32 mg/g on PP with STP using F 1. This was followed by PP and BP i.e. 11.23 U/g of tannase activity and 1.87 mg/g of gallic acid (Table 5).

Two combinations were proposed in the third stage i.e. PP with STP and BP with STP in the ratio of 3:1. This experiment was designed to check whether the production in these FAR combinations was enhanced by STP/BP/PP. The results indicated that PP is most supporting FAR for the production as compared to other FAR but only at the optimum concentration. The increase in PP concentration resulted in decline in tannase as well as gallic acid production (Table 5). This could be due to higher tannin content and sugar content. It has been reported that high tannin and sugar content will have inhibitory effect on tannase production [32,34-37]. The decrease in activity beyond optimum substrate concentration i.e. 3:1 could be due to increase in total sugar and tannin content with the increase in concentration of substrate might lead to substrate toxicity in the medium. The results obtained were in agreement with the reported literature [38]. FAR used in the present study are rich in carbon sources and increase in substrate concentration leads to increase in total sugar content. It is observed that the increase in carbon sources creates osmotic stress to depress enzyme synthesis and which could be a reason for the reduction in tannase production at higher substrate concentration [39]. Concentrations of other nutrient supply of the FAR also contribute to the inhibitory effect on the tannase production such as Ca^{+2}, Zn^{+2}, Fe^{+3}.

Conclusions

Our study was the first attempt to formulate combination of FAR as an alternate substrate for microbial production of tannase and gallic acid. The FAR selected in the study were PP, STP, TSP, BP, CC and CH. FAR formulation for SSF experiment was designed in three stages. This study revealed that the FAR formulation of PP with STP (1:1 ratio) was the best among the six formulations attempted for all the microbial isolates. Maximum production of tannase by F1 on this FAR formulation was 19.02 U/g and gallic acid was 5.32 mg/g. Further the bacterial isolates also have shown equally good amount of production. Among the bacterial isolates B 2.2 has shown maximum production of tannase (13.21 U/g) and gallic acid (3.51 mg/g) on the same FAR formulation. The study revealed that these FAR and their combinations surely contribute as cheap and efficient raw material for production of tannase and gallic acid by native isolates. The results also indicate that there is an ample of scope for further research to optimize the process parameters of SSF to achieve enhanced production. However, there is a need to check tannin utilized by these isolates on all possible FAR combinations before it can replace synthetic tannic acid in industrial processes.

Acknowledgement

The author Dr. K.E. Nandini gratefully acknowledges Department of Biotechnology, Govt. of India, New Delhi for fellowship of DBT-RA and financial support to carry out the research work. Authors express their sincere thanks to the

FAR	Tannase activity (U/g)			Gallic acid (mg/g)		
	F1	B 2.2	B 2.7	F1	B 2.2	B 2.7
CH	3.22 ± 0.04^a	-	-	1.89 ± 0.03^a	-	-
CC	3.43 ± 0.49^a	2.90 ± 0.49^a	1.98 ± 0.32^a	0.98 ± 0.12^b	0.80 ± 0.16^a	0.77 ± 0.08^a
BP	4.01 ± 0.50^a	4.44 ± 0.99^a	3.77 ± 0.76^b	1.13 ± 0.08^b	0.86 ± 0.06^a	0.87 ± 0.06^a
TSP	5.75 ± 0.34^b	3.10 ± 0.53^a	2.57 ± 0.50^b	1.07 ± 0.05^b	0.84 ± 0.05^a	0.82 ± 0.04^a
STP	7.62 ± 0.37^c	4.98 ± 0.33^b	3.66 ± 0.33^c	2.52 ± 0.07^c	1.76 ± 0.09^b	1.29 ± 0.07^b
PP	8.49 ± 0.21^d	6.88 ± 0.48^c	4.05 ± 0.36^c	2.88 ± 0.04^d	2.06 ± 0.40^b	1.58 ± 0.03^c

Data with different superscripts (a, b, c, d and e) differ significantly at the probability level p ≤ 0.05.

Table 4: Production of tannase and gallic acid by SSF on FAR

FAR (1:1)	Tannase activity (U/g)			Gallic acid (mg/g)		
	F1	B 2.2	B 2.7	F1	B 2.2	B 2.7
PP+CC	8.96 ± 0.40^a	3.97 ± 031^a	2.74 ± 0.85^a	1.65 ± 0.06^a	1.26 ± 0.07^a	1.04 ± 0.08^a
PP+TSP	10.94 ± 0.80^b	6.15 ± 0.47^b	4.74 ± 0.31^b	1.80 ± 0.02^b	1.60 ± 0.05^b	1.40 ± 0.04^b
PP+BP	11.23 ± 0.34^b	5.87 ± 0.41^b	4.33 ± 0.42^b	1.87 ± 0.04^b	1.60 ± 0.05^b	1.40 ± 0.04^b
PP+STP	19.21 ± 0.40^c	13.21 ± 0.43^c	9.15 ± 0.24^c	5.32 ± 0.05^c	3.51 ± 0.06^c	3.36 ± 0.03^c
FAR (3:1)						
BP +STP	10.24 ± 0.24^a	5.25 ± 0.07^a	4.69 ± 0.20^a	2.98 ± 0.04^a	0.71 ± 0.01^a	0.57 ± 0.01^a
PP+STP	14.52 ± 0.34^b	7.54 ± 0.13^b	6.45 ± 0.20^b	3.57 ± 0.05^b	1.97 ± 0.01^b	1.63 ± 0.01^b

Data with different superscripts (a, b, c and d) differ significantly at the probability level p ≤ 0.05

Table 5: Production of tannase and gallic acid by SSF on different formulations of FAR

Honorable Vice Chancellor, JIIT, NOIDA and Head of the Department and staff, Biotechnology Department, JIIT, NOIDA for their kind support.

References

1. Li S, Yang X, Yang S, Zhu M, Wang X (2012) Technology prospecting on enzymes: application, marketing and engineering. Comput Struct Biotechnol J 2: e201209017.

2. Gurung N, Ray S, Bose S, Rai V (2013) A broader view: microbial enzymes and their relevance in industries, medicine, and beyond. Biomed Res Int 2013: 329121.

3. Pinto GAS, Leite SGF, Terzi SC, Couri S (2001) Selection of tannase-producing Aspergillus niger strains. Braz J Microbiol 32: 24-26.

4. Lekha PK, Lonsane BK (1997) Production and application of tannin acyl hydrolase: state of the art. Adv Appl Microbiol 44: 215-260.

5. Mukherjee G, Banerjee R (2006) Effects of temperature, pH and additives on the activity of tannase produced by a co-culture Rhizopus oryzae and Aspergillus foetidus. World J Microbiol Biotechnol 22: 207-212.

6. Aguilar CN, Rodríguez R, Gutiérrez-Sánchez G, Augur C, Favela-Torres E, et al. (2007) Microbial tannases: advances and perspectives. Appl Microbiol Biotechnol 76: 47-59.

7. Beniwal V, Kumar A, Sharma J, Chhokar V1 (2013) Recent advances in industrial application of tannases: a review. Recent Pat Biotechnol 7: 228-233.

8. Batra A, Saxena RK (2005) Potential tannase producers from the genera Aspergillus and Penicillium. Process Biochem 40: 1553-1557.

9. Purohit JS, Dutta JR, Nanda RK, Banerjee R (2006) Strain improvement for tannase production from co-culture of Aspergillus foetidus and Rhizopus oryzae. Bioresour Technol 97: 795-801.

10. Ayed L, Hamdi M (2002) Culture conditions of tannase production by Lactobacillus plantarum. Biotechnol Lett 24: 1763-1765.

11. Beniwal V, Chhokar V, Singh N, Sharma J (2010) Optimization of process parameters for the production of tannase and gallic acid by Enterobacter cloacae MTCC 9125. J American Sci 6: 389-397.

12. Lokeshwari N, Reddy RD (2010) Microbiological production of gallic acid by a mutant strain of Aspergillus oryzae using cashew husk. Pharmacophore 1: 112-122.

13. Clesceri LS, Greenberg AE, Trussel RR (1989) Standard methods for the analysis of water and wastewater, 17th edition, American Public Health Association.

14. Kumar R, Kumar A, Nagpal R, Sharma J, Kumari A (2010) A novel and sensitive plate assay for screening of tannase-producing bacteria. Ann Microbiol 60: 177-179.

15. Garrity GM, Bell JA, Lilburn TG (2004) Taxonomic outline of the prokaryotes. Bergey's manual of systematic bacteriology. Second edition. Volume 3. Springer, New York, Berlin, Heidelberg.

16. Tejano EA (1985) State of the Art of Coconut Coir Dust and Husk Utilization (General Overview). National Workshop on Waste Utilization, Coconut Husk 1984, Philippine Coconut Authority, Diliman, Quezon City, Philippines.

17. Tartrakoon T, Chalearmsan N, Vearasilp T, Meulen U (1999) The nutritive value of banana peel (Musa sapieutum L.) in growing pigs, In Sustainable technology development in animal agriculture. Deutscher Tropentag, Berlin.

18. El-Siddig, Gunasena HPM, Prasad BA, Pushpakumara DKNG, Ramana KVR, et al. (2006) Tamarind, Tamarindus indica. Southampton Centre for under-utilized crops, Southampton, UK.

19. Aguilar CN, Aguilera-Carbo A, Robledo A, Ventura J, Belmares R, et al. (2008) Production of antioxidant nutraceuticals by solid-state cultures of pomegranate (Punica granatum) peel and creosote bush (Larreatri dentata) leaves. Food Technol Biotechnol 46: 218-222.

20. Sabu A, Augur C, Swati C, Pandey A (2006) Tannase production by Lactobacillus sp. ASR-S1 under solid-state fermentation. Process Biochem 41: 575-580.

21. Viuda-Martos M, Fernández-López J, Pérez-Álvarez JA (2010) Pomegranate and its many functional components as related to human health: A Review. Comp Rev Food Sci Food Safety 9: 635-654.

22. Israel AU, Ogali RE, Obot IB (2011) Extraction and characterization of coconut (Cocos nucifera L.) coir dust. Songklanakarin J Sci Technol 33: 717-724.

23. Sharma S, Bhat TK, Dawra RK (2000) A spectrophotometric method for assay of tannase using rhodanine. Anal Biochem 279: 85-89.

24. Mondal KC, Banerjee R, Pati BR (2000) Tannase production by Bacillus licheniformis. Biotechnol Lett 20: 767-769.

25. Mondal KC, Banerjee D, Banerjee R, Pati BR (2001) Production and characterization of tannase from Bacillus cereus KBR9. J Gen Appl Microbiol 47: 263-267.

26. Selwal MK, Yadav A, Selwal KK, Aggarwal NK, Gupta R, et al. (2010) Optimization of cultural conditions for tannase production by Pseudomonas aeruginosa IIIB 8914 under submerged fermentation. World J Microbiol Biotechnol 26: 599-605.

27. Belur PD, Gopal M, Nirmala KR, Basavaraj N (2010) Production of novel cell-associated tannase from newly isolated Serratia ficaria DTC. J Microbiol Biotechnol 20: 732-736.

28. Bradoo S, Gupta R, Saxena RK (1996) Screening of extracellular tannase producing fungi: development of a rapid and simple plate assay. J Gen App Microbiol 42: 325-329.

29. Nandini KE, Gaur A, Sundari SK (2013) The suitability of natural tannins from food and agricultural residues (FAR) for producing industrially important Tannase and Gallic acid through microbial fermentation. Int J Agric Food Sci Technol 4: 999-1010.

30. Sabu A, Pandey A, Daud MJ, Szakacs G (2005) Tamarind seed powder and palm kernel cake: two novel agro residues for the production of tannase under solid state fermentation by *Aspergillus niger* ATCC 16620. Bioresour Technol 96: 1223-1228.

31. Jana A, Maity C, Halder SK, Mondal KC, Pati BR, et al. (2012) Tannase production by *Penicillium purpurogenum* PAF6 in solid state fermentation of tannin-rich plant residues following OVAT and RSM. Appl Biochem Biotechnol 167: 1254-1269.

32. Saad H, Charrier-el Bouhtoury F, Pizzi A, Rode K, Charrier B, et al. (2012) Characterization of pomegranate peels tannin extractives. Industrial crops and products 40: 239-246.

33. Lekha PK, Lonsane BK (1994) Comparative titres, location and properties of tannin acyl hydrolase produced by *Aspergillus niger* PKL 104 in solid-state, liquid surface and submerged fermentations. Process Biochem 29: 497-503.

34. Banerjee R, Mukherjee G, Patra KC (2005) Microbial transformation of tannin-rich substrate to gallic acid through co-culture method. Bioresour Technol 96: 949-953.

35. Rodrigues TH, Dantas MA, Pinto GA, Gonçalves LR (2007) Tannase production by solid state fermentation of cashew apple bagasse. Appl Biochem Biotechnol 137-140: 675-88.

36. Srivastava A, Kar R (2009) Characterization And Application Of Tannase Produced By *Aspergillus niger* ITCC 6514.07 On Pomegranate Rind. Braz J Microbiol 40: 782-789.

37. Raaman N, Mahendran B, Jaganathan C, Sukumar S,Chandrasekaran V (2010) Optimisation of extracellular tannase production from *Paecilomyces variotii*. World J Microbiol Biotechnol 26:1033-1039.

38. Kumar R, Sharma J, Singh R (2007) Production of tannase from *Aspergillus ruber* under solid-state fermentation using jamun (*Syzygium cumini*) leaves. Microbiol Res 162: 384-390.

39. Lokeswari N, Raju KJ, Pola S, Bobbarala V (2010) Tannin acyl hydrolase from *Trichoderma viride*. Int J Chem Anal Sci 1: 106-109.

18

Diethylaminoethyl Cellulose Immobilized Pointed Gourd (*Trichosanthes dioica*) Peroxidase in Decolorization of Synthetic Dyes

Farrukh Jamal[1]* and Tanvi Goel[2]

[1]*Department of Biochemistry, Dr. Ram Manohar Lohia Avadh University, Faizabad-224001, U.P., India*
[2]*Division of Life Sciences, Research Centre, Nehru Gram Bharti University, Jhunsi, Allahabad-221505, U.P., India*

Abstract

Diethylaminoethyl (DEAE) cellulose adsorbed Pointed Gourd Peroxidase (PGP) was employed in decolorization of synthetic dyes. The expressed activity of immobilized preparation on fifth repeated use was ~50% and decolorization achieved for synthetic dyes DR19 and dye mixture (DR19+DB9) was 64.9% and 61.5% respectively. Immobilized enzyme could effectively decolorize up to 88.2% and 77.4% of DR19 and dye mixture respectively in stirred batch process at 40°C whereas dye color removal monitored at 30°C and 50°C was comparatively low under similar conditions. Immobilized enzyme in the packed column used for the continuous removal of dye color could successfully decolorize DR19 and dye mixture to 69.4% and 51.4% after 50 d of operation. Thus, DEAE immobilized PGP is a simple, economical and effective preparation to remove color of synthetic dyes.

Keywords: Pointed gourd peroxidase; Diethylaminoethyl; Immobilization; Dye decolorization; Continuous reactor

Introduction

The waste water from dye industries is rich in compounds that are toxic and detrimental to different life forms including humans. Color is usually the first contaminant to be recognized which affects the aesthetic merit, transparency and gas solubility of water bodies. The unused synthetic dyes in textile effluent mostly go untreated in the river and water bodies. Disperse dyes constitute the largest group of colorants used in the industry and are difficult to remove by chemical treatment. The processes involving physical and chemical treatment for decolorization of textile wastewater have numerous operational problems, and involve high cost [1]. On the other hand, most of the synthetic dyes are xenobiotic compounds which are poorly removed by the use of conventional biological aerobic treatments [2].

The dye color removal from wastewater is an area of technological innovations. In recent times, the approach has shifted towards enzyme based treatment of colored wastewater/industrial effluents. Enzymes are vital to new processes as they are environmental friendly and are capable of specifically reducing hazardous wastes. Peroxidases (EC 1.11.1.7) are a group of heme-containing enzymes that have been isolated from diverse sources including plants, animals and microorganisms [3,4]. These enzymes have the ability to act on a number of aromatic compounds in the presence of hydrogen peroxide. The function of the latter is to oxidize the enzyme into its catalytically active form which in turn is capable of reacting with the phenolic contaminants [4]. However, during the reaction process these enzymes get inactivated and this inactivation might be due to the free radical formation during enzymatic reaction. These radicals are adsorbed on the enzymes active site thus, blocking the substrate binding sites [5].

Peroxidases are very useful in either removal of recalcitrant toxic compounds or transforming them into innocuous products. They can change the characteristics of a given waste rendering it more amenable for treatment [6]. Their catalytic action is extremely efficient and selective as opposed to chemical catalysts due to higher reaction rates, milder reaction conditions (relatively low temperature and in the entire aqueous phase pH range) and greater stereo specificity [7]. Though much attention has been paid in the utilization of biocatalysts in several fields, their involvement has been felt very recently in solving the environmental problems [8,9]. The use of free enzymes poses inherent limitations as the stability and catalytic ability of free enzymes decreases with the complexity of the effluents [10]. Some of these limitations are overcome by the use of enzymes in immobilized form which can be used as catalysts with long lifetime [10,11].

An extensive work has been done on the enzymatic removal of aromatic compounds from wastewater by using peroxidase and hydrogen peroxide [12]. Enzymatic method has some advantages over conventional methods of treatment which include: applicability over a broad range of pH, temperature, salinity and contaminant concentration, action on recalcitrant materials and simplicity in controlling the process [13]. However, an effective use of enzymes was hampered due to their non-reusability, sensitivity to various denaturants and high cost [14]. Some of these constraints may be overcome by immobilizing enzymes on various supports [15,16].

Work in the area of enzyme technology has provided significant clues that facilitate using enzymes optimally at large scale by cross-linking, entrapping and immobilizing [17,18]. The current study demonstrates a simple, inexpensive and high yield procedure for immobilization of Pointed Gourd Peroxidase (PGP) on DEAE cellulose for the dye color removal of disperses dyes. The decolorization of disperse dyes was done in batch process as well as in a continuous vertical bed-reactor. The reusability of the I-PGP for dye color removal was measured.

Material and Methods

Materials

Bovine serum albumin, *o*-dianisidine HCl, Disperse Red 19

---

*Corresponding author: Farrukh Jamal, Assistant Professor, Department of Biochemistry (DST-FIST & UGC-SAP Supported), Dr. Ram Manohar Lohia Avadh University, Faizabad-224001, U.P., India

(DR19), Disperse Black 9 (DB9), DEAE cellulose, glutaraldehyde was procured from Sigma Chemical Co. (St. Louis, MO, USA). All other chemicals were of analytical grade. The pointed gourds were procured from local market.

Partial purification of PGP by ammonium sulphate precipitation

The protein was extracted from pointed gourd using ammonium sulphate as described previously [19]. This preparation of protein was aliquoted and stored for further use.

Protein estimation and assay of peroxidase (PGP) activity

Protein was estimated using the procedure of Lowry et al. [20]. The salt purified protein was assayed for peroxidase activity as describe by Jamal et al. [21,22]. Peroxidase activity was measured by the change in the optical density (λ=460 nm) at 37°C, by estimating the initial rate of oxidation of 6.0 mM o-dianisidine HCl in presence of 18.0 mM H_2O_2 in 0.1 M sodium acetate buffer (pH 5.6) for 15 min. Immobilized enzyme preparation was continuously agitated for entire duration of assay. One unit (1.0 U) of peroxidase activity was defined as the amount of enzyme protein that catalyzes the oxidation of 1mmol of o-dianisidine HCl in the presence of H_2O_2 per min at 37°C into colored product ($\varepsilon_m = 30$ 000 $M^{-1} L^{-1}$)

Treatment and activation of DEAE cellulose

The activation of DEAE cellulose was done using method described elsewhere [17,23]. Briefly 6.0 g of DEAE cellulose was gently stirred and allowed to swell overnight in 150 mL of distilled water. The swollen DEAE cellulose was filtered with a Buchner Funnel and incubated with 120 mL of 0.5 N HCl for 1 h. Acid treated DEAE cellulose was collected by filtration on Buchner Funnel and was washed with distilled water continuously till it attained pH 7.0. 125 mL of 0.5 N NaOH was added to HCl treated DEAE cellulose and it was stirred on a magnetic stirrer for 1 h at 4°C. The treated ion exchanger was washed again with distilled water till it attained neutral pH. Further, it was suspended and stored in 100 mL of distilled water at 4°C.

Cross-linking of DEAE cellulose adsorbed PGP

PGP (1300 units) was added to 1.2 g of DEAE cellulose and stirred overnight at 4°C. The preparation was treated with 0.3% (v/v) glutaraldehyde for 2 h at 4°C with constant stirring. Cross-linking was performed in presence of o-dianisidine HCl. Ethanolamine was added to a final concentration of 0.01% (v/v) to stop cross-linking. The solution was allowed to stand for 90 min at room temperature and the pellet was collected by centrifugation at 3000 × g for 30 min on a cooling centrifuge 4°C [24,25].

Preparation of synthetic dye solutions and calculation of percent dye decolorization

The synthetic solutions of disperse dyes (30-50 mg/mL) were prepared in distilled water to examine their decolorization by soluble and immobilized PGP. A mixture of disperse dyes consisting of DR19 and DB9 was prepared by mixing each dye in equal proportion of color intensity [26]. To compare various experiments, the decolorization was calculated for each dye or mixture of dyes. Dye decolorization was monitored by measuring the difference at the maximum absorbance for each dye as compared with control experiments without enzyme on UV-visible spectrophotometer (JASCO V-550, Japan). Untreated dye solution (inclusive of all reagents except the enzymes) was used as control for calculation of percent decolorization. The dye decolorization

was calculated as the ratio of the difference of absorbance of treated and untreated dye to that of treated dye and converted in terms of percentage. Three independent experiments were carried out in duplicate and the mean was calculated.

Reusability of immobilized PGP in the decolorization of disperse dye

DR19 and mixture of dyes (DR19+DB9) were incubated with immobilized PGP for 2 h as mentioned earlier. The enzyme was separated by centrifugation and stored in the assay buffer for over 12 h. Experiments was repeated upto 10 times with the same preparation of PGP and each time with a fresh batch of dye solution. Dye decolorization was monitored at specific wavelength maxima of the dye solutions. The percent decolorization was calculated by taking untreated dye or mixture of dyes as control (100%).

Decolorization of dye solution by soluble and immobilized peroxidase in batch processes

The dye solution (200 mL) was treated with soluble and immobilized PGP (180 U) in 100 mM sodium acetate buffer, pH 5.6 in the presence of 0.2 mM riboflavin as redox mediator and 0.8 mM H_2O_2 for different durations and at varying temperatures [21] . The treated samples were centrifuged at 3000 × g for 15 min. The residual dye concentration was measured spectrophotometrically at specific wavelength maxima of the dye. Untreated dye solution was considered as control (100%) for the calculation of percent decolorization.

Continuous dye decolorization using a vertical PGP-immobilized column

A packed bed-reactor system was developed for the continuous removal of dyes from solutions. A column (15 × 2.0 cm) was filled with DEAE cellulose immobilized PGP (1500 U) and equilibrated with 100 mM sodium acetate buffer pH 5.6. The flow rate was maintained at 12 mL h^{-1} and the feed solution contained either DR19 or mixture of DR19 and DB9 in two independent reactor systems. Dye solutions were run under the same experimental conditions. Reactors were operated at room temperature (37°C) for a period of 60 days. Treated samples were collected at an interval of 10 d and the absorbance of each sample was recorded.

Results

Reusability and storage stability of DEAE-PGP complex in dye decolorization

Table 1 shows the expressed activity of DEAE-PGP complex. Immobilized PGP (I-PGP) expressed ~ 50% activities after being repeatedly use for five times. Thereafter, the expressed peroxidase activity declined progressively to 15.5% in the tenth use. I-PGP preparation was used to remove the dye color of disperse dye DR19 and dye mixture (DR19 and DB9). The decolorizing ability of I-PGP was 51.4% for DR19 in the eight repeated use whereas the dye mixture (DR19+DB9) showed 51.7% decolorization in the seventh repeated use. Thereafter, decolorizing potential of I-PGP underwent downfall progressively and was parallel to the decline in the expressed peroxidase activity of the I-PGP preparation.

Storage stability along with the potential to remove the dye color was monitored upto 60 d with an interval of 5 to 10 d. Table 2 shows the remaining activity in the stored preparations and their ability to remove the dye color. The 35 d stored I-PGP preparation expressed 53.4% peroxidase activity and decolorized DR19 and dye mixture (DR19+DB9) to 73.5% and 59.9% respectively. Thereafter the decolorization potential underwent downfall with the 60 d stored preparation expressing poor peroxidase activity (23.8%) and removed dye color to 29.3% (DR19) and 22.6% (DR19+DB 9).

Dye treatment in a stirred batch process at varying time and temperature

The decolorization of dyes by soluble and DEAE immobilized PGP was carried out at varying time durations and temperatures as shown in Table 3. It was observed that maximum dye color removal was

No of uses of DEAE-PGP complex	Expressed peroxidase activity (%)	Percent dye decolorization	
		DR19 (λ_{495nm})	DR19 and DB9 (λ_{460nm})
1	89.7	91.5	82.2
2	78.8	87.4	80.6
3	69.7	74.6	72.7
4	58.4	69.8	65.4
5	50.7	64.9	61.5
6	42.4	60.6	59.4
7	36.3	58.9	51.8
8	29.7	51.4	43.2
9	18.9	33.6	28.5
10	15.5	21.4	19.6

Table 1: Re-usability of DEAE-immobilized PGP in decolourization of dye and dye mixture. Immobilized PGP was independently incubated with DR19 and mixture of dyes (DR19+DB9) (150 mL) at 40°C. Dye decolourization was determined after incubation period of 2 h. The immobilized enzyme was collected by centrifugation (3000 × g) and stored in assay buffer at 4°C overnight. The similar experiment was repeated 10 times. Each value represents the mean for three independent experiments performed in duplicate.

Storage stability of I-PGP complex		Dye colour removal (%)	
Days (d)	Remaining Activity (%)	DR19 (λ_{495nm})	DR19 and DB9 (λ_{460nm})
5	91.5	89.8	76.7
10	76.8	88.7	74.3
15	74.6	87.4	73.6
20	69.8	85.9	73.2
25	67.5	83.3	71.4
30	61.9	77.9	69.7
35	53.4	73.5	59.9
40	45.3	67.2	56.2
50	34.5	53.8	41.7
60	23.8	29.3	22.6

Table 2: Storage stability of DEAE- immobilized PGP preparations. Expressed activity is the activity achieved after immobilization. Original activity is the activity of soluble counterparts and taken as 100%.

achieved at 40°C with an exposure time of 2 h. Soluble PGP decolorized 80.9% and 74.1% of DR19 and mixture of dyes after 2 h of incubation, respectively. However, DEAE immobilized PGP was more effective and sustainable as compared to its soluble counterpart in the decolorization of both DR19 and mixture of dyes. Decolorization by I- PGP under similar conditions was 88.2% and 77.4% for DR19 and dye mixture respectively after 2 h of treatment. At lower temperature of 30°C, the performance of soluble PGP were lower with 76.9% and 55.1% dye color removal achieved after an incubation period of 2 h. However under similar conditions the I-PGP was much effective and decolorized DR19 and dye mixture to an extent of 78.4% and 65.4% respectively. Further, at 50°C and under similar conditions of incubation soluble PGP effectively removed the dye color of DR19 (55.9%) and dye mixture (51.2%). On increasing the incubation time (160 min), the dye color removal by soluble PGP was lowered to 31.3% (DR19) and 29.4% (DR19+DB9) whereas the performance of I-PGP was sufficiently higher 48.1% (DR19) and 60.2% (DR19+BB9). It was also observed that although, the optimum temperature was 40°C, yet I-PGP still retained sufficient potential at 50°C as compared to that recorded at 30°C.

Dye decolorization in a packed vertical I- PGP column

Table 4 shows the dye color removal of DR19 and mixture of dyes (DR19+DB9) by passing through I-PGP packed vertical bed reactor. About 88.5% and 74.5% of dye decolorization was achieved even after 10 d of its continuous operation. With the increase in the duration of continuous operation there was decrease in the decolorization ability of the I-PGP packed column. On 30 d of continuous use I-PGP decolorized DR19 (77.4%) and dye mixture (69.5%). This decrease continued to 60 d with only 50.5% and 53.4% dye color recorded for DR19 and dye mixture respectively.

Discussion

Immobilizing enzymes directly from crude homogenate is relatively a cost-effective and seemingly a feasible approach [27]. Although the immobilized form of bio-molecules holds commercial importance, protocols available for such preparations are limited. Immobilization by adsorption is an effective procedure for binding enzymes directly from partially purified preparations or even from crude homogenates [24]. The present study was aimed to immobilize PGP on a support which could be used to treat disperse dyes and dye mixtures which are generally present in textile effluents.

DEAE cellulose adsorbed PGP was cross-linked with glutaraldehyde and this preparation was found to be very efficient in removing the dye colors of disperse dyes viz., DR19 and dye mixture (DR19+DB9). I-PGP

Time (min)	Percent Dye decolourization at varying temperatures at											
	30°C				40°C				50°C			
	DR19 (λ_{495nm})		DR19+ DB9 (λ_{460nm})		DR19 (λ_{495nm})		DR19+ DB9 (λ_{460nm})		DR19 (λ_{495nm})		DR19+ DB9 (λ_{460nm})	
	S- PGP	I-DEAE	S-PGP	I-DEAE	S- PGP	I-DEAE	S-PGP	I-DEAE	S- PGP	I-DEAE	S-PGP	I-DEAE
20	70.8	72.5	59.2	62.5	74.8	76.5	66.2	70.5	61.8	62.5	46.2	61.5
40	70.6	72.6	58.7	63.6	74.6	78.4	67.7	72.6	61.6	62.6	48.7	62.6
60	72.8	74.4	57.4	65.5	78.8	81.3	69.4	73.5	60.8	64.4	49.4	65.5
80	73.6	76.4	54.5	65.5	79.9	86.4	71.5	75.5	60.6	66.4	53.5	67.5
100	73.7	76.4	54.6	65.4	79.9	86.9	72.4	77.4	58.7	66.4	55.6	67.9
120	76.9	78.4	55.1	65.4	80.9	88.2	74.1	77.4	55.9	58.4	51.2	63.4
140	60.4	72.2	33.2	64.3	65.4	88.1	45.2	69.3	47.4	58.2	36.2	63.3
160	50.3	69.1	30.4	60.2	52.7	88.3	39.8	69.1	31.3	48.1	29.4	60.2

Table 3: Dye decolourization in batch processes. DR 19 or mixture of dyes (DR19+DB9) (200 mL) was independently treated with soluble and immobilized PGP (180 U) for varying times and temperatures in batch processes. Aliquots were taken from containers at different time intervals for measuring the dye colour removal. Each value represents the mean for three independent experiments.

No of days (d)	DR19 (λ_{495nm})	DR19+ DB9 (λ_{460nm})
	I-PGP	I-PGP
10	88.5	74.5
20	85.6	73.6
30	77.4	69.5
40	77.8	68.5
50	69.4	51.4
60	50.5	43.4

Table 4: Continuous removal of dye colour in a vertical bed reactor. Dye / dye mixture were independently passed through the reactor (15 × 2 cm) filled with I-PGP (1500 U) at room temperature continuously. Samples from the column outlet were collected and after centrifugation were analyzed spectrophotometrically for the dye colour removal.

could be used seven times to achieve dye color removal of DR19 and dye mixture to over 50% which is considerably effective than soluble peroxidases. Immobilized preparation of enzymes has advantage over free enzyme as it could be reused several times and may also provide a better environment for the catalytic activity of enzyme [14,28]. Moreover, such enzymes in immobilized states can be stored for longer durations and relatively easier to handle and use. The data of storage stability of I-PGP suggest that immobilized preparations confer long lasting peroxidase activity which could be used in reactors for the treatment of effluents containing phenolic and other aromatic pollutants including dyes which are primarily represented in textile effluents. The reusability and storage experiments further supported that the use of such a cheaper source of enzyme and support will definitely minimize the cost of immobilization and provide a suitable approach for the treatment of huge volumes of wastewater in batch processes as well as in continuous reactors.

In our earlier studies we have shown that the dye color removal was poor when these dyes were treated without enzyme or only in the presence of redox mediator or the enzyme alone. Thus it's essential that the enzyme catalytic activity is effective in conjunction with redox mediators and optimum concentration of hydrogen peroxide [4,19]. Further, the enzyme works well in acidic pH (usually in the range of 3 to 6) and therefore the study was performed at pH 5.6 (data not shown). The decolorization data indicates that the enzyme retained sufficient activity in the immobilized state at the operational pH.

The dyes tested may have a narrow redox potential range and perhaps a correlation exists between redox potential and dye reduction rates. It is known that the closer the redox potential is between dye and redox mediator, the faster is dye reduction, because electron transfer is facilitated due to the low potential difference. Such behavior explains the better catalytic properties of riboflavin. However, dye reduction rate is not only determined by redox potential, but also by other factors such as chemical structure, environmental conditions and anaerobic sludge affinity and concentration. The chromophore cleavage by PGP in conjunction with redox mediators was favorable for azo dyes, because the reduction occurs in the nitrogen bonds, which have more affinity to receive electrons, based on electronegative properties, as compared to carbon-carbon bond chromophore of the anthraquinone dyes. Therefore, the effect of redox mediators on dye reduction is related to the molecular structure, being more evident for azo dyes with low decolorization rates in the absence of these compounds, and ineffective for anthraquinone dyes because of the structural stability of the latter [26].

Our studies showed that immobilized peroxidase was much more effective in removing dye color as compared to soluble enzyme in a batch process. It may be due to the reason that immobilization shielded the number of reactive free amino groups, which are not protected in soluble case and hence, were more susceptible to reaction with the reactive products like free radicals [29]. Our findings are in accordance to our earlier studies using immobilized PGP-Concanavalin A complex on calcium alginate pectin gel in decolorization of synthetic dyes [24].

To evaluate the efficiency of immobilized PGP on a large scale for the removal of dye color, a vertical continuous reactor system was designed and operated continuously with a flow rate of 15 mL h^{-1}. In this work, a flow rate of 15 mL h^{-1} was maintained to run the reactor without any operational problem like clogging which may be of concern since precipitate is formed during the enzymatic reaction. Both the reactors worked for more than 60 d approximately, thus explaining their efficiency towards dye decolorization. A significant loss of color appeared when DR19 or mixture of dyes was treated with I- PGP in the presence of redox mediator, riboflavin in a continuous reactor system. It has earlier been reported that the disappearance of peak in visible region was either due to the breakdown of chromophoric groups present in dyes or the removal of pollutants in the form of insoluble products [23].

One of the remarkable properties of the immobilized enzyme over its soluble counterpart was that it can be separated from the reaction mixture and hence, can be reused repeatedly to transform its substrate. Although PGP immobilized on DEAE cellulose was losing its activity over repeated uses, the decrease in efficiency of the enzyme after few cycles may be due to the binding of the active sites of the enzyme by the product produced during the enzymatic reactions [30].

Conclusion

The preparation and application of DEAE-immobilized pointed gourd peroxidase in decolorization of synthetic dyes was investigated. The data of the present work reveals the effectiveness of the immobilized peroxidase in sustainable dye color removal. The DEAE immobilized PGP expressed remarkable peroxidase activity, can be repeatedly used and stably stored for long periods. The system in this study is developed with a cheaper biocatalyst that is quite effective in treating dyes continuously in a small laboratory reactor. The enzymes in soluble form cannot be exploited on large scale due to their limitations of stability and reusability. Consequently, the use of immobilized enzymes has significant advantages over soluble enzymes. In the near future, cost effective, eco-friendly technologies based on the enzymatic approach for treatment of dyes present in the industrial effluents/wastewater will play a vital role. By immobilization using adsorption on DEAE, the apparent rate of enzyme inactivation was reduced which allowed a significant reduction in enzyme requirements for treatment. This increases enzyme lifetime which represents a very significant saving in terms of treatment costs.

Acknowledgement

We are thankful to the Department of Science and Technology (DST-FIST) under the Ministry of Science and Technology for providing financial assistance towards infrastructure development for carrying out this work. There is no conflict of interest whatsoever regarding this manuscript.

References

1. Jamal F, Qidwai T, Singh D, Pandey PK (2012) Biocatalytic activity of immobilized pointed gourd (*Trichosanthes dioica*) peroxidase-concanavalin A complex on calcium alginate pectin gel. J Mol Cat B: Enzy 74: 125-131.

2. Husain Q (2006) Potential applications of the oxidoreductive enzymes in the decolorization and detoxification of textile and other synthetic dyes from polluted water: a review. Crit Rev Biotechnol 26: 201-221.

3. Agostini E, Hernández-Ruiz J, Arnao MB, Milrad SR, Tigier HA, et al. (2002) A peroxidase isoenzyme secreted by turnip (Brassica napus) hairy-root cultures: inactivation by hydrogen peroxide and application in diagnostic kits. Biotechnol Appl Biochem. 35: 1-7.

4. Jamal F, Qidwai T, PandeyPK, Singh D (2011) Catalytic potential of cauliflower (Brassica oleracea) bud peroxidase in decolorization of synthetic recalcitrant dyes using redox mediator. Catal Commun 15: 93-98.

5. Bratkovskaja I, Vidziunaite R, Kulys J (2004) Oxidation of phenolic compounds by peroxidase in the presence of soluble polymers. Biochemistry (Mosc) 69: 985-992.

6. Karam J, Nicell JA (1997) Potential application of enzymes in waste treatment. J Chem Tech Biotechnol 69: 141-153.

7. Mohan SV, Prasad KK, Prakasham RS, Sarma PN (2002) Enzymatic pretreatment to enhance the biodegradability of industrial wastewater. Chem Wkly 23:163-168.

8. Mohan SV, Prasad KK, Rao NC, Sarma PN (2005) Acid azo dye degradation by free and immobilized horseradish peroxidase (HRP) catalyzed process. Chemosphere 58: 1097-1105.

9. Nicell JA (1994) Kinetics of horseradish peroxidase-catalysed polymerization and precipitation of aqueous 4-chlorophenol. J Chem Technol Biotechnol 60: 203-215.

10. Zille A, Tzanov T, Gübitz GM, Cavaco-Paulo A (2003) Immobilized laccase for decolourization of Reactive Black 5 dyeing effluent. Biotechnol Lett 25: 1473-1477.

11. Rogalski J, Jozwik E, Hatakka A, Leonomicz A (1995) Immobilization of laccase from Phlebia radiata on controlled porosity glass. J Mol Catal B: Enzy 95: 99-108.

12. Wright H, Nicell JA (1999) Characterization of soybean peroxidase for treatment of aqueous phenols. Bioresource Technology 70: 69-79.

13. Maki F, Yugo U, Yasushi M, Isao I (2006) Preparation of peroxidase-immobilized polymers and their application to the removal of environment-contaminating compounds. Bulletin of the Fiber and Textile Research Foundation 15: 15-19.

14. Husain M, Husain Q (2008) Application of redox mediators in the treatment of organic pollutants by using oxidoreductive enzymes: A review. Crit Rev Environ SciTech 38:1-42.

15. BayramoÄŸlu G, Arica MY (2008) Enzymatic removal of phenol and p-chlorophenol in enzyme reactor: horseradish peroxidase immobilized on magnetic beads. J Hazard Mater 156: 148-155.

16. Vasileva N, Godjevargova T, Ivanova D, Gabrovska K (2009) Application of immobilized horseradish peroxidase onto modified acrylonitrile copolymer membrane in removing of phenol from water. Int J Biol Macromol 44: 190-194.

17. Kulshrestha Y, Husain Q (2006) Direct immobilization of peroxidase on DEAE cellulose from ammonium sulphate fractionated proteins of bitter gourd (Momordica charantia). Enz Microb Technol 38: 470-477.

18. Jamal F, Singh S, Khatoon S, Mehrotra S (2013) Application of Immobilized Pointed Gourd (Trichosanthes dioica) Peroxidase-Concanavalin A Complex on Calcium alginate Pectin Gel in decolorization of synthetic dyes using batch processes and continuous two reactor system. J Bioprocess Biotech 3: 131.

19. Jamal F, Pandey PK, Qidwai T (2010) Potential of peroxidase enzyme from Trichosanthes dioica to mediate disperse dye decolorization in conjunction with redox mediators. J Mol Catal B: Enzy 66: 177-181.

20. Lowry OH, Rosebrough NJ, Farr Al, Randall RJ (1951) Protein measurement with the Folin phenol reagent. J Biol Chem 193: 265-275.

21. Jamal F, Singh S, Qidwai T, Pandey PK, Singh D (2012) Optimization of internal conditions for biocatalytic dye color removal and a comparison of redox mediator's efficiency on partially purified Trichosanthes dioica peroxidase. J Mol Catal B: Enzy 74: 116-124.

22. Akhtar S, Khan AA, Husain Q (2005) Simultaneous purification and immobilization of bitter gourd (Momordica charantia) peroxidases on bioaffinity support. J Chem Technol Biotechnol 80: 198-205.

23. Ashraf H, Husain Q (2009) Removal of α-naphthol and other phenolic compounds from polluted water by white radish (Raphanus sativus) peroxidase in the presence of an additive, polyethylene glycol. Biotechnol Bioproc E 14: 536-542.

24. Musthapa MS, Akhtar S, Khan AA, Husain, Q (2004) An economical, simple and high yield procedure for the immobilization/stabilization of peroxidases from turnip roots. J Sci Ind Res 63: 540-547.

25. Haider T, Husain Q (2007) Calcium alginate entrapped preparations of Aspergillus oryzae β galactosidase: its stability and applications in the hydrolysis of lactose. International Journal of Biological Macromolecules 41: 72-80.

26. Jamal F, Qidwai T, Pandey PK, Singh R, Singh S (2011) Azo and anthraquinone dye decolorization in relation to its molecular structure using soluble Trichosanthes dioica peroxidase supplemented with redox mediator. Catal Commun 12: 1218-1223.

27. Gupta MN, Mattiasson B (1992) Unique applications of immobilized proteins in bioanalytical systems. Methods Biochem Anal 36: 1-34.

28. Qayyum H, Maroof H, Yasha K (2009) Remediation and treatment of organopollutants mediated by peroxidases: a review. Crit Rev Biotechnol 29: 94-119.

29. Tatsumi K, Wada S, Ichikawa H (1996) Removal of chlorophenols from wastewater by immobilized horseradish peroxidase. Biotechnol Bioeng 51: 126-130.

30. Cheng J, Ming Yu S, Zuo P (2006) Horseradish peroxidase immobilized on aluminium-pillared inter-layered clay for the catalytic oxidation of phenolic wastewater. Water Res 40: 283-290.

19

Enhancing Biosorption Characteristics of Marine Green Algae (*Ulva lactuca*) for Heavy Metals Removal by Alkaline Treatment

Laura Bulgariu[1]* and Dumitru Bulgariu[2,3]

[1]*Technical University Gheorghe Asachi of Iaşi, Faculty of Chemical Engineering and Environmental Protection, Department of Environmental Engineering and Management, D. Mangeron Street, 71A, 700050-Iaşi, Romania*
[2]*Alexandru Ioan Cuza University of Iaşi, Faculty of Geography and Geology, Department of Geology, Carol I Street, no. 20A, 700506-Iaşi, Romania*
[3]*Romanian Academy, Filial of Iaşi, Collective of Geography, Carol I Street, no. 18, 700506-Iaşi, Romania*

Abstract

The biosorption characteristics of alkaline treated marine green algae (*Ulva lactuca*) have been studied for the removal of Pb(II), Zn(II) and Co(II) from aqueous solution, at room temperature. Batch experiments were performed to examine the effect of NaOH concentration used for treatment, initial heavy metals concentration and contact time, in comparison with untreated algae. The biosorption capacity of alkaline treated marine green algae increases with increasing of NaOH concentration, up to 0.6 mol L^{-1}, when an improvement of biosorption capacity with 11.75% for Pb(II), 60.64% for Zn(II) and 62.53% for Co(II) respectively, was obtained. The Langmuir model provides best correlation of equilibrium experimental data, and the pseudo-second order describes well the biosorption kinetics of considered heavy metals. The heavy metal ions could be easily desorbed from loaded biosorbent, and this may be reuse at least in three biosorption/desorption cycles.

Keywords: Heavy metals; Alkaline treatment; Marine green algae (*Ulva lactuca*); Biosorption

Introduction

Industrial activities often produce wastewater containing large amounts of heavy metals that are discharged into environment, and that become an important source of pollution. Due to their toxicity, mobility and accumulation tendency, the contamination of aqueous environments with heavy metals is an important issue with serious ecological and human health consequences [1,2]. Therefore, it is desirable to eliminate the heavy metals from industrial wastewaters, and this could be also important from economical considerations [3]. Among heavy metals, Pb(II), Zn(II) and Co(II) are the most common contaminants of wastewater, due to their varied uses in different industrial activities, and have priority for removal from aqueous waste stream [4].

Conventional methods for heavy metals removal from wastewater include chemical precipitation, membrane filtration, adsorption on activated carbon and biological techniques [5-9]. Unfortunately, many of these methods are limited because are expensive, or have some disadvantages such as low selectivity, moderate removal yields, high energy consumption or generates large amounts of sludge, that are also difficult to treat [10].

Biosorption provide potential alternative to overcome the disadvantages of conventional methods for wastewater treatment containing heavy metals. Different materials of biologic origin, such as fungi, bacteria, yeasts, moss, aquatic plants, algae, etc. [10-13], can be efficient used for heavy metals removal from aqueous solution, due to their low-cost and minimization of secondary wastes [3].

Marine green algae could be especially useful as they are fairly abundant in many regions of the world, have a greatly metal removal potential and large surface area [14,15]. The metal binding capacity of marine green algae is determined by the presence of various functional groups of polysaccharides, proteins and lipids on the cell wall surface [16], but also by the small and uniform distribution of binding sites [17].

Nevertheless, there are still many aspects that need to be solved before that the marine green algae to be used in real applications in wastewater treatment. The most important of these are: (i) the biosorption capacity of marine green algae should be improved, mainly because low biosorption capacity may cause large amounts of wastes loaded with heavy metals [18], and (ii) minimization of secondary pollution, which is determined by the dissolution of organic compounds from biomass, during of biosorption process [19]. According to the studies from literature, the biosorption capacities of marine green algae for different heavy metals varied from 0.01 to 1.9 mmol g^{-1}, which is significant lower that those obtained for commercial adsorbents, such as activated carbon. In consequence, a new preparation methods of biosorbents derived from marine green algae should be developed.

In order to improve the biosorption characteristics of marine green algae (*Ulva lactuca*) for heavy metals, a simple alkaline treatment was used as a preparation method of biosorbent. This treatment implied the mixing of marine green algae with well known concentration of NaOH solution, and the main advantage of this procedure is that the enhancement of biosorption characteristics of marine green algae is done without using expensive additives, and thus the cost of biosorbent preparation remains low.

In this study, the biosorption characteristics of alkaline treated marine green algae have been investigated for the removal of Pb(II), Zn(II) and Co(II) ions from aqueous solution, at laboratory scale. The effect of NaOH concentration used for alkaline treatment, initial heavy

*Corresponding author: Laura Bulgariu, Technical University Gheorghe Asachi of Iaşi, Faculty of Chemical Engineering and Environmental Protection, Department of Environmental Engineering and Management, D. Mangeron Street, 71A, 700050-Iaşi, Romania

metals concentration and contact time was studied at room temperature (22 ± 0.5°C) in batch experiments, in comparison with untreated marine green algae. Three isotherm models (Langmuir, Freundlich and Dubinin-Radushkevich) and three kinetic models (pseudo-first order, pseudo-second order and intra-particle diffusion model) were used for the mathematical description of single component biosorption on untreated and alkaline treated marine green algae, and various models parameters have been calculated in each case.

Experimental

Preparation of biosorbents

The marine green algae (*Ulva lactuca*) were collected from Romanian Black Sea coast, during of summer (July-August, 2009). Before use, the raw marine green algae were washed several times with double distilled water, to remove impurities, dried in air at 70°C for 10 hours, and the grounded and sieved until the granulation reach less that 1.0-1.5 mm. The obtained biomass was stored in desiccators for further use. The chemical composition and some physical-chemical properties of this biomass have been presented in a previous study [20].

The alkaline treatment was done by mixing the marine green algae samples with aqueous solutions of NaOH. Thus, 5.0 g of algae biomass was mechanically shaken for 1 hour with 100 mL of NaOH solution, with known concentration (range from 0.2 to 1.0 mol L^{-1}). After 24 hour of stand-by, the alkaline treated marine green algae samples were filtrated, washed with double distilled water until a neutral pH, dried in air and then mortared.

Reagents

All chemical reagents were of analytical degree and were used without further purifications. In all experiments, double distilled water, obtained from a commercially distillation system, was utilized for the preparation and dilution of solutions.

The stock solution of heavy metals (Pb(II), Zn(II) and Co(II), respectively), containing 10^{-2} mol M(II) L^{-1} were prepared by dissolving metal nitrate salts (purchased from Aldrich) in double distilled water. Fresh dilution were prepared and used for each experiment. The initial pH value of working solutions was obtained using 10^{-3} mol L^{-1} HNO_3 solution.

Biosorption experiments

The biosorption experiments were performed for a single component, by batch technique at room temperature (22 ± 0.5°C), mixing samples of 0.2 g of untreated and alkaline treated marine green algae with 25 mL solution of known heavy metal concentration (0.20-3.39 mmol L^{-1}), in 150 mL conical flasks, with intermittent stirring. All experiments were run in duplicate, in optimum experimental conditions (pH=5.0, biosorbent dose=8.0 g L^{-1}), establish previously. For kinetics experiments a constant amount of untreated and alkaline treated marine green algae of 8.0 g L^{-1} was mixed with 25 mL of 0.85 mmol L^{-1} heavy metals solution, at various time intervals, between 5 and 180 min. At the end of biosorption procedure, the solid and liquid phases were separated by filtration, and the heavy metals concentration in filtrate was determined spectrophotometrically (Digital Spectrophotometer S 104 D, 1 cm glass cell) [21], using a prepared calibration graph.

The feasibility of alkaline treated marine green algae was assessed using desorption experiments, performed with 0.1 mol L^{-1} HCl solution, by batch technique. The heavy metals-loaded biosorbent was washed with double distilled water and dried in air. The obtained samples (1.0 g)

were treated with 10 mL of 0.1 mmol L^{-1} HCl solution. Each sample was intermittently stirred for 2 hours, then filtrated and the concentration of heavy metals in filtrate was determined. The described procedure was repeated for three cycles, using the same biosorbent sample.

FT-IR spectra measurements

The FT-IR spectra of untreated and alkaline treated marine green algae were recorded with a Bio-Rad FT-IR Spectrometer, in a 400-4000 cm^{-1} spectral domain, 4 cm^{-1} resolution and 32 scans, by KBr pellet technique. The analysis of FT-IR spectra was carried out by examining the spectral bands that are modified.

Data evaluation

The biosorption efficiency of marine green algae, before and after alkaline treatment, for studied heavy metals was quantitatively evaluated from experimental results, with a relative standard deviation less than ± 0.5%, using:

(a) amount of heavy metals retained by mass unit of biosorbent (q, mmol g^{-1}), calculated from the mass balance expression:

$$q = \frac{(c_0 - c) \cdot (V/1000)}{m} \qquad (1)$$

(b) percent of heavy metals removal (R,%), that can be obtained from:

$$R,\% = \frac{c_0 - c}{c_0} \cdot 100 \qquad (2)$$

where: c_0 is the initial concentration of heavy metals solution (mmol L^{-1}); c is the equilibrium concentration of heavy metals solution (mmol L^{-1}); V is the volume of solution (mL) and m is the biosorbent mass (g).

The difference between the biosorption capacity of untreated and alkaline treated marine green algae was calculated according to expression [22]:

$$\Delta q_e = \frac{q_{e,treated} - q_{e,untreated}}{q_{e,untreated}} \cdot 100 \qquad (3)$$

where: $q_{e, treated}$ and $q_{e, untreated}$ are the amounts of heavy metals retained on mass unit of alkaline treated and untreated marine green algae respectively, (mmol L^{-1}).

The desorption efficiency (Desorption,%) was evaluated using the bellow mathematical equation [16]:

$$Desorption,\% = \frac{q_{desorbed}}{q_{retained}} \cdot 100 \qquad (4)$$

where: $q_{desorbed}$ is the amount of heavy metals desorbed on mass unit of biosorbent (mmol g^{-1}); $q_{retained}$ is the amount of heavy metals retained on mass unit of biosorbent (mmol g^{-1}).

Results and discussion

Effect of NaOH concentration used for alkaline treatment

The effect of NaOH concentration used for alkaline treatment of marine green algae, for the biosorption of Pb(II), Zn(II) and Co(II) ions is presented in Figure 1. The obtained results indicate that the alkaline treatment improves the biosorption capacity of marine green algae for all studied heavy metals, in comparison with untreated biomass (cNaOH=0.0 mol L^{-1}).

In addition, the amounts of heavy metals retained on mass unit of biosorbent (q, mmol g^{-1}) increases with increasing of NaOH concentration up to 0.6 mol L^{-1}, after that remains almost constant. This means that a NaOH concentration of 0.6 mol L^{-1} is sufficient for the efficient alkaline treatment of marine green algae, and this value of NaOH concentration was used for the preparation of alkaline treated marine green algae required in the further experiments. The enhancing of biosorption capacity obtained under these conditions (Δq_e) was 6.36% for Pb(II), 34.50% for Zn(II) and 40.60% for Co(II) respectively, in comparison with untreated biomass.

Generally, the alkaline treatment enhances the biosorption characteristics of biological materials, mainly due to the hydrolysis reactions [23,24]. The hydrolysis reaction can lead to the formation of more carboxylic (–COOH) and hydroxyl groups (–OH), both in un-dissociated and dissociated forms, that improve the metal-binding properties of algae biomass [25].

FT-IR spectra (Figure 2) of untreated and alkaline treated marine green algae were used to characterize the biosorbents. Several important peaks, corresponding to the essential functional groups of the algae cell wall are observed in the spectra of untreated marine green algae (spectra 1).

The broad and strong band from 3296 cm^{-1} is attributed to the overlapping of O–H and N–H stretching vibrations. Peak at 2939-2908 cm^{-1} correspond to carboxylic/phenolic vibrations. The peaks at 1535 cm^{-1} and 1417 cm^{-1} can be attributed to the HO- bonds of quinine, and the peak of 1656 cm^{-1} to the C=N and C=O stretching. The peaks at 1338 cm^{-1}, 1055 cm^{-1} and 1031 cm^{-1} can be assigned to N–H bond, –CH3 release and C–OH stretching vibrations, due to different functional groups on algae cell. The bands from 574 cm^{-1} and 464 cm^{-1} corresponds

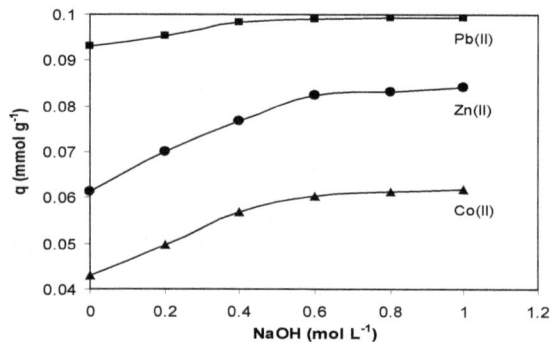

Figure 1: Effect of NaOH concentration used for alkaline treatment on the biosorption efficiency of studied heavy metals.

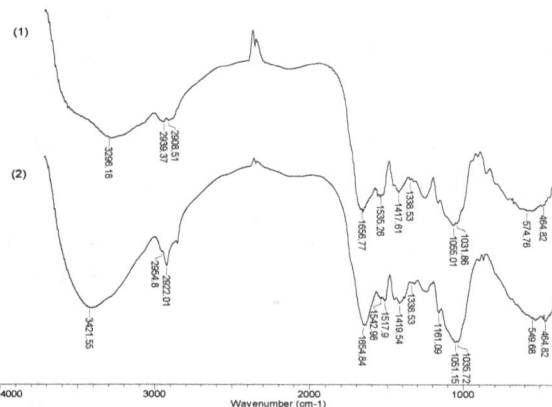

Figure 2: FT-IR spectra of untreated (1) and alkaline treated (2) marine green algae (*Ulva lactuca*).

to the C–N–S shearing due to polypeptides structure of algae cells.

After alkaline treatment of marine green algae with 0.6 mol L^{-1} NaOH solution, the most significant modifications can be observed in case of absorption bands corresponding to carboxyl and hydroxyl groups (Figure 2 – spectra 2). Thus, the absorption bands from 3296 cm^{-1}, 1656 cm^{-1} and 1055^{-1}031 cm^{-1} are shifted to 3421 cm^{-1}, 2954 cm^{-1}, 1654 cm^{-1} and 1051^{-1}035 cm^{-1}, respectively. These modifications suggest that after alkaline treatment, the carboxyl and hydroxyl groups became more available for the interactions with heavy metals, without to be significant affected the chemical structure of marine green algae.

Effect of initial heavy metals concentration

It was previously showed [20] that the untreated marine green algae can be efficient use for the removal of heavy metal ions from aqueous solution, and that the maximum removal efficiency is obtained at initial solution pH of 5.0 and 8.0 g L^{-1} biomass dose. In order to compare the biosorption characteristics of untreated and alkaline treated marine green algae, the experimental conditions were maintained the same.

The effect of initial concentration of Pb(II), Zn(II) and Co(II) ions on the biosorption efficiency both on untreated and alkaline treated marine green algae, in optimum experimental conditions, was examined in the initial concentration range from 0.20 to 3.39 mmol M(II) L^{-1}, and the obtained results are illustrated in Figure 3.

An increase of the amounts of heavy metals retained on mass unit of biosorbent with increasing of initial concentration can be observed for both untreated and alkaline treated marine green algae. This variation is mainly determined by the increase of interaction probability between heavy metals and functional groups from biosorbent surface, with increasing of initial concentration of studied metal ions.

On the other hand in the studied concentration range, the alkaline treated marine green algae have better biosorptive characteristics than untreated marine green algae (Figure 3), for all heavy metals. This is more evident at high initial concentrations. Thus, for an initial concentration of 3.39 mmol M(II) L^{-1}, the biosorption capacity is 0.3432 mmol g^{-1} for Pb(II), 0.2086 mmol g^{-1} for Zn(II) and 0.1177 mmol g^{-1}Co(II) respectively in case of untreated marine green algae, while in case of alkaline treated marine green algae the obtained values of biosorption capacities were 0.3835 mmol g^{-1} for Pb(II), 0.3352 mmol g-1 for Zn(II) and 0.1913 mmol g^{-1} for Co(II), respectively. These differences are mainly determined by changing the availability of functional superficial groups after alkaline treatment. When marine green algae are treated with alkaline solution, the hydrolysis processes occur, and these will transform inactive superficial groups into available functional groups for interaction with heavy metals, and so the biosorption capacity increase greatly. The change of availability of superficial functional groups by this simple chemical treatment determine the increase of the biosorption capacity with 11.75% for Pb(II), 60.64% for Zn(II) and 62.53% for Co(II) respectively, and thus the economical feasibility of this biosorbent was increased.

In addition, the oxidability index (CCO, mg O$_2$ L^{-1}), which is a measure of the total content of organic compounds from certain aqueous solution, determined according with the standard procedure [26], decrease from 117.3 mg O$_2$ L^{-1} in case of aqueous solution separated on untreated marine green algae, to 85.04 mg O$_2$ L^{-1}in case of aqueous solution separated on alkaline treated marine green algae. The decrease of this parameter indicates that after alkaline treatment the dissolution of organic compounds from biomass during of biosorption process is significantly reduced, and so the secondary pollution is

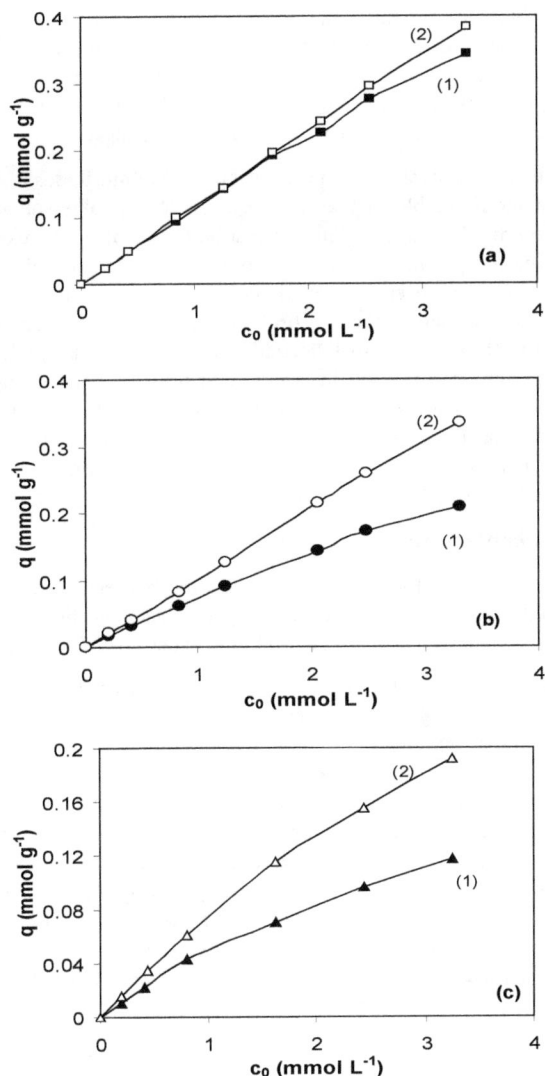

Figure 3: Effect of initial heavy metals concentration on their removal onto untreated (1) and alkaline treated (2) marine green algae: (a) Pb(II); (b) Zn(II); (c) Co(II) (pH=5.0; 8 g biosorbent L^{-1}, temperature=22°C; contact time=24 hours).

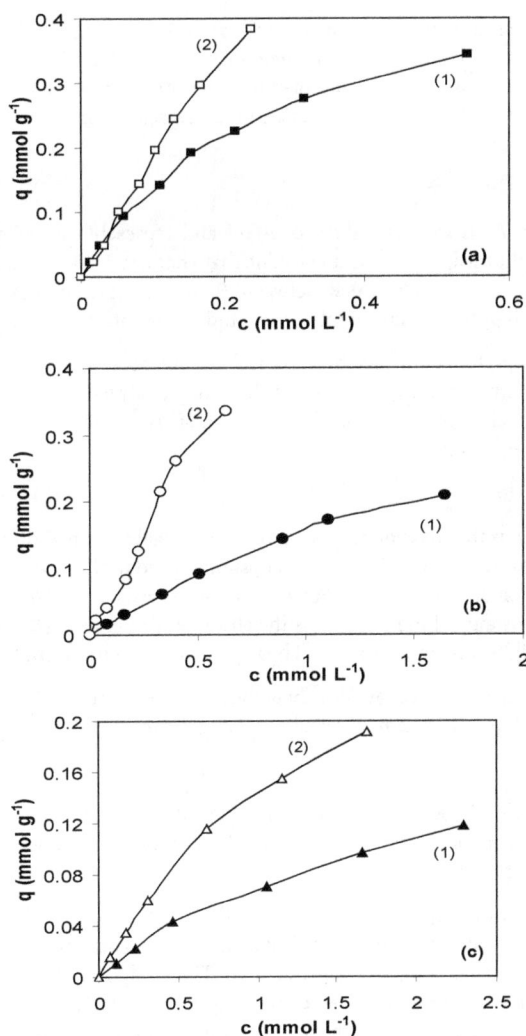

Figure 4: Biosorption isotherms of (a) Pb(II), (b) Zn(II) and (c) Co(II) respectively, onto untreated (1) and alkaline treated (2) marine green algae (pH=5.0; 8 g biosorbent L^{-1}, temperature=22°C; contact time=24 hours).

minimized. This can be also another advantage of the use of alkaline treated marine green algae as biosorbent for the removal of heavy metals from wastewaters.

Equilibrium modelling

The analysis of equilibrium data by fitting them to different isotherm models is important to asses the practical biosorption capacity and optimization of the biosorption system design. In this study, the experimental results were fitted using three isotherm models (Langmuir, Freundlich and Dubinin-Radushkevich), and the best fit isotherm model was selected based on the linear regression correlation coefficients.

The Langmuir isotherm model is probably the most widely applied isotherm model, which is based on the monolayer biosorption onto homogeneous surface [3,27]. This model assumed that the biosorption forces are similar to the forces in chemical interactions, and is used to estimate the maximum biosorption capacity (q_{max}, mmol g^{-1}),

corresponding to the biosorbent surface saturation. The linear form of Langmuir isotherm model is:

$$\frac{1}{q} = \frac{1}{q_{max}} + \frac{1}{q_{max} \cdot K_L} \cdot \frac{1}{c} \tag{5}$$

where: q_{max} is the maximum biosorption capacity (mmol g^{-1}), corresponding to the complete monolayer coverage of the surface, c is the concentration of heavy metals at equilibrium (mmol L^{-1}) and K_L is the Langmuir constant, related to the biosorption/desorption energy (L g^{-1}).

The essential feature of the Langmuir isotherm model can be expressed in terms of R_L, a dimensionless constant referred to as separation factor or equilibrium parameter, defined by the equation:

$$R_L = \frac{1}{1 + K_L \cdot c_0} \tag{6}$$

where: c_0 is the highest initial concentration of heavy metals in aqueous solution (mmol L^{-1}).

The value of RL indicate the type of isotherm, thus the biosorption process is irreversible when R_L=0, favourable when $0<R_L<1$, linear when

$R_L=1$, or unfavourable when $R_L>1$.

The Freundlich isotherm model is derived from multilayer biosorption and biosorption on heterogeneous surface [3,28], and was chosen to estimate the biosorption intensity of heavy metals on untreated and alkaline treated marine green algae. The linear form of this model is:

$$\lg q = \lg K_F + \frac{1}{n}\lg c \tag{7}$$

where: K_F is the Freundlich constant and represent an indicator of biosorption capacity, n is a constant that characterized the surface heterogeneity. The $1/n$ value between 0 and 1 indicate that the biosorption process is favourable under studied conditions.

In order to appreciate the physical or chemical nature of the biosorption process, the equilibrium data were analyzed by Dubinin-Radushkevich isotherm model [29,30], that is expressed by the following equation:

$$\ln q = \ln q_{max}^{D-R} - \beta \cdot \varepsilon^2 \tag{8}$$

where: is the maximum amount of heavy metals retained on mass unit of biosorbent (mmol g^{-1}), β is a constant related to the biosorption energy (mol^2 kJ^{-2}) and ε is the Polanyi potential ($\varepsilon=RT \ln(1+1/c)$), R is the gas constant (kJ mol^{-1} K^{-1}), T is the absolute temperature (K) and c is the equilibrium concentration of heavy metals in solution (mol L^{-1}).

The values of β gives an idea about the mean free energy (E, kJ mol^{-1}) of biosorption, which can be calculated using the relation:

$$E= \frac{1}{\sqrt{-2\beta}} \tag{9}$$

If E value is between 8 and 16 kJ mol^{-1} the biosorption process follows by chemical ion exchange, while E<8 kJ mol^{-1} is characteristic of physical retention.

The biosorption isotherms of Pb(II), Zn(II) and Co(II) ions on untreated and alkaline treated marine green algae, used for the equilibrium modelling are illustrated in Figure 4, and the values of Langmuir, Freundlich and Dubinin-Radushkevich isotherm parameters, evaluated from the slope and intercept of corresponding linear dependencies, are summarized in Table 1.

The values of correlation coefficients (R^2) show that the biosorption isotherm data of studied heavy metals are very well represented by the Langmuir isotherm model and indicate the formation of monolayer coverage of heavy metal ions on the outer surface of biomass. The maximum biosorption capacity (q$_{max}$, mmol g^{-1}), which is a measure to form a monolayer, calculated from Langmuir isotherm equation increases in the order Zn(II)>Pb(II)>Co(II), which suggests that Zn(II) ions have higher affinity for the functional groups of the biosorbents, and can be largely retained.

On the other hand, it can also observe from Table 1 that the obtained values of maximum biosorption capacities were higher in case of alkaline treated marine green algae, than in case of untreated biomass. The increase of maximum biosorption capacities, calculated according with equation (3) was by 43.70% in case of Pb(II), 37.40% in case of Zn(II) and 94.31% in case of Co(II), respectively. This is another argument which sustains the hypothesis that after alkaline treatment more many superficial functional groups become available for interaction with heavy metal ions. In consequence the amount of heavy metals required to form a complete monolayer on alkaline treated biomass surface is higher.

The calculated R$_L$ values range between 0 and 1 in all cases, indicating

that the biosorption of Pb(II), Zn(II) and Co(II) ions from aqueous solution, both on untreated and alkaline treated marine green algae is a favourable process. The fractional values of 1/n from Freundlich isotherm model, suggests the heterogeneity of both biosorbents surface, and also indicate a favourable biosorption of Pb(II), Zn(II) and Co(II) ions on untreated and alkaline treated marine green algae.

The higher values of q_{max}^{D-R} parameter from Dubinin-Radushkevich isotherm model (Table 1) than the experimental q values obtained under optimized conditions indicate that both untreated and alkaline treated marine green algae have a porous structure. The obtained values of biosorption energy (E, kJ mol^{-1}) range between 11.45 and 14.12 kJ mol^{-1} and shows that biosorption is mainly a chemical process that occurs through electrostatic interactions, for all studied heavy metals, on untreated and alkaline treated marine green algae. A slightly decrease of the mean biosorption energy values can be observed in case of alkaline treated marine green algae, which is probably determined by the increment of the availability of functional groups from biosorbent surface for heavy metals biosorption.

Effect of contact time

Figure 5 shows the effect of contact time between biosorbents (untreated and alkaline treated marine green algae) and heavy metal ions from aqueous solution, under optimum experimental conditions.

It can be observed that the efficiency of biosorption process increases with increasing of contact time in each case. The biosorption process is very fast during the initial stage, when in the first 30 min around 80% of heavy metals (79.88% for Pb(II), 79.53% for Zn(II) and 78.90% for Co(II), respectively) are retained in case of untreated marine green algae, while in case of alkaline treated marine green algae in the first 5 min the removal percents were higher than 85% (88.92% for Pb(II), 88.82% for Zn(II) and 85.19% for Co(II), respectively). After this fast initial step, the rate of biosorption process become slower near to equilibrium, which is practically obtained after 60 min in case of untreated biomass and 30 min in case of alkaline treated marine green algae, for all studied heavy metals.

The results presented above shows that in case of alkaline treated marine green algae, the required time for the biosorption process is significant lower. The very fast biosorption on alkaline treated marine green algae make this biosorbent more suitable for continuous flow rate treatment systems.

Kinetics modelling

The biosorption kinetics is significant in the design of wastewater treatment and allows the selection of optimum conditions for operating in full-scale batch process, as it provide valuable insights into the process pathways and mechanism of biosorption.

In order to investigate the biosorption kinetics of Pb(II), Zn(II) and Co(II) respectively on untreated and alkaline treated marine green algae, three kinetic models, namely pseudo-first order, pseudo-second order and intra-diffusion particle model, were used to analyze the experimental data.

The linear form of the pseudo-first order kinetic model [31,32] is given by the following relation:

$$\lg(q_e - q_t) = \lg q_t - \frac{k_1}{2.303}\cdot t \tag{10}$$

where: q_e and q_t are the amounts of heavy metals retained on mass unit of biosorbent (mmol g^{-1}) at equilibrium and at time t respectively, and k_1 is the rate constant of pseudo-first order kinetic model (min^{-1}).

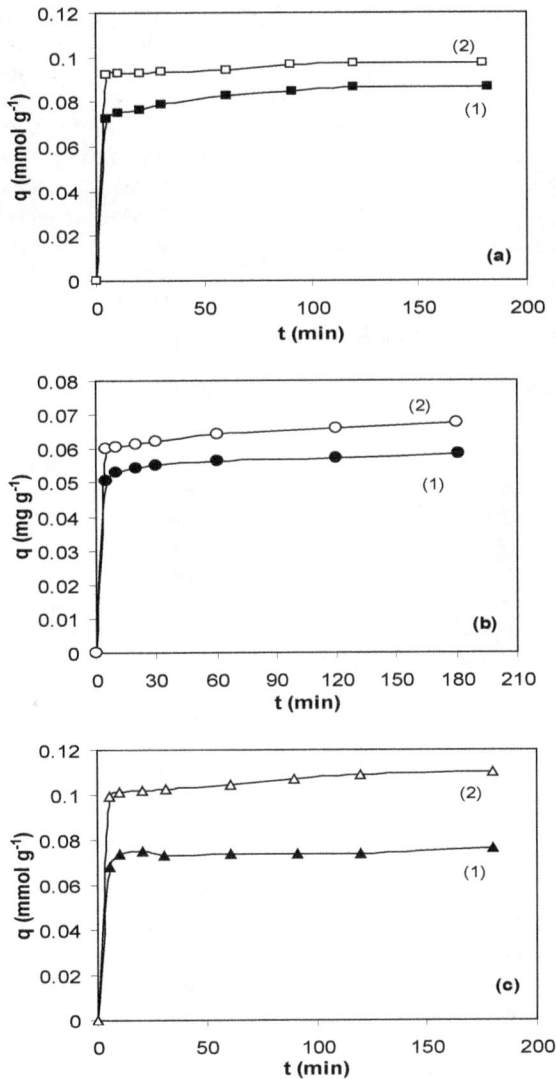

Figure 5: Effect of contact time on the heavy metals biosorption onto untreated (1) and alkaline treated (2) marine green algae: (a) Pb(II); (b) Zn(II); (c) Co(II) (pH=5.0; 8 g biosorbent L^{-1}, temperature=22°C).

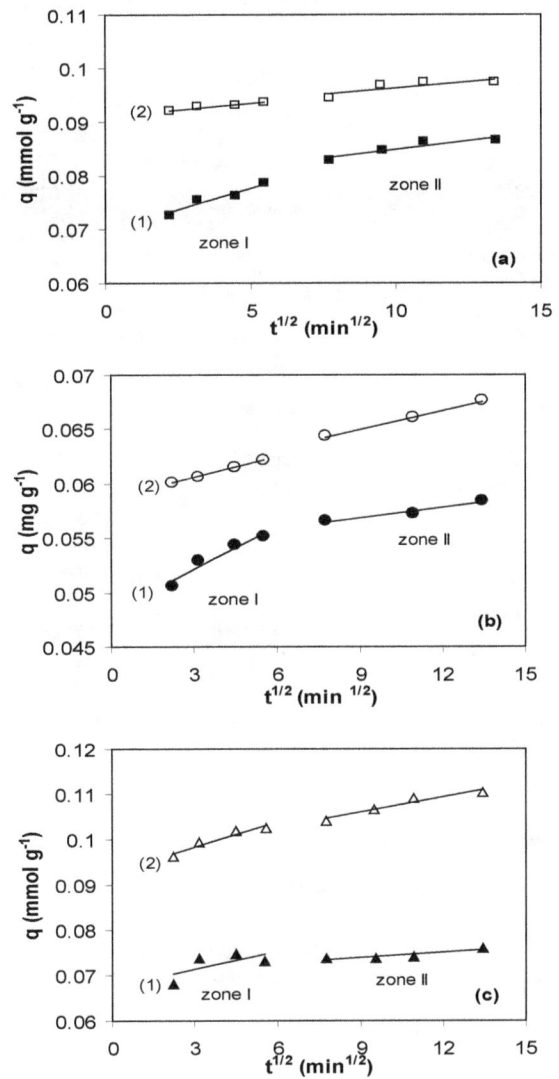

Figure 6: The plots of intra-particle diffusion model for the biosorption of (a) Pb(II), (b) Zn(II) and (c) Co(II) respectively, onto untreated (1) and alkaline treated (2) marine green algae.

In comparison with the pseudo-first order kinetic model, the pseudo-second order kinetic model is based on the assumption that the rate controlling step in the biosorption mechanism is chemical interaction between metal ions and superficial functional groups of biosorbent. The mathematical expression of the pseudo-second order kinetic model can be written in linear form as [31,33]:

$$\frac{t}{q_t} = \frac{1}{k_2 \cdot q_e^2} + \frac{t}{q_e} \qquad (11)$$

where: k_2 is the pseudo-second order rate constant (g mmol^{-1} min^{-1}).

The intra-particle diffusion model was used to determine the participation of diffusion process in the biosorption of considered heavy metals on untreated and alkaline treated marine green algae. The kinetic equation of intra-particle diffusion model is given by [34]:

$$q_t = k_{diff} \cdot t^{1/2} + c \qquad (12)$$

where: k_{diff} is the intra-particle diffusion rate constant (mmol g^{-1} min$^{-1/2}$) and c is the concentration of heavy metals from solution, at equilibrium (mmol L^{-1}).

The kinetic parameters of the pseudo-first order and pseudo-second order models, calculated from the linear dependences lg(q$_e$-q$_t$) vs. t and t/q$_t$ vs. t respectively, together with the corresponding correlation coefficients (R^2) are presented in Table 2.

The obtained results showed that the equilibrium biosorption capacities (q$_e$, mmol g^{-1}) calculated from the pseudo-first order equation is very different from the experimental values (q$_e^{exp}$, mmol g^{-1}), for all studied heavy metals and both biosorbents. This means that the pseudo-first order kinetic model is not adequate for to describe the kinetics data of Pb(II), Zn(II) and Co(II) biosorption on untreated and alkaline treated marine green algae.

The experimental data were further fitted by the pseudo-second order kinetic model, and better fitting was obtained with high

Isotherm model	Parameter	Untreated marine green algae			Alkaline treated marine green algae		
		Pb(II)	Zn(II)	Co(II)	Pb(II)	Zn(II)	Co(II)
Langmuir	R^2	0.9998	0.9991	0.9991	0.9981	0.9957	0.9996
	q_{max} mmol·g⁻¹	0.4538	0.5012	0.2406	0.6521	0.6886	0.4675
	K_L L·g⁻¹	4.4374	0.4262	0.4371	3.5563	1.1075	0.4734
	R_L	0.0623	0.4144	0.4127	0.0765	0.2140	0.3936
Freundlich	R^2	0.9850	0.9954	0.9702	0.9874	0.9778	0.9893
	1/n	0.7197	0.8611	0.7692	0.6067	0.7535	0.8007
	K_F mmol·g⁻¹·(L·mmol⁻¹)$^{1/n}$	0.6447	0.1523	0.0667	0.4549	0.5436	0.1412
Dubinin-Radush-kevich	R^2	0.9877	0.9336	0.9478	0.9960	0.9577	0.9562
	$q_{max}$$^{D-R}$ mmol·g⁻¹	0.5205	0.5632	0.2586	0.7089	0.6916	0.4874
	E kJ·mol⁻¹	12.0712	14.1271	14.1287	11.4549	13.9057	13.4515

Table 1: Isotherm parameters of the biosorption of Pb(II), Zn(II) and Co(II) respectively, on untreated and alkaline treated marine green algae (*Ulva lactuca*).

Kinetics model	Parameter		Untreated marine green algae			Alkaline treated marine green algae		
			Pb(II)	Zn(II)	Co(II)	Pb(II)	Zn(II)	Co(II)
Pseudo-first order model	q_{exp} mmol·g⁻¹		0.0865	0.584	0.0759	0.0974	0.0676	0.1101
	R^2		0.9270	0.8932	0.7774	0.8450	0.9969	0.8933
	q_e mmol·g⁻¹		0.0193	0.0061	0.0028	0.0690	0.0079	0.0813
	k_1 min⁻¹		6.1244·10⁻³	2.8658·10⁻³	5.2106·10⁻⁴	4.5158·10⁻³	2.6053·10⁻³	3.9097·10⁻⁴
Pseudo-second order model	R^2		0.9998	0.9998	0.9996	0.9999	0.9997	0.9996
	q_e mmol·g⁻¹		0.0875	0.0586	0.0755	0.0978	0.0678	0.1108
	k_2 g·mmol⁻¹·min⁻¹		4.7626	11.0509	14.5476	10.6541	17.4956	16.4462
Intra-particle diffusion model	zone I	R^2	0.9445	0.9401	0.4898	0.9431	0.9402	0.9242
		c mmol·L⁻¹	0.0694	0.0539	0.0670	0.0910	0.0481	0.0925
		k_{diff}I mmol·g⁻¹·min⁻¹	0.0017	0.0003	0.0014	0.0005	0.0013	0.0019
	zone II	R^2	0.8453	0.9977	0.7407	0.6805	0.9971	0.9431
		c mmol·L⁻¹	0.0782	0.0587	0.0704	0.0914	0.0599	0.0960
		k_{diff}II mmol·g⁻¹·min⁻¹	0.0007	0.0007	0.0004	0.0005	0.0006	0.0011

Table 2: Kinetics parameters of the biosorption of Pb(II), Zn(II) and Co(II) respectively, on untreated and alkaline treated marine green algae (*Ulva lactuca*).

correlation coefficients ($R^2 > 0.9996$), in all cases. More, the values of biosorption capacities calculated from pseudo-second order kinetic model (q_e, mmol g⁻¹) are close to the experimental values (q_e^{exp}, mmol g⁻¹), for all studied heavy metals and both biosorbents. These indicate clear that the biosorption of Pb(II), Zn(II) and Co(II) ions on untreated and alkaline treated marine green algae comply with the pseudo-second order kinetic model, which is based on the assumption that the rate controlling step in biosorption mechanism is the chemical interaction, and similar behaviour have been reported for various types of algae used as biosorbents [11,35].

In addition, the higher values of the rate constants obtained in case of alkaline treated marine green algae in comparison with untreated biomass suggests that the rate of biosorption process is limited by the availability of superficial functional groups to interact with metal ions from aqueous solution. When, marine green algae are treated with alkaline solution, some hydrolysis and dissociation processes occur, and these make that the chemical interactions to occur easier. In consequence, the rate of biosorption process increase (as well as the rate constants are higher), than in case of untreated marine green algae.

According to intra-particle diffusion model, if the diffusion process is the rate controlling step in studied systems, the graphical representations of q_t vs. $t^{1/2}$ should yield a straight line, passing through the origin [34]. However, in this study, the plots q_t vs. $t^{1/2}$ obtained in case of Pb(II), Zn(II) and Co(II) biosorption on untreated and alkaline treated marine green algae (Figure 6) does not go through the origin, and two separated zones exist in all cases. The parameters of intra-particle diffusion model (k_{diff} and c) evaluated for each zone, are also presented in Table 2.

The deviation of straight line from the origin indicate that the intra-particle diffusion process is not only rate controlling step [2,34], and the boundary diffusion controls the biosorption up to a certain degree. The first zone (zone I) is attributed to mass transfer of heavy metals from the bulk solution to biosorbent surface (k_{diff}^{I}), while the second zone (zone II) indicates the intra-particle diffusion (k_{diff}^{II}). The significant difference between these two zones is given by their slope, when the first zone has a pronounced slope, the slope of second zone is much lower. This indicates that the binding sites are located on biosorbent surface and are readily accessible for the heavy metal ions, or at external intra-layer surface. Hence, the obtained experimental data indicate a limited contribution of mass transfer and boundary layer diffusion in the biosorption process of Pb(II), Zn(II) and Co(II) respectively, on untreated and alkaline treated marine green algae, and that the intra-particle diffusion influenced the biosorption process up to certain degree.

Desorption and reusability

In order to make the biosorption process more economical feasible, it is necessary to regenerate the loaded biosorbent. Generally, the regeneration process of loaded biosorbent is performed by desorption, using chemical reagents that must have low cost, highly efficiency and do not damage the structure of biosorbent.

In this study, the desorption of heavy metals (Pb(II), Zn(II) and Co(II), respectively) from loaded alkaline treated marine green algae was tested under batch conditions, using different concentrations of aqueous solution of hydrochloric acid. The experimental results have indicate that both studied heavy metals can be readily eluted with 0.1

mol L^{-1} HCl solution, when 97.5% of Pb(II), 98.02% of Zn(II) and 98.13% of Co(II) ions were recovered, for a ratio alkaline treated biosorbent mass:acid volume=1:10. The biosorbent samples, washed with double distilled water and dried in air, were reuse in three biosorption/desorption cycles, and the loss in the biosorption capacity was less than 3.5% for Pb(II), 5.7% for Zn(II) and 6.2% for Co(II), respectively.

Conclusions

In this study, the biosorptive characteristics of alkaline treated marine green algae (*Ulva lactuca*) have been investigated for the removal of Pb(II), Zn(II) and Co(II) ions form aqueous solution, in comparison with untreated biomass. The experiments were examined the effect of NaOH concentration used form marine green algae treatment, initial heavy metals concentration and contact time, in optimum experimental conditions, at room temperature. The experimental results have indicate that the alkaline treated marine green algae have better biosorption characteristics than untreated biomass, and have potential for serving as biosorbent for removal of heavy metals from aqueous solution.

Acknowledgement

This work was supported by a grant of the Romanian National Authority for Scientific Research, CNCS-UEFISCDI, project number PN-II-ID-PCE-2011-3-0559.

References

1. Freitas OMM, Martins RJE, Delerue-Matos CM, Boaventura RAR (2008) Removal of Cd(II), Zn(II) and Pb(II) from aqueous solutions by brown marine macro algae: Kinetic modelling. J Hazard Mater 153: 493-501.

2. Sulaymon AH, Mohammed AA, Al-Musawi TJ (2013) Removal of lead, cadmium, copper, and arsenic ions using biosorption: equilibrium and kinetic studies. Desalination and Water Treatment 51: 4424-4434.

3. Montazer-Rahmati MM, Rabbani P, Abdolali A, Keshtkar AR (2011) Kinetics and equilibrium studies on biosorption of cadmium, lead, and nickel ions from aqueous solutions by intact and chemically modified brown algae. J Hazard Mater 185: 401-407.

4. Mendoza CA, Cortes G, Munoz D (1998) Heavy metal pollution in soils and sediments of rural developing district 063, Mexico. Environ Toxic Water 11: 327-336.

5. Davis TA, Volesky B, Mucci A (2003) A review of the biotechnology of heavy metal biosorption by brown algae. Water Res 37: 4311-4330.

6. Wang J, Chen C (2009) Biosorbents for heavy metals removal and their future. Biotechnol Adv 27: 195-226.

7. Llanos J, Williams PM, Cheng S, Rogers D, Wright C, et al. (2010) Characterization of a ceramic ultrafiltration membrane in different operational states after its use in a heavy-metal ion removal process. Water Res 44: 3522-3530.

8. Kumar PS, Ramalingam S, Abhinaya RV, Kirupha SD, Murugesan A, et al. (2012) Adsorption of Metal Ions onto the Chemically Modified Agricultural Waste. CLEAN – Soil, Air, Water 40: 188-197.

9. Mudhoo A, Garg VK, Wang S (2012) Removal of heavy metals by biosorption. Environ Chem Lett 10: 109-117.

10. Arief VO, Trilestari K, Sunarso J, Indraswati N, Ismadji S (2008) Recent Progress on Biosorption of Heavy Metals from Liquids Using Low Cost Biosorbents: Characterization, Biosorption Parameters and Mechanism Studies. CLEAN – Soil, Air, Water 36: 937-962.

11. Donmez G, Aksu Z, Ozturk A, Kutsal T (1999) A comparative study on heavy metal biosorption characteristics of some algae. Process Biochem 34: 885-892.

12. Feng D, Aldrich C (2004) Adsorption of heavy metals by biomaterials derived from the marine alga Ecklonia maxima. Hydrometallurgy 73: 1-10.

13. Klos A, Rajfur M, Waclawek M, Waclawek W (2005) Ion exchange kinetics in lichen environment. Ecol Chem Eng 12: 1353-1365.

14. Hamdy AA (2000) Biosorption of heavy metals by marine algae. Curr Microbiol 41: 232-238.

15. Pavasant P, Apiratikul R, Sungkhum V, Suthiparinyanont P, Wattanachira S,

et al. (2006) Biosorption of Cu^{2+}, Cd^{2+}, Pb^{2+} and Zn^{2+} using dried marine green macroalga *Caulerpa lentillifera*. Bioresource Technol 97: 2321-2329.

16. Deng L, Su Y, Su H, Wang X, Zhu X (2007) Sorption and desorption of lead (II) from wastewater by green algae *Cladophora fascicularis*. J Hazard Mater 143: 220-225.

17. Kızılkaya B, Dogan F, Akgul R, Turker G (2012) Biosorption of Co(II), Cr(III), Cd(II), and Pb(II) Ions from Aqueous Solution Using Nonliving Neochloris Pseudoalveolaris: Equilibrium, Thermodynamic, and Kinetic Study. J Disper Sci Technol 33: 1055-1065.

18. Hu JL, He XW, Wang CR, Li JW, Zhang CH (2012) Cadmium adsorption characteristic of alkali modified sewage sludge. Bioresource Technol 121: 25-30.

19. Karthikeyan S, Balasubramanian R, Iyer CSP (2007) Evaluation of the marine algae Ulva fasciata and Sargassum sp. for the biosorption of Cu(II) from aqueous solutions. Bioresource Technol 98: 452-455.

20. Lupea M, Bulgariu L, Macoveanu M (2012) Biosorption of Cd(II) from aqueous solution on marine green algae biomass. Environ Eng Manag J 11: 607-615.

21. Dean JA (1995) Analytical Chemistry Handbook. Mc-Grow Hill.

22. Yazici H, Kilic M, Solak M (2008) Biosorption of Copper(II) by Marrubium globosum sub sp. Globosum leaves powder: effect of chemical pretreated. J Hazard Mater 151: 669-675.

23. Chen JP, Yang L (2005) Chemical modification of Sargassum sp. for prevention of organic leaching and enhancement of uptake during metal biosorption. Ind Eng Chem Res 44: 9931-9942.

24. Bulgariu L, Bulgariu D, Macoveanu M (2011) Adsorptive performances of alkaline treated peat for heavy metals removal. Separ Sci Technol 46: 1023-1033.

25. Gupta VK, Rastogi A (2008) Biosorption of lead from aqueous solutions by green algae Spirogyra species: kinetics and equilibrium studies. J Hazard Mater 152: 407-414.

26. Fresenius W, Schneider W, Quentin KE (1998) Water Analysis. A Practical Guide to Physico-Chemical and Microbiological Water Examination and Quality Assurance. Springer-Verlag, Berlin, Germany.

27. Chong KH, Volesky B (1995) Description of two-metal biosorption equilibria by Langmuir-type models. Biotechnol Bioeng 47: 451-460.

28. Hameed BH, Ahmad AA, Aziz N (2007) Isotherms, kinetics and thermodynamics of acid dye adsorption on activated palm ash. Chem Eng J 133: 195-203.

29. Lodeiro P, Barriada JL, Herrero R, Sastre de Vicente ME (2006) The marine macroalga Cystoseira baccata as biosorbent for cadmium(II) and lead(II) removal: Kinetic and equilibrium studies. Environ Pollut 142: 264-273.

30. Sari A, Tuzen M (2008) Biosorption of cadmium(II) from aqueous solution by red algae (Ceramium virgatum): Equilibrium, kinetic and thermodynamic studies. J Hazard Mater 157: 448-454.

31. Gerente C, Lee VKC, Lee P, McKay G (2007) Application of chitosan for the removal of metals from wastewaters by adsorption - mechanisms and models review. Crit Rev Env Sci Tech 37: 41-127.

32. Febrianto J, Kosasih AN, Sunarso J, Ju YH, Indrawati N, et al. (2009) Equilibrium and kinetic studies in adsorption of heavy metals using biosorbent: A summary of recent studies. J Hazard Mater 162: 616-645.

33. Ho YS, Porter JF, McKay G (2002) Equilibrium isotherm studies for the sorption of divalent metal ions onto peat: copper, nickel and lead single component systems. Water Air Soil Poll 141: 1-33.

34. Wu Y, Zhang S, Guo X, Huang H (2008) Adsorption of chromium(III) on lignin. Bioresource Technol 99: 7709-7715.

35. Romera E, Gonzalez F, Ballester A, Blazquez ML, Munoz JA (2007) Comparative study of biosorption of heavy metals using different types of algae. Bioresource Technol 98: 3344-3353.

20

Response Surface Methodology for Mycoprotein Production by *Fusarium Venenatum* ATCC 20334

S. M. Hosseini[1] and K. Khosravi-Darani[2]*

[1]*Department of Food Science and Technology, Shaheed Beheshti University, M.C., Tehran, Iran*
[2]*Department of Food Technology Research, National Nutrition and Food Technology Research Institute, Shaheed Beheshti University, M.C., P. O. Box: 19395-4741, Tehran, Iran*

Abstract

In this research, the effect of process variable on yield (%w/w protein per biomass) of mycoprotein production by *Fusarium venenatum* ATCC 20334 in surface culture evaluated. A face centered central composite design (FCCD) was employed to determine maximum protein production at suitable initial concentration of date juice (as a carbon and energy source), nitrogen concentration and seed size. Analysis of variance showed that the contribution of a quadratic model was significant for the response. The optimal condition for mycoprotein production contains 20 g/l of date juice, 4.48 g/l of nitrogen source and 12.97% (v/v) of seed size. In these conditions, 46.48 ± 0.2% (w/w) protein was obtained in the dried cell weight. Heat treatment of fungal biomass at 64 -65°C for 20-30 min reduced the RNA content to an acceptable level for human food grade products. Finally, after reduction of ribonucleic acid contents of mycoprotein, the amino acids and fatty acids profiles of product were determined.

Keywords: *Fusarium venenatum*; Mycoprotein; Central Composite Design (CCD); Surface culture; Date juice

Introduction

Fusarium venenatum has been cultured as a mycoprotein source for human consumption in England for over a decade under the trade name of "Qourn". This product with a fibrous texture is a rich source of high quality protein including essential amino acids. It is also less energy dense than equivalent meat products and does not have animal fats and cholesterol [1]. Mycoprotein shows satiety and satiation properties which can be a solution for overweight by enabling people to achieve a healthier diet (low fat and high fiber) [2].

The most commonly used medium for biomass production by *F. venenatum* is Vogel medium with glucose as a carbon source [3]. Several economic substrate like agricultural wastes are been introduced as the carbon and energy sources for mycoprotein production, e.g. wheat starch (the substrate chosen by RHM company in England), potatoes (in Ireland), cassava, rice or cane juice (in tropical countries) [4].

Date fruit with high content of carbohydrates, minerals and vitamins, is produced in Middle East. Unfortunately a large amount of this product is wasted, while it is rich in carbohydrates and other required metabolites for microbial growth and production. Use of date as carbon and energy sources results cheap fermentation process due to less required pretreatment and low cost substrate. Date has been used for the production of baker's yeast biomass [5], lactic acid [6,7], alkaline protease [8], xantan [9], cultivation of mushrooms [10] and some other microbial metabolites.

Many studies have been conducted for mycoprotein production. Wiebe used *F. venenatum* A3/5 to produce mycoprotein in 150,000 liter pressure reactors in continues flow process on glucose and ammonium (as the carbon and nitrogen sources) [11]. Ahangi were used *F. oxysporum* for production of mycoprotein while this fungus shows allergic and toxic symptoms in consumer. The results showed that in optimum condition the dried fungal biomass contained 42%w/w protein and the productivity was still low (5g/l biomass contain of 42% protein) [12]. Also the results indicated that mechanical agitation damages mycelial biomass and reduces the yield of production [13]. Therefore, application of surface culture method for production of mycoprotein can be proposed to fill gap of research in this field.

The optimization of variables in mycoprotein production (as like as other fermentation processes), is of primary importance due to the impact on the feasibility and efficiency of the process. The first step is identification of the main process variables. The selection of the variables in this study was based on our prior experience by Plackett–Burman design (PBD) [14]. In addition, the rheological properties of fungal biomass from *F. venenatum* were determined in different conditions of temperature and shear [15]. Since the PBD is typically used as a preliminary optimization technique, more accurate quantitative analysis of the effect of variables for mycoprotein production is required [16]. Further optimization can be conducted by response surface methodology (RSM), a factorial base design introduced by G.E.P. Box in the 1950s [17]. RSM mainly includes central composite design (CCD), Box-Behnken design, one-factor design, D-optimal design, user defined design, and historical data design [18].

In this study, a face centered composite design (FCCD) was adapted to optimize the levels of medium variables (date juice and nitrogen source concentrations as well as seed size) on production of mycoprotein. This is the first time that date juice is used for protein production by *F. venenatum* ATCC 20334 in surface culture. The biomass was produced under determined condition. Finally, after reduction of ribonucleic acid contents of mycoprotein, the amino acids and fatty acids profiles of product were determined.

*Corresponding authors: Dr. Kianoush Khosravi-Darani, Department of Food Technology Research, National Nutrition and Food Technology Research Institute, Shaheed Beheshti University, M.C., P. O. Box: 19395-4741, Tehran, Iran

Materials and Methods

Microrganism

F. venenatum ATCC 20334 was used throughout this investigation. Strain was maintained at 4°C on agar-solidified Vogel slants. The components of modified Vogel medium is described elsewhere [14].

Inoculum and media preparation

For preparation of date juice, the wastage of date (prepared from Dombaz Company) was added to water and boiled in a 50 L tank for 30 min. The components of filtrate was determined [8] and used as carbon and energy source in inoculum development medium. Inocula were prepared in 250 conical flasks containing 50 mL Vogel medium, which were inoculated and incubated on a rotary shaker at 30°C and 200 rpm for 72 h. Production medium also contained date juice as carbon source and other medium components were the same as in seed medium. Fermentation was conducted in 500 mL flasks each containing 100 mL of production medium. The culture medium was inoculated with fungal suspension, and incubated in surface culture at 30°C.

After fermentation process, biomass was harvested by filtration of 100 mL cultivation medium through pre-dried Whatman No.1 filter papers. Then, clarified suspension was passed throw a 0.45 μm membrane, was washed twice with cold distilled water and dried using an oven at 60ºC to a constant weight. The cell dry weight was quantified gravimetrically.

RNA reduction

The RNA content of biomass was reduced in order to meet required safety standards [11] by subjecting the biomass to heat shock at 64 -65 °C for 20-30 min.

Response surface methodology

The main and interaction of three variables which influence the mycoprotein production were analyzed and optimized by FCCD (α = 1) in three levels Table 1. A total of 20 experimental runs with different combination of variables (consisting 14 experimental runs and 6 additional runs at the center point) to check reproducibility were carried out. Protein production was taken as the response (Y). The general form of second order polynomial, which its coefficients were analyzed by Minitab 14, is as Equation1:

$$Y_i = \beta_0 + \sum \beta_i X_i + \sum \beta_{ii} X_i^2 + \sum \beta_{ij} X_i X_j \qquad \text{Eq (1)}$$

where Y_i is the predicted response, $X_i X_j$ are input variables, which influence the response variable Y; β_0 is the offset term; β_i is the ith coefficient; β_{ii} the jth quadratic coefficient and β_{ij} is the ij^{th} interaction coefficient.

Statistical analysis of the model was performed to evaluate the analysis of variance (ANOVA). The optimum levels of variables (within the experimental range) for maximum protein production were determined, and maximum protein production was confirmed by running trial number 10.

variables	Coded levels		
	-1	0	+1
Date juice (g/l)	10	15	20
$(NH_4) H_2PO_4$ (g/l)	3	4	5
Seed size (% v/v)	8	12	16

Table 1: Levels of independent variables in the experimental design for mycoprotein production.

Analytical methods

The crude protein content was measured using the Kjeldahl technique [19]. The crude protein content of the biomass was calculated by multiplying 6.25 to total nitrogen. The protein content was represented by %w/w protein per total dry weight. Determination of RNA content of biomass was performed by spectrophotometeric method [20]. Amino acid profile of *F. venenatum* was determined by Pico-Tag method. This method involves hydrolysis of biomass with HCL; derivatized with phenylisothiocyanate to produce phenylthiocarbamyl amino acids and analysis by reverse phase HPLC [21]. Fatty acid composition of biomass was determined by a Younglin ACME 6000M gas chromatograph (Korea) using a capillary column (Technokroma TR-CN 100) (60 m per 0.25 mm ID, film thickness 0.2 μm) operated with hydrogen as the carrier gas (0.2 ml/min). The operating conditions were as follows: injector temperature 250°C; detector temperature 260°C; split ratio 80:1, oven temperature program was 5 min at 150°C, then increasing by 5°C per min up to 175°C then 3 min in this temperature followed by 3°C per min up to 190°C, and finally 15 min at 190°C [22].

Results and Discussion

Regression model and statistical testing

The experimental central composite design was applied to analyze the main and interaction effects of date juice, nitrogen concentration and seed size. The optimum condition was obtained for mycoprotein production, as the design and results of trials are given in Table 2. Multiple regressions were used to analyze the data and thus polynomial equation was derived from regression analysis as follows:

$$Y = 45.5063 + 0.9559\ X_1 + 0.4129\ X_2 + 0.3448\ X_3 + 0.0062\ X_1^2 - 0.5978\ X_2^2 - 0.6153\ X_3^2 + 0.0931\ X_1X_2 - 0.6706\ X_1X_3 + 0.8356\ X_2X_3 \quad (2)$$

Regression analysis of the experimental data showed that date juice, nitrogen and seed size had positive linear effect on mycoprotein production (P< 0.05). Among the three variables date juice had highest impact on protein production as given by highest linear coefficient (0.9559), followed by nitrogen source (0.4127) and seed size (0.3448). Elimination of insignificant terms was done step by step. Date juice showed insignificant quadratic effect on protein production with maximum *P* value among other terms. Hence, in the first step this term was excluded from the regression and conclusions were repeated Table 3. All rest terms had significant effects on protein production by low *P* values (<0.05) as showed in Table 3. So the model Equation 2 was modified to reduced fitted model Equation 3:

$$Y = 45.5071 + 0.9590\ X_1 + 0.4127\ X_2 + 0.3448\ X_3 - 0.5954\ X_2^2 - 0.6129\ X_3^2 + 0.0931\ X_1X_2 - 0.6706\ X_1X_3 + 0.8356\ X_2X_3 \quad (3)$$

Analysis of variance (ANOVA) for the protein production obtained from this design is shown in Table 4. ANOVA gives the value of the model and can explain whether the model adequately fits the variation observed in protein production with the designed variable level. F test for regression was significant at a P<5%. The coefficient of determination (R^2) for production of mycoprotein was 99.5%. This value showed good agreement between experimental observations and predicted values. The coefficient of variation indicates the high degree of precision. The higher reliability of the experiment is usually indicated by low value of S. In the present case, a low value of S at 0.1146 denotes that the experiments performed are highly reliable. The *P* value for lack of fit was 0.236. This amount indicated that the experimental data obtained fitted well with the model and explained the effect of date juice and nitrogen

Run	Date Juice (g/l)(X_1)	$(NH_4) H_2PO_4$ (g/l)(X_2)	Seed size (% v/v) (X_3)	Protein (g/100)(Y)	
				experimental	predicted
1	-1	-1	-1	42.910	42.843
2	1	-1	-1	45.875	46.003
3	-1	1	-1	41.775	41.811
4	1	1	-1	45.290	45.251
5	-1	-1	1	43.210	43.203
6	1	-1	1	43.670	43.587
7	-1	1	1	45.595	45.513
8	1	1	1	46.250	46.270
9	-1	0	0	44.441	44.551
10	1	0	0	46.405	46.463
11	0	-1	0	44.378	44.499
12	0	1	0	45.260	45.324
13	0	0	-1	44.515	44.549
14	0	0	1	44.088	44.962
15	0	0	0	45.720	45.507
16	0	0	0	45.505	45.507
17	0	0	0	45.530	45.507
18	0	0	0	45.501	45.507
19	0	0	0	45.640	45.507
20	0	0	0	45.500	45.507

Table 2: Face centered composite design matrix for evaluation of the three variables and experimental and predicted mycoprotein production.

Variable	Coef.	SE Coef.	t-value	P-value[a]
constant	45.5071	0.03842	1184.423	0.000
X_1	0.9559	0.03622	26.389	0.000
X2	0.4127	0.03622	11.398	0.000
X3	0.3448	0.03622	9.519	0.000
X22	-0.5954	0.06404	-9.299	0.000
X32	-0.6129	0.06404	-9.572	0.000
X1X2	0.0931	0.04050	2.299	0.042
X_1X_3	-0.6706	0.04050	-16.559	0.000
X_2X_3	0.8356	0.04050	20.633	0.000

[a]Significant at P<0.05
S = 0.1146 R^2 = 99.5% R^2(adj) = 99.1%

Table 3: Coefficients and t-values of process variables for mycoprotein production using face centered composite design.

Source	Degree of freedom (DF)	Sum of squares (SS)	Mean of squares (MS)	$S_{tatistics}$	P-value[a]
Model	8	27.1240	3.39049	258.39	0.000
Linear	3	12.0295	4.00984	305.59	0.000
Square	2	5.8410	2.92049	222.57	0.000
Interaction	3	9.2534	3.08448	235.07	0.000
Residual error	11	0.1443	0.01312		
Lack of fit	6	0.1015	0.01692	1.98	0.236
Pure error	5	0.0428	0.00856		
Total	19	27.2683	-	-	

[a]Significant at P<0.05

Table 4: Analysis of variance for mycoprotein production using face centered composite design.

source contents and seed size on mycoprotein production by *F. venenatum*. The insignificant value of lack of fit (more than 0.05) showed that the quadratic model was valid for the present study [23].

Interaction between factors influencing mycoprotein production

The interaction effects and optimal levels of the variables were determined by plotting the response surface curves. The effect of interaction of date juice and nitrogen on protein production is illustrated in Figure 1. The surface plot showed increasing trend for mycoprotein production with increased date juice and a moderate value of nitrogen. The maximum protein production was achieved by adding 20 and 4 g/L of sugar syrup and $NH_4H_2PO_4$ as carbon and nitrogen sources. In fact, increased concentration of produced protein is a result of increased carbon and nitrogen sources concentrations. Further increase of nitrogen concentration causes decreased production of mycoprotein. Carbon is the main component of cellular structure and energy storage. Nitrogen is a key factor in protein and nucleic acid production. In addition, $NH_4H_2PO_4$ plays a buffer role in medium. The initial sugar requirement for growth varies from species to species and from strain to strain [24].

Figure 2 represents the interaction between date juice and seed size. This Fig showed that intermediate level of seed size and high date juice favored mycoprotein production. Results indicate that an inoculum amount of 12% is the suitable inoculums size for protein production. Increase of inoculum size results increased protein production from date juice. Obviously, this observation is due to decreased lag phase and promoted cell growth in early stage of fermentation. Similar results has been reported by Jin et al. for protein production from starch waste by *Asperglilus oryzae*, who determined 7.5% (v/v) is optimum level of seed size in data range of 1 to 15.5% (v/v) [25]. Further increase of inoculum size with constant concentration of substrate may cause decreased production due to limitation of medium components and resulted compe-

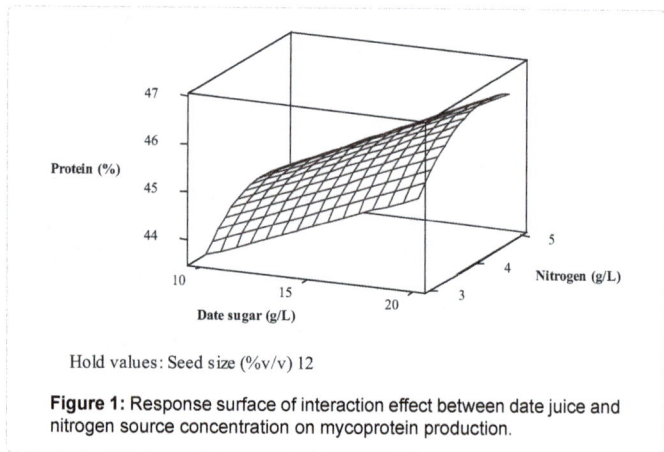

Hold values: Seed size (%v/v) 12

Figure 1: Response surface of interaction effect between date juice and nitrogen source concentration on mycoprotein production.

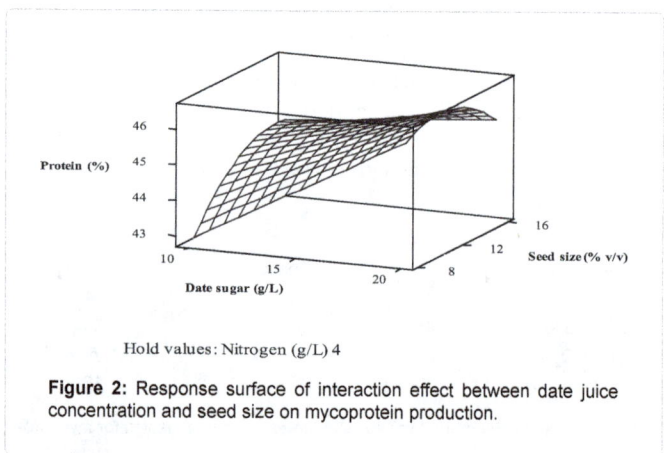

Hold values: Nitrogen (g/L) 4

Figure 2: Response surface of interaction effect between date juice concentration and seed size on mycoprotein production.

tition between cells.

Figure 3 shows effect of nitrogen and seed size on mycoprotein production. High production of protein was observed at higher level of nitrogen and seed size. Therefore, increase of both variables enhances protein production.

Optimization and verification

The surface plots in Figure1-3 indicated that the produced protein in some regions could be more than 46.5%. Hence, the response optimization based on desirability function was carried with Minitab 14. The parameters given by response optimization for date juice, nitrogen and seed size consist of 20 g/L, 4.48 g/L and 12.97%, respectively. Under this condition protein production was 46.55%. Therefore, these conditions were used for confirmation of the predicted value of protein output. Maximum yield of protein production 46.48 ± 0.2%w/w was obtained under optimized experimental conditions.

Heat treatment at 64 -65°C for 20-30 min reduces the RNA content from 7.88 to 0.9% which is a safe level for human consumption (maximum limit is 2 g/day). This result is in agreement with previous studies about F. venenatum [11] (Wiebe, 2002). Amino and fatty acids composition of fungal biomass are presented in Table 5 and 6. Table 5 shows that the amino acid profile of biomass include all the essential amino

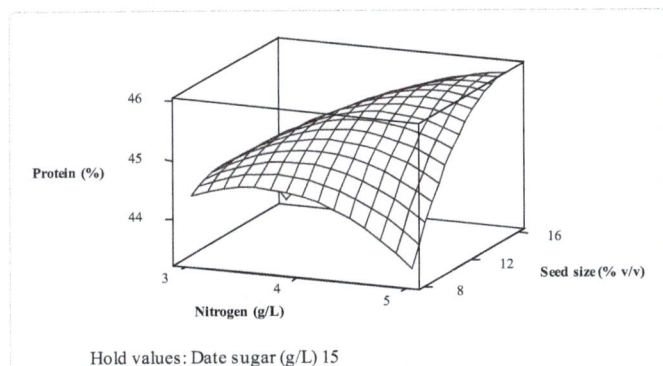

Hold values: Date sugar (g/L) 15

Figure 3: Response surface of interaction effect between nitrogen juice concentration and seed size on mycoprotein production.

Amino acid	Content (% w/w)
Alanine	5.250±1.061
Arginine	6.125±0.187
Aspartic	5.750±0.353
Cystine	3.250±1.061
Glutamic	13.125±0.177
Glycine	4.750±0.353
Histidine	3.000±0.707
Isolucine	4.250±0.353
Leucine	2.750±0.353
Lysine	7.250±1.061
Methionine	3.000±0.707
Phenylalanine	4.500±0.707
Proline	2.250±0.353
Serine	5.750±0.353
Theronine	3.500±0.707
Tyrosine	4.500±0.707
Valine	5.000±0.707

Table 5: Amino acid composition of fungal biomass produced by F. venenatum.

Fatty acid	Content (% w/w)
Myristic (C14:0)	0.384±0.047
Pentadecanoic (C15:0)	0.152±0.014
Palmitic (C16:0)	14.253±0.945
Palmitoleic (C16:1)	0.450±0.007
Margaric (C17:0)	0.205±0.002
Heptadec-9enoic (C17:1)	0.121±0.019
Stearic (C18:0)	6.967±0.146
Elaideic (C18:1t)	0.079±0.007
Oleic (C18:1)	18.439±0.653
Linelaideic (C18:2t)	0.076±0.002
Linoleic (C18:2)	30.859±0.638
γ-Linolenic (C18:3)	0.328±0.002
α-Linolenic (C18:3)	25.180±0.748
Arashidic (C20:0)	0.519±0.048
Gadoleic (C20:0)	0.194±0.003
Behenic (C22:0)	0.421±0.034
Erucic (C22:1)	0.142±0.017
Linoceric (C24:0)	0.739±0.69

Table 6: Fatty acid composition of fungal biomass produced by F. venenatum.

acids. This is in agreement with results reported by Rodger in 2001 [26]. Analysis of fatty acids profile indicated that the ratio of unsaturated to saturated fatty acid was 3.21 to 1 Table 6. Rodger also has reported ratio of 3.5 to 1 for unsaturated to saturated fatty acids. High amounts of unsaturated fatty acids may cause several benefits for human health.

Conclusions

In this study, process variables in surface culture of F. venenatum for protein production from date juice were investigated by response surface methodology. Optimization of variables of date juice and nitrogen source concentrations as well as seed size on mycoprotein production was done by applying FCCD. The protein content of biomass was obtained 46.48% under optimum conditions (date juice 20 g/L, (NH_4) H_2PO_4 4.48 g/L and seed size 12.97% v/v). The results suggest that date juice can be used to produce fungal protein without substantial modification. The scale up of mycoprotein production on modified vogel medium based on date juice is the future trend of this research.

Acknowledgements

We would like to thank the National Nutrition and Food Technology Research Institute (NNFTRI) of the Iran for financial support of this research project.

References

1. Turnbull WH, Leeds AR, Edwards DG (1992) American Journal of Clinical Nutrition 55: 415-419.

2. Williamson DA, Geiselman PJ, Lovejoy J, Greenway F, Volaufova J, et al. (2006) Effects of consuming mycoprotein, tofu or chicken upon subsequent eating behaviour, hunger and safety. Appetite 46: 41-48.

3. Vogel HJ (1956) Microbial Genetics Bulletin 13: 42-44.

4. Trinci APJ (1992) Biotechnology and Bioengineering 40: 1181-1189.

5. Nancib N, Nancib A, Boudrant J (1997) Bioresource Technology 60: 67-71.

6. Chauhan K, Trivedi U, Patel KC (2007) Bioresource Technology 98: 98-103.

7. Nancib N, Nancib A, Boudjelal A, Benslimane C, Blanchard, et al. (2001) Bioresource Technology 78: 149-153.

8. Khosravi-darani K, Falahatpishe HR, Jalali M (2008) African Journal of Biotechnology 7: 1536-1542.

9. Khosravi-Darani K, Farhadi Gh, Mohammadifar MA, Hadian Z, Seyed Ahmadian, et al. (2009) Iranian Journal of Nutrition Science and Food Technology 4: 49-56.

10. Jwanny EW, Rashad MM, Abdu HM (1995) Applied Biochemistry and Biotechnology 50: 71-78.

11. Wiebe MG (2002) Myco-protein from Fusarium venenatum: a well-established product for human consumption. Applied Microbiology and Biotechnology 58: 421-427.

12. Ahangi Z, Shojaosadati SA, Nikoopour H (2008) Pakistan Journal of Nutrition 7: 240-243.

13. Moo-Young M, Chisti Y, Vlach D (1993) Biotechnology Advances 11: 469-479.

14. Hosseini SM, Khosravi-Darani K, Mohammadifar MA, Nikoopour H (2009) Asian Journal of Chemistry 21: 4017-4022.

15. Hosseini SM, Mohammadifar MA, Khosravi-Darani K (2010) International Research. Journal of Biotechnology 1: 13-18.

16. Khosravi-Darani K, Zoghi A (2008) Bioresource Technology 99: 6986-6993.

17. Bruns RE, Scarminio IS, De Barros Neto B (2006) Statistical design – Chemometrics, First ed. Amsterdam: Elsevier.

18. Li Q, Cheng KK, Zhang JA, Li JP, Wang GH (2010) Applied Biochemistry and Biotechnology 160: 604-612.

19. AOAC (2005) AOAC Official Methods of Analysis, AOAC International, Gaithersburg, MD.

20. Chomczynski P, Sacchi N (1987) Analytical Biochemistry 162: 156-159.

21. Bidlingmeyer BA, Cohen SA, Tarvin TL (1984) Pico-Tag method for amino acid determination 336: 93-104.

22. Korean Standard Association (2003) Animal and vegetable fats and oils analysis by gas chromatography of methyl esters of fatty acids (KS H ISO 5508). Geneva: ISO.

23. Zulkali MM, Ahmad AL, Norulakmal NH (2006) Bioresource Technology 97: 21-25.

24. Khosravi-darani K, Zoghi A, Fatemi SSA (2008) Iranian Journal of Chemistry and Chemical Engineering 27: 91-104.

25. Jin B, Van Leeuwen HJ, Patel B, Yu Q (1998) Bioresource Technology 66: 201-206.

26. Rodger G (2001) Food Technology 55: 36-41.

21

Application of Immobilized Pointed Gourd (Trichosanthes dioica) Peroxidase-Concanavalin A Complex on Calcium Alginate Pectin Gel in Decolorization of Synthetic Dyes Using Batch Processes and Continuous Two Reactor System

Farrukh Jamal[1]*, Sangram Singh[1], Sadiya Khatoon[1] and Sudhir Mehrotra[2]

[1]Department of Biochemistry, Dr. Ram Manohar Lohia Avadh University, Faizabad-224001, U.P, India
[2]Department of Biochemistry, University of Lucknow, Lucknow, U.P, India

Abstract

Ammonium sulphate fractionated pointed gourd (*Trichosanthes dioica*) peroxidase- concanavalin A (PGP-Con A) complex was entrapped into calcium alginate-pectin gel. Catalytic performance of immobilized PGP-Con A complex in dye decolorization was examined on repeated use and reusing after a prolonged period of storage. Immobilized and entrapped peroxidase preparation retained 59.6% of the original activity after a period of 50 d. Entrapped PGP-Con A complex decolorized 91.2% and 82.1% of the initial color from DR19 and dye mixture [DR19+DB9] after 20 d, respectively. Considerable color removal was found even after 120 d and 80 d respectively, of operation of two reactor system and total organic carbon analysis was quite comparable to color loss. This study shows the efficacy, durability and sustainability of using immobilized *T. dioica* peroxidase in batch and continuous two reactor catalytic system for the removal of synthetic dyes from industrial effluents.

Keywords: Concanavalin A; Pointed gourd peroxidase (PGP); Immobilization; Biocatalysis; Continuous reactor; Disperse dyes

Abbreviations

Pointed Gourd Peroxidase- concanavalin A (PGP-Con A); Disperse Red (DR 19); Disperse Black (DB 9); *Trichosanthes dioica* (*T. dioica*); Days (d); Weight/volume (w/v); Concanavalin A (Con A); Percent (%); U (units); TOC (total organic carbon); λ (wavelength); nm (nanometre)

Introduction

Enzymes are not only environmental friendly but are also capable of specifically reducing hazardous wastes and hence key to new processes. Due to considerable increase in the level of phenolic compounds and dyes contamination in waste water; approaches that are cheaper, sustainable and eco-friendly are highly desired. Peroxidases are heme-containing monomeric glycoproteins that oxidize a wide variety of molecules [1,2]. These enzymes find wide range of applications in detoxification, dye decolorization and removal of various toxic organic pollutants which contaminate water and industrial effluents [3-5]. Extensive research has focused on developing processes in which enzymes are employed to remove dyes from polluted water [6-8]. Several factors like shorter treatment period; operation of high and low concentrations of substrates; reduced lag phase of biomass, reduction in sludge volume and ease of controlling the process makes enzymatic treatment potentially useful as compared to microbial treatment. Although these enzymes catalyze a variety of aromatic compounds in the presence of hydrogen peroxide, several limitations prevent the use of liquid form of enzymes as their stability and catalytic ability decreases with the complex nature of the effluents [9]. Some of these limitations may be overcome by the use of immobilized enzymes that can be used as catalysts with long lifetime. Immobilization with different polymeric materials is studied for enzyme encapsulation along with their application in treatment of various pollutants. However, appropriate selection of encapsulation material specific to the enzyme and optimization of process conditions is still in early stage of development [10]. The immobilized form of enzymes has several merits over the soluble enzymes such as enhanced stability, easier product recovery and purification, protection of enzymes against denaturants, proteolysis and reduced susceptibility to contamination [9].

The methodologies employed for the immobilization of peroxidases are limited and among them bioaffinity-based physical adsorption strategy is useful and economical [11]. This process can immobilize enzyme directly from crude homogenate overcoming the high cost of purification which employs commercially available enzyme/expensive supports [12]. Besides conferring ease to immobilize proteins, other advantages of such protocols include lack of chemical modification, proper orientation of enzyme on the support, high yield and enhanced stability of glycoenzymes/enzymes [13].

Novel research in the area of enzyme technology has provided significant evidence and strategies that facilitate using enzymes optimally at large scale by entrapping and immobilizing [14,15]. Although enzymes entrapped in porous polymeric matrices pose inherent limitations of enzyme leaching; however by controlling the pore dimensions such leaching can be minimized. Alternatively, entrapping cross-linked or pre-immobilized enzyme preparations could be a better and pragmatic option [16].

Immobilization of enzymes is a tricky and articulated approach with regard to proper orientation, as involvement of key amino acids must be avoided to prevent loss of enzymatic activity. In case of glycoenzymes, glycosyl moieties can safely be used in immobilizations as they do not participate in catalysis. Lectins are proteins which recognize and interact with exposed carbohydrate moieties of glycoproteins and glycoenzymes. These proteins are useful in characterizing glycoproteins

*Corresponding author: Farrukh Jamal, Assistant Professor, Department of Biochemistry, Dr. Ram Manohar Lohia Avadh University, Faizabad-224001, U.P, India

and certain glycoenzymes have been immobilized on concanavalin A affinity matrices or as Con A-glycoenzyme complexes [13,17].

The current study demonstrates a simple, inexpensive and high yield procedure for immobilization of glycosylated *Trichosanthes dioica* peroxidase. *T. dioica* popularly known as pointed gourd is widely planted in tropical areas and consumed as vegetables. Salt fractionated pointed gourd peroxidase immobilized with lectin Con A as insoluble PGP-Con A were entrapped into calcium alginate-pectin beads. A comparative study on the reusability and storage stability of soluble and immobilized form of PGP (PGP-Con A complex and PGP-Con A-calcium alginate-pectin complex) has been presented for using such enzymes in effectively removing/minimising the color in industrial effluent contaminated with dyes.

Materials and Methods

Chemicals

Sodium alginate, Bovine serum albumin (BSA), Concanavalin A, *o*-dianisidine HCl, Disperse Red 19 and Disperse Black 9 was procured from Sigma Chemical Co. (St. Louis, MO, USA). Pectin was obtained from SRL Chemicals, Mumbai, India. All other chemicals were of analytical grade. The pointed gourds were procured from Narendra Dev University of Agriculture and Technology, Faizabad, U.P., India. The samples were aseptically transferred into sterilized plastic bags.

Ammonium sulphate purification of PGP proteins and protein estimation

The homogenate of pointed gourd (~300 gm) was homogenized in 600 mL of 100 mM sodium acetate buffer, pH 5.6, filtered through multi-layers of cheesecloth and centrifuged at 10,000×g on a Remi C-24 cooling centrifuge for 25 min at 4°C. The supernatant was salt fractionated by continuously stirring at 4°C overnight using 10% to 80% (w/v) $(NH_4)_2SO_4$. The precipitate was centrifuged at 10,000×g on a Remi C-24 cooling centrifuge, dissolved in 100 mM sodium acetate buffer, pH 5.6 and dialyzed against the assay buffer (100 mM glycine HCl buffer, pH 5.6 [4,7]. This preparation of protein was aliquoted and stored at -20°C for further use.

Protein concentration was estimated using BSA as a standard protein and following the procedure of Lowry et al. [18].

Preparation of insoluble PGP-Con A complex and entrapment in calcium alginate-pectin beads

The peroxidase proteins (1100 U) were mixed with an increasing concentration of Con A (0.1-1.0 mL) in a series of tubes. Final volume of each tube was adjusted to 5 mL with 100 mM phosphate buffer (pH 5.6). The reaction mixtures were incubated overnight at 37°C. The precipitates were collected after centrifugation at 3000xg for 20 min at room temperature and washed twice with the same buffer. Finally precipitates were suspended in 2 mL assay buffer and each precipitate was analyzed for enzyme activity. The precipitate (PGP-Con A complex) exhibiting maximum activity was used for further studies.

PGP-Con A complex (1250 U) was mixed with sodium alginate (2.5%) and pectin (2.5%) in 10 mL of 100 mM sodium acetate buffer (pH 5.6). Using a syringe the mixture was slowly extruded as droplets. Further, beads were gently stirred for 2 h in calcium chloride solution (1.5% w/v), washed and stored in 100 mM sodium acetate buffer (pH 5.6) at 4°C for further use [7,19].

Measurement of peroxidase activity and effect of enzyme loading

Peroxidase activity was determined by measuring a change in optical density (A_{460} nm) at 37°C of initial rate of oxidation of 6.0 mM *o*-dianisidine HCl in presence of 18.0 mM H_2O_2 in 0.1 M sodium acetate buffer (pH 5.6) for 15 min. Immobilized enzyme preparation was continuously agitated for entire duration of assay [5].

One unit (1.0 U) of enzyme activity is the amount of enzyme protein that catalyzes oxidation of 1.0 μmole of *o*-dianisidine HCl per min at 37°C into colored product. An increasing concentration of enzyme (110-1000 U) was mixed to calcium alginate-pectin gel in a series of tubes. Expression of loaded enzyme was monitored by assaying the peroxidase activity.

Storage stability and reusability of soluble and immobilized PGP

Soluble and immobilized PGP were stored at 4°C in 50 mM sodium acetate buffer (pH 5.6) for over 50 d. The aliquots from each preparation (1.20 U) were taken out in triplicates at the gap of 5 d and were analyzed for the remaining enzyme activity. The catalytic activity measured on day 1 was considered as control (100%) for the calculation of remaining storage activity.

Immobilized PGP was taken in triplicates for assaying peroxidase activity. After each assay the immobilized enzyme preparation was taken out, washed and stored overnight in 50 mM sodium acetate buffer, pH 5.6 at 4°C. The activity was assayed for ten successive days. The catalytic activity determined for the first time was considered as control (100%) for the calculation of remaining percent activity after each use.

Preparation of synthetic dye solutions and calculation of percent dye decolorization

The synthetic solutions of disperse dyes (25-50 mg/mL) were prepared in distilled water to examine their decolorization by PGP. A mixture of disperse dyes consisting of DR19 and DB9 was prepared by mixing each dye in equal proportion of color intensity [5]. To compare various experiments, the decolorization was calculated for each dye or mixture of dyes. Dye decolorization was monitored by measuring the difference at the maximum absorbance for each dye as compared with control experiments without enzyme on UV-visible spectrophotometer (JASCO V-550, Japan). Untreated dye solution (inclusive of all reagents except the enzymes) was used as control for calculation of percent decolorization. The dye decolorization was calculated as the ratio of the difference of absorbance of treated and untreated dye to that of treated dye and converted in terms of percentage. Five independent experiments were carried out in duplicate and the mean was calculated.

Decolorization of dye solution by soluble and immobilized peroxidase in batch processes

The dye solution (200 mL) was treated with soluble and immobilized PGP (27.6 U) in 100 mM sodium acetate buffer, pH 5.6 in the presence of 0.2 mM riboflavin as redox mediator and 0.8 mM H_2O_2 for 1 h at 37°C. The treated samples were centrifuged at 3000 x g for 15 min. The residual dye concentration was measured spectrophotometrically at specific wavelength maxima of the dye. Untreated dye solution was considered as control (100%) for the calculation of percent decolorization.

Reusability of immobilized PGP in the decolorization of disperse dye

DR19 and mixture of dyes [DR19+DB9] were incubated with immobilized PGP for 1h as mentioned earlier. The enzyme was separated by centrifugation and stored in the assay buffer for over 12 h. Experiments was repeated upto 10 times with the same preparation of PGP and each time with a fresh batch of dye solution. Dye decolorization was monitored at specific wave length maxima of the dye solutions. The percent decolorization was calculated by taking untreated dye or mixture of dyes as control (100%).

Continuous dye decolorization using a two-reactor system

For the continuous removal of dyes from solutions two-reactor system was developed. A column (15x2.0 cm) was filled with 10 g entrapped Con A-PGP complex (1268 U) connected to second column (15x2.0 cm) containing activated silica. Silica was activated by heating in an oven (120°C) for overnight and then washed thrice with 30 mL distilled water. The flow rate was maintained at 6.4 mL/h and the feed solution contained either DR19 or mixture of DR19 and DB9 in two independent reactor systems. Dye solutions were run under the same experimental conditions. Reactors were operated at room temperature (37°C) for a period of 4 months. Immobilized PGP treated and activated silica gel adsorbed samples were collected at an interval of 20 d and the absorbance of each sample was recorded.

Procedure for the dye decolorization was followed by centrifugation and clear diluted solution was considered for total organic carbon determination. TOC was evaluated by using a TOC analyzer. Each control and treated DR19 or mixture of dyes (DR19+DB9) was diluted 10-fold before determining their TOC content.

Results and Discussion

Purification of PGP and preparation of PGP-Con A complex

The crude extract of pointed gourd exhibited an initial specific activity of 96 U/mg of protein. Peroxidase was partially purified by ammonium sulphate precipitation and specific activity of preparation was increased 3.5 fold over crude enzyme. This enzyme preparation was used for direct immobilization as enzyme-Con A complex.

Peroxidases from *T. dioica* are usually glycosylated proteins. These enzymes in soluble states may leak out of beads over prolonged retention or repeated use. Thus to prevent the leaching of enzymes from porous gel beads, these enzyme molecules were complexed with lectin concanavalin A. The insoluble PGP-Con A complex was subsequently entrapped into calcium alginate-pectin gel. With 0.2 mL of Con A the PGP-Con A complex expressed an activity of 79% which on entrapment into calcium alginate-pectin gel resulted in further decrease of peroxidase activity (Table 1).

Entrapment of PGP-Con A complex in calcium alginate-pectin beads

Immobilization by means of entrapment is a simple and effective technique. Partially purified PGP precipitated with Con A was used for direct immobilization onto calcium alginate pectin beads. Entrapped PGP-Con A peroxidase complex retained only 56% of the original activity (Table 1). Further, the effect of enzyme loading on entrapped activity was evaluated by entrapping increasing concentration of enzyme. Optimum concentration (418 U/mL) was sufficient for maximum expression of peroxidase activity by entrapped preparation.

Cross-linked enzymes or pre-immobilized enzymes that could remain inside polymeric matrices for longer duration than soluble enzymes provide higher mechanical and operational stability to enzymes [16,20]. It indicated that enzymes with high molecular mass could stay for longer period inside polymeric matrix. Pre-immobilization increases molecular dimensioning of the enzyme and thus prevents its leaching from alginate beads.

Reusability and Storage stability of immobilized PGP

Reusability of preparation has been shown in Figure 1. After sixth repeated use, PGP-Con A retained 56.6% of the original activity whereas entrapped exhibited 71.1% peroxidase activity. However, thereafter progressive decline in activity was recorded and on day 10 PGP-Con A complex and entrapped PGP-Con A complex retained only 19.6% and 38.7% activity respectively.

Enzyme reuse provides a number of cost effective advantages that are often an essential prerequisite for establishing an economically viable enzyme catalyzed process [9]. Immobilized PGP (entrapped PGP-Con A complex) retained 53.6% of its original activity even after its ninth successive uses (Figure 1). PGP-Con A complex exhibited loss in enzymatic activity after sixth repeated use as compared to entrapped PGP. The loss in activity upon reusability may be due to several factors, one of which could be due to inhibition of enzyme by its reaction product [21]. Another factor causing profound loss in activity of PGP-Con A complex could be the loss in recovery while monitoring its reusability.

Storage stability of soluble and immobilized preparations of PGP was monitored at a gap of 5 d for 50 d at 4°C (Figure 2). The immobilized peroxidase preparation [entrapped PGP-Con A complex] retained 59.6% of the original activity after a period of 50 d. The PGP-Con A complex exhibited 43.5% peroxidase activity and on the contrary the soluble form of PGP lost upto 84.7% activity under similar storage conditions. Thus, immobilized PGP exhibited remarkable stability on prolonged storage. On the basis of results obtained in the present work, it can be concluded that the stability offered by immobilized

Type of Immobilization	Original Activity*	Expressed Activity*
PGP-Con A Complex	100%	79%
PGP-Con A complex entrapped on calcium-alginate pectin gel	100%	56%

*Original activity is the activity of soluble PGP and the expressed activity is the activity achieved after immobilization.

Table 1: Activities of PGP immobilized with lectin Con A and calcium-alginate pectin gel.

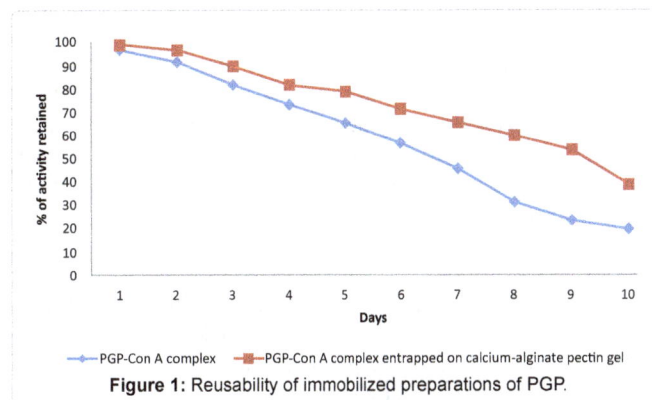

Figure 1: Reusability of immobilized preparations of PGP.

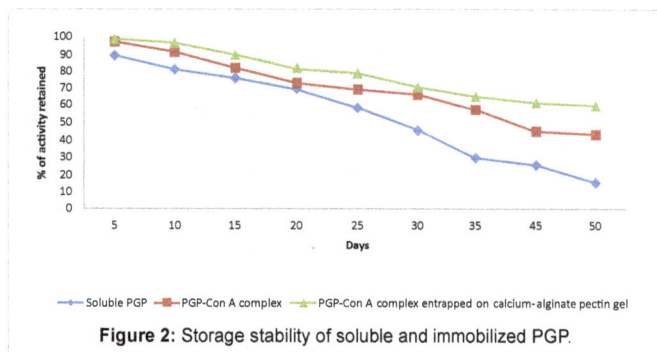

Figure 2: Storage stability of soluble and immobilized PGP.

PGP preparation could successfully be employed in reactors for the treatment of effluents containing phenolic and other aromatic pollutants including dyes which are primarily represented in textile effluents. The reusability and storage experiments further supported that the use of such a cheaper source of enzyme and support will definitely minimize the cost of immobilization and provide a suitable approach for the treatment of huge volumes of wastewater in batch processes as well as in continuous reactors.

Reusability of immobilized PGP in dye decolorization

In order to make immobilized PGP (entrapped PGP-Con A complex) use more practical in a reactor, it was necessary to investigate its reusability on repeated use in the decolorization of disperse dyes. The dye decolorizing reusability of entrapped Con A-PGP complex continuously decreased upto its tenth repeated use (Table 2). Immobilized PGP retained 51.4% and 35.6% for DR19 and dye mixture [DR19+DB9] respectively, after tenth repeated use.

Dye decolorization by immobilized PGP in batch process and a two-reactor system

The decolorization of dyes by PGP has been illustrated in Table 3. Soluble PGP decolorized 81.9% and 59.1% of DR19 and mixture of dyes after 2 h of incubation, respectively. However, immobilized PGP was more effective as compared to its soluble counterpart in the decolorization of both DR19 and mixture of dyes. Decolorization by immobilized PGP under similar conditions was 78.4% and 69.5% for DR19 and dye mixture respectively after 1 h of incubation. On the contrary the performance of the two-reactor system in terms of dye decolorization is shown in Table 4. Entrapped Con A-PGP complex decolorized 91.2% and 82.1% of the initial color from DR19 and dye mixture [DR19+DB9] after 20 d, respectively. Considerable color removal from DR19 (59.4%) and mixture of disperse dyes (53.2%) was found even after 120 d and 80 d respectively of operation of the two-reactor system.

The treatment of DR19 and mixture of dyes [DR19+DB9] by passing through a double reactor system provided almost the water free from dyes. The dyes treated by immobilized PGP present in the first column, got adsorbed in the second column, which contained activated silica. Both the reactors worked for more than 120 d approximately, thus explaining their efficiency towards dye decolorization. A significant loss of color appeared when DR19 or mixture of dyes was treated with entrapped PGP-Con A complex in the presence of redox mediator, riboflavin in a continuous reactor system (Table 4). It has earlier been reported that the disappearance of peak in visible region was either due to the breakdown of chromophoric groups present in dyes or the removal of pollutants in the form of insoluble products [22].

The level of TOC in case of continuous reactors was significantly decreased in the presence of immobilized PGP treated polluted water. However, the PGP treated dye solutions exhibited great loss of TOC from the wastewater (Table 4), which suggested that the major toxic compounds could have been removed out of the treated samples. Immobilized HRP removed 88% of TOC from model wastewater containing mixture of chlorophenols [15]. It has been reported that a significant amount of TOC from polluted water containing dyes/dye-mixtures and dyeing effluent was removed when treated with soluble and immobilized bitter gourd peroxidase [23]. These evidences strongly proved that immobilized PGP-Con A complex could be successfully used for the removal of dye effluents loaded with recalcitrant synthetic dyes.

Conclusions

The stability exhibited by calcium alginate-pectin entrapped Con A-PGP preparation was significantly higher compared to soluble PGP and Con A-PGP complex. Thus, immobilized PGP preparations could be exploited for developing bioreactors for the treatment of phenolic and other aromatic pollutants including synthetic dyes present in industrial effluents. A two-reactor system with simple operational protocol for decolorization/degradation of disperse dyes has been focused for the

No of Uses	DR19 (λ495nm)	DR19 and DB9 mixture (λ460nm)
	Percent Dye decolorization	Percent Dye decolorization
1	93.7	88.2
2	89.4	84.6
3	78.6	74.7
4	72.8	69.4
5	67.9	63.5
6	65.6	60.4
7	63.9	51.4
8	57.4	43.2
9	53.6	38.5
10	51.4	35.6

Table 2: Dye/dyes mixture decolorization and reusability of immobilized PGP. Immobilized PGP was independently incubated with DR19 and mixture of dyes [DR19+DB9] (300 mL) for 2 h at 40°C. Dye decolorization was determined after incubation period of 2 h. The immobilized enzyme was collected by centrifugation and stored in assay buffer at 4°C overnight. Next day, the similar experiment was repeated. This procedure was repeated 10 times. Each value represents the mean for three independent experiments performed in duplicate.

Time (min)	Percent Dye decolorization			
	DR19(λ495nm)		DR19 and DB9 mixture (λ460nm)	
	Soluble PGP	Entrapped PGP-Con A complex	Soluble PGP	Entrapped PGP-Con A complex
20	69.8	78.5	61.2	69.5
40	74.6	78.5	60.7	69.6
60	78.8	78.4	60.4	69.5
80	79.9	78.4	59.5	69.5
100	79.9	78.4	59.4	69.4
120	81.9	78.4	59.1	69.4
140	62.4	78.4	38.2	69.3
160	59.6	78.4	28.5	69.4
180	42.2	78.4	25.6	69.4

Table 3: Dye decolorization in batch processes. DR19 or mixture of dyes [DR19+DB9] (250 mL) was treated with soluble and immobilized PGP (27.6 U) for varying times in batch processes. Each value represents the mean for five independent experiments performed in duplicate.

No of days	DR19 (λ495nm)		Mixture of DR19 and DB9 (λ460nm)	
	Percent Dye decolorization	Percent TOC Removal	Percent Dye decolorization	Percent TOC Removal
20	91.2	93.2	82.1	83.4
40	79.4	80.1	76.9	77.3
60	77.7	78.2	64.3	67.8
80	65.8	66.7	53.2	56.3
100	63.4	62.8	33.7	37.4
120	59.4	60.2	23.2	25.1

Table 4: Continuous removal of color and TOC from DR19 and a mixture of [DR19+DB9] by immobilized PGP in a two-reactor system. DR19 / mixture of dyes was treated with PGP in a two-reactor system as described in text. Each value represents the mean for five independent experiments performed in duplicate.

potential future use of immobilized peroxidases. Interestingly, the described system is developed with a cheaper biocatalyst that is quite effective in treating dyes continuously in a small laboratory reactor.

Future directions

With the recent focus on enzyme based treatment of colored wastewater/ industrial textile effluent, the enzymes in soluble form cannot be exploited on large scale due to their limitations of stability and reusability. Consequently, the use of immobilized enzymes has significant advantages over soluble enzymes. In the near future, cost effective, eco-friendly technologies based on the enzymatic approach for treatment of dyes present in the industrial effluents/wastewater will play a vital role. Comprehensive studies on the chemical modification of pointed gourd peroxidase would provide us lead to exploit novel versions of peroxidases and thereby increase their resistance to withstand high concentrations of phenolic compounds/ dyes in effluent wastewater. Thus, our aim would be to characterize and express putative peroxidase from *Trichosanthes dioica* (Pointed Gourd) whose peroxidase activity has recently been shown by us to effectively decolorize several recalcitrant synthetic dyes as well as non textile dyes. This would go a long way in expressing an effective version of peroxidase gene and subsequently use its products in bioreactors to treat textile effluents. Such peroxidase system would retain high remaining activity in a recycle processing unit deploying immobilized enzyme.

Acknowledgement

We are thankful to the Department of Science and Technology (DST-FIST) under the Ministry of Science and Technology for providing financial assistance towards infrastructure development for carrying out this work. Grant received from CSTUP, Lucknow is gratefully acknowledged.

References

1. Azevedo AM, Martins VC, Prazeres DM, VojinoviÄ‡ V, Cabral JM, et al. (2003) Horseradish peroxidase: a valuable tool in biotechnology. Biotechnol Annu Rev 9: 199-247.

2. Chanwun T, Muhamad N, Chirapongsatonkul N, Churngchow N (2013) Hevea brasiliensis cell suspension peroxidase: purification, characterization and application for dye decolorization. AMB Express 3: 14.

3. Husain Q (2006) Potential applications of the oxidoreductive enzymes in the decolorization and detoxification of textile and other synthetic dyes from polluted water: a review. Crit Rev Biotechnol 26: 201-221.

4. Jamal F, Pandey PK, Qidwai T (2010) Potential of peroxidase enzyme from Trichosanthes dioica to mediate disperse dye decolorization in conjunction with redox mediators. J Mol Cat B: Enzym 66: 177-181.

5. Jamal F, Qidwai T, Pandey PK, Singh R, Singh S (2011) Azo and anthraquinone dye decolorization in relation to its molecular structure using soluble Trichosanthes dioica peroxidase supplemented with redox mediator. Catal Commun 12: 1218-1223.

6. López C, Mielgo I, Moreira MT, Feijoo G, Lema JM (2002) Enzymatic membrane reactors for biodegradation of recalcitrant compounds. Application to dye decolourisation. J Biotechnol 99: 249-257.

7. Jamal F, Qidwai T, Singh D, Pandey PK (2012) Biocatalytic activity of immobilized pointed gourd (Trichosanthes dioica) peroxidase-concanavalin A complex on calcium alginate pectin gel. J Mol Cat B: Enzy 74: 125-131.

8. Silva MC, Correa AD, Amorim MTSP, Parpot P, Torres JA, et al. (2012) Decolorization of the phthalocyanine dye reactive blue 21 by turnip peroxidase and assessment of its oxidation products. J Mol Cat B: Enzym 77: 9-14.

9. Zille A, Tzanov T, Gübitz GM, Cavaco-Paulo A (2003) Immobilized laccase for decolourization of Reactive Black 5 dyeing effluent. Biotechnol Lett 25: 1473-1477.

10. Gholami-Borujeni F, Mahvi AH, Naseri S, Faramarzi MA, Nabizadeh R, et al. (2011) Application of immobilized horseradish peroxidase for removal and detoxification of azo dye from aqueous solution. Res J Chem Environ 15, 217-222.

11. Akhtar S, Khan AA, Husain Q (2005) Simultaneous purification and immobilization of bitter gourd (Momordica charantia) peroxidases on bioaffinity support. J Chem Technol Biotechnol 80: 198-205.

12. Gupta MN, Mattiasson B (1992) Unique applications of immobilized proteins in bioanalytical systems. Methods Biochem Anal 36: 1-34.

13. Mislovicová D, Gemeiner P, Sandula J, Masárová J, Vikartovská A, et al. (2000) Examination of bioaffinity immobilization by precipitation of mannan and mannan-containing enzymes with legume lectins. Biotechnol Appl Biochem 31 : 153-159.

14. Gomez JL, Bodalo A, Gomez E, Bastida J, Hidalgo AM, et al. (2006) Immobilization of peroxidases on glass beads: An improved alternative for phenol removal. Enz Microb Technol 39: 1016-1022.

15. Husain M, Husain Q (2007) Applications of Redox Mediators in the Treatment of Organic Pollutants by Using Oxidoreductive Enzymes. Crit Rev Environ Sci Technol 38: 1-42.

16. Betancor L, López-Gallego F, Hidalgo A, Fuentes M, Podrasky O, et al. (2005) Advantages of the pre-immobilization of enzymes on porous supports for their entrapment in sol-gels. Biomacromolecules 6: 1027-1030.

17. Kulshrestha Y, Husain Q (2006) Direct immobilization of peroxidase on DEAE cellulose from ammonium sulphate fractionated proteins of bitter gourd (Momordica charantia). Enz Microb Technol 38: 470-477.

18. Lowry OH, Rosebrough NJ, Farr AL, Randall RJ (1951) Protein measurement with the Folin phenol reagent. J Biol Chem 193: 265-275.

19. Nigma SC, Tsao I-F, Sakoda A, Wang HY (1988) Techniques for preparing hydrogel membrane capsule. Biotechnol Techn 2: 271-276.

20. Musthapa SM, Akhtar S, Khan AA, Husain Q (2004) An economical, simple and high yield procedure for the immobilization/stabilization of peroxidases from turnip roots. J Sci Ind Res 63: 540-547.

21. Gåserød O, Sannes A, Skjåk-Braek G (1999) Microcapsules of alginate-chitosan. II. A study of capsule stability and permeability. Biomaterials 20: 773-783.

22. Moreira MT, Palma C, Mielgo I, Feijoo G, Lema JM (2001) In vitro degradation of a polymeric dye (Poly R-478) by manganese peroxidase. Biotechnol Bioeng 75: 362-368.

23. Akhtar S, Husain Q (2006) Potential applications of immobilized bitter gourd (Momordica charantia) peroxidase in the removal of phenols from polluted water. Chemosphere 65: 1228-1235.

22

Improvement and Immobilization of a new Endo-β-1,4-xylanases KRICT PX1 from *Paenibacillus* sp. HPL-001

Hee Kyung Lim[1], No-Joong Park[1], Young Kyu Hwang[1,2], Kee-In Lee[1,2] and In Taek Hwang[1,2]*

[1]*Nanocatalysis Center, Green Chemistry Division, Korea Research Institute of Chemical Technology, Yuseong PO Box 107, Daejon 305-600, Republic of Korea*
[2]*Department of Green Chemistry and Environmental Biotechnology, University of Science and Technology, 217 Gajungro, Yuseong- gu, Daejon, 305-350, Republic of Korea*

Abstract

A new endo-1,4-beta-xylanase gene (FJ380951), KRICT PX1, isolated from *Paenibacillus* sp. HPL-001, was expressed in *E. coli*. Enzyme purification, mutation, immobilization, and molecular simulation were conducted. The P1H1 mutant of one amino acid substitution (168Leu→Gln) showing the strongest xylanase activity was selected among 250 mutant clones. The specific activity of the P1H1 xylanase was 53.3 U/mg protein at 50°C, an approximately five-fold increase compared to that of the original KRICT PX1 xylanase (10.25 U/mg protein) at the same condition of pH 5~10, and enhanced activity from 20.1 to 36.6 U/mg proteins at 60°C. The structural dynamics of the mutant P1H1 became stronger than that of the wild type, which could be eliminated by a helix deformation with the H-bond missing between Leu_{168} and Asp_{164} after mutation. Xylanase could be reused for 5 batches without significant loss of activity after being immobilized on surface functionalized silica-based mesoporous cellular foam. However, xylanase activity declined to 60% at the 10th batch. Further research on practical applications will be necessary for industrial usage.

Keywords: Biocatalyst; Error-Prone PCR; Hemicelluloses; Mutation; Xylanase

Introduction

Endo-β-1,4-xylanases catalyze the hydrolysis of the backbone of xylan to produce xylooligosaccharides, which in turn can be converted to xylose by β-xylosidase [1]. Endo- xylanase (1,4-β-D-xylan xylohydrolase, EC 3.2.1.8) is initiates the degradation of xylan into xylose and xylo-oligosaccharides of different sizes [2]. This enzyme also has been widely used in baking and industrial processes such as the clarification of fruit juices, waste treatment, and the bio-bleaching of paper pulp [3]. An increasing number of reports and articles mentioning the isolation of newer microbial species for xylanase production reveal growing interest by the scientific community in this field. Owing to the rising biotechnological importance of xylanases, interest in industrially applicable xylanases has markedly increased. Accordingly, many attempts are being made to isolate new strains and to discover more relevant xylanases [4,5].

Before using xylanase at industrial levels, several criteria have to be fulfilled. Pilot scale processes are generally carried out at high temperatures; therefore bacterial xylanase having broad ranges of pH and temperature stability are preferred in industrial applications [6]. Similarly, xylanases extracted from actinomycetes are also operational over a broad range of reaction parameters [7,8], whereas fungal xylanases are stable under acidic pH conditions, ranging from pH 4 to 6 only [9]. However, some fungal species produce xylanases that are active at highly alkaline pH, but their number is few and they are less efficient in comparison to the acidic fungal xylanases. In order to obtain the maximal effect of xylanase, some of the reaction parameters, such as enzyme dose, retention time, pH, and temperature, should be optimized. Xylan-degrading enzymes of microorganisms are potentially important in various industrial processes for use under high temperatures and a broad range of pH conditions. A large proportion of research goes into finding or creating enzymes with specific properties, but the inherent activity, stability, and enantioselectivity of enzymes are not always suitable for industrial applications; there is consequently a need to engineer natural enzymes [10,11].

Enhancing the characteristics of the enzyme to lend it more suitable characteristics such as thermo-stability, pH optima, or substrate specificity can be accomplished using a number of approaches. The most commonly used random mutagenesis method is error-prone PCR [12], which introduces random mutations during PCR by reducing the fidelity of DNA polymerase. The fidelity of DNA polymerase can be reduced by adding manganese ions or by biasing the dNTP concentration. The use of a compromised DNA polymerase causes disincorporation of incorrect nucleotides during the PCR reaction, yielding randomly mutated products.

Immobilization of enzymes is an alternative to facilitate downstream processing such as recycling the enzymes as well as enhancing activity, stability, and selectivity. Silica-based mesoporous materials are attractive supports for enzyme immobilization because they have large surface areas and high pore volumes, with well-defined pore sizes. They also offer the possibility of enzyme immobilization at mild conditions. Among mesoporous materials used as enzymes supports; mesocellular foams (MCF) have shown better performance in terms of retaining activity and mass transfer feasibility [13].

We have recently discovered a new xylanase, KRICT PX1, from a strain of *Paenibacillus* sp. HPL-001 (Korean Collection for Type Culture: KCTC 11365BP) and expressed in *Escherichia coli*. Also, some of the biochemical properties of the purified enzyme were reported [5]. In the present study, we immobilized the enzyme on silica-based Mesoporous Cellular Foam (MCF) and enhanced the thermo-stability of the enzyme to make it more suitable for industrial needs using error-prone PCR.

Materials and Methods

Isolation and mutation of *KRICT PX1* gene

Xylanase KRICT PX1 has been isolated from a *Paenibacillus* sp.

***Corresponding author:** In Taek Hwang, Department of Green Chemistry and Environmental Biotechnology, University of Science and Technology, 217 Gajungro, Yuseong- gu, Daejon, 305-350, Republic of Korea*

HPL-001 (KCTC11365BP) and the sequences of nucleotides or deduced amino acids were analyzed with CLC Free Workbench, Ver. 3.2.1 (CLC bio A/S, www.clcbio.com). Related sequences were obtained from database searches (SwissPort and GenBank). The genome sequence of KRICT PX1 was submitted to GenBank, and assigned as Accession Number FJ380951. The xylanase KRICT PX1 gene consisting of 996 bp encoding a protein of 332 amino acids (38.1 kDa) has been cloned and expressed in *Escherichia coli*. The xylanase KRICT PX1 was purified with an immobilized glutathione column and cleaved with Factor Xa before elution through a benzamidine column.

Site-directed mutagenesis was performed on KRICT PX1 using an Error-prone PCR kit from Clontech according to the protocol recommended by the manufacturer and using the following primers: PX1M FP-*Bam*HI (forward), AGGGATCCCATGAGGTTGCG, PX1M RP-*Hind*III(reverse), CCAAGCTTGGCCTAAAGAGTCAAG. The PCR conditions used were as follows: 1 min at 95°C, 25 cycles (30 sec at 94°C, 30 sec at 55°C, 1 min at 68°C), and 5 min at 68°C. Products were digested with *Bam*HI and *Hind*III and cloned into the pSTV28 plasmid vector (Takara Bio Inc.). The obtained plasmid library was used to transform competent *E. coli* JM109 (Promega).

Selection and Expression of the KRICT PX1 Mutant Gene in *E. coli*

E. coli JM109 transformants were picked with a sterile tooth pick and re-suspended in separate wells of a 96-well flat-bottom block containing 200 μl of LB broth (Difco) and ampicillin in each well. After growing the cells for 24 h at 37°C, the cells were collected by centrifugation and then suspended in 200 μl of 1 × phosphate-buffered saline (Sigma-Aldrich). Lysis was directly performed with a 96 pin probe-attached sonic disrupter (Sonic vibra-cells, Sonics & Materials). Activity was determined by the DNS method [14]. An aliquot of 50 μl of lysate from each well was mixed in 50 μl of 2% birch wood xylan (Fluka). After incubation at 50°C for 30 min, two volumes of stop DNA solution (1% (w/v) DNS, 30% (w/v) potassium sodium tartarate, and 0.4 N NaOH) were added. The mixture was boiled at 99°C for 10 min, and activity was determined from the increase of absorbance at 540 nm on a Bench Mark Plus plate reader (Bio-rad).

The four xylanase-active clones were selected from the library screening, and the nucleotide sequence of each insert was determined by automated sequencing under BigDye™ terminator cycling conditions. The reacted product was purified using ethanol precipitation and analyzed with an Automatic Sequencer 3730 × l (Applied Biosystems, Weiterstadt, Germany). Nucleotide and deduced amino acid sequences of selected *KRICT PX1* mutant clones were analyzed with *KRICT PX1* using the ClustalW2 tool (http://www.ebi.ac.uk/). The *KRICT PX1* mutant gene was inserted into the pIVEXGST fusion vector and transformed into *E. coli* BL21 (Roche Applied Science) to produce a recombinant fusion protein.

Purification and analysis of recombinant enzyme

The transformed *E. coli* with GST-fused xylanase were grown overnight in 100 ml of LB medium containing ampicillin (100 μg/ml) at 37°C and 200 rpm in a shaking incubator. After 4 h, the cultures were induced with 1 mM IPTG and incubated under the same conditions for 4 h longer. The induced bacteria were collected by centrifugation and suspended in 10 ml of ice-cold 1X phosphate-buffered saline (Sigma-Aldrich), and this process was performed three times. The cells from the final washing process were resuspended in the lysis buffer (pH 7.0, 200 mM Tris-HCl, 10 mM NaCl, 10 mM β-mercaptoethanol, 1 mM

EDTA), and treated with a sonic disruptor (CosmoBio Co., LTD). After cell disruption, the lysate was centrifuged at 10,000 g for 20 min at 4°C and the supernatant was eluted through a GST binding resin column (Novagen, Madison WI, USA).The column was previously equilibrated with a washing buffer (pH 7.0, 50 mM Tris-HCl, 15 mM NaCl), and the lysate was then applied to the equilibrated column with a flow rate at 10-15 cm/h. After washing the GST resin with 20 bed volumes of cold washing buffer, the fusion protein was eluted and fractionated with 10 bed volumes of freshly prepared elution buffer (pH 7.0, 20 mM Glutathione, 50 mM Tris-HCl, 15 mM NaCl). All fractions were examined with Bradford's protein determination, a PAGE analysis, and a DNS assay. Active fractions were pooled and treated with Restriction Protease Factor Xa (Roche Applied Science), and eluted through a *p*-aminobenzamidine-agarose column (Sigma- Aldrich) according to the manufacturer's instruction. The elute was precipitated with 70% ammonium sulfate, solubilized in phosphate-buffered saline, and dialyzed to concentrate at 4°C with a dialysis membrane (Spectra/Por CE, MWCO 10,000), and then the protein content was determined prior to being stored at -20°C. Sodium dodecyl sulfate-polyacrylamide gel electrophoresis (SDS-PAGE) on 15% polyacrylamide was performed by the method reported by Laemmli [15]. The protein fractions were boiled for 3 min and applied to the gel. Proteins were visualized by Coomassie brilliant blue R 250 staining. The protein concentration was determined by a Bradford [16] assay using bovine serum albumin as a standard.

Xylanase activity was measured according to a method slightly modified from Saha [17] using 50 μl of a 1% (w/v) solution of birch wood xylan (Fluka) and 200 mM each pH buffer (50 mM citric buffer for pH 2-6.5, phosphate buffer for pH 7-9, and glycine buffer for pH 9.5-12) incubated with 30 μl of an appropriately diluted enzyme (3.3 mg/ml) for 20 min at different temperatures and pHs. The activity of the xylanase according to the different substrates was assessed at the same conditions of 1% xylan from beech wood, birch wood, and oat spelt less than 50°C at pH 7.0. The released reducing sugars were assayed using the DNS method [14]. One unit of xylanase activity was defined as the amount of enzyme that liberated 1 μmol of xylose equivalents per minute under the assay conditions. The optimal temperature and pH conditions for the xylanase activity of recombinant KRICT PX1 and KRICT P1H1 protein were examined in 96-well micro plates with a DNS assay at various temperatures (50°C, 60°C) and pH conditions (ranging from pH 5 to 10). The effect of the birch wood xylan concentration on the xylanase activity was evaluated under optimal assay conditions. A diluted enzyme solution (100 μg protein in 30 μl) was incubated with 0.5 ml of various concentrations (0-10 mg/ml) of xylan in 50mM citrate buffer (pH 5.5) and glycine buffer (pH 9.5) at 50°C for 20 min. Xylanase activity was measured with a DNS assay at 540 nm, as described above. The kinetic parameters (Michaelis-Menten constant, K_m, and maximal reaction velocity, V_{max}) were estimated by a linear regression from double-reciprocal plots according to Lineweaver and Burk [18]. The effects of metallic ions and other chemicals on the xylanase activity of KRICT PX1 and KRICT P1H1 proteins were studied as described above at pH 7 with addition of 1 mM NaCl, LiCl, KCl, NH_4Cl, $CaCl_2$, $MgCl_2$, $MnCl_2$, $CuSO_4$, $ZnSO_4$, $FeCl_3$, $CsCl_2$, ethylenediamine tetra-acetic acid (EDTA), 2-mercaptoethanol (2-ME), dithiothreitol (DTT), phenylmethane sulphonyl fluoride (PMSF), acetate, and furfural, respectively.

Structure modeling and molecular dynamics simulation analysis

The three-dimensional structures of xylanases KRICT PX1 and

KRICT P1H1 were modeled using the Discovery Studio version 3.0 program at the ExPASy server. The RCSB Protein Data Bank entries for these proteins are 2q8x and 2fgl. The modeled structures were visualized and analyzed using Swiss-Pdb Viewer version 3.51. The Figures were created with MOLSCRIPT version 2.1. A molecular dynamics simulation analysis was conducted with the calculation protocol of minimization 2,000 step (steepest descent, fixed backbone constraint, 1st stage only), minimization 2,000 step (steepest descent), heating 100,000 step (0 → 300K), equilibration 500,000 step (300K), production 400,000 step (300K, NVT, 200 structures), and implicit solvent (distance dependent: dielectric constant 1).

Synthesis of Mesoporous Cellulose Foam (MCF) and enzyme immobilization

MCF was prepared by a hydrothermal method that has been reported elsewhere [19]. MCF synthesis: 1.62 g (0.279 mmol) of Pluronic P123 (triblock copolymer, chemical formula:

HO(CH$_2$CH$_2$O)$_{20}$(CH$_2$CH(CH$_3$)O)$_{70}$(CH$_2$CH$_2$O)$_{20}$H, molecular weight: 5,800) was transferred into a 100 ml volume of a polypropylene bottle and dissolved in a mixture of 33.33 g (1,852 mmol) of deionized water containing 0.8 g (13.32 mmol) of acetic acid. After clearly dissolved, the solution was heated to 60°C and kept for 1 hr in an oil bath. Another solution was prepared with 2.67 g (11 mmol) of sodium silicate as a silica source in 33.33 g of water. The molar composition of the final solution was Na$_2$SiO$_3$:P123:H$_2$O:acetic acid=1:0.025:336.4:121. The latter solution was dropped into the solution, and then aged at 60°C for 1h and reacted at 100°C for 12 h. The bottle was subsequently cooled to room temperature. A white precipitated product was filtered with water and ethanol. The final product was obtained by heat treatment at 550°C for 6 hrs. The surface area (Braunauer Emmett Teller, BET) of MCF and a Transmission Electron Microscope (TEM) image of MCF are presented in Figure 1.

Amine grafted MCF: Amine grafted MCF was prepared by post grafting with 3- aminopropyl triethoxysilane as an amine source. 1 g of MCF dried at 200°C for 12 h was added to 80 ml toluene under N$_2$ gas with a flow rate of 5 cc/min. 1.11 g (5 mmol) of (3- aminopropyl) triethoxylsilane was slowly dropped to the solution. After mixing for 5 min, the mixture was heated to 110°C and maintained at that temperature for 24 hrs under reflux condition. Final product was recovered by filtration with water and ethanol.

Aldehyde grafted MCF: 0.5 g amine functionalized MCF was added to 50 ml water and then 0.2 g (2 mmol) of glutaraldehyde was injected to the mixture. After stirring at room temperature for 5min, the mixture was heated to 60°C and maintained at that temperature for 24 hrs. The final product was filtered with water and ethanol and FT-IR spectrum was presented at Figure 2. Enzyme immobilization was conducted at pH 7, MCF or alkyl MCF (2 mg) was added to 4 ml

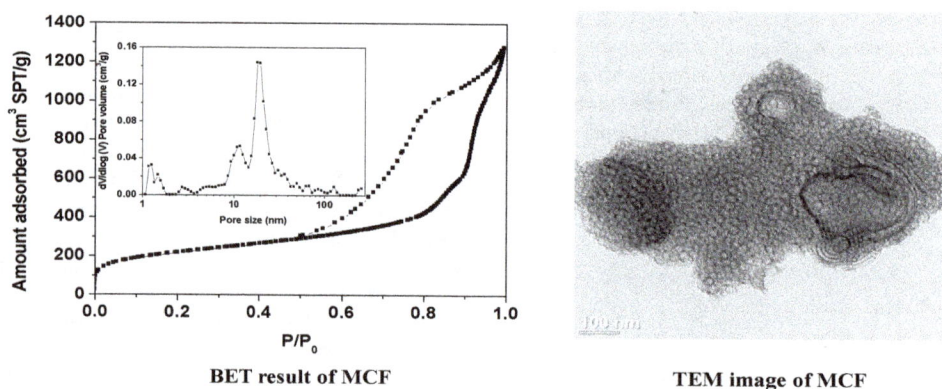

BET result of MCF

TEM image of MCF

Figure 1: Surface area (BET) and Transmission Electron Microscope image (TEM) of Mesocellular Foams (MCF).

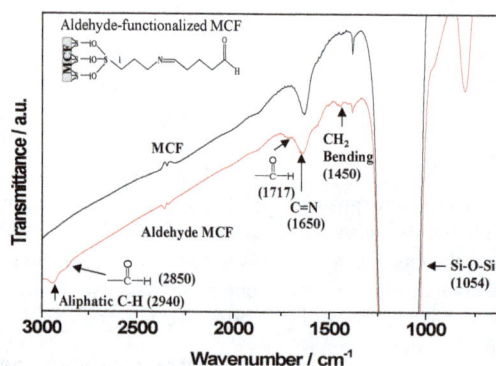

Figure 2: Aldehyde-functionalized MCF was confirmed by Fourier Transform Infrared Spectroscopy (FT-IR). Black line: standard MCF, Red line: aldehyde-functionalized MCF.

buffer solution (50 mM of Tris-HCl, pH 7.0) containing xylanase (0.5 mg/ml) in a 15-ml micro tube. The micro tube was shaken at 85 strokes per minute using a shaking incubator model SI-300 overnight at 20°C. The suspension was centrifuged at 2,000 rpm for 2 min and then the supernatant was taken from each micro tube. The solid (the MCF or alkyl-MCF with immobilized xylanase) was washed twice with 4 ml of buffer, centrifuged and the liquid removed. The supernatants were collected and checked for residual protein concentration according to the Bradford protein assay using bovine serum albumin as standard. The adsorption ratio (%) of xylanase protein on the supports was defined as $(m_i-\Sigma\ m_f)/m_i$, where m_i and Σm_f are the masses of xylanase protein of initial xylanase solution and supernatants, respectively. The xylanase protein loading on the supports was defined as $(m_i-\Sigma m_f)/W\ support$, where m_i and Σm_f are the masses of xylanase protein of in the initial xylanase solution and supernatants, respectively. $W_{support}$ indicates the weight of the support. Repeated batch reaction was conducted with xylanase immobilized on MCF using the same reaction mixture as for batch mode reaction of xylanase activity assay after wash enzyme and filtering with membrane filter (ø 100 μm).

Results and Discussion

Enzyme improvement with error-prone PCR

We screened a random mutation library generated by error prone PCR of the xylanase gene *KRICT-PX1* for enhanced thermo-stability and to make it more suitable for industrial needs. A DNS assay in a 96-well micro-plate was used to select mutants with enhanced xylanase activity. Several mutants showed xylanase activity at pH 9.0. From the results, we identified 6 transformants that expressed a library of 250 random mutants. Table 1 lists the top six mutants selected from the pH 9.0 conditions, together with their DNA and amino acid changes, and activity at 50°C and 60°C. DNA and amino acid sequence analyses showed that the mutant of P1F10 had weak xylanase activity and a DNA change of $_{45}$Cytosine to Thymine without an amino acid change. The mutant of the P1C11 clone had weak xylanase activity and a DNA sequence change of $_{841}$Adenine to Thymine with a $_{281}$Threonine to Serine amino acid change. Mutant P1H1 showed three DNA sequences ($_{183}$Thymine to Adenine, $_{503}$Thymine to Adenine, $_{534}$Thymine to Cytosine) and one amino acid ($_{168}$Leucine to Glycine) changes with remarkable improvement of xylanase activity. Mutant P2A2 showed two DNA ($_{386}$Cytosine to Thymine, $_{177}$Adenine to Guanine) and one amino acid ($_{123}$Alanine to Valine) change with moderate xylanase activity. P2C1 showed only one DNA ($_{102}$Cytosine to Thymine) change without an amino acid mutation and weak xylanase activity. P2D3 showed one DNA ($_{800}$Guanine to Adenine) and one amino acid ($_{267}$Glycine to Serine) change. Among the 250 mutant clones, the P1H1 clone was selected; it showed the greatest xylanase activity, a maximum of 53.3 U/mg protein at pH 7.0 and 50°C, which was approximately a 5-fold increase compared to the original KRICT PX1 xylanase activity of 10.25 at the same conditions. Also, the xylanase activity of P1H1 was enhanced from 20.1 to 36.6 U/mg proteins at pH 7.0 and 60°C (Figure 3). A molecular dynamics simulation study showed that the structural dynamics of the mutant P1H1 became stronger than that of the wild type, which could be eliminated by a helix deformation with the H-bond missing between $_{168}$Leu and $_{164}$Asp after mutation (Figure 4).

Clone ID*	DNA change**	Amino acid change***	Activity****		
			pH 6.0, 50°C	pH 9.0, 50°C	pH 9.0, 60°C
P1F10	45 C → T	None	★★	★	-
P1C11	841 A → T	281 Thr → Ser	★	★★	★
	183 T → A	None	★★★★★	★★★★★	★★★★
P1H1	503 T → A	168 Leu → Gln			
	534 T → C	None			
P2A2	386 C → T	123 Ala → Val	★★	★★★	★★
	177 A → G	None			
P2C1	102 C → T	None	★	★	-
P2D3	800 G → A	267 Gly → Ser	★★	★★★	★

*Clone ID means clone position among mutant library in 96-well plate.
**Position and base in DNA sequence of xylanase krict PX1; Adenin (A), Timin (T), Sitosin (C), Guanin (G).
***Position and amino acid in natural xylanase and mutant; Threonine (Thr), Leucine (Leu), Serine (Ser), Glutamine (Gln), Alanine (Ala), Valine (Val), Glycine (Gly).
****The activity was arbitrarily expressed with results from 96-well micro-plate DNS assay.

Table 1: Changes of DNA, amino acid, and activity in selected mutants.

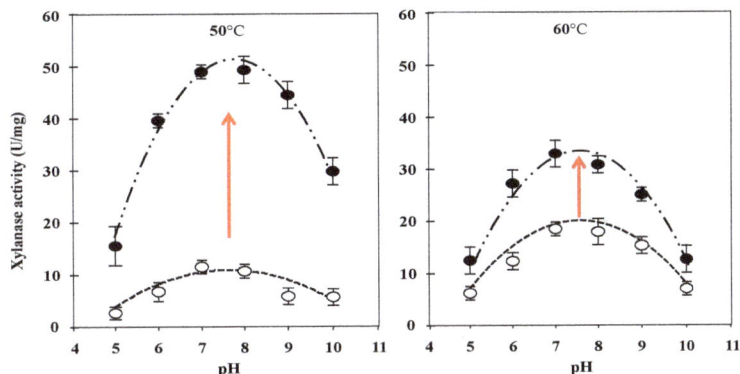

Figure 3: Improvement of the xylanase activity of mutant P1H1 compared to the original KRICT PX1 xylanase under 50 and 60°C.

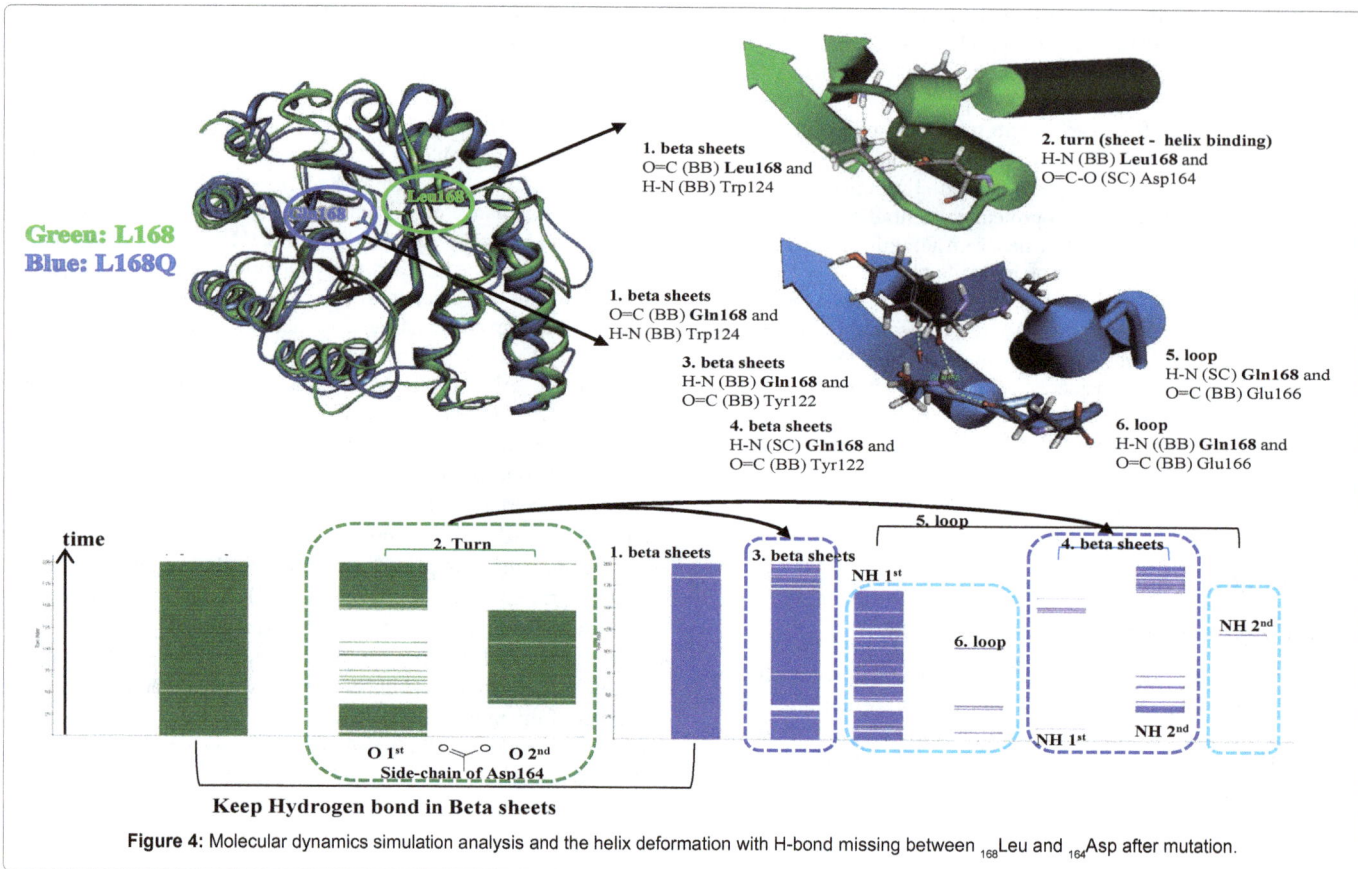

Figure 4: Molecular dynamics simulation analysis and the helix deformation with H-bond missing between [168]Leu and [164]Asp after mutation.

There is a need to engineer natural enzymes [10,11] because the inherent activity, stability, and enantioselectivity of existing enzymes are not always suitable for industrial applications. The most commonly used random mutagenesis method is error-prone PCR [12], which introduces random mutations during PCR by reducing the fidelity of DNA polymerase (Figure 5). The use of compromised DNA polymerase causes misincorporation of incorrect nucleotides during the PCR reaction, yielding randomly mutated products. Determining which mutation positions are critical in gaining the ability to enhance practical applications. No dramatic enhancement of activity was found for xylan

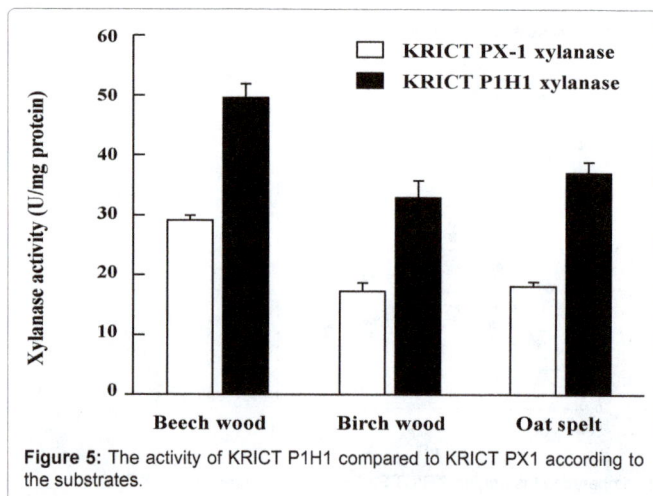

Figure 6: The enhanced birchwoodxylancatalyzing Property of P1H1 compared to the original KRICT PX1 xylanase. X1: xylose, X2: xylobiose, X3: xylotriose, X4: xylotetrose, X5: xylopentose.

hydrolysis among the 250 mutations. However, one of the 6 selected mutants, P1H1, showed good thermostability and enhanced xylanase activity. Nevertheless, further research on practical applications will be necessary for industrial usage. The activity of mutant P1H1 was improved roughly two-fold relative to that of KRICT PX1 when compared to different substrates, that is, beech wood, birch wood, and oat spelt xylan by these two xylanases revealed that the hydrolysis of birchwood xylan by P1H1 xylanase was significantly greater than that of KRICT PX1, and xylobiose was the major product with smaller amounts of xylotriose and xylose (Figure 6).

One of the exciting applications of xylanases is the production of xylo-oligosaccharides. Xylo-oligosaccharides have stimulatory effects on the selective growth of human intestinal *bifidiobacteria* and are frequently defined as probiotics. Xylo-oligosaccharides (xylobiose,

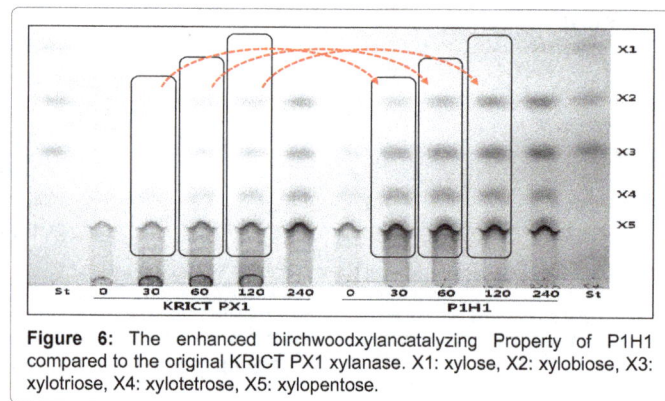

Figure 5: The activity of KRICT P1H1 compared to KRICT PX1 according to the substrates.

xylotriose, xylotetraose, etc.) prepared from various sources of xylans such as wheat bran, birchwood, or corncob can be utilized selectively by the beneficial intestinal microflora, viz. *bifdiobacteria*, are thus expected to be used as a valuable food additive [20]. Further studies on the production of xylo-oligosaccharides from economical agricultural residues such as wheat bran and rice bran are currently underway.

Another change in the characteristics of the P1H1 mutant was that the effects of additives on the xylanase activity were weakened by $CaCl_2$, $MgCl_2$, $MnCl_2$, and $FeCl_3$ (Table 2). Xylanase activity of *KRICT PX1* was completely inhibited by 1 mM $CaCl_2$, $MgCl_2$, $MnCl_2$, and $FeCl_3$ whereas

Additives	Relative activity (%) at 1 mM	
	KRICT P1H1	KRICT PX1
None	100	100
NaCl	100	100
LiCl	100	100
KCl	100	100
NH_4Cl	100	100
$CaCl_2$	100	0
$MgCl_2$	100	1
$MnCl_2$	83	3
$CsCl_2$	99	100
$CuSO_4$	21	8
$ZnSO_4$	30	2
$FeCl_3$	79	6
EDTA	54	100
2-ME	100	100
DTT	100	100
PMSF	100	100
Acetate	72	100
Furfural	100	100

Table 2: The improved activity of P1H1 compared to the original KRICT PX1 xylanase effective of additives.

these metal chloride ions did not have any dramatic effects on the P1H1 xylanase. Acetate and furfural are the major residual components of biomass pretreatment involving enzyme deactivation or inhibition. However, P1H1 xylanase was safe up to 1 mM of acetate, furfural, phenylmethanesulfonylfluoride (PMSF), dithiothreitol (DTT), and 2-mercaptoethanol (ME), respectively.

Enzyme immobilization

To investigate the potential of surface functionalized MCF spheres for immobilization with a new xylanase KRICT PX1 from *Paenibacillus* sp. HPL-001, enzymes functionalized with on the three supports regardless of surface functionalization, but the residual xylanase activity was higher in the case of immobilization on MCF-C_3-NH_2 and MCF-C_3-RC(=O)H than standard MCF. However, the residual enzyme activity was decreased to 70, 80, and 50% of the original activity after immobilization on MCF-C_3-NH_2, MCF-C_3-RC(=O)H, and standard MCF, respectively (Figure 7A). This might be due to the decreasing opportunity of contact between the enzyme and the substrate, because the immobilized body became bigger, and pores could impede enzyme action.

Among these results, MCF-C_3-NH_2 immobilized xylanase was conducted repeat-batch reaction as using the same reaction mixture of xylanase activity assay after washing the enzyme and filtering with a membrane filter (ø 100 μm). The residual activity was decreased with the number of recycling steps; however, the immobilized enzyme showed only a slight reduction of residual activity after 6 cycles. At the 10th batch reaction, about 40% of the original activity was lost. This loss of specific activity suggests a diffusion limitation on the substrate or product flux due to the association of the enzyme within the pores of the carriers (Figure 7B).

Similar to our work, Mukesh and Ramesh [20] reported that the xylanase from *Bacillus pumilus* strain MK001 showed the maximum xylanase immobilization efficiency on different matrices of entrapment

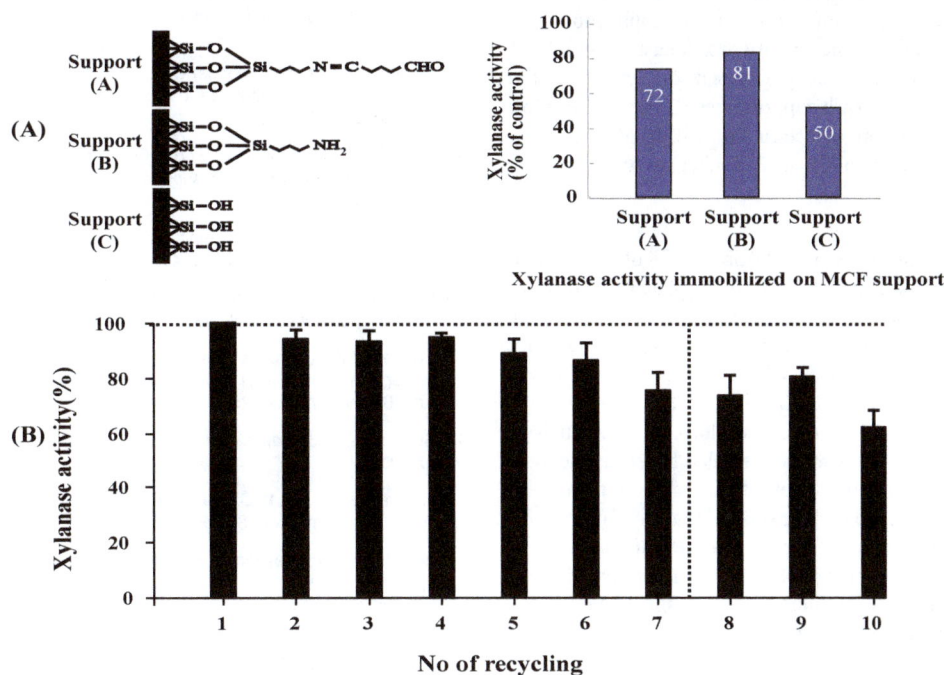

Figure 7: Repeated-batch MCF-immobilized xylanase KRICT PX1.

using gelatin (40%, GE), physical adsorption on chitin (35%, CH), ionic binding with Q-sepharose (45%, Q-S), and covalent binding with HP-20 beads (42%, HP-20). The reason for low immobilization efficiency may be the crowding of other proteins on the support with a direct effect on the accessibility of enzyme molecules to the adsorption material. MCF has high mechanical stability and rigidity, regenerability, and ease of preparation in different geometrical configurations, providing the system with permeability and suitable surface area for a given biotransformation and reducing the system cost.

Xylanases are an important group of carbohydrases, with a worldwide market of approximately 200 million dollars, and are being routinely used in various industrial processes such as animal feed digestion, waste treatment, energy generation, production of chemicals, and paper manufacturing. One of the exciting applications of xylanases is the production of xylo-olygosaccharides. They also show remarkable potential for utilization in pharmaceuticals, in feed formulations, in agricultural applications, and as food additives. In industrial applications, the free forms of enzymes pose difficulties, e.g., instability of enzyme structure and sensitivity under harsh process conditions, non-recovery of the active form of the enzyme from the reaction mixture for reuse, and contamination of the final product [21]. The strategies used for improving the stability of proteins include the use of additives, introduction of disulfide bonds [22], site-specific mutagenesis [23], and chemical modification or cross-linking [24]. However, in all these strategies, the yield and the reusability of free enzymes as industrial catalysts are quite limited. Increased attention therefore has been paid to immobilization techniques, which offer advantages over free enzymes in terms of choice of batch or continuous processes, rapid termination of reactions, controlled product formation, ease of enzyme recovery from the reaction mixture, and adaptability to various engineering designs. The extent of stabilization depends on the enzyme structure, the immobilization method, and the type of support [21]. During the last decade, numerous supports for xylanase immobilization have been investigated [25-30].

The reuse of an immobilized enzyme is an important factor considering cost effectiveness for commercial applications. A continuous assay of residual enzyme activity of xylanase immobilized on surface functionalized-MCF was performed to determine the retention of xylanase activity by each support over 10 enzyme reaction cycles. Xylanase immobilized on surface functionalized-MCF showed better retention and retained up to 70% activity after seven cycles.

Conclusion

A mutation library was produced by error-prone PCR of the xylanase gene *KRICT-PX1*. A DNS assay in a 96-well micro-plate used to select mutants with enhanced xylanase activity from mutant clones. P1H1 clone was selected on the basis of it having the highest xylanase activity of 53.5 U/mg protein at pH 7 and 50°C, which is approximately a five-fold increase compared to that of the original KRICT-PX1 xylanase with activity of 10.25 U/mg protein at the same conditions. The TLC analysis of the products catalyzed by two xylanases revealed faster degradation of birchwood xylan and larger production of small xylo-oligomers by P1H1. DNA sequencing of the P1H1 clone showed that it had one amino acid substitution (168Gly → Gln) caused by a point mutation of 503rd thymine to adenine. Further research on the biochemical properties and a structural analysis of this point-mutated xylanase should be carried out to shed light on the catalysis mechanism of xylanase. The overall performance of the immobilized xylanase in terms of catalytic activity, thermal and pH stability reuse, and xylo-oligosaccharide production

is more promising than that of the free enzyme. The higher efficiency observed for xylanase immobilized on cost effective supports such as MCF favors its commercial utilization for various industrial processes such as production of xylo- oligosaccharides, animal feed digestion, pharmaceuticals, waste treatment, energy generation, production of chemicals, and paper manufacturing, as well as in agricultural applications and as food additives.

References

1. Zhang GM, Huang J, Huang GR, Ma LX, Zhang XE (2007) Molecular cloning and heterologous expression of a new xylanase gene from *Plectosphaerella cucumerina*. Appl Microbiol Biotechnol 74: 339-346.

2. Collins T, Gerday C, Feller G (2005) Xylanases, xylanase families and extremophilic xylanases. FEMS Microbiol Rev 29: 3-23.

3. Kulkarni N, Shendye A, Rao M (1999) Molecular and biotechnological aspects of xylanases. FEMS Microbiol Rev 23: 411-456.

4. Dhiman SS, Sharma J, Battan B (2008) Industrial applications and future prospects of microbial xylanases: A review. BioResources 3: 1377-1402.

5. Hwang IT, Lim HK, Song HY, Cho SJ, Chang JS (2010) Cloning and characterization of a xylanase, KRICT PX1 from the strain *Paenibacillus* sp. HPL-001. Biotechnol Adv 28: 594-601.

6. Kulkarni N, Rao N (1996) Application of xylanases from alkalophilic thermophilic *Bacillus* sp. NCIM 59 in biobleaching of bagasse pulp. J Biotechnol 51: 167-173.

7. Garg AP, Roberts JC, McCarthy AJ (1998) Bleach boosting effect of cellulose free xylanase of *Streptomyces thermoviolaceus* and its comparison with two commercial enzyme preparation on birchwood kraft pulp. Enzyme Microb Technol 22: 594-598.

8. Beg QK, Bhushan B, Kapoor M, Hoondal GS (2000) Enhanced production of a thermostable xylanase from *Streptomyces* sp. QG-11-3 and its application in biobleaching of eucalyptus kraft pulp. Enzyme Microb Technol 27: 459-466.

9. Christov LP, Szakacs G, Balakrishnam H (1999) Production, partial characterization and use of fungal cellulose free xylanases in pulp bleaching. Process Biochem 34: 511-517.

10. Turner NJ (2009) Directed evolution drives the next generation of biocatalysts. Nat Chem Biol 5: 567-573.

11. Choudhury D, Biswas S, Roy S, Dattagupta JK (2010) Improving thermostability of papain through structure-based protein engineering. Protein Eng Des Sel 23: 457-467.

12. Leung DW, Chen E, Goeddel DW (1989) A method for random mutagenesis of a defined DNA segment using a modified polymerase chain reaction. Techniques 1: 11-15.

13. Lei C, Shin Y, Magnuson JK, Fryxell G, Lasure LL, et al. (2006) Characterization of functionalized nanoporous supports for protein confinement. Nanotechnology 17: 5531-5538.

14. Miller GL (1959) Use of dinitrosalicyclic acid reagent for determination of reducing sugar. Anal Chem 31: 426-428.

15. Laemmli UK (1970) Cleavage of structural proteins during the assembly of the head of bacteriophage T4. Nature 227: 680-685.

16. Bradford MM (1976) A rapid and sensitive method for the quantitation of microgram quantities of protein utilizing the principle of protein-dye binding. Anal Biochem 72: 248-254.

17. Saha BC (2003) Hemicellulose bioconversion. J Ind Microbiol Biotechnol 30: 279-291.

18. Lineweaver H, Burk D (1934) The determination of enzyme dissociation constants. J Am Chem Soc 56: 658-666.

19. Seo YK, Suryanarayana I, Hwang YK, Shin N, Ahn DC, et al. (2009) Swift synthesis of hierarchically ordered mesocellular mesoporous silica by microwave-assisted hydrothermal method. J Nanosci Nanotechnol 8: 3995-3998.

20. Kapoor M, Kuhad RC (2007) Immobilization of xylanase from *Bacillus pumilus* strain MK001 and its application in production of xylo-oligosaccharides. Appl Biochem Biotechnol 142: 125-138.

21. Krajewska B (2004) Application of chitin- and chitosan-based materials for enzyme immobilizations: a review. Enzyme and Microbial Technology 35: 126-139.

22. Perry LJ, Wetzel R (1984) Disulfide bond engineered into T4 lysozyme: stabilization of the protein toward thermal inactivation. Science 226: 555-557.

23. Imanaka T, Shibazaki M, Takagi M (1986) A new way of enhancing the thermostability of proteases. Nature 324: 695-697.

24. Braxton S, Wells JA (1992) Incorporation of a stabilizing Ca(2+)-binding loop into subtilisin BPN'. Biochemistry 31: 7796-7801.

25. Rogalski J, Szczodrak J, Dawidowicz A, Ilczuk Z, Leonowica A (1985) Immobilization of cellulase and d-xylanase complexes from Aspergillus terreus F-413 on controlled porosity glasses. Enzyme and Microbial Technology 7: 395-400.

26. Abdel-Naby MA (1993) Immobilization of Aspergillus niger NRC 107 Xylanase and beta-Xylosidase, and Properties of the Immobilzed Enzymes. Appl Biochem Biotechnol 38: 69-81.

27. Tyagi R, Gupta MN (1995) Immobilization of Aspergillus niger xylanase on magnetic latex beads. Biotechnol Appl Biochem 21: 217-222.

28. Dumitriu S, Chornet E (1997) Immobilization of xylanase in chitosan-xanthan hydrogels. Biotechnology Progress 13: 539-545.

29. Gouda MK, Abdel-Naby MA (2002) Catalytic properties of the immobilized Aspergillus tamarii xylanase. Microbiol Res 157: 275-281.

30. Ai Z, Jaing Z, Li L, Deng W, Kusakabe I, et al. (2005) Immobilization of Streptomyces olivaceoviridis E-86 xylanase on Eudragit S-100 for xylo-oligosaccharide production. Proc Biochem 40: 2707-2714.

23

Thermodynamic Studies of *Trans*-4-Hydroxy-L-Proline at Different Ionic Medium, Ionic Strength and Temperature

Emilia Furia*, Anna Napoli, Antonio Tagarelli and Giovanni Sindona

Department of Chemistry and Chemical Technologies, University of Calabria, 87036 Arcavacata di Rende (CS), Italy

Abstract

The aim of this work was to determine the solubility as well as the acidicconstants of *trans*-4-hydroxy-L-proline in 0.16 and 3.5 Mol.kg^{-1} NaClO$_4$ and in 0.16 and 3.18 Mol.kg^{-1} NaCl at two different temperature, *i.e.* 298.15 and 310.15 K. *trans*-4-Hydroxy-L-proline, HL, is one of the most abundant amino acids present in collagen. As a major part of this protein, the measurement of hydroxyproline levels can be used as an indicator of collagen content. The protonation constants of ligand, K_1 and K_2, combined with the salting effects on the ligand in the ionic media were treated by the specific ion interaction theory, SIT, to give equilibrium constants at the infinite dilution reference state at 298.15 K and at 310.15 K, as well as specific interaction coefficients of the cation and anion of *trans*-4-hydroxy-L-proline with the media ions at the standard temperature, *i.e.* 298.15 K.

Keywords: *Trans*-4-Hydroxy-L-proline; Acidic constants; Constant ionic mediummethod; Infinite dilution reference state; SIT model

Introduction

In our on-going investigation of metals with biological ligands, we were interested in the metal–*trans*–4–hydroxy–L–proline complex formation equilibria. *trans*–4–Hydroxy–L–proline, HL (Figure 1), a constituent of several major structural proteins and other biodegradable synthetic products, is one of the main nonessential amino acid which contribute to the formation of collagen, the most abundant protein in humans as a primary constituent of bone, skin and connective tissues.

Defects in collagen synthesis lead to easy bruising, internal bleeding, breakdown of connective tissue of the ligaments and tendons, and increased risk of blood vessel damage. Studies on the properties of *trans*–4–hydroxy–L–proline in aqueous solution are useful in understanding the details of the mechanism of its action in aqueous systems. Previous works [1,2] reported thermodynamic data about equilibrium constants of HL at 0.16 Mol.kg^{-1}at two different temperature (*i.e.*298.15 and 310.15 K). By contrast, other available values [3] were obtained in non–aqueous solvents. In the present work the dependence of equilibrium constants of the ligand on ionic strength and media as well as on temperature was studied. It seems reasonable to predict different values of acidic constants at different ionic strength, medium and temperature. Thus, we have evaluated acidic constants K_1 ($L^- + H^+ \rightleftharpoons HL$) and K_2 ($L^- + 2H^+ \rightleftharpoons H_2L^+$) of the lig and at two different ionic strength, I, in two distinct media, i.e.0.16 and 3.5 Mol.kg^{-1} NaClO$_4$ and 0.16 and 3.18 Mol.kg^{-1}NaCl, as well as at the two different temperatures 298.15 and 310.15 K. The adoption of the constant ionic medium method, proposed by Biedermann and Sillén [4], was necessary in order to minimize activity coefficient variation in spite of the change of the reagent concentrations. By this approach, it was possible to replace in the calculations activities with concentrations and to minimize the liquid junction potential due to the hydrogen ion concentration. This was varied in order to determine acidic constants of ligand, taking into account that protonation takes place at different acidities. The potentiometric results and the solubility data obtained in the ionic media were processed by the specific ion interaction theory, SIT, [5,6] to yield acidic constants at the infinite dilution reference state and specific interaction coefficient of the cation and anion of HL with the media ions.

Materials and Methods

Apparatus and reagents

The cell arrangement was as formerly described [7]. The test solutions, stirred during titrations, were purified with a slow stream of nitrogen gas as reported in a previous work [8]. Glass electrodes were the same used in preceding evaluations [9]; they attained a constant potential within few minutes after the addition of the reagents and remained unchanged for several hours to within ± 0.1 mV. The titrations were carried out as described in a previous paper [10]. The electromotive force values were recorded with a precision of ± 10^{-5} V by an OPA 111 low–noise precision DIFET operational amplifier. The cell assembly was retained in a thermostat kept at (298.1 ± 0.1) and (310.1 ± 0.1) K.

A perchloric acid and an hydrochloric acid stock solutions were prepared and standardized as described previously [9,11]. Sodium perchlorate stock solutions were prepared and standardized according to Biedermann [12]. Sodium hydroxide titrant solutions were prepared and standardize as reported formerly [10]. Purissimum grade (100

*Corresponding author: Emilia Furia, Department of Chemistry and Chemical Technologies, University of Calabria, 87036 Arcavacata di Rende (CS), Italy

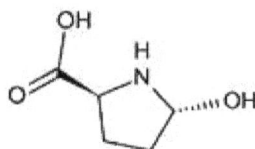

Figure 1: Structure of *trans*–4–hydroxy–L–proline, HL.

% Aldrich p.a.) *trans–4–hydroxy–L–proline* product was kept in a desiccator over silica gel and it was utilized without purification. All solutions were prepared with twice distilled water.

Solubility measurements

Solubility studies are of both theoretical and practical interest and they allow the determination of activity coefficients for nonelectrolyte solutes in aqueous solutions containing a large excess of salts [13-15]. The knowledge of the activity coefficients of neutral species is necessary when modeling the dependence of protonation constants on ionic strength according to the SIT equations [5,6]. Saturated HL were prepared as already described in a previous work [16]. Solid HL was wrapped up in a highly retentive filter paper (Whatman 42) bag and then was kept in a glass cylinder containing pure water as well as sodium perchlorate and sodium chloride aqueous solutions at pre-established ionic strength values (0.16 and 3.5 Mol.kg^{-1}, and 0.16 and 3.18 Mol.kg^{-1}, respectively), under continuous stirring with a magnetic bar (Figure 2).

The cylinder was then placed in a thermostatic water bath at (298.1 ± 0.1) and (310.1 ± 0.1) K and the ligand concentration was detected over time, until it reached a constant value, which usually occurred in about 3 up to 5 days. Finally, the absorption spectra in the UV region were recorded on a series of HL solutions. The absorbance, A_λ, may be expressed as equation 1:

$$A_\lambda = \iota\varepsilon[HL] \qquad (1)$$

Where ι is the optical path and ε is the molar absorptivity. To find suitable conditions for determining the solubility, S, of HL, A_λ were measured between 200 and 300 nm taking as a blank the ionic medium. Three replicates were run for each point. A typical spectrum of ligand recorded is reported in Figure 3.

Absorption spectrum of HL shows one single intense band centered at 193.5 nm. The solubility, S, was deduced by interpolation on a calibration curve, based on standard solutions. The reproducibility of the solubility data was of 1%.

Taking into account that $S°$ is the solubility of the neutral species

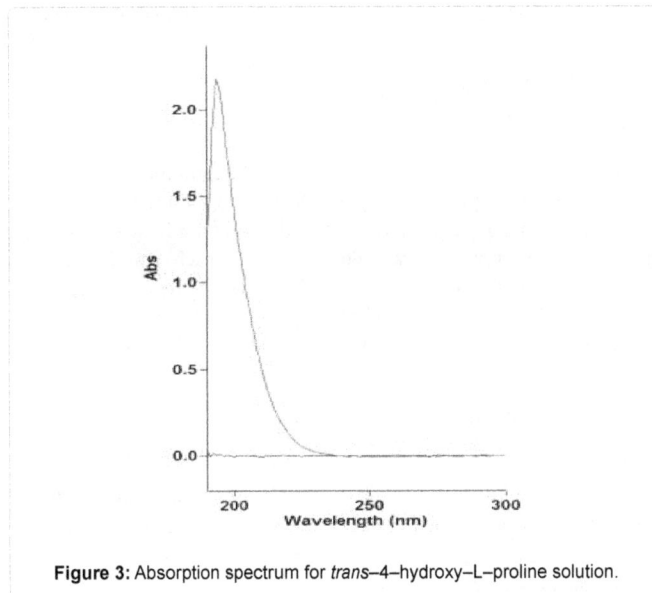

Figure 2: The apparatus for preparation of saturated *trans–4–hydroxy–L–proline* solutions.

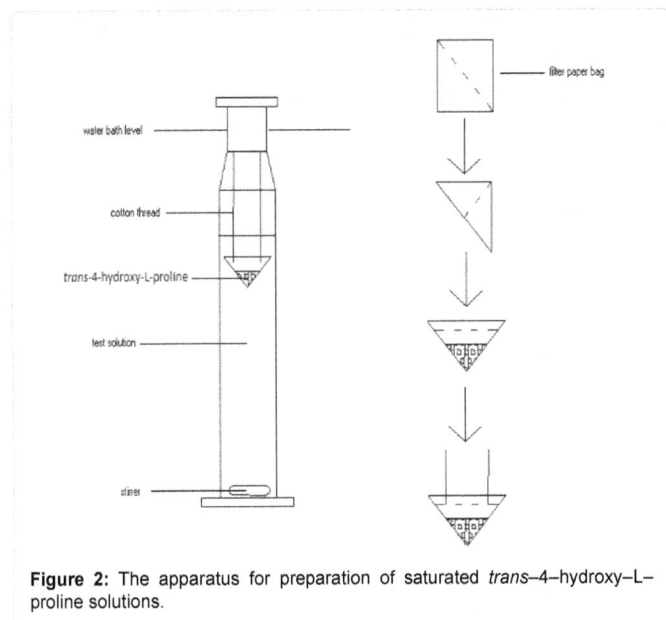

Figure 3: Absorption spectrum for *trans–4–hydroxy–L–proline* solution.

in the electrolyte solutions, the total solubility, S, of HL can be written as follows:

$$S = S^o \left\{ 1 + \left(K_1[H^+] \right)^{-1} + \left(K_1 K_2 \left[H^+ \right]^2 \right)^{-1} \right\} \qquad (2)$$

where the acidic constants of HL to the equilibria 3 and 4:

$$L^- + H^+ \rightleftharpoons HL \, K_1 (3)$$

$$L^- + 2H^+ \rightleftharpoons H_2L^+ \, K_2 (4)$$

The solubility of the HL, valid in the molal concentration scale, is related to activity coefficient [15], γ, by the equation 5:

$$\log \gamma_{HL} = -\log \left(\frac{S_0^{\,o}}{S^o} \right) \qquad (5)$$

Where $S_0°$ is the solubility of HL at the infinite dilution reference state. The results are summarized in Table 1.

Trans–4–Hydroxy–L–proline exhibits salting–in [17] behaviour, the solubility increasing monotonically with increasing ionic strength.

Potentiometric measurements

The protonation equilibria of *trans–4–hydroxy–L–proline* were investigated by potentiometric titrations with an alkaline glass electrode, GE, at 298.15 and at 310.15 K with cell (G)

RE / Test Solution / GE (G)

in which RE stands for the silver reference electrode = Ag/AgCl/0.0105 Mol.kg^{-1} AgClO$_4$, (I – 0.0105) molkg^{-1} NaClO$_4$/IMol.kg^{-1} NaClO$_4$, when NaClO$_4$ was inert salt at two different ionic strength I (*i.e.* 0.16 and 3.5 Mol.kg^{-1}) and = Ag / AgCl / IMol.kg^{-1}NaCl saturated with AgCl / IMol.kg^{-1}NaCl, when NaCl was inert salt at two different ionic strength I (*i.e.* 0.16 and 3.18 Mol.kg^{-1}). Test solution had the general composition: C_LMol.kg^{-1}HL, C_AMol.kg^{-1}H$^+$, C_BMol.kg^{-1}NaOH, (I – C_A – C_B) Mol.kg^{-1}Na$^+$, where C_L were between (5 10^{-3} and 10 10^{-3}) Mol.kg^{-1} and I was 0.16 and 3.5 Mol.kg^{-1}NaClO$_4$, and 0.16 and 3.18 NaCl Mol.kg^{-1}. The equilibrium constants at the infinite dilution reference state can be readily evaluated from results obtained in this ionic strength range according to the SIT [5].

At a given I the electromotive force of cell (G) can be written, in mV, at the temperatures of 298.15 and 310.15 K as equations 6 and 7, respectively:

$$E = E^o + 59.16\log\left[H^+\right] + E_j \quad (6)$$

$$E = E^o + 61.54\log\left[H^+\right] + E_j \quad (7)$$

Where E^o is constant in each series of measurements and E_j is the liquid junction potential [4] which is a linear function of $[H^+]$, $E_j = -j[H^+]$. The j parameters, at the different ionic strengths, are known from a previous evaluation [9,18]. In each run E^o values, constant within 0.1 mV, were calculated from measurements in solutions of $(10^{-3} \geq [H^+] \geq 10^{-4})$mol kg^{-1} in the absence of HL. $[H^+]$ was decreased stepwise by coulometric generation of OH$^-$ ions with the circuit (C)

– Pt / Test Solution / AE + (C)

where AE, auxiliary electrode,=IMol.kg^{-1} inert salt/0.1 Mol. kg^{-1}NaCl, $(I$-0.1) Mol.kg^{-1} Na$^+$/Hg$_2$Cl$_2$/Hg. In the test solution of a given volume V dm^3, C_B=($\mu F\, 10^{-6}$/V) mol dm^{-3} where μF stands for the micro–faradays passed through the cell, according to the assumption that at the cathode the only reactions that occur are

$$H^+ + e^- \rightarrow \tfrac{1}{2}\,H_2 \text{ and } H_2O + e^- \rightarrow \tfrac{1}{2}\,H_2 + OH^-.$$

After the introduction of a known amount of HL, dissolved in a known excess of CBMol.kg^{-1}NaOH, the acidification was achieved, in the pH range from 12 to 2, by adding H+ standardized solutions, according to the equilibria reported in equations 3 and 4:

$$L^- + H^+ \rightleftharpoons HL\, K_1 \,(3)$$

$$L^- + 2H^+ \rightleftharpoons H_2L^+ K_2 \,(4)$$

The primary C_L, C_A, C_B and [H$^+$] data form the basis of the treatment to obtain the equilibrium constants.

Results and Discussion

The protonation constants of *trans*–4–hydroxy–L–proline, K_1 and K_2, were calculated from the data acquired by performing two titrations for each of the involved equilibria for the different ionic media and strengths and for the two different temperatures (298.15 and 310.15 K). In particular, a data set of 192 experimental points was used. An overview of the working conditions of all titrations carried out in NaClO$_4$ and NaCl media is reported in Table 2 and Table 3, respectively.

The experimental data (C_L, C_A, C_B, [H$^+$]) were processed by numerical procedures. For the numerical treatment we employed the least–squares computer program Superquad [19] to seek the minimum of the function:

$$U = \sum (E_i^{obs} - E_i^{cal})^2 \quad (8)$$

Where $E^{obs} = E^o + 59.16\log[H^+]$ and $E^{obs} = E^o + 61.54\log[H^+]$ at 298.15 and 310.15

I, Mol.kg⁻¹		(S.10³, Mol.kg⁻¹) 298.15 K	(S.10³, Mol.kg⁻¹) 310.15 K	(logγ_HL) 298.15 K	(logγ_HL) 310.15 K
Pure water		31.61 ± 0.01	40.16 ± 0.01	0	0
NaClO₄	0.16	32.75 ± 0.02	43.80 ± 0.02	0.01 ± 0.02	0.04 ± 0.02
	3.5	35.02 ± 0.04	46.25 ± 0.04	0.04 ± 0.04	0.06 ± 0.04
NaCl	0.16	32.68 ± 0.02	43.29 ± 0.02	0.01 ± 0.02	0.03 ± 0.02
	3.18	34.93 ± 0.04	49.72 ± 0.04	0.04 ± 0.04	0.09 ± 0.04

Table 1: Solubility,S, of *trans*–4–hydroxy–L–proline in /molkg⁻¹NaClO₄, in /Mol. kg⁻¹NaCl and in pure water at 298.15 and 310.15 K.

I, Mol.kg⁻¹	NaClO₄	C_L, Mol.kg⁻¹	pH_range
298.15 K	0.16	5.10⁻³	2.79–11.1
		10.10⁻³	2.97–10.8
	3.5	5.0.10⁻³	2.1–12.1
		9.94.10⁻³	2.2–11.8
310.15 K	0.16	5.10⁻³	2.88–11.3
		10.10⁻³	3.03–10.7
	3.5	5.2.10⁻³	2.2–12.1
		10.1.10⁻³	2.1–11.7

Table 2: Summary of the working conditions of 8 titrations for protonation constants K_1 and K_2 in /Mol.kg⁻¹NaClO₄ at two different temperature.

I, Mol.kg⁻¹	NaClO₄	C_L, Mol.kg⁻¹	pH_range
298.15 K	0.16	5.1.10⁻³	3.55–12.09
		9.97.10⁻³	3.27–11.12
	3.18	5.0.10⁻³	23.43–11.33
		9.96.10⁻³	3.1–11.45
310.15 K	0.16	5.3.10⁻³	3.4–11.78
		10.10⁻³	3.23–11.69
	3.18	5.2.10⁻³	3.1–12.06
		10.10⁻³	2.9–11.17

Table 3: Summary of the working conditions of 8 titrations for protonation constants K_1 and K_2 in /Mol.kg⁻¹NaCl at two different temperature.

I, Mol.kg⁻¹		Log K₁log K₂ 298.15 K		log K₁log K₂ 310.15 K	
NaClO₄	0.16	9.63 ± 0.02	11.46 ± 0.04	9.79 ± 0.02	11.93 ± 0.04
	3.5	9.93 ± 0.02	12.18 ± 0.04	10.14 ± 0.02	12.78 ± 0.04
NaCl	0.16	9.63 ± 0.02	11.46 ± 0.04	9.78 ± 0.02	11.93 ± 0.04
	3.18	9.37 ± 0.02	12.00 ± 0.04	10.03 ± 0.02	12.56 ± 0.04

Table 4: Survey of the log K_n values, molal scale, by numerical methods.

K, respectively, while E_{cal} is a value calculated for a given set of parameters. In the numerical treatment the ion product of water has been taken from the literature [20] for different temperature and different ionic strength and media. Results are given in Table 4 and the uncertainties on equilibrium constants represent 3σ.

As it can be seen in Table 4 the value of the acidic constants increases as temperature increases, thus supporting that reactions are thermodynamically favored.

Dependence on ionic medium, ionic strength and temperature by SIT approach

The acidic constants at the infinite dilution reference state, $°K_1$ and $°K_2$, of HL were determined by the numerical values reported in Table 4, by assuming the validity of the SIT [5]. According to this theory, constants and other quantities in the following treatment are expressed on the molal scale by using the conversion factors from Grenthe et al. [21]. The activity coefficient, Z_i, of the species i with charge z_i can be expressed in aqueous solution as stated by the SIT:

$$\log \gamma i = -z_i^2 D + \sum \varepsilon(i,k)m_k \quad (9)$$

Where D is the Debye-Hückel term

$$D = \frac{A\sqrt{I}}{1+b\sqrt{I}} \quad (10)$$

Where the b value is arbitrarily chosen, generally $1 \leq b \leq 1.5$. In equation 9 ε is the specific ion interaction coefficient of i with species k of molality m_k. The SIT model is based on the assumption that interaction coefficients of ions with the same charge type are nearly zero. Interaction coefficients are the result of short range forces and depend on the ionic strength but their variation in the range $0.5 \leq I \leq 3.5$molal is sufficiently low that they may be assumed as constants.

Parameters A and b in equation 10 can be taken from the literature. In particular, $b = 1.5$ [22,23], while A varies with temperature [24,25]. In the range $273 \leq T \leq 348$ K A_T can defined as follows [22]:

$$A_T = 0.510 + 76.286 f_1(T) + 1.4189 f_2(T) \, (11)$$

Where

$$f_1(T) = \left(\frac{1}{\Theta} - \frac{1}{T}\right) (12)$$

$$f_2(T) = \left(\frac{\Theta}{T} - 1 + \ln\frac{T}{\Theta}\right) (13)$$

and is the standard temperature, *i.e.* 298.15 K.

According to the specific interaction theory [5,6] the protonation constants of *trans*-4- hydroxy-L-proline, K_1 and K_2, combined with the salting effects of inert salts (i.e. $NaClO_4$ and $NaCl$) on HL deduced from the solubility determinations, were processed to give equilibrium constants at the infinite dilution reference state, $^\circ K_1$ and $^\circ K_2$. The effect of the ionic media at different ionic strength (*i.e.* 0.16 and 3.5 Mol.kg^{-1} $NaClO_4$, and in 0.16 and 3.18 Mol.kg^{-1} $NaCl$) on the various equilibrium constants (according to equilibria 3 and 4) determined in this work can be expressed as:

$$\log {^\circ K_1} = \log K_1 + 2D + \log \gamma_{HL} - \left[\varepsilon(H^+, ClO_4^-) + \varepsilon(L^-, Na^+)\right] m \, (14)$$

$$\log {^\circ K_2} = \log K_2 + 2D + \left[\varepsilon(H_2L^+, ClO_4^-) - 2\varepsilon(H+, ClO_4^-) - \varepsilon(L^-, Na^+)\right] m \, (15)$$

when inert salt is $NaClO_4$, and as:

$$\log {^\circ K_1} = \log K_1 + 2D + \log \gamma_{HL} - \left[\varepsilon(H^+, Cl^-) + \varepsilon(L^-, Na^+)\right] m \, (16)$$

$$\log {^\circ K_2} = \log K_2 + 2D + \left[\varepsilon(H_2L^+, Cl^-) - 2\varepsilon(H^+, Cl^-) - \varepsilon(L^-, Na^+)\right] m \, (17)$$

When inert salt is NaCl. The activity coefficients of HL, $\log \gamma_{HL}$, were deduced bysolubility measurements in the electrolyte solutions, and from Ciavatta [5] at 298.15 K $\varepsilon (H^+, ClO_4^-) = 0.14$ and $\varepsilon (H^+, Cl^-) = 0.12$. Hence, plots on the known terms of equations 14-17 as a function of I result in straight lines in which the intercept correspond to the constants at zero ionic strength and the slope are the interaction coefficients, ε (i, k), between HL and the counter ions of the ionic media (Table 5).

The interaction coefficients between Na^+ ion and the anion of the acid as well as between $ClO4^-$ ion and H_2L^+ agree with those expected for small ion [21]. As concerning $\varepsilon (H_2L^+, Cl^-)$ the value obtained is negligible.

To calculate acidic constants of HL at the infinite dilution reference state at the temperature of 310.15 K, it was necessary to evaluate D, which is dependent on ionic strength I as well as on temperature by parameter A_T (equation 11). Similarly to what has been done at 298.15 K, it was possible to calculate $^\circ K_1$ and $^\circ K_2$ at 310.15 K, by assuming the validity of the SIT and taking into account equilibria 3 and 4. However in this case it was possible to deduce just the algebraic sum of interaction coefficients ε (i,k) since $\varepsilon (H^+, ClO_4^-)$ and $\varepsilon (H^+, Cl^-)$ were unknown at 310.15 K. Results of extrapolation are reported in Table 6.

The trend of two acidic constants at two different temperatures was, accordingly, matched at the two selected ionic media (Figure 4).

Log $^\circ K_1$ = 9.87 ± 0.05	$\varepsilon (L^-, Na^+)$ = 0.02 ± 0.05
Log $^\circ K_2$ = 11.67 ±0.05	$\varepsilon (H_2L^+, ClO_4^-)$ = -0.02 ± 0.05
	$\varepsilon (H_2L^+, Cl^-)$ = 0.001 ± 0.05

Table 5: Results of extrapolation to zero ionic strength at 298.15 K.

log $^\circ K_1$ = 10.02 ± 0.05	log $^\circ K_2$ = 12.14 ± 0.05

Table 6: Results of extrapolation to zero ionic strength at 310.15 K.

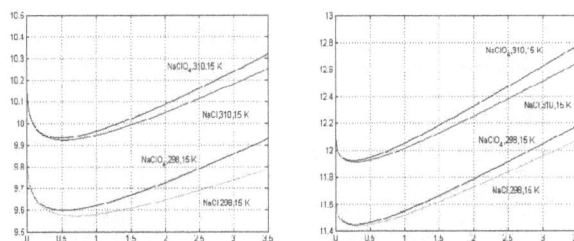

Figure 4: log K_1 (a) and log K_2(b) at different temperature and ionic medium.

I, Mol.kg^{-1} NaClO$_4$	298.15 K log K_1	log K_2	310.15 K log K_1	log K_2	References
0.16	9.63	11.46	9.79	11.93	This work
	9.81	11.68	9.16	10.82	Zielinski et al.,[1] Makaret al.,[2]

Table 7: Comparison between acidic constants of *trans*-4–hydroxy–L–proline reported in literature at the same ionic strength in NaClO$_4$ (i.e. 0.16 Mol.kg^{-1}) reported in this work.

A common behaviour was observed, since acidic constants increase as temperature increases. By contrast, a distinct effect of the ionic medium was clearly observed, whereby K_1 decreases with different inert salt, while K_2 results unaffected by the ionic medium.

Conclusions

The solubility and the acidic constants of *trans*-4–hydroxy–L–proline have been determined, at 298.15 and 310.15 K, in 0.16 and 3.5 Mol.kg^{-1} NaClO$_4$ and in 0.16 and 3.18 mol kg^{-1} NaCl solutions and at the infinite dilution reference state. Results have been used to evaluate the salting effect of NaClO$_4$ as well as of NaCl on the neutral molecule and interaction coefficients ε (i, k). The acidic constants obtained in this work and those taken from literature [1,2] are collected in Table 7 for a comparison.

The reactions are thermodynamically favored since acidic constants increase as temperature increases. The agreement between the data sets is just satisfactory, especially as concerning log K_2 by Makaret al. [2] at 310.15 K.

The available values from other authors [3] were obtained in non–aqueous solvents, for this reason it was not possible to make any comparison.

The results obtained in this work can be used for further studies regarding complex formation equilibria between HL and metal cations.

References

1. Zielinski S, Lomozik L, Wojciechowska A (1981) Potentiometric studies on the complex formation of lanthanides with proline and hydroxyproline. MonatshChem 112: 1245-1252.

2. Makar GKR, Touche MLD, Williams DR (1976) Thermodynamic considerations in co-ordination. Part XXIII.1 Formation constants for complexes of protons, zinc(II), and acid anions and their use in computer evaluation of a better zinc therapeutical. J ChemSoc Dalton *Trans* 11: 1016-1019.

3. Pettit GL (2006) IUPAC: Stability Constant Data Base. Tools of the Trade.

4. Biedermann G, Sillén LG (1953) Studies on the hydrolysis of metal ions. IV. Liquid junction potentials and constancy of activity factors in NaClO4–HClO4 ionic medium. Ark Kem 5: 425-440.

5. Ciavatta L (1980) The specific interaction theory in evaluating ionic equilibria. Ann Chim 70: 551-567.

6. Ciavatta L (1990) The specific interaction theory in equilibrium analysis. Some empirical rules for estimating interaction coefficients of metal ion complexes. Ann Chim 80: 255-263.

7. Furia E, Porto R (2008) 2–Hydroxybenzamide as ligand. Complex formation with dioxouranium(VI), aluminium(III), neodymium(III) and nickel(II) ions. J ChemEng Data 53: 2739-2745.

8. Furia E, Porto R (2004) The hydrogen salicylate ion as ligand. Complex formation equilibria with dioxouranium (VI), neodymium (III) and lead (II). Ann Chim 94: 795-804.

9. Furia E, Porto R (2002) The effect of ionic strength on the complexation of copper (II) with salicylate ion. Ann Chim 92: 521-530.

10. Furia E, Sindona G (2010) Complexation of L-cystine with metal cations. J ChemEng Data 55: 2985-2989.

11. Porto R, De Tommaso G, Furia E (2005) The second acidic constant of salicylic acid. Ann Chim 95: 551-558.

12. Biedermann G (1964) Study on the hydrolysis equilibria of cations by emf methods.Ft. Belvoir Defense Technical Information Center.

13. Long FA, McDevit WF (1952) Activity coefficients of nonelectrolyte solutes in aqueous salt solution. Chem Rev 51: 119-169.

14. Randal M, Failey CF (1927) The activity coefficient of the undissociated part of weak electrolytes. Chem Rev 4: 291-318.

15. Setschenow JZ (1889) Über die konstitution der salzlosungenaufgrundihresverhaltenszukohlensaure. Z PhysChem 4: 117-125.

16. Furia E, Nardi M, Sindona G (2010) Standard potential and acidic constants of Oleuropein. J ChemEng Data 55: 2824-2828.

17. Osol A, Kilkpatrick M (1933) The "salting-out" and "salting-in" of weak acids. I. The activity coefficients of the molecules of ortho, meta and parachlorobenzoic acids in aqueous salt solutions. J Am ChemSoc 55: 4430-4440.

18. Bottari E, Porto R (1986) Protolyticequilibria in aqueous sodium deoxycholate solutions. MonatshChem117: 589-597.

19. Gans P, Sabatini A, Vacca A (1985) SUPERQUAD: an improved general program for computation of formation constants from potentiometric data. J ChemSoc Dalton *Trans* 6: 1195-1200.

20. Baes CF, Mesmer RE (1976) The Hydrolysis of Cations. A Wiley– Interscience Publication, New York.

21. Grenthe I, Fuger J, Konigs RJM, Lemire RJ, Muller AB, et al., (2004) Chemical thermodynamics of Uranium. Nuclear Energy Agency, France.

22. Bretti C, Foti C, Sammartano S (2004) A new approach in the use of SIT in determining the dependence on ionic strength of activity coefficients. Application to some chloride salts of interest in the speciation of natural fluids. Chem Spec Bioavailab 16: 105-110.

23. Scatchard G (1976) Equilibrium in solution: surface and colloid chemistry. Harvard University Press, Cambridge, Massachusetts.

24. Robinson RA, Stokes RH (1959) Electrolyte solutions. Butterworths scientific publications, London.

25. Helgenson HC, Kirkham DH, Flowers GC (1981) Theoretical prediction of the thermodynamic behavior of aqueous electrolytes at high pressures and temperatures: IV. Calculation of activity coefficients, osmotic coefficients, and apparent molal and standard and relative partial molal properties to 600°C and 5kb. Am J Sci 281: 1249-1516.

24

Enhancement of Biogas Production from Laying Hen Manure via Sonolysis as Pretreatment

Duygu Karaalp[1], Kubra Arslan[2] and Nuri Azbar[2,3]*

[1]*Ege University, Graduate School of Natural and Applied Sciences, Biotechnology Department, Izmir, Turkey*
[2]*Ege University, Faculty of Engineering, Bioengineering Department, Izmir, Turkey*
[3]*Ege University, Center for Enviromental Studies, Izmir, Turkey*

Abstract

Aim of this study is to investigate the effect of sonication as a pretreatment on the efficiency of anaerobic digestion of Laying Hen Manure (LHM) under mesophilic conditions. In this study, the pretreatment studies were carried out using two different sonotrode (BS2d22 and BS2d40) and two different types of booster (B2-1,4 and B2-1,8) at four different amplitudes (9, 31, 40, 81 μm) at two time settings (5 and 15 min duration). Biochemical Methane production Potential (BMP) test protocol was also employed in order to evaluate the effect of sonication under varying amplitudes on the biogas production for 50 days. The BMP results obtained in this work suggested that sonication significantly enhanced the biogas productivity of chicken manure which was much lower if it was digested alone. Between 12-70% increase in methane production depending on the sonolysis matrix used was obtained in comparison to control group which had no sonication as pretreatment. The best result was obtained at an amplitude of 81 μm and 5 min sonication duration which is a combination of BS2d22/B2-1,4. The results of experiment demonstrated that use of this pretreatment technology could significantly enhance the biogas production from chicken manure.

Keywords: Anaerobic digestion; Biogas; Chicken manure; Methane; Pretreatment; Ultrasound

Introduction

Ultrasonication (US) has been considered as an environmentally and economically sound pretreatment strategy [1]. Several studies found in the literature have shown that sonication is an effective pretreatment for the enhancement of anaerobic biodegradation of organic wastes [1-3]. Sonication is the application of ultrasound waves to a confined liquid resulting in violent collapse and turbulence which then drives thermal destruction and forming of radicals [4]. Sonication is reported to result in increased solubilisation and reduction of particle size allowing for greater biodegradability [5]. Anaerobic Digestion (AD) processes are frequently preferred for biomethane production from organic wastes, especially animal manure. However, efficiency of anaerobic conversion for biogas production have always been questioned because of the fact that some potential wastes for bioconversion are relatively problematic including chicken manure [1,2]. The prime drawbacks of the conventional AD technology is the need for extremely long hydraulic retention times and large bioreactor volumes due to the lower microbial conversion rates. The non-availability of the readily biodegradable soluble and organic matters and lower digestion rate constant necessitates the pretreatment of biomass to be digested [6]. Some pretreatment strategies are capable of separating the lignin from the readably degradable cellulose fibers, allowing for greater AD efficiency. Therefore, applying a pre-treatment process are preferred in order to improve the rate limiting step of the AD process as a result of the existence of the lignocellulose and hemicellulose and to shorten the required reaction time for biomethane production [6-11]. The efficiency of disintegration depends on the sonication parameters and also on biomass characteristics, therefore the evaluation of the optimum parameters varies with the type of sonicator and biomass to be treated. With this inspiration from literature, it was decided to test the hypothesis that whether or not the sonication would enhance the anaerobic conversion of laying hen manure to biomethane and up to what extent.

Materials and Methods

The laying hen manure used in this study was kindly obtained from a local farm located in Izmir, Turkey. All the Chicken Manure (CM) was stored in a refrigerator at +4°C until used. Some characteristic parameters of stock CM were 85% DM (organic dry matter) and 30% DM (dry matter), this stock manure was then diluted down to 5% DM before sonication experiments.

The anaerobic consortium was obtained from the anaerobic sludge digester of a local yeast factory and was used as inoculum in BMP (Biochemical Methane Production Potential) assay.

Pretreatment process

All sonication experiments were carried out in 1000 ml glass container at room temperature. The ultrasonic processor used in this study was a Hielsher UIP 1000HD which has a maximum operating power of 1000 W, operational frequency of 20 kHz ± 1 kHz and maximum amplitude value up to 150 μm depending on sonotrode and booster (Figure 1).

The ultrasonic pretreatment studies were carried out using two different sonotrode (BS2d22 and BS2d40) and booster (B2-1,4 and B2-1,8) at three different amplitudes (9, 31, 81 μm) and at two time settings (5 and 15 min) as shown in Table 1. 800 ml of manure mixture having 5% DM content in 1000 ml glass container was sonicated according to

***Corresponding author:** Nuri Azbar, Professor, Faculty of Engineering, Bioengineering Department, Ege University, Izmir, Turkey*

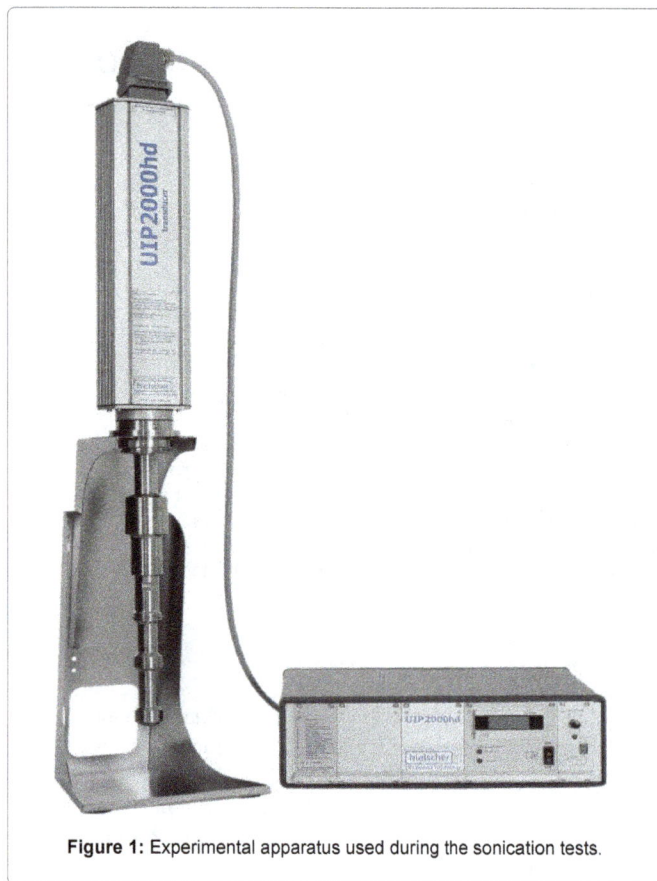

Figure 1: Experimental apparatus used during the sonication tests.

UIP1000hd	Frontal area [cm²]	Max. Amplitude* at 100% [μm]				
		Booster as Reducer			Booster as Booster	
Sonotrode		B2-1,8	B2-1,4	No Booster	B2-1,4	B2-1,8
BS2d22 (F)	3,8	26	**40**	57	**81**	106*
BS2d40 (F)	12,5	**9**	14	17	26	**31**

*With amplitudes >100 μm, the lifetime of sonotrodes will be reduced drastically

Table 1: Experimental conditions during sonication.

the experimental conditions given in table 1 (bold figures) for 5 and 15 min of reaction times.

Biochemical Methane Potential (BMP) Assay

Biochemical Methane Potential assay (BMP) was used to monitor the anaerobic biodegradability of chicken manure [12]. The BMP assay was performed in serum bottles (working volume is 100 ml) and these test was applied to both raw and pre-treated samples for comparison purpose in parallel. Each BMP bottle was seeded with anaerobic stock culture (40 ml), basal medium (15 ml) and chicken manure (20 ml; 5% DM). The control serum bottles that contained only anaerobic stock culture and basal medium but no organic wastes were also run in all experiments to determine the background gas production.After seeding the BMP test bottles, they were purged with N_2 gas for 2-3 min to maintain anaerobic conditions and then sealed with natural rubber stoppers and plastic screw-caps. The serum bottles were then incubated at 36 ± 1°C and were shaken at 150 rpm throughout the study. Total gas production recorded daily for at least 50 days by a glass syringe and the methane content of biogas in the headspace was determined by Gas Chromatography (GC). All experiments were run as triplicate and the mean values of net biogas production were reported in this study.

Analytical methods

The pH measurements were taken with a pH meter (WTW pH meter). Dry matter and Suspended Solids (SS) values were measured by following the standard methods. The methane and carbon dioxide content of the biogas was measured on an Agilent gas chromatograph equipped with a flame ionization detector and a DB-FFAP 30 m × 0.32 mm × 0.25 mm capillary column (J&W Scientific, USA).

Results and Discussions

Figure 2 shows the results of batch anaerobic digestion tests applied to both raw and pretreated chicken manure under various sonication conditions having amplitude values of 9, 31,40 and 81 μm (peak-to-peak measurement, producer information) as given in Table 1. The aim of these tests was to comparatively evaluate the effect of various combinations of sonotrode and booster couples on the enhancement of anaerobic conversion of chicken manure tested. Figure 2 also displays the volumetric daily methane production over time for all pretreatments and a control set that had no pretreatment at all.All BMP tests were carried out for at least 50 days. Each test group was tested against the control group which had no US pretreatment. As shown in Figure 2, in most instances the methane content could be increased by acoustic cavitation in comparison to the untreated sample. Best results could be reached on the one hand with high amplitudes and low reaction time and on the other hand with low amplitudes and long reaction time. The cumulative biomethane measurements for all tests (except for one condition) showed that sonication resulted in a higher biomethane conversion (obvious indication of better biodegradability) with respect to quantity of biomethane produced in comparison to control group (untreated raw sample). If the final biomethane gas production values at day 50 are carefully evaluated, the sonication pretreatment resulted in varying amount of percentage increase biomethane production varying between 0% and 74%. Lower left corner in Figure 2 shows the results of sonication experiments with amplitude of 9 μm and varying reaction time between 5 to 15 min. When the amplitude was increased to 31 μm as seen in the upper left corner of Figure 2, both reaction time resulted in almost equal increase in biomethane production which is better than the control one. It is quite obvious from the lower right corner of the Figure 2 that 40 μm amplitude was the best one in terms of biomethane production but needing longer reaction time. On the other hand, the use of amplitude of 81 μm required only 5 min reaction time to achieve the same high biomethane production. It is seen from the Figure 2 that even though sonication pretreatment resulted in higher biogas production than the control one and an increase in ultrasound amplitude had a positive effect on the biomethane yield in general, furthermore the effect of reaction time is also significant. It is known that increasing reaction time and amplitude leads to a higher number of cavitation bubbles and therefore to increased cavitational effects. Perez-Elvira (2009) studied the effect of ultrasound pretreatment for anaerobic digestion improvement in both batch experiments and continuously fed plot bioreactor [13]. The authors reported up to 42% increase in biogas production with batch tests. Their data from continuously operating digester with a volume of 100 L also indicated that pretreatment of sludge resulted in higher volatile solids removal and increased biogas production varying between 25-37%. These results were achieved at HRT value of 15 d which could actually be obtained in a bioreactor fed with non-pretreated at HRT=20 d. This result indicated that pretreatment also allowed use of higher organic loading rates resulting shorter HRT values

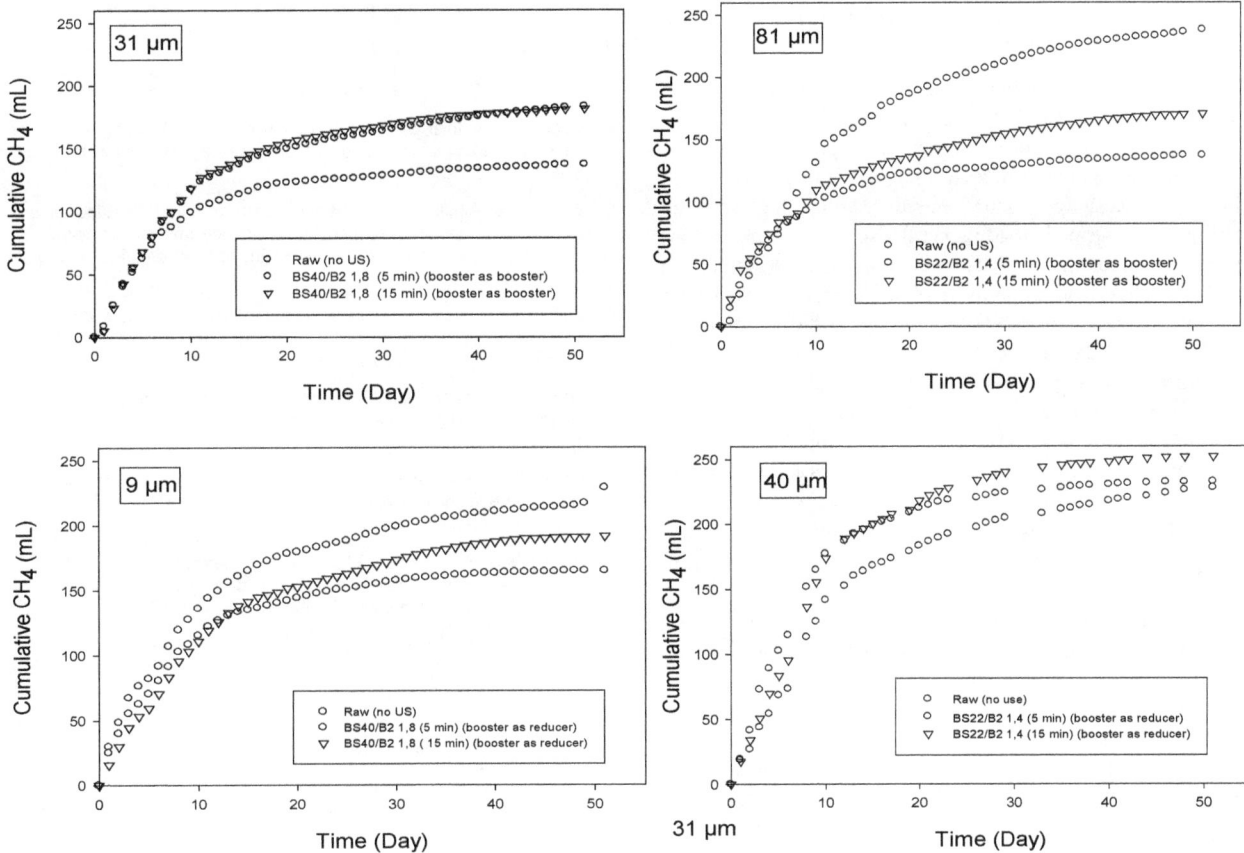

Figure 1: The results of batch anaerobic digestion tests.

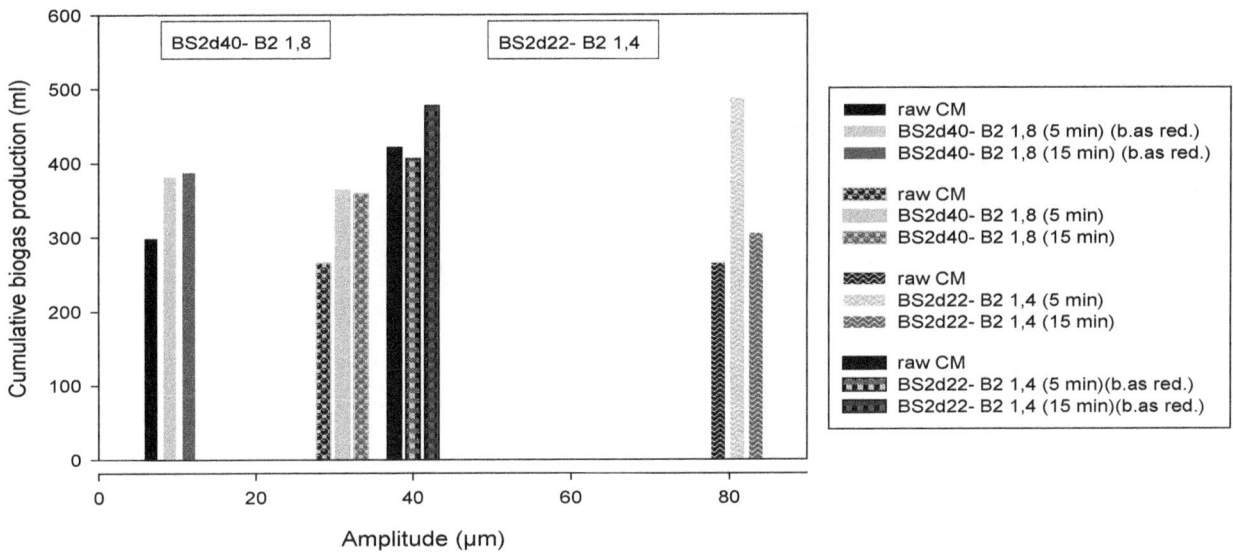

Figure 3: The effect of amplitude values on BMP test for each operational condition.

Figure 3 shows the effect of amplitude values on BMP test for each operational condition. It is obvious from Figure 3 that both amplitude and reaction time are significant factors. Up to the amplitude of 40 μm in general (except for 31 μm), increasing biogas volume was observed in comparison to control group that had no pretreatment. While the best result in terms of highest biogas production was obtained at an amplitude of 81 μm (BS2d22/B2-1, 4 as booster), longer reaction time (>5 min) has detrimental effect resulting in reducing biogas production.

It is important to note that the experiment using the amplitude of 40 μm (BS2d22/B2-1, 4 as reducer) showed high biogas production but needed 3 times longer reaction time (15 min).

Sonication has been increasingly used for enhancing the digestion of wastewater treatment plant sludge, animal manures in both Europe and USA. According to Hogan et al. sonication provides improved solids destruction, substantially increased biogas production, enhanced dewatering, decreased sludge production, reduced operating costs, and a reasonable payback period [14]. Muller et al. also reported that various disintegration pretreatment methods such as maceration, sonication and ozonation at full scale significantly increased improvement of biogas production due to increased solubilisation of degradable material [15]. Cesaro et al. reported 24% higher biogas production than untreated one when sonication was applied to solid waste samples. Cesaro et al. explained the mechanism that sonolysis significantly improved the solubilisation of organic waste [16]. Personal communication of the authors with large scale user of sonolysis technology as a pre-treatment in biogas production indicated higher biogas production up to 20% in comparison to those which had no similar pre-treatment.

Conclusions

Various pre-treatment strategies for AD including sonication were tested before which was quite less for laying hen manure. Disintegration via ultrasonication improves biomass digestibility by disrupting biosolids and bacterial cells, releasing intracellular components due to the solubilisation of the particulate matter, decreasing Solid Retention Time (SRT) and improving the overall performance of anaerobic digestion. Laboratory test resulted in promising results for the applications of full-scale systems indicating that pre-treatment is capable of increasing the efficiency, productivity and applicability of AD systems. The full-scale installations of ultrasonication also provided promising results which would be as high as 50% increase in the biogas generation.

Even though, each pretreatment would increase the capital cost of AD plant, obvious benefits such as increase in biomethane volume per unit mass of manure would make sonication more affordable and preferable. Literature reports indicate that the average ratio of the net energy gain to electric consumed by the ultrasound device is 2.5. In addition, strict environmental regulations and legislations have great impact on agricultural operations forcing farmers to rethink environmentally sound and economically tolerable novel solutions for the safe disposal of animal manure. In this manner, animal farms, especially chicken farms, have been experiencing significantly increasing waste management stress that needs the incorporation of the technologies that will comply with strict regulations while being capable of marinating profitability. Last but not least, profitability will drive the agricultural sector implement these technologies and exploit the benefits.

Acknowledgement

The authors would like to express appreciation for the support of the Seyyidoglu Ultrasonic Solutions Corporation, Turkey. The authors also wish to thank TUBITAK-CAYDAG under the grant No 111Y019 for the financial support of this study. The data presented in this article was produced within the project above, however it is only the authors of this article who are responsible for the results and discussions made herein.

References

1. Kameswari KSB, Kalyanaraman C, Thanasekaran K (2011) Effect of ozonation and ultrasonication pretreatment processes on co-digestion of tannery solid wastes. Clean Technologies and Environmental Policy 13:517-525.

2. Cesaro A, Belgiorno V (2013) Sonolysis and ozonation as pretreatment for anaerobic digestion of solid organic waste. UltrasonSonochem 20: 931-936.

3. Quiroga G, Castrillón L, Fernández-Nava Y, Marañón E, Negral L, et al. (2014) Effect of ultrasound pre-treatment in the anaerobic co-digestion of cattle manure with food waste and sludge. Bioresour Technol 154: 74-79.

4. Gonze E, Gonthier Y, Boldo P, Bernis A (1998) Standing waves in a high frequency sonoreactor: Visualization and effects. Chemical Engineering Science 53: 523-532.

5. Bougrier C, Albasi C, Delgenes JP, Carrere H (2006) Effect of ultrasonic, thermal and ozone pre-treatments on waste activated sludge solubilisation and anaerobic biodegradability. Chemical Engineering and Processing: Process Intensification 45:711-718.

6. Shimizu T, Kudo K, Nasu Y (1993) Anaerobic waste-activated sludge digestion-a bioconversion mechanism and kinetic model. Biotechnol Bioeng 41: 1082-1091.

7. Angelidaki I, Ahring BK (2000) Methods for increasing the biogas potential from the recalcitrant organic matter contained in manure. Water Sci Technol 41: 189-194.

8. Eastman JA, Furguson JF (1981) Solubilization of particulate organic carbon during the acid phase of anaerobic digestion. J Water Pollut Control Fed53:352-366.

9. Elbeshbishy E, Nakhla G (2011) Comparative study of the effect of ultrasonication on the anaerobic biodegradability of food waste in single and two-stage systems. Bioresour Technol 102: 6449-6457.

10. Elbeshbishy E, Aldin S, Hafez H, Nakhla G, Ray M (2011) Impact of ultrasonication of hog manure on anaerobic digestability. Ultrason Sonochem 18: 164-171.

11. Pilli S, Bhunia P, Yan S, LeBlanc RJ, Tyagi RD, et al. (2011) Ultrasonic pretreatment of sludge: a review. Ultrason Sonochem 18: 1-18.

12. Owen WF, Stuckey DC, Healy Jr JB, Young LY, McCarty PL (1979) Bioassay for monitoring biochemical methane potential and anaerobic toxicity. Water Research 13:485-492.

13. Pérez-Elvira S, Fdz-Polanco M, Plaza FI, Garralón G, Fdz-Polanco F (2009) Ultrasound pre-treatment for anaerobic digestion improvement. Water Sci Technol 60: 1525-1532.

14. Hogan F, Mormede S, Clark P, Crane M (2004) Ultrasonic sludge treatment for enhanced anaerobic digestion. Water Sci Technol 50: 25-32.

15. Müller JA, Winter A, Strünkmann G (2004) Investigation and assessment of sludge pre-treatment processes. Water Sci Technol 49: 97-104.

16. Cesaro A, Naddeo V, Amodio V, Belgiorno V (2012) Enhanced biogas production from anaerobic codigestion of solid waste by sonolysis. Ultrason Sonochem 19: 596-600.

25

Effect of Media Sterilization Time on Penicillin G Production and Precursor Utilization in Batch Fermentation

Kishore Kumar Gopalakrishnan and Swaminathan Detchanamurthy*

Research Scholar, Department of Chemical & Process Engineering, University of Canterbury, New Zealand

Abstract

Benzyl Penicillin (Penicillin G) is a secondary metabolite (Antibiotic) derived from *Penicillium chrysogenum*. Batch fermentation is widely used in the production of Penicillin G in laboratory scale and as well as in Industrial scale. The fermentation media for the production is sterilised at 121o C for 30 minutes as a usual practise in both scales of production. In our present study the production media used to produce Penicillin G was sterilized at various time intervals and the change in penicillin production along with the level of precursors utilized by the *P. chrysogenum* were analysed. The change in sterilization time varied the proportion of fermentation media converted from complex to simple. Studies were carried out in both shake flask level and also in laboratory scale fermenter (3 litres) with media containing PAA (Phenyl acetic acid) as precursor. The fermentation media used in this study contained K_2SO_4, KH_2PO_4, $(NH4)_2SO_4$, corn steep liquor (N_2 source), Lactose (Carbon source) and $CaCO_3$. The steam batch sterilisation at 121o C was attempted with different time intervals between 25 to 50 minutes with 5 minutes increment. It was observed that change in the sterilization time increased the Penicillin-G production by 30 % upto 30 minute and only 6% upto 45 minute and then it started to drop. HPLC method was used to carry out quantitative analysis of the product Penicillin G and the precursor Phenyl acetic acid. The results further concluded that though the rise in the sterilization temperature increased the Penicillin G production rate, it was cost effective as more energy required to rise the sterilization temperature which in turn increased the cost of production of Penicillin G.

Keywords: Benzyl Penicillin; Penicillin G; Fermentation; Sterilization; Phenyl acetic acid

Abbreviations: LCSB: Lactose Corn Steep Broth; PAA: Phenyl Acetic Acid; POAA: Phenoxy Acetic Acid

Introduction

First and foremost important unit operation required for carrying out fermentation successfully is media sterilisation [1]. In the fermentation industry the vegetative inoculum was prepared in seed tank and then transferred to fermentation tank with sterilised media. Nutrient media sterilization is an important process made for maintaining pure culture during the fermentation process. Nutrient media can be sterilized by 1) heating 2) filtration 3) irradiation 4) sonic vibration and 5) exposure to chemical agents. Throughout the fermentation industry, heating using steam is widely used as a sterilization method, since it is reliable and easy to control [2].

Nutrient media sterilisation by autoclaving was primarily used for the eradication of indigenous microbes, another important change which can occurs on the nutrient media simultaneously due to autoclaving is change in nutrient composition. Nutrient media widely used for the fermentation is complex mixtures of amino acids, vitamins, proteins, lipids, carbohydrates, nucleotides, nucleosides and minerals. Autoclaving of the fermentation media can lead to change in the composition of the nutrients by various mechanisms such as hydrolysis, degradation, conjugation and the formation of insoluble salts such as magnesium or calcium salts of phosphate. These salts are essential for the growth of the microbes and the production of the secondary metabolites [3]. During the sterilisation process before fermentation, indigenous microbial population were destroyed. At the same time simultaneous chemical changes in media can also be expected [4].

Primary Metabolites are found in all plant cells. These include sugars (carbohydrates), fats, oils and proteins, which are involved in the fundamental biochemical reactions common to all life. Secondary metabolites are the organic compounds which are produced in biological systems. These secondary metabolites would not involve directly in growth or reproduction of the organism [5] but they are not formed without a reason.

In an industrial production of penicillin by fermentation, the yield is augmented by the addition of a precursor such as phenyl acetic acid (PAA) or phenoxyacetic acid [6]. In order to avoid toxic effects, the precursor must be added repeatedly in small amounts during the fermentation. The penicillin side chain precursors, phenylacetic acid (PAA) and phenoxyacetic acid (POAA) are aromatic protonophores which are transported across the plasma membrane by free diffusion [7].

Among the secondary metabolites produced, penicillin is the most commercial antibiotics over the last 50 years. The total world market for β-lactum antibiotics is estimated to be about 15 billion US$. The β-lactum antibiotics are now accounted for over 65% of the world antibiotics market [8]. The mode of penicillin fermentation is fed batch. The precursor phenoxy acetic acid (POAA) is added to the fermenter throughout the process in order to avoid substrate inhibition. For the production of optimal antibiotic the fermentation parameters such as temperature, pH, dissolved oxygen, carbon dioxide, sugar, precursor, ammonia, etc. are closely monitored and controlled by computer programs [9].

*Corresponding author: Swaminathan Detchanamurthy, Research Scholar, Department of Chemical & Process Engineering, University of Canterbury, New Zealand

After the formulation of industrial fermentation media, it was sterilized for killing the total microorganisms present before the inoculation of the desired culture. Mostly in fermentation industry the steam sterilization method is used. Once media was autoclaved at 15 psi pressure for 15 minutes it was extended under streaming steam for continuous three days, which increases the hydrolysis of lactose and maltose to the greatest extend which makes maximum utilization of media by the microorganisms [10]. In steam sterilization the heating and cooling process-control to maintain steady temperature profiles is difficult. To avoid the risk of incomplete sterilization the "over kill" heating is frequently employed [11].

Materials and Methods

Culture and chemicals

The culture used in this study was penicillin-G producing *Penicillium chrysogenum* strain from SPIC pharmaceuticals division, kudikadu village, Cuddalore, India. The *Penicillium chrysogenum* spores were provided in the ampoules, to which 1mL of sterile water is mixed in sterile condition. 0.1 mL of the dilution was plated to LCSB plate (Lactose Corn Steep Broth) and kept for incubation for about 13 to 15 days. One mL was taken from this and serial dilution is made up to 10^{-6}. The colony selection was based on the following morphological characteristics 1) Green colour 2) 3-12 mm in diameter 3) Uniform sporulation 4) Circular slightly enlonged with raised centres and 5) White margin.

F1 Suspension preparation

The selected colony was diluted in 2 mL of sterile water, and then the suspension was added into a conical flask which contained 100 g of rice. It was kept for 13 days incubation at 25° C with 60 to 65% of relative humidity. To the above suspension, 200 mL of 0.05% Tween 60 was added and mixed. Then the strain was collected along with Tween 60 by rice sediment back. Tween 60 surfactant was used to separate spores from rice. The 100 mL of this suspension was taken as the F_1.

Viability test

From F_1 suspension one mL was taken and made serial dilutions till 10^{-6} in Tween 60 solution. It was plated in a viability count media and then incubated at 25°C for 5 days . Then colony count was made for analysing the activity of spores. The F_1 suspension showed viability count of 1.5 to $2x10^8$ cells/mL was taken for F_2 stage. The number of colonies present in the viability count plate with less than $1.5x10^8$ cells/mL was not taken into consideration and that stain was not taken to the F_2 stage. Because the number of colonies in the viability count media plates will reflect the activity of that stain.

Seed and fermentation medium

The seed media used in this work contains cotton seed meal 10 (g/l); corn steep powder 20 (g/l); sucrose 40 (g/l); ammonium sulphate 2 (g/l); KH_2PO_4 0.5 (g/l); $CaCO_3$ 5 (g/l); (pH 6.3) and sterilised at 121°C for 15 min. The seed media was inoculated after autoclaving at 121°C for 15 min with 10 mL of inoculum which contained $9x10^5$ mL of spores per mL and the culture was incubated in a rotary shaker at 25o C and 230 rpm for 54 hours. . Fermentation media contained the following nutrient composition: phenoxy acetic acid 10 (g/l); potassium sulphate 5 (g/l); KH_2PO_4 1.5 (g/l); ammonium sulphate 12 (g/l); corn steep meal 27.5 (g/l); lactose 120 (g/l); calcium carbonate 10 (g/l) and corn oil 10 mL. The media pH was adjusted to 6.5.

Sterilization

Vertical autoclaves employed for media sterilisation was stainless steel and which can withstand pressure up to 22 psi, equipped with temperature and pressure display (macro scientific works Pvt Ltd). The media flasks are plugged with cotton and covered with aluminium foil and sterilised as batches at various time intervals of 25, 30, 35, 40, 45, 50 and 55 min at constant pressure of 22 psi.

Inoculation and fermentation

Sterilised media (pH 6.5) in 3 batch flasks was inoculated with 3 mL of inoculum under laminar air flow and kept in orbital shaking incubator (REMI Instruments) at 220 rpm for 164 hours.

Analytical measurements

The samples from all 3 batch flasks were examined at 41st, 82nd, 123rd and 164th minutes for finding the level of activity of Penicillin G production using RP-HPLC (Shimadzu Corporation). The column used in the analysis was C18–Hypersil Octadodecyl Silane (length 25 cm, diameter 0.46 cm). The polar mobile phase was contained potassium dihydrogen orthophosphate and 45% of acetonitrile (80:20) at pH 4.6. and flow rate was used at 1.4 mL/min. In the case of standard fermentation procedures methanol was used as the secondary mobile phase instead of acetonitrile (60:40). The detector used was UV detector. The detector was of fixed type and the Light source was a D_2 (Deuterium) Lamp.

Results and Discussion

On penicillin fermentation the media will be sterilised at 121°C for 30 min, but in our study the media was sterilised at 121°C for various intervals of time between 25 and 55 minutes and the analysis was done at various fermentation time intervals, 41, 82, 123 and 164 minutes, respectively. The analysis was included the detection of penicillin G production and unutilization of precursor phenoxyaceticacid (POAA) in the media. The media sterilised at 121°C for 45 minutes has shown the maximum production of penicillin G and utilization of POAA at all times (41, 82, 123 and 164 minutes) of analysis. The analysis was carried out in HPLC and the elution peak for POAA is obtained around third minute and the elution peak for Penicillin G around fifteenth minute.

41st minute analysis

The 5 set of flask undergoing fermentation containing different times sterilised media were taken from shaker after 41 minutes and analysed the level of penicillin G activity in that and the amount of unutilized POAA present in that. The media sterilised for 45 minute shown Table 1 activity 9.63% more than the activity at 30 min (Usual sterilisation time). After 45 min, prolonged sterilisation up to 55 min leads to drop in the activity. The activity shown by 55 min sterilised media was only 32.39% compared to the activity 45 min sterilised media fermentation. The residual precursor POAA content was more in 25 minute sterilised media, which decreased steadily still 45 minute sterilised media. The flasks sterilised above 45 minutes showed an increase in POAA accumulation. 45 minute sterilised media showed 98.3 % POAA less accumulated while compared with the residual POAA present in fermentation at 55 minutes sterilised media Figure 1.

82nd minute analysis

The 5 set of flask undergoing fermentation containing different time sterilised media were taken from shaker after 82nd minute and

S. No	Media sterilisation time (minutes)	Penicillin activity (1μ/mL)	Residual POAA (mg/mL)
1	25	1021	8.01
2	30	1329	4.52
3	35	1406	2.191
4	40	1438	2.005
5	45	1457	0.331
6	50	911	10.04
7	55	472	16.76

Table 1: Penicillin activity and residual POAA content in various sterilization time intervals of 41 minute fermentation media.

S. No	Media sterilisation time (minutes)	Penicillin activity (1μ/mL)	Residual POAA (mg/mL)
1	25	10682	6.32
2	30	14019	3.62
3	35	14883	1.65
4	40	15812	1.38
5	45	16004	0.221
6	50	9912	6.71
7	55	5178	10.81

Table 2: Penicillin activity and residual POAA content in various sterilization time intervals of 82 minutes fermentation media.

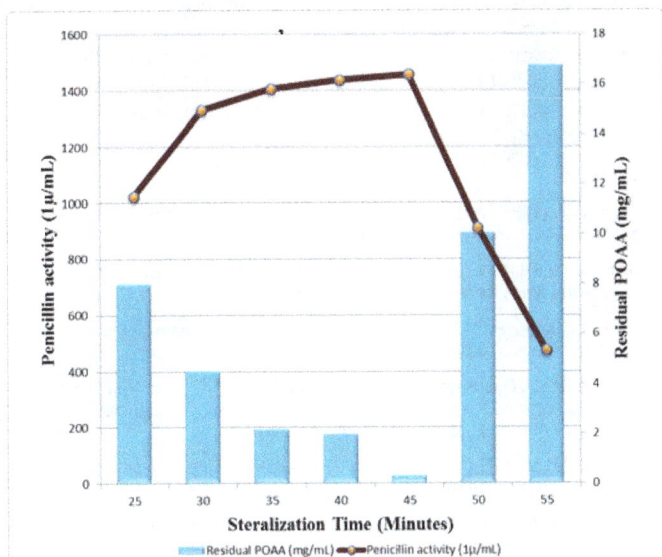

Figure 1: Penicillin activity and residual POAA content in various sterilization time intervals of 41 minute fermentation media.

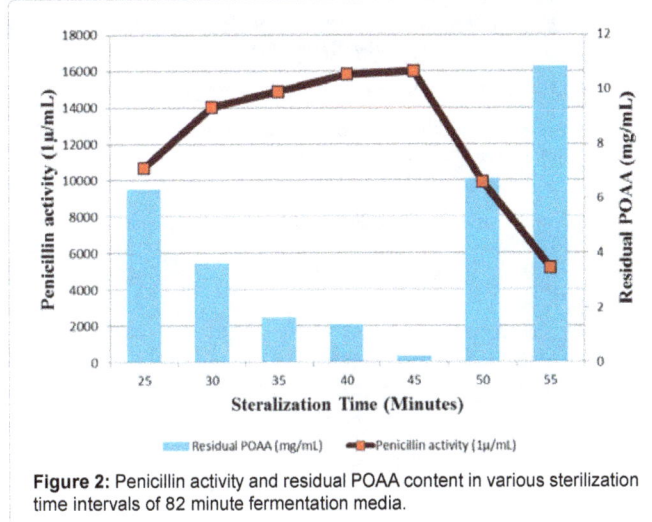

Figure 2: Penicillin activity and residual POAA content in various sterilization time intervals of 82 minute fermentation media.

S. No	Media sterilisation time (minutes)	Penicillin activity (1μ/mL)	Residual POAA (mg/mL)
1	25	29081	5.213
2	30	37791	2.91
3	35	38934	1.33
4	40	39195	1.1
5	45	39806	0.171
6	50	19861	5.15
7	55	9851	8.12

Table 3: Penicillin activity and residual POAA content in various sterilization time intervals of 123 minutes fermentation media.

analysed the level for penicillin activity in that and the amount of unutilized POAA present in that. The media sterilised for 45 minute showed Table 2 activity 14.16% more than the activity at 30 minute (Usual sterilisation time). After 45 minute of prolonged sterilisation up to 55 minute lead to drop in activity of penicillin G. The activity showed by 55 minute sterilised media was only 32.74% when compared to the activity 45 minute sterilised Media fermentation. The residual precursor POAA content was more in 25 minute sterilised, which decreased steadily till 45 minutes sterilised media. The flasks sterilised more than 45 minute showed increase in POAA accumulation. 45 minute sterilised media showed 87.24% POAA less accumulated while compared with the residual POAA present in fermentation 55 minutes sterilised media Figure 2.

123rd minute analysis

The 5 set of flask undergoing fermentation containing different time sterilised media were taken from shaker after 123rd minutes and analysed for the level for penicillin activity in that and the amount of unutilized POAA present in that. The media sterilised for 45 minutes Table 3 showed activity of 5.332% more than the activity at 30 minute (Usual sterilisation time). After 45 minute of prolonged sterilisation up to 55 minute lead to drop in activity of penicillin G. The activity showed by 55 minute sterilised media was only 24.74% when compared to the activity 45 minute sterilised Media fermentation. The residual precursor POAA content was more in 25 minute sterilised, which decreased steadily till 45 minute sterilised media. The flasks sterilised more than

45 minute showed increase in POAA accumulation. 45 minute sterilised media showed 97.89% POAA. Less accumulated while compared with the residual POAA present in fermentation 55 minutes sterilised media Figure 3.

164th minute analysis

The 5 set of flask undergoing fermentation containing different time sterilised media were taken from shaker after 164th minutes and analysed the level for penicillin activity in that and the amount of unutilized POAA present in that. The media sterilised for 45 minute Table 4 showed activity of 8.315% which was more than the activity at 30 minute (Usual sterilisation time). After 45 minute, prolonged sterilisation up to 55 minute lead to drop in activity. The activity showed by 55 minute sterilised media was only 26.24% when compared to the activity of 45 minute sterilised fermentation media. The residual precursor POAA content was more in 25 minute sterilised, which decreased steadily till 45 minute sterilised media. The flasks sterilised more than 45 minute showed increase in POAA accumulation. 45 minute sterilised media

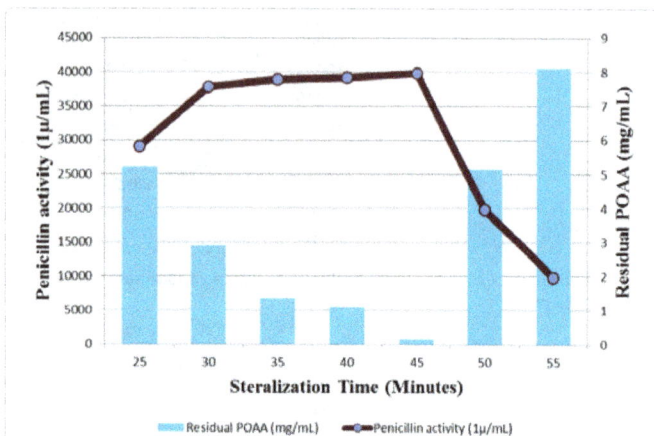

Figure 3: Penicillin activity and residual POAA content in various sterilization time intervals of 123 minute fermentation media.

S. No	Media sterilisation time (minutes)	Penicillin activity (1µ/mL)	Residual POAA (mg/mL)
1	25	35071	4.21
2	30	45291	2.32
3	35	48195	1.05
4	40	48954	0.91
5	45	49057	0.14
6	50	30672	4.02
7	55	12876	6.21

Table 4: Penicillin activity and residual POAA content in various sterilization time intervals of 164 minutes fermentation media.

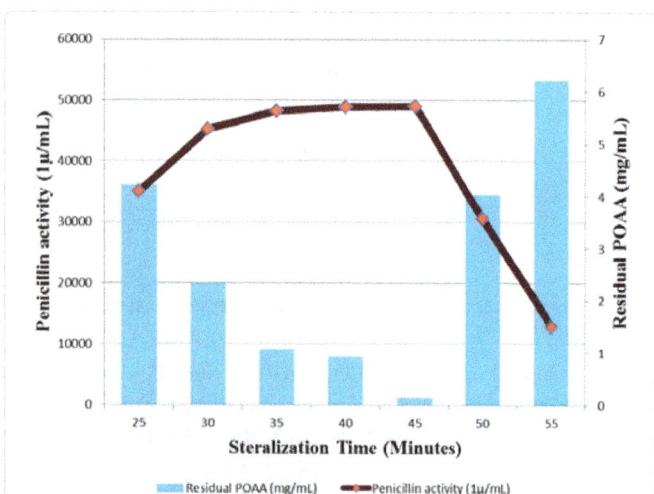

Figure 4: Penicillin activity and residual POAA content in various sterilization time intervals of 164 minute fermentation media.

show 97.74% POAA less accumulated while compared with the residual POAA present in fermentation 55 minutes sterilised media. The maximum Penicillin G fermentation is obtained in the 45 minutes sterilized media, whereas the media sterilised more than 45 minutes showed a decline in the Penicillin G fermentation Figure 4.

In order to maintain the unicellular fermentation, the media used should be free of microorganisms before inoculation of the *Penicillium chrysogenum*. Steam sterilisation is one of the best methods in sterilizing the fermentation media in industrial scale. The unknown benefit in sterilisation is making the media to be easily and maximum utilised

by the microorganisms. The number and type of chemical reaction that can occur in fermentation media during heat sterilization are too complex for definition. The temperature and time profile of the sterilization process could theoretically affect the resulting component concentrations. Long sterilisation also has negative impact on the production on some cases; inhibitory products also can be produced or charring of the media which leads to decline the production of the fermentation product. The studies carried out by us proved that increasing the sterilization time has increased the Penicillin G production by 30% upto 30 minute and above 30 minute the increase in sterilization time didn't increase the Penicillin G production much though there was a little increase. Moreover, it was observed that though the productivity of Penicillin G was increased as the effect of rise in Sterilization time, the cost involved in increasing the duration of sterilization was more than the cost of the product. Hence currently we are working on an alternative cheaper approach to increase the duration of sterilization without affecting the quality of the product and to get more profit from our Penicillin G production process.

References

1. Cooney CL (1985) Media sterilization. In: Cooney CL, Humphrey AE (eds) Pergamon Press, London. pp. 209-211.

2. Deindoerfer FH (1957) Microbiological Process Discussion - Calculation of heat sterilization times for fermentation media. Applied microbiology 5: 221-228.

3. Doolin LE, Mertz FP (1973) The effect of inorganic phosphate on the biosynthesis of vancomycin. Can. J. Microbiol. 19: 263-270.

4. Laverne Boeck D, Joseph Alford S, Jr Richard Pieper L, Floyd Huber M (1989) Interaction of media components during bioreactor sterilization: definition and importance of R0. J Industrial Microbiol 4: 247-252.

5. Fraenkel Gottfried S (1959) The raison d'Etre of secondary plant substances. Science 129: 1466-1470.

6. Eriksen SH, Jensen B, Schneider I, Kaasgaard S, Olsen J (1994) Utilization of side chain precursors for penicillin biosynthesis in a high-producing strain of Penicillium chrysogenum. Appl. Microbiol. Biotechnol. 40: 883-887.

7. Hillenga DJ, Versantvoort HJ, van der Molen S, Driessen AJM, Konings WN (1995) Penicillium chrysogenum takes up the penicillin G precursor phenylacetic acid by passive diffusion. Appl. Environ. Microbiol 61: 2589-2595.

8. Elander RP (2003) Industrial production of b-lactam antibiotics. Appl. Microbiol. Biotechnol. 61: 385-392.

9. Rowlands RT (1991) Industrial strain improvement and the Panlabs penicillin club In: Kleinkauf H, von Dohren H (eds) 50 years of penicillin applications: history and trends. Public, Czech Republic.

10. Mudge CS (1917) The effect of sterilization upon sugars in culture media. J. Bact. 2: 403-415.

11. Enzinger M, Goodsir S, Korozynski M, Parham F, Schneier M (1978) Validation of stream sterilization cycles Technical monograph No 1, Parenteral Drug Association, Inc., Philadelphia, USA.

12. Anderson TM, Bodie EA, Goodman N, Schwartz RD (1986) Inhibitory effect of autoclaving whey-based medium on propionic acid production by Propionibacterium shermanii. Appl. Environ. Microbiol 51: 427-428.

26

Automatic Sleep Stage Detection and Classification: Distinguishing Between Patients with Periodic Limb Movements, Sleep Apnea Hypopnea Syndrome, and Healthy Controls Using Electrooculography (EOG) Signals

Emad Malaekah*, SobhanSalari Shahrbabaki and Dean Cvetkovic

School of Electrical and Computer Engineering, RMIT University, Australia

Abstract

Background: To improve the diagnostic and clinical treatment of sleep disorders, the first important step is to identify or detect the sleep stages. Utilizing the conventional method-known as visual sleep stage scoring-is tedious and time-consuming. Therefore, there is a significant need to create or develop a new automatic sleep stage detection system to assist the sleep physician in evaluating the sleep stages of patients or non-patient subjects. The first aim of this study is to develop an algorithm for automatic sleep stage detection based on Electrooculography (EOG) signals. The second aim is to utilize sleep quality parameters to classify and screen Periodic Limb Movements of Sleep (PLMS) patients and Sleep Apnea Hypopnea Syndrome (SAHS) patients, as distinct from healthy control subjects.

Methods: 10 patients with Periodic Limb Movements of Sleep (PLMS), 10 patients with Sleep Apnoea Hypopnea Syndrome (SAHS), and 10 healthy control subjects were utilised in this study. Several features were extracted from EOG signals such as cross-correlation, energy entropy, Shannon entropy and maximal amplitude value. K-Nearest Neighbour was used for the classification of sleep stages. Several polysomnographical (PSG) features were measured for screening and classification of the sleep disorders, such as the percentage of the sleep stages over the total time of sleep, the duration of the sleep stages, Sleep Latency (SL), and sleep efficacy. A decision tree analysis was utilised for identifying the three groups of subjects.

Results: The overall accuracy, sensitivity and specificity of automatic sleep stage detection were 80.5%, 81.3% and 88.8%, respectively. The Cohen's Kappa was 0.73. The performance of the classified sleep disorders showed an overall accuracy of 90%. The sensitivity and specificity were 90% and 95%. The Cohen's Kappa was 0.85.

Conclusion: One advantage of the automatic sleep stage detection method based on Electrooculography (EOG) signals is that it can be utilized with portable sleep stage recording instead of using a multichannel signal. Classification of sleep disorders based on the automatic system is an improvement, in that it can make the screening or diagnostic processes much faster and easier than with other methods.

Keywords: Sleep stage; Sleep scoring; Polysomnography; Electrooculography; Sleep quality

Introduction

The sleep phenomenon has gained reasonable scientific interest for an extended time. Sleep refers to a behavioral state that varies from wakefulness by a loss of reactivity, readily and reversibly, in relation to events within one`s environment [1]. Sleep can be categorized into two primary and distinct behaviors: NREM (Non-Rapid Eye Movement) sleep and REM (Rapid Eye Movement) sleep [2]. NREM is categorized as light sleep, termed N1 and N2 (the latter is further broken down into S1, S2), and deep sleep, which is termed N3 (and further broken down into S3 and S4) [3]. Deep sleep is also known as Slow-Wave Sleep (SWS).The abbreviations W, N1, N2, N3, and R are derived from the new standard of Iber and colleagues [4].

Performing Polysomnography (PSG) entails a comprehensive sleep study assessing numerous physiological signals such as an Electroencephalogram (EEG), an Electrooculogram (EOG), an Electromyogram (EMG), respiratory effort, an Electrocardiogram (ECG), and others. It is the gold standard for measuring sleep states [5], sleep quality, and sleep quantity. The manual scoring of sleep stages based on EEG, EOG and EMG is a subjective and time-consuming process; hence the need for comprehensive and more accurate automatic techniques that are easy to apply and can be used in experimental and clinical ambulatory research.

Several attempts have been made to utilise the EEG or EOG signals

only for detecting sleep stages, or to detect only one particular sleep stage, such as Slow-Wave Sleep (SWS) [6,7].

EEG automatic detection has been employed for detecting sleep stages. The method was comprised of four steps: segmentation, extraction of parameters, analysis of cluster, and classification. The parameters compared included the harmonic parameters, Hjorth, and relative band energy [8]. An automatic algorithm used by Liang [9] for detection of SWS utilized one or two EOG/EEG channels. The result of this study obtained 80% sensitivity, and a Cohen's kappa value of 0.755.

"Sleep disorder" refers to a medical condition in the patterns of sleep of an animal or human being, also known as somnipathy [10]. The classification of sleep disorders is essential in order to differentiate

---

*Corresponding author: Emad Malaekah, School of Electrical and Computer Engineering, RMIT University, 376-392 Swanston Street, GPO Box 2476V, Melbourne, VIC 3001, Australia

between disorders, and to enhance understanding of etiology, pathophysiology and symptoms, thus enabling appropriate treatment [11]. The Pittsburgh Sleep Quality Index (PSQI), which was developed by Buysse [12], has been used as a standardised subjective measure to evaluate sleep quality. PSQI is based on several questions relating to the evaluation of psychometric properties of sleep quality for the duration of one month. The periodic limb leg movement disorder is defined as a nearly irresistible urge to move legs while asleep [13]. Studies indicate that PLMS occurs in stages 1 or 2 of the sleep period before REM sleep. On the other hand, Obstructive Sleep Apnea Hypopnea Syndrome (OSAHS) has been on the increase in the last fifty years, with significant morbidity rates in both developing and developed countries. OSAHS also causes daylight sleepiness [14]. Sleep apnea hypopnea syndrome leads to fragmentation of sleep and limits the quantity of time spent in the deeper sleep stages 3 and 4.

In this study we focused on two main objectives. The first aim was to develop an automatic sleep stage detection method based on two EOG signals, as compared with the manual sleep stage detection system based on EEG, EOG, and EMG signals. The second important aim was to develop an automatic system for classifying different sleep disorders based on the sleep quality extracted from the sleep stages.

Materials and Methods

Participants and data collection

The PSG data was downloaded from the online database [11]. The PSG signals included three EEG signals (C3-A1, FP1-A1 and O1-A1), two EOG signals, and one submental EMG channel. The Right EOG (REOG) and Left EOG (LEOG) signals were the only signals utilised from this PSG dataset. This PSG data was then recorded from 10 healthy controls: 7 females aged 20-65 years (average age: 40 years), and 3 males aged 20-27 years (average age: 23.5 years). Next, these signals were taken in 10 patients with Periodic Limb Movements of Sleep (PLMS): 8 adult males aged 31-71 years (average age: 51 years), and 2 adult females aged 27-69 years (average age: 48 years). Finally, signals were recorded from 10 patients with Sleep Apnea Hypopnea Syndrome (SAHS): 6 adult males aged 38-73 years (average age: 55 years), and 4 adult females aged 52-74 years (average age: 60 years). The collected data was acquired in a Belgian sleep hospital using a digital 32-channel polygraph (Brainnet System of MEDATEC, Brussels, Belgium). The sample frequency was 200 Hz. The visual sleep stage was scored by an expert according to the AASM criteria.

EOG signal Processing

Pre-processing: The EOG data was segmented into 5-second epochs. The entire EOG vector was processed utilising a zero-phase band pass filter with a Cascaded Integrator-Comb (CIC) filter of order six, for the following different frequency bands: delta (0.5-2 Hz), delta1 (2-4 Hz), theta (4-8 Hz), alpha (8- 12 Hz), sigma (12-16 Hz), beta1 (16-20 Hz), and beta2 (20-30 Hz) (Figure 1). Since an EOG signal might be affected by EEG, EMG, and ECG artifacts, launching a suitable algorithm in order to remove the artifacts and noises is necessary. A cascade of three adaptive filters based on a least mean square algorithm was employed, and by this means ECG, EEG, and EMG artifacts were duly eliminated (more information in [15,16]).

Feature extraction: Several features were extracted from the EOG signal in the time and frequency domain, such as variance, Maximal Peak Amplitude Value (MPAV), Minimum Peak Amplitude Value (MPAV), total power, energy entropy, Shannon entropy, and cross-correlation. In order to select the best feature that classified variations

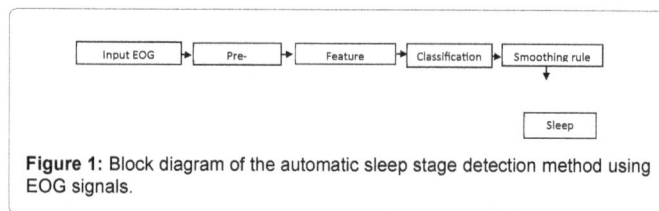

Figure 1: Block diagram of the automatic sleep stage detection method using EOG signals.

in sleep stages and wakefulness, the Sequential Feature Selection method (SFS) was used.

Classification

In this study, the K-Nearest Neighbor was used for classification of sleep and wakefulness stages. The KNN is based on a nonparametric method for different pattern classification approach, which represents as one robust classifier. The KNN classifier works based on a comparison between a new sample (testing data) and baseline (training data). It attempts to find out the K-Nearest Neighbor within the baseline, and indicates a class which seems more normally in the nearest neighbor of K. The value of K might need to be diverse in order to detect the corresponding class between the training and testing data. In this paper, the value of K varies from 1 to 5. The Euclidean distance metric is utilized for calculating the distance between two points. The training and testing data was evaluated based on 10-fold cross-validation.

Smoothing rule

The smoothing rule is one of the common methods for increasing the accuracy of detecting the sleep stages. This rule is used as in the following example: three consecutive readings of N1, N2, and N2 were replaced as sequence N1, N1 and N1.

Classification of sleep disorders

In order to classify the PLSM, SAHS, and healthy control subjects, different PSG sleep stage parameters were first measured, such as Sleep Latency (SL), Sleep Efficiency (SE), number of times subject woke up (NW), Total Sleep Time (TST), Waking After Sleep Onset (WASO), Slow-Wave Sleep (SWS), and Rapid Eye Movement sleep (REM). During the first REM period, characteristics of the sleep stages (N1, N2, N3 and R) were based on the automatic detection system detailed in the previous section.

A decision tree analysis was used to classify the three groups of subjects based on the following rules:

- Rule (1) used the percentage of the sleep stage N1 parameter to separate SAHS patients from PLMS patients and healthy control subjects. If N1(%) was more than 7% and less than 9%, then a subject would be classified as an SHAS patient; if N1(%) was less than 4%, then a subject would be classified as a healthy control; if N1(%) was more than 10%, then a subject would be classified as a PLMS patient.

- Rule (2) used the percentage of the sleep stage N2 parameter to separate the SHAS from PLMS patients. If N2(%) was more than 80%, a patient would be classified as an SAHS patient; If N2(%) was less than 80% and more than 60%, a patient would be classified as a PLMS patient.

- Rule (3) utilised Slow-Wave Sleep Duration (SWSD) in minutes and the percentage of sleep stage N3 to distinguish between PLMS and healthy control subjects. If SWSD was more than 70 min and N3(%) more than 20%, a subject would be classified as a healthy control. All these rules were based on some percentage of sleep stage and the time duration of sleep parameters, such as N1%, N2%, and SWSD. The automatic classification algorithm is described in Figure 2.

A statistical analysis was conducted using post-hoc tests to ascertain whether there were significant differences between the three groups. Sensitivity and specificity tests, as well as Cohen's Kappa, were conducted to evaluate the automatic classification algorithm for the three groups.

Results

Automatic sleep stage detection

In this pilot study, we utilised the EOG signal only for detection of the sleep stages of 30 subjects, comprising 10 healthy controls, 10 PLMS patients, and 10 SAHS patients. Several features were extracted from the EOG signal based on different frequency bands as mentioned in the previous section. The overall agreement, sensitivity, and specificity of the detection of sleep stages for healthy control subjects were 83.5%, 85%, and 88% respectively. The Cohen's Kappa was 0.79. Table 1 shows the confusion matrix with sensitivity and specificity after applying the smoothing rule to one healthy subject as an example. The results show that the best detection was in wakefulness, and in sleep stage N3 (by 91%). The detection of sleep stage N1 after utilizing the smoothing rule was significantly improved.The overall agreement, sensitivity, and specificity for detection of the sleep stages of PLMS patients were 80%, 82%, and 86%, respectively. The Cohen's Kappa was 0.71, which was lower than the Cohen's Kappa of the healthy controls. The reason for this is that the normal distribution of sleep stages with healthy controls was much more consistent than that of the PLMS patients. Table 2 shows the confusion matrix, sensitivity, and specificity of the sleep stages of a PLMS patient. It is obvious that the total number of sleep stage N2s was higher than other sleep stages, which increased the detection of stages N1 and R. On the other hand the overall agreement, sensitivity, and specificity for detection of sleep stages with SAHS patients were 78%, 77%, and 80%, respectively, while the Cohen's Kappa represented was lower than in the other two groups by 0.67. Table 3 shows the confusion matrix, sensitivity, and specificity of sleep stages of an SAHS patient. It is clear that the lower sensitivity was in the wakefulness stages, with an improvement in the detection of sleep stage N1.

Figures 3, 4 and 5 show the hypnograms of visual sleep stage scoring vs. automatic scoring for a healthy control, a PLMS and an SAHS patient, respectively, as opposed to with automatic sleep stage detection. It can be observed that due to some sleep stages was scored as sleep stage N2 or N3 which made the view of hypnogram included some non-corrected classification. However, all of the visual scores were consistent with 84% for a healthy subject, 86% for a PLMS patient, and 79% for SAHS patients.

Automatic classification of the PLMS, SAHS, and healthy control subjects

Figure 2 shows the automatic classification algorithm used to classify the patients with PLMS, the patients with SAHS, and the healthy control subjects. The significant sleep parameters, such as N1(%), N2(%), and SWSD were used to identify the three groups on the basis of the thresholds as described in the previous section. Table 4 shows the post-hoc test analysis for the three groups of participants. There were significant differences between the PLMS patients and the healthy control participants, particularly in sleep stages N1, N2, and SWS duration. Furthermore, there were significant differences between the SAHS patients and the healthy control participants in the following

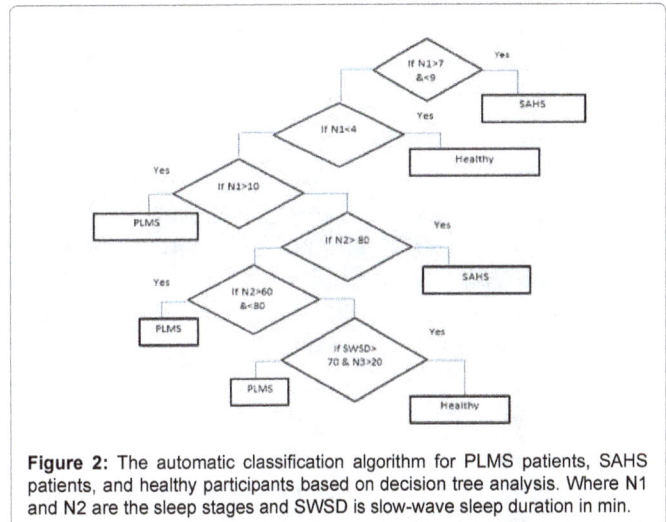

Figure 2: The automatic classification algorithm for PLMS patients, SAHS patients, and healthy participants based on decision tree analysis. Where N1 and N2 are the sleep stages and SWSD is slow-wave sleep duration in min.

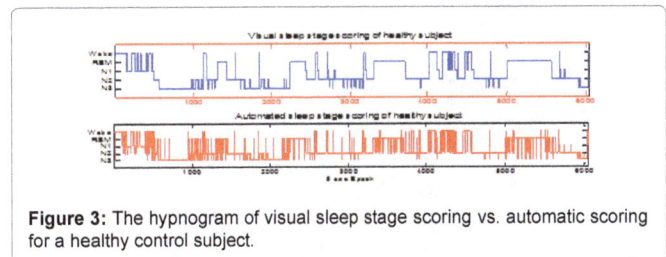

Figure 3: The hypnogram of visual sleep stage scoring vs. automatic scoring for a healthy control subject.

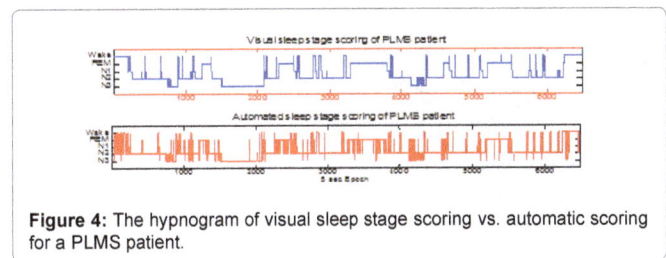

Figure 4: The hypnogram of visual sleep stage scoring vs. automatic scoring for a PLMS patient.

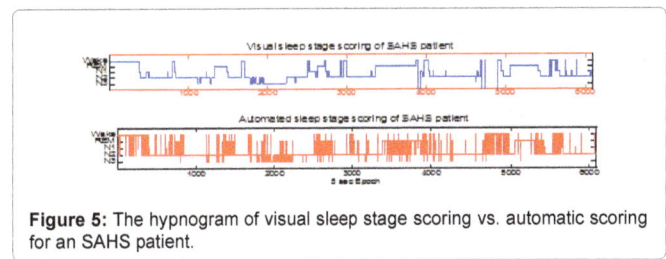

Figure 5: The hypnogram of visual sleep stage scoring vs. automatic scoring for an SAHS patient.

sleep parameters: SWS duration, REM duration, and sleep stages N2, N3, and R. The participants with PLMS differed from the SAHS patients within some sleep parameters, such as in sleep stage N2. Figure 4 shows the bar plots of the three groups. The SL, WASO, and NW were all significant sleep stage parameters for the PLMS patients. Sleep stage N2 was a significant sleep parameter for the SAHS patients. The sensitivity and specificity of identification in the PLMS patients, SAHS patients, and healthy controls were 90% and 95%, respectively (Table 5). The level of accuracy and Cohen's kappa were 90% and 0.85, respectively.

Discussion

In this pilot study, we aimed to use EOG signals for automatic sleep

		Automatic detection					
		W	N1	N2	N3	R	Sensitivity (%)
Visual detection	W	575	18	120	5	16	91
	N1	22	238	108	5	29	78
	N2	15	10	2280	39	62	80.6
	N3	3	1	168	1018	46	91
	R	11	3	150	43	1065	87.4
	Specificity (%)	90	85	70	87.4	81.2	

Table 1: The confusion matrix of a healthy control subject.

		Automatic detection					
		W	N1	N2	N3	R	Sensitivity (%)
Visual detection	W	524	6	163	5	28	87
	N1	50	150	163	9	36	70
	N2	25	5	3091	13	44	80
	N3	10	0	191	224	19	85
	R	186	11	277	16	836	78
	Specificity (%)	88	88	60	87.6	82.2	

Table 2: The confusion matrix of a PLMS patient.

		Automatic detection					
		W	N1	N2	N3	R	Sensitivity (%)
Visual detection	W	547	10	153	17	43	66
	N1	29	134	181	11	59	72
	N2	38	13	2893	68	138	79.5
	N3	3	2	104	731	24	84
	R	11	12	255	32	968	85.7
	Specificity (%)	85	75	66	96	85	

Table 3: The confusion matrix of an SAHS patient.

stage detection, and then used the data to classify PLMS patients, SAHS patients, and healthy control subjects. The overall inter-rate agreement between the visual sleep scoring and automatic sleep stage scoring was 80.5%, with a Cohen's Kappa of 0.73. On the other hand, the accuracy level of automatic classification of sleep disorders was 90%, and the Cohen's Kappa was 0.85.

We employed different features which extracted from the EOG signals and then utilized the KNN classifier for detection of wakefulness and the sleep stages. Some studies use the decision rule based on various thresholds for predicting the sleep stages [17], however due to the contract with the threshold from subject to subject particular to the EOG signal, the resultant accuracy reached 72%. Therefore, we used the KNN classifier due to its simplicity and strength in detecting the sleep stages. Several studies employ signals in addition to EOG signals for automatic sleep stage detection, such as Electroencephalography (EEG) and Electromyogram (EMG) signals [18-20]. These require more electrodes and more complicated algorithms to increase the accuracy level which has been observed. On the other hand some studies use only one EEG signal for automatic sleep detection [21,22].

Since the number of occurrences of sleep stage N2 in PLMS and SAHS patients was more than in the healthy control subjects, this was a distinct difference between these three groups. This led the overall accuracy of sleep stage N2 to be very low, which means the KNN classifier was predicated the other sleep stages as sleep stage N2. In Table 2, for example, it was obvious that the total number of occurrences of sleep stage N2 was higher than the other sleep stages, which caused increased overall detection of the other sleep stages or wakefulness stage.

Similar studies have utilised the EOG signal for detection of the sleep stages, or of one particular sleep stage such as Slow-Wave Sleep

Sleep parameters	PLMS vs Healthy			PLMS vs SAHS			SAHS vs Healthy		
	t	SD	p	t	SD	p	t	SD	p
Sleep latency	0.37	43.2	0.71	0.2	40.2	0.81	0.28	21.5	0.77
Sleep efficiency	-0.62	22.4	0.54	-0.57	18.3	0.57	-0.24	14.5	0.81
WASO	0.61	94.9	0.55	0.50	81.04	0.62	0.29	59.6	0.77
Number of waking	0.58	1.13	0.57	0.48	193	0.64	0.28	143	0.78
REM Latency	1.9	48.36	0.07	-0.43	101.7	0.67	1.4	95.2	0.17
First REM Period	-2.07	0.08	0.06	-1.0	0.60	0.34	0.69	0.60	0.5
Total Sleep Time	-0.41	118.9	0.68	-0.51	90.6	0.61	-0.03	77.3	0.97
SWS duration	-2.56	70.5	'0.03	50.2	0.51	0.62	-6.01	34.38	'0.01
REM duration	-1.8	41.7	0.10	1.0	32.9	0.33	-3.25	33.4	'0.01
W(%)	0.30	23.2	0.76	0.30	18.8	0.76	0.08	15.38	0.93
N1(%)	2.07	6.1	0.06	1.2	5.7	0.25	1.34	4.23	0.21
N2(%)	2.95	16.6	'0.01	-2.93	10.2	'0.01	5.25	15.06	'0.01
N3(%)	-3.04	15.2	'0.01	0.80	11.7	0.44	-7.61	7.30	'0.01
R(%)	-1.38	10.3	0.2	1.75	6.1	0.11	-2.78	9.03	'0.02

Table 4: The post-hoc t-tests of the differences in the sleep parameters between the three groups.

		Automatic Classification		
		PLMS	SAHS	Healthy
True Classification	PLMS	9	1	0
	SAHS	0	9	1
	Healthy	1	0	9
	Sensitivity		90%	
	Specificity		95%	
	Cohen's kappa		0.85	

Table 5: Confusion matrix, sensitivity, and specificity of the three groups.

(SWS) [17,22]. An automatic method was previously developed for detection of SWS based on two EOG channels [22]. This study employed the amplitude criterion for detecting SWS, and beta power [18-23] was utilised to reduce the artefact. The result shows inter-rater reliability between the visual and the developed automatic method of 96%, with a Cohen's kappa value of 0.70. The sensitivity and specificity were 75% and 96%, respectively. Another study employed two-channel electrooculography for automatic sleep stage classification particular to the left mastoid (M1) [17]. The synchronous Electroencephalographic (EEG) activity during SWS and S2 were detected by calculating peak-to-peak and cross-correlation amplitude differences in the 0.5 to 6 Hz range, and between the two EOG channels. The result indicated epoch by epoch agreement between the visual and the developed automatic method of 72%, with a Cohen's kappa value of 0.63.

The second aim of this study was to utilise the sleep stages that were detected based on an automated system for the purpose of classifying PLMS patient, SAHS patients, and healthy control subjects. We provide significant evidence that supports the use of the PSG sleep stage features-such as sleep stage N1(%), N2(%), and SWSD-for an automatic classification of PLMS patients, SAHS patients, and healthy control subjects. The percentage of sleep stage N1 was the most significant feature distinguishing patients with SAHS from healthy control subjects. The study found that sleep stage N(%) was between 7% and 9% for 7 SAHS patients. Conversely, 8 healthy subjects had a percentage of sleep stage N1 for less than 5% of total sleep duration. However, some patients with PLMS show higher percentages in sleep stage N1, which led the mean average of sleep stage N1(%) to be higher than in the other two groups. Figure 6 presents evidence that the mean average of sleep stage N1(%) for the healthy control group was lower compared with the other groups.

The percentage of sleep stage N2 was used to distinguish between SAHS and PLMS patients. This study found that most of the SAHS patients had a higher percentage of sleep stage N2 (above 80%) than PLMS patients. In Figure 2, we show the threshold that was used to distinguish between these two groups of patients. The SWSD and the percentage of sleep stage N3 were used to separate the PLMS patients from the healthy control subjects. We found that most of the patients with PLMS had less SWSD compared to healthy subjects. The longest SWSD of the healthy controls was above 70 min. The reason for using the percentage of sleep stage N3 is because some patients with PLMS had similar SWSD to the healthy control group. Figure 2 shows the threshold that was used to separate these two groups. The overall accuracy was 90%, and the Cohen's kappa was 0.85.

Conclusion

In conclusion, this paper aimed to develop an automatic method for detection of the sleep stages based on EOG signals, and then utilised these sleep stages for classification of PLMS patients, SAHS patients, and healthy control subjects. There was a significant advantage which supports utilising automatic sleep stage detection based on only EOG signals on ambulatory sleep recordings. The sensitivity of identifying PLMS, SAHS and healthy control participants was 90%, 90%, and 90%, respectively. This study suggests that using an automatic classification system in screening processes is more effective and efficient compared to some standards such as PSQI.

Acknowledgements

The authors gratefully acknowledge the financial support of the Ministry of Higher Education of the Kingdom of Saudi Arabia.

References

1. Verhulst SL, Schrauwen N, Haentjens D, Suys B, Rooman RP, et al. (2007) Sleep-disordered breathing in overweight and obese children and adolescents: prevalence, characteristics and the role of fat distribution. Arch Dis Child 92: 205-208.

2. Sadock BJ, Sadock VA, Belkin GS (2010) Kaplan & Sadock's pocket handbook of clinical psychiatry. Lippincott Williams & Wilkins, Philadelphia, USA.

3. Miano S, Paolino MC, Castaldo R, Villa MP (2010) Visual scoring of sleep: A comparison between the Rechtschaffen and Kales criteria and the American Academy of Sleep Medicine criteria in a pediatric population with obstructive sleep apnea syndrome. Clin Neurophysiol 121: 39-42.

4. Nishida M, Pearsall J, Buckner RL, Walker MP (2009) REM sleep, prefrontal theta, and the consolidation of human emotional memory. Cereb Cortex 19: 1158-1166.

5. Davis SD, Eber E, Koumbourlis AC (2014) Diagnostic tests in pediatric pulmonology: Applications and Interpretation. Humana Press, USA.

6. Chan M, Estève D, Escriba C, Campo E (2008) A review of smart homes-present state and future challenges. Comput Methods Programs Biomed 91: 55-81.

7. Berthomier C, Drouot X, Herman-Stoïca M, Berthomier P, Prado J, et al. (2007) Automatic analysis of single-channel sleep EEG: validation in healthy individuals. Sleep 30: 1587-1595.

8. Nakano H, Tanigawa T, Furukawa T, Nishima S (2007) Automatic detection of sleep-disordered breathing from a single-channel airflow record. Eur Respir J 29: 728-736.

9. Hang LW, Su BL, Yen CW (2013) Detecting Slow Wave Sleep via One or Two Channels of EEG/EOG Signals. International Journal of Signal Processing Systems 1: 84-88.

10. Derry CP, Davey M, Johns M, Kron K, Glencross D, et al. (2006) Distinguishing sleep disorders from seizures: diagnosing bumps in the night. Arch Neurol 63: 705-709.

11. Tsuno N, Besset A, Ritchie K (2005) Sleep and depression. J Clin Psychiatry 66: 1254-1269.

12. Doi Y, Minowa M, Okawa M, Uchiyama M (2000) Prevalence of sleep disturbance and hypnotic medication use in relation to sociodemographic factors in the general Japanese adult population. J Epidemiol 10: 79-86.

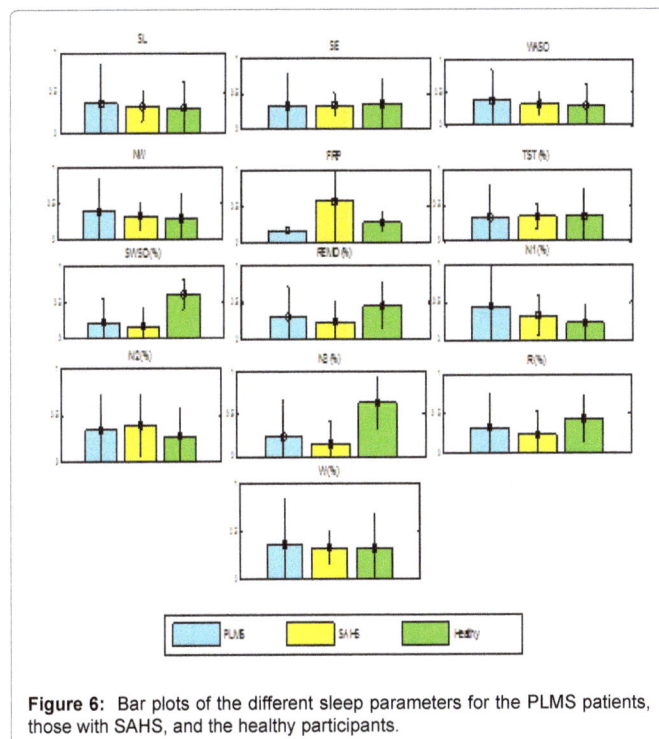

Figure 6: Bar plots of the different sleep parameters for the PLMS patients, those with SAHS, and the healthy participants.

13. Roth T (2007) Insomnia: definition, prevalence, etiology, and consequences. J Clin Sleep Med 3: S7-10.

14. Pack AI, Maislin G, Staley B, Pack FM, Rogers WC, et al. (2006) Impaired performance in commercial drivers: role of sleep apnea and short sleep duration. Am J Respir Crit Care Med 174: 446-454.

15. The DREAMS Subjects Database.

16. Correa AG, Laciar E, Patiño HD, Valentinuzzi ME (2007) Artifact removal from EEG signals using adaptive filters in cascade. J Phys Conf Ser 90: 012081.

17. Virkkala J, Hasan J, Värri A, Himanen SL, Müller K (2007) Automatic sleep stage classification using two-channel electro-oculography. J Neurosci Methods 166: 109-115.

18. Agarwal R, Gotman J (2001) Computer-assisted sleep staging. IEEE Trans Biomed Eng 48: 1412-1423.

19. Anderer P, Gruber G, Parapatics S, Woertz M, Miazhynskaia T, et al. (2005) An E-health solution for automatic sleep classification according to Rechtschaffen and Kales: validation study of the Somnolyzer 24 × 7 utilizing the Siesta database. Neuropsychobiology 51: 115-133.

20. Álvarez-Estévez D, Fernández-Pastoriza JM, Hernández-Pereira E, Moret-Bonillo V (2013) A method for the automatic analysis of the sleep macrostructure in continuum. Expert Systems with Applications 40: 1796-1803.

21. Liang SF, Kuo CE, Hu YH, Pan YH, Wang YH (2012) Automatic stage scoring of single-channel sleep EEG by using multiscale entropy and autoregressive models. Instrumentation and Measurement, IEEE Transactions on 61: 1649-1657.

22. Malaekah E, Cvetkovic D (2014) Automatic sleep stage detection using consecutive and non-consecutive approach for elderly and young healthy subject. Biosignals and Biorobotics Conference: Biosignals and Robotics for Better and Safer Living (BRC), 5th ISSNIP-IEEE: 1-6.

23. Virkkala J, Hasan J, Värri A, Himanen SL, Müller K (2007) Automatic detection of slow wave sleep using two channel electro-oculography. J Neurosci Methods 160: 171-177.

27

A Genetically Engineered STO Feeder System Expressing E-Cadherin and Leukemia Inhibitory Factor for Mouse Pluripotent Stem Cell Culture

Masanobu Horie, Akira Ito, Yoshinori Kawabe and Masamichi Kamihira*

Department of Chemical Engineering, Faculty of Engineering, Kyushu University, 744 Motooka, Nishi-ku, Fukuoka 819-0395, Japan

Abstract

Embryonic stem (ES) cells and induced pluripotent stem (iPS) cells are usually cultured on mouse embryonic fibroblasts (MEFs) isolated from fetuses. MEFs are primary cells and can only be cultured for several passages before senescence. Although the STO mouse stromal cell line has been used as a MEF substitute, performance of a STO feeder layer for ES/iPS cell culture is inferior to a MEF feeder layer. Thus, the development of effective feeder systems may be beneficial for advancing stem cell technology using ES/iPS cells. We established a STO feeder cell line expressing mouse leukemia inhibitory factor (LIF) and mouse E-cadherin (designated as STO/EL cells). ES/iPS cells were cultured on STO/EL feeder cells without LIF addition to the medium, while maintaining the expression of stem cell markers, Oct3/4, Nanog and Rex1. Quantitative evaluation of feeder performance by an alkaline phosphatase-positive colony-forming assay revealed that the colony forming efficiency was comparable to that of the conventional and most reliable culture method using MEFs with LIF addition. STO/EL cells can be used as an efficient feeder system for supporting ES/iPS cell culture in an undifferentiated state. The STO/EL feeder system may contribute toward medical and biological research of ES/iPS cells.

Keywords: Embryonic stem cell; Induced pluripotent stem cell; Feeder cell; E-cadherin; LIF; STO

Abbreviations: AP: Alkaline Phosphatase; CM: Conditioned Medium; ES: Embryonic Stem; iPS: induced Pluripotent Stem; LIF: Leukemia Inhibitory Factor; LPA: Lysophosphatidic Acid; MEF: Mouse Embryonic Fibroblasts; ROCK: Rho-associated Coiled-Coil Forming Kinase

Introduction

Pluripotent cell lines such as ES cells derived from embryos and iPS cells generated from somatic cells have become important tools for medical and biological research [1,2]. These pluripotent cells are characterized by self-renewal and a differentiation potential for all cell types of the adult organism. A robust culture system for maintaining the self-renewal of ES/iPS cells and differentiation potential is crucial for research. Undifferentiated ES/iPS cell culture is usually supported by a feeder cell layer in the presence of anti-differentiation factors such as LIF. MEFs isolated from fetuses are often used as feeder cells for mouse and human ES/iPS cell culture. However, MEFs are primary cells and can only be cultured for several passages before senescence. Although the STO mouse stromal cell line has been used as a substitute for MEFs, the performance as a feeder for ES/iPS cell culture is inferior to MEFs. The detailed mechanism for supporting the undifferentiated state of ES/iPS cells on a MEF feeder layer remains unclear. In addition, STO cells appear to produce few or less growth factors to support ES/iPS cell culture. Thus, we theorized that close contact between STO and ES cells would improve the performance of STO cells as feeders for ES cell culture. E-cadherin, a Ca^{2+}-dependent cell-cell adhesion molecule, plays important roles in intercellular adhesion, colony formation and differentiation of ES cells [3]. Previously, we generated E-cadherin-expressing STO cells that promote cell-cell interactions with ES cells inherently expressing E-cadherin, and demonstrated improved performance comparable to that of a MEF feeder layer [4].

In our E-cadherin gene-engineered feeder system, LIF addition to the culture medium was essential. In the present study, we constructed genetically engineered STO cells co-expressing E-cadherin and LIF to further improve the performance of STO cells as feeders for ES/iPS cell culture.

Materials and Methods

Cell culture

Mouse ES cell lines, H-1 (Riken BioResource Center, Tsukuba, Japan) and 129/Sv (Chemicon, Pittsburgh, PA, USA), and a mouse iPS cell line, iPS-MEF-Ng-20D-17 [2] (Riken BioResource Center) were maintained on mitotically inactivated feeder cells that were treated with mitomycin C for 2 h. Cells were cultured on 0.1% gelatin- (Nacalai Tesque, Kyoto, Japan) coated tissue culture dishes (Greiner Bio-one, Frickenhausen, Germany) in ES/iPS medium consisting of Knockout-DMEM™ (Invitrogen, Carlsbad, CA, USA) supplemented with 4 mM L-glutamine (Wako Pure Chemical Industries, Osaka, Japan), non-essential amino acids (NEAA; Invitrogen), 100 μM 2-mercaptoethanol (Millipore, Billerica, MA, USA), 100 U/ml penicillin G potassium (Wako Pure Chemical Industries), 50 μg/ml streptomycin sulfate (Wako Pure Chemical Industries) and 15% Knock-out-serum-replacement (Invitrogen). For routine ES/iPS cell culture, 1×10^3 U/ml LIF (ESGRO™; Millipore) was added to the medium. Culture medium was changed every day, and cells were passaged every 2-3 d.

MEFs were isolated from the fetuses of 14 d pregnant BALB/c mice and cultured in Dulbecco's modified Eagle's medium (DMEM; Sigma-Aldrich, St. Louis, MO, USA) supplemented with 10% fetal bovine serum (FBS; Biowest, Miami, FL, USA) and 4 mM L-glutamine. STO

*Corresponding author: Masamichi Kamihira, Department of Chemical Engineering, Faculty of Engineering, Kyushu University, 744 Motooka, Nishi-ku, Fukuoka 819-0395, Japan

cells were cultured in DMEM supplemented with 10% FBS, 100 U/ml penicillin G potassium, 50 µg/ml streptomycin sulfate and NEAA. Cells were cultured at 37°C in a 5% (v/v) CO2 incubator. Animal experimentation was approved by the Ethics Committee for Animal Experiments of the Faculty of Engineering, Kyushu University (A21-098-1).

Establishment of E-cadherin and LIF expressing STO cells

The expression plasmid vector for E-cadherin, pcDNA4/E-cad-IRES-EGFP [5] was transfected into STO cells using a lipofection reagent (Lipofectamine2000; Invitrogen). Cells were selected in medium containing 1 mg/ml zeocin (Invitrogen), and stable E-cadherin-expressing clones (designated as STO/E cells) were established by a limiting dilution method [4]. For construction of an expression plasmid vector for LIF, LIF cDNA in pMFGmLIF [6] (Riken BioResource Center) was PCR amplified using the primers, 5'-CCG CTC GAG ACC ATG AAG GTC TTG GCC GCA G-3' and 5'-GGA ATT CCT AGA AGG CCT GGA CCA CCA C-3' to insert XhoI and EcoRI digestion sites (underlined), respectively. PCR was performed using KOD plus DNA polymerase (Toyobo, Osaka, Japan) at an initial denaturation of 94°C for 2 min, followed by 30 cycles of amplification at 94°C for 15 s, 56°C for 30 s and 68°C for 37 s. The PCR product was digested with the relevant restriction enzymes and ligated into a XhoI and EcoRI-digested pIRES2 vector from pIRES2-DsRed-Express (Clontech, Palo Alto, CA, USA) to generate pmLIF. STO and STO/E cells were transfected with pmLIF by electroporation using a Neon transfection system (Invitrogen). Cells were selected by culture in medium containing 800 µg/ml G418 (Sigma-Aldrich). Stable LIF-expressing clones (designated as STO/L cells), as well as E-cadherin and LIF co-expressing clones (designated as STO/EL cells) were established by a limiting dilution method.

RT-PCR analysis

After plating ES/iPS cells onto gelatin-coated dishes to remove feeder cells, total RNA was extracted from ES/iPS cells using RNAiso Plus reagent (Takara Bio, Otsu, Japan). RNA was reverse-transcribed into cDNA from 1 µg total RNA using a ReverTra Ace First Strand cDNA synthesis kit (Takara Bio). Specific gene sequences were PCR amplified using the primers shown in Table 1.

Western blot analysis

Cell lysates (50 µg protein) of STO, STO/E, STO/L and STO/EL cells were subjected to SDS-PAGE in a 10% (w/v) polyacrylamide gel. Proteins were then transferred onto a polyvinylidene difluoride membrane (GE Healthcare, Buckinghamshire, UK). After blocking with a 5% (w/v) skim milk solution, membranes were incubated with a rabbit anti-E-cadherin monoclonal antibody (Santa Cruz Biotechnology, Santa Cruz, CA, USA) or rabbit anti-mouse LIF polyclonal antibody (Santa Cruz Biotechnology) for 1 h, and probed with a peroxidase-labeled anti-rabbit antibody (Santa Cruz Biotechnology) for 1 h. Proteins were detected with an ECL detection system (GE Healthcare).

Alkaline phosphatase (AP) staining

Cells were fixed in 4% paraformaldehyde for 5 min at room temperature and treated with a solution containing naphthol AS-MX phosphate (Sigma-Aldrich) as a substrate and Fast Violet B Salt (Sigma-Aldrich) as a coupler for 20 min at 37°C. Cells showing AP activity stained dark brown and were observed with a phase-contrast microscope (Olympus, Tokyo, Japan).

Target gene (product size)	Primer sequence
Actc1 (124 bp)	FW: 5'-CCA GAT CAT GTT TGA GAC CTT CAA-3' RV: 5'-GAA CAT TAT GAG TTA CAC CAT CGC-3'
BMP2 (249 bp)	FW: 5'-GGG ACC CGC TGT CTT CTA GTG TTG C-3' RV: 5'-TGA GTG CCT GCG GTA CAG ATC TAG CA-3'
α-Fetoprotein (173 bp)	FW: 5'-TCG TAT TCC AAC AGG AGG-3' RV: 5'-AGG CTT TTG CTT CAC CAG-3'
Gata4 (207 bp)	FW: 5'-CTG GAG GCG AGA TGG GAC GGG ACA CTA C-3' RV: 5'-CCG CAG GCA TTA CAT ACA GGC TCA CC-3'
GAPDH (150 bp)	FW: 5'-CTA CCC CCA ATG TGT CCG TC-3' RV: 5'-GCT GTT GAA GTC GCA GGA GAC-3'
Nanog (163 bp)	FW: 5'-GCG GCT CAC TTC CTT CTG ACT T-3' RV: 5'-GAC CAG GAA GAC CCA CAC TCA T-3'
Neurod3/ngn1 (405 bp)	FW: 5'-CAT CTC TGA TCT CGA CTG C-3' RV: 5'-CCA GAT GTA GTT GTA GGC G-3'
Oct3/4 (459 bp)	FW: 5'-CTG AGG GCC AGG CAG GAG GAG CAC GAG-3' RV: 5'-CTG TAG GGA GGG CTT CGG GCA CTT-3'
Zfp42/Rex1 (287 bp)	FW: 5'-ACG AGT GGC AGT TTC TTC TTG GGA-3' RV: 5'-TAT GAC TCA CTT CCA GGG GGC ACT-3'

Table 1: Primer sequences for RT-PCR analysis.

ES/iPS cell culture using conditioned medium (CM)

STO, STO/E, STO/L and STO/EL cells treated with mitomycin C for 2 h were seeded at 5×10^5 cells/well in 6 well culture plates. After 1 d, medium was replaced with fresh LIF-free ES/iPS medium and cultures were continued for a further 1 d. Culture supernatants were then collected and filtered through a 0.45 µm cellulose acetate filter (Advantec, Tokyo, Japan). CM was prepared by mixing each culture supernatant with fresh LIF-free ES/iPS medium at a ratio of 1:1. 129/Sv cells were cultured for 10 d on a MEF feeder layer in each CM without LIF and then 1×10^4 cells were re-plated onto a fresh MEF feeder layer. After 3 d culture in ES/iPS medium containing 1×10^3 U/ml LIF, AP-positive colonies were counted.

Hanging drop assay

Embryoid bodies (EBs) were induced using the hanging drop method [7]. After harvesting ES/iPS cells, the cells were re-suspended in LIF-free ES/iPS medium at 7×10^3 cells/ml. Droplets of the cell suspension (15 µl) were placed on the lid of a bacterial grade 100 mm plastic dish (AsOne, Osaka, Japan). The lid was inverted and placed on the bottom half of a dish filled with phosphate buffered saline (PBS) and then incubated at 37°C in a 5% CO$_2$ incubator. After 2 d, cells were transferred to gelatin-coated dishes and cultured in DMEM supplemented with 15% FBS, NEAA, 100 U/ml penicillin G potassium and 50 µg/ml streptomycin sulfate.

Colony forming assay

ES/iPS cells were cultured for 10 d on the various feeder layers in ES/iPS medium with or without LIF and then 1×10^4 cells were re-plated onto a MEF feeder layer. After 2 d culture in ES/iPS medium containing 1×10^3 U/ml LIF, AP staining was performed as described above. AP-positive colonies were counted using microscope images from five fields of view in three separate wells per sample.

Magnetic force-based culture

Magnetically-labeled ES/iPS cells and a magnetic force were used to enhance the physical contact between ES/iPS cells and feeder layers [4]. Magnetite cationic liposomes (MCLs) were prepared from colloidal magnetite nanoparticles (Fe_3O_4, 10 nm average particle size; Toda Kogyo, Hiroshima, Japan) and a lipid mixture consisting of N-(α-trimethylammonioacetyl)-didodecyl-D-glutamate chloride, dilaur+oylphosphatidylcholine and dioleoylphosphatidyl-ethanolamine at a molar ratio of 1:2:2 as described elsewhere [8]. For magnetic labeling of ES/iPS cells, MCLs were added to ES/iPS cell cultures at a net magnetite concentration of 100 pg/cell. In our previous study, ES cells rapidly took up MCLs and reached 13 pg/cell at 2 h after MCL addition, and 95% of cells were magnetically captured, suggesting that almost all cells were labeled with MCLs and the uptake amount was sufficient for magnetic cell attraction [9]. After a 2 h incubation, 1×10^5 ES/iPS cells/well were seeded onto feeder layers in 24 well plates. A cylindrical neodymium magnet (30 mm diameter; 0.4 T magnetic induction) was then placed under the wells to apply a vertical magnetic force. Cells were cultured for 7 d, including two passages, under the applied magnetic force. Then, 1×10^4 cells/well were re-plated onto MEFs in 24 well plates. After 3 d culture, AP-positive colonies were counted.

ROCK inhibitor and Rho activator treatments

For ROCK inhibitor and Rho activator treatments, Y-27632 (Wako Pure Chemical Industries) or lysophosphatidic acid (LPA; Enzo Life Science, Farmingdale, NY, USA) were respectively added at 10 μM to ES/iPS medium. ES/iPS cells were then plated onto feeder layers in ES/iPS medium containing Y-27632 or LPA. Cells were cultured for 10 d, 1 $\times 10^4$ cells/well were then re-plated onto MEFs in 24 well plates. After 3 d culture, AP-positive colonies were counted.

Statistical analysis

Data were expressed as the means ± standard deviation (SD). Statistical comparisons were evaluated using a one-way analysis of variance (ANOVA), and $P < 0.05$ was considered significant.

Results

Establishment of genetically engineered STO cells expressing E-cadherin and LIF

We previously established E-cadherin-expressing STO cells (STO/E cells) [4]. STO/E feeder layers exhibited excellent performance for supporting undifferentiated culture of mouse ES cells comparable to that of MEF feeder layers in the presence of LIF [4] Figure 1A. However, the effectiveness of STO/E feeder layers was greatly reduced without LIF Figure 1A. In the present study, the LIF gene was introduced into STO and STO/E cells. Transgene expression was analyzed by RT-PCR Figure 1B, and protein level expression by western blot analysis Figure 1C. Genetically engineered STO cells expressed elevated levels of E-cadherin (STO/E cells), LIF (STO/L cells) and both E-cadherin and LIF (STO/EL cells) compared with those of MEFs and parental STO cells. E-cadherin expression was almost identical between STO/E and STO/EL cells. From the image analysis of RT-PCR Figure 1B, LIF expression of STO/L cells was 1.7-fold higher than that of STO/EL cells. Western blot analysis Figure 1C revealed that LIF protein level in STO/L cell lysate was 1.2-fold higher than that of STO/EL cells. To evaluate the activity of LIF expressed by genetically modified cells, ES cells were cultured on MEF feeder layers in CM prepared from STO/L and STO/EL cell culture supernatants. AP-positive ES cell colonies were counted after 10 d culture (3 passages) Figure 1D. As compared with conventional ES cell cultures using ES/iPS medium containing 1×10^3 U/ml LIF, the number of AP-positive colonies decreased in CM from STO and STO/E cells. However, the efficiency of the medium was improved

using CM from STO/L and STO/EL cells. For ES/iPS culture in STO/L CM, the number of AP-positive colonies was comparable to that of ES/iPS culture in medium containing 1×10^3 U/ml LIF, while AP-positive colonies in STO/EL CM was slightly lower compared with that of using STO/L CM. These results indicated that LIF expression in STO/EL cells was lower compared with that of STO/L cells, and was consistent with RT-PCR results Figure 1B.

Stem cell marker expression and pluripotency analyses of ES/iPS cells cultured on STO/EL feeders

Two mouse ES cell lines (H-1 and 129/Sv cells) and a mouse iPS

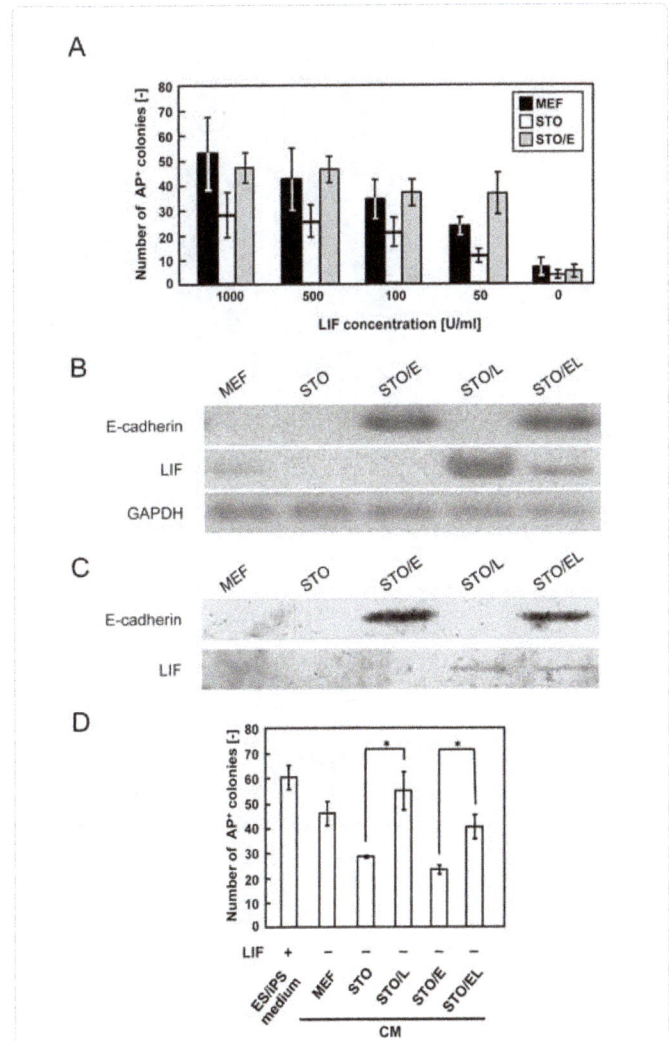

Figure 1: STO/EL cell establishment.
(A) Effect of LIF concentration on maintaining ES cell cultures in an undifferentiated state. 129/Sv cells were cultured in various LIF concentrations (0–1,000 U/ml) on MEF (black column), STO (white column) and STO/E (gray column) feeder layers. ES cells were re-plated onto a MEF feeder layer and cultured in medium containing 1×10^3 U/ml LIF for 2 d. Then, AP-positive ES colonies were counted.
(B) RT-PCR analysis of E-cadherin and LIF expression in STO/EL cells.
(C) Western blot analysis of E-cadherin and LIF expression in STO/EL cells.
(D) Effect of CM on the number of AP-positive ES cell colonies. 129/Sv cells were cultured on a MEF feeder layer in CM prepared from the various feeder layers. The control experiment was performed by culturing 129/Sv cells on a MEF feeder layer in ES/iPS medium containing 1×10^3 U/ml LIF. Experiments were performed in triplicate, and data are the means ± SD, $*P < 0.05$.

cell line (iPS-MEF-Ng-20D-17) were used to investigate whether STO/EL cells could support the self-renewal of pluripotent stem cells. ES and iPS cells cultured on MEF, STO, STO/E and STO/L feeder layers in LIF-free medium changed morphology and a decrease in the number of AP-positive cells was observed Figure 2A. iPS cells maintained a relatively undifferentiated state irrespective of which cells were used as feeder layer. However, ES and iPS cells cultured on a STO/EL feeder layer exhibited a typical ES/iPS cell morphology with tightly packed cell colonies and smooth borders Figure 2A. Cell morphology and the

number of AP-positive cells were very similar to those of cells cultured on a MEF feeder layer in LIF-containing medium. To further investigate the undifferentiated state of ES/iPS cells cultured on each feeder layer for 10 d (3 passages), stem cell marker expression was analyzed by RT-PCR Figure 2B. The expression levels of Oct3/4, Zfp42/Rex1 and Nanog were detected in ES/ iPS cells cultured on a STO/EL feeder layer. The stem cell marker expression levels were almost identical to those of cells cultured on a MEF feeder layer in LIF-containing medium as the control. However, the expression of stem cell marker genes were low or

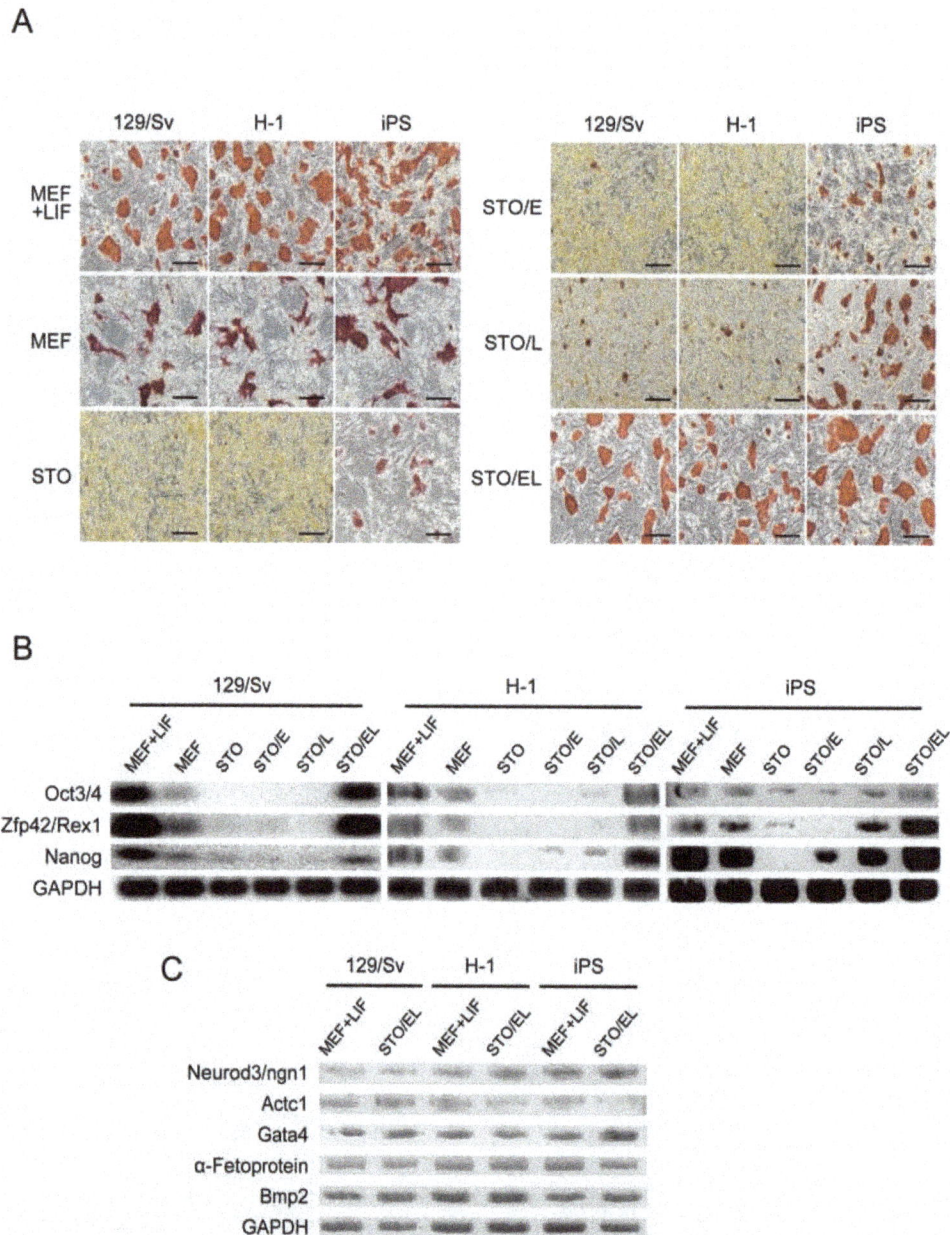

Figure 2: Expression of stem cell marker and pluripotency genes.
(A) AP staining of ES/iPS cells cultured on MEF, STO, STO/E, STO/L and STO/EL feeder layers in medium with (MEF+LIF) or without (MEF, STO, STO/E, STO/L and STO/EL) LIF. Scale bars: 500 μm.
(B) RT-PCR analysis of genetic markers of the undifferentiated state. ES/iPS cells were cultured on the various feeder layers in LIF-free medium (MEF, STO, STO/E, STO/L and STO/EL). The control experiment (MEF+LIF) was performed by culturing ES/iPS cells on a MEF feeder layer in medium containing 1 × 10³ U/ml LIF.
(C) RT-PCR analysis of marker genes related to formation of the three germ layers. ES/iPS cells were cultured on a STO/EL feeder layer for 10 d in LIF-free medium. The control experiment (MEF+LIF) was performed by culturing ES/iPS cells on a MEF feeder layer in medium containing 1 × 10³ U/ml LIF. ES/iPS cells were then cultured in suspension with differentiation medium to form EBs. After 11 d, the expression of marker genes was analyzed by RT-PCR.

absent in cells cultured on MEF, STO, STO/E or STO/L feeder layers in LIF-free medium.

The pluripotency of ES/iPS cells cultured on STO/EL feeders in LIF-free medium was investigated by analyzing differentiation capability via EB formation. For EB formation, ES/iPS cells cultured on STO/EL feeders in LIF-free medium and MEF feeders in LIF-containing medium, as the control, were transferred to suspension culture in differentiation medium. Expression of marker genes related to the formation of the three germ layers was analyzed by RT-PCR Figure 2C. Cells in EBs expressed the marker genes of all three germ layers (Neurod3/ngn1, Actc1, Gata4, α-fetoprotein and BMP-2), indicating that in vitro pluri-

potency was maintained in ES/iPS cells cultured on STO/EL feeders in LIF-free medium. Moreover, ES/iPS cells cultured for 10 d on STO/EL feeders in LIF-free medium, which were transplanted into the femurs of SCID mice developed into teratomas at the injection sites (data not shown). These results indicate that STO/EL cells as feeders support the undifferentiated state and pluripotency of ES/iPS cells.

Colony forming assay of ES/iPS cells cultured on STO/EL feeders

For quantitative evaluation of feeder performance, an AP-positive colony-forming assay was performed on ES/iPS cells after 10 d culture on the various feeder layers. As shown in Figure 3A, ES/iPS cells cul-

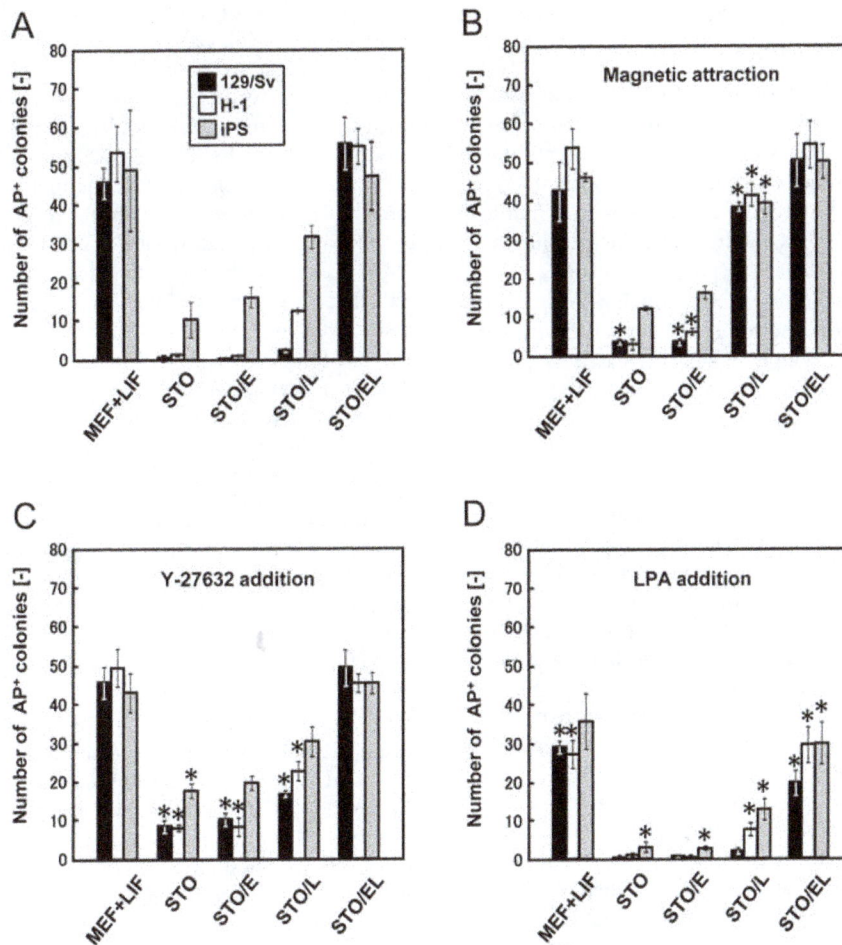

Figure 3: Quantitative analysis of the maintenance of ES/iPS cells in an undifferentiated state by an AP-positive colony forming assay.
(A) ES/iPS cells were cultured on the various feeder layers for 10 d in LIF-free medium. Then, ES/iPS cells were re-plated onto a MEF feeder layer and cultured in medium containing 1×10^3 U/ml LIF for 2 d. The control experiment (MEF+LIF) was performed by culturing ES/iPS cells on a MEF feeder layer in medium containing 1×10^3 U/ml LIF.
(B) Effect of the physical interaction of ES/iPS cells with feeder cells using magnetic culture. Magnetically-labeled ES/iPS cells were cultured on the various feeder layers in LIF-free medium under an applied magnetic force for 10 d. The control experiment (MEF+LIF) was performed by culturing magnetically-labeled ES/iPS cells on a MEF feeder layer in medium containing 1×10^3 U/ml LIF under an applied magnetic force for 10 d. ES/iPS cells were then re-plated onto a MEF feeder layer and cultured in medium containing 1×10^3 U/ml LIF for 2 d. *$P<0.05$ vs the respective data shown in Figure 3A.
(C) Effect of a ROCK inhibitor on ES/iPS cell culture on the various feeder layers. ES/iPS cells were cultured on the various feeder layers in LIF-free medium containing a ROCK inhibitor (Y-27632) for 10 d. ES/iPS cells were then re-plated onto a MEF feeder layer and cultured in medium containing 1×10^3 U/ml LIF for 2 d. The control experiment (MEF+LIF) was performed by culturing ES/iPS cells on a MEF feeder layer in medium containing 1×10^3 U/ml LIF and Y-27632 for 10 d. *$P<0.05$ vs the respective data shown in Figure 3A.
(D) Effect of a Rho activator on ES/iPS cells cultured on the various feeder layers. ES/iPS cells were cultured on the various feeder layers in LIF-free medium containing a Rho activator (LPA) for 10 d. ES/iPS cells were then re-plated onto a MEF feeder layer and cultured in medium containing 1×10^3 U/ml LIF for 2 d. The control experiment (MEF+LIF) was performed by culturing ES/iPS cells on a MEF feeder layer in medium containing 1×10^3 U/ml LIF and LPA for 10 d. *$P<0.05$ vs the respective data shown in Figure 3A. Experiments were performed in triplicate, and data are the means ± SD. Black column, 129/Sv; white column, H-1; gray column, iPS-MEF-Ng-20D-17.

tured on a STO/EL feeder layer in LIF-free medium formed a higher number of AP-positive colonies compared with those of STO, STO/E or STO/L feeder layers, and the AP level was comparable to that of a MEF feeder in LIF-containing medium.

To enhance the physical interaction between ES/iPS and feeder cells, magnetically-labeled ES/iPS cells were attracted to the feeder cells by magnetic force Figure 3B. The number of AP-positive ES/iPS cell colonies that were magnetically attracted to the MEF feeder layer in the presence of LIF was almost identical to that of cultures without magnetic attraction Figure 3A. Similarly, the number of AP-positive ES/iPS cell colonies cultured on a STO/EL feeder layer did not change in response to magnetic attraction. The ability to support undifferentiated culture of ES cells was slightly improved on STO/E and STO feeder layers in the absence of LIF, but was still relatively low. However, the performance of STO/L feeders was greatly improved by the magnetic attraction.

To investigate whether E-cadherin-mediated signal transduction was involved in maintaining ES/iPS cells in an undifferentiated state via modulation of the Rho-ROCK cascade, an inhibitor and activator of the Rho-ROCK signaling pathway were added to the medium. ES cell culture on STO, STO/E or STO/L feeder layers in LIF-free medium containing the ROCK inhibitor, Y-27632 resulted in significantly increased AP-positive colonies for both cell lines Figure 3C, compared with that of culture without Y-27632 Figure 3A. ROCK inhibitor treatment did not significantly affect iPS cells. However, addition of the Rho and Erk1/2 activator, LPA to the medium resulted in decreased AP-positive colonies for all cell lines and feeder conditions Figure 3D.

Discussion

Undifferentiated culture of pluripotent stem cells is supported by a complex network of soluble and membrane-bound cytokines, as well as the extracellular matrix provided by stromal feeder cells. The most reliable protocols require freshly isolated MEFs as feeders for the maintenance of ES/iPS cells. However, MEFs exhaust their stem cell supportive properties and undergo senescence after several passages. There have been numerous reports on substitutes for MEFs, such as mouse mesenchymal cell lines STO [10] and 3T3 [4], human placenta cells [11], human amniotic epithelial cells [12] and the mouse testicular stromal cell line JK1 [13]. These feeder cells produce growth factors including LIF, activin A, transforming growth factor β (TGF-β), basic fibroblast growth factor (bFGF), Wnts and bone morphogenetic protein-4 (BMP-4) [14,15], which are important for the self-renewal of ES/iPS cells. In an analysis of conditioned medium from MEF cultures, 85 proteins were identified and classified into categories such as differentiation and growth factors, and extracellular matrix and remodeling [16]. Importantly, the pluripotency of mouse ES cells is dependent on intracellular signaling including phosphorylation by the Janus family of tyrosine kinases (JAK) via LIF and the LIF receptor, which leads to activation of the signal transducer protein STAT3 [17]. The combination of IL-6 and soluble IL-6 receptor also interacts with and activates a homodimer of gp130, and the gp130-mediated signaling pathway maintains mouse ES cells without involvement of the LIF receptor [18,19]. STAT3 activation sufficiently maintains the undifferentiated state of mouse ES cells, and is enhanced by inhibition of the mitogen-activated protein (MAP) kinase (Erk1/2) pathway [17,20]. Human ES cells also express LIF, IL-6 and gp130 receptors, but STAT3 activation is not essential for maintenance of human ES cells in an undifferentiated state. Thus, it has been considered that human ES cells are distinct from mouse ES cells. Recently, Hanna et al. reported that LIF-dependent "naïve" human ES cells, similar to mouse ES cells, could be derived from conventional human ES

cells by ectopic induction of Oct4, Klf4 and Klf2 combined with LIF and inhibitors of the glycogen synthase kinase 3b (GSK3b) and Erk1/2 pathways [21]. Therefore, LIF/STAT3 signals may be required to maintain a naïve state of pluripotency even in human ES cells.

In the present study, we developed a STO/EL feeder system to activate JAK/STAT signaling in ES/iPS cells to maintain pluripotency via LIF produced by the feeder cells in LIF-free medium. We demonstrated that the STO/EL feeder layer maintains the pluripotency of ES/iPS cells in LIF-free medium. LIF-expressing STO cells are available from the ECACC cell bank (SNL; EC07032801) [22], and the cells have been used as feeders for ES/iPS cell culture [2,23]. In the present study, the STO/L feeder layer did not sufficiently support undifferentiated culture of ES/iPS cells Figure 3A in LIF-free medium, albeit STO/L cells produced sufficient LIF and the expression level was higher compared with that of STO/EL cells Figure 1C. These results suggest that E-cadherin expression plays a pivotal role in the performance of STO cells as feeders. We hypothesized two possibilities regarding the improved performance of the STO/EL feeder layer; 1) enhanced physical contact between STO and ES/iPS cells facilitates LIF signaling, and/or 2) direct signal transduction via E-cadherin modulation of the Rho-ROCK cascade. The number of AP-positive colonies significantly increased by culturing magnetically-labeled ES/iPS cells that were attracted to STO/L feeder cells by magnetic force Figure 3B. This observation suggests that close physical contact with ES/iPS cells is important for STO feeder layers to perform efficiently. Notably, the feeder performance of STO and STO/E cells in LIF-free ES cell cultures also slightly improved in magnetic culture. Although these results suggest that STO cells produce anti-differentiation factors other than LIF, the production of LIF by feeder cells or medium supplementation is essential to maintain ES/iPS cell cultures in an undifferentiated state.

Thus far, several signaling pathways other than the JAK/STAT3 pathway have been identified as regulators of self-renewal in ES/iPS cells. Harb et al. reported that the Rho-ROCK-myosin signaling axis determines the cell-cell integrity of self-renewing ES cells [24]. A ROCK inhibitor blocks apoptosis and supports human ES cell proliferation without affecting pluripotency following dissociation into single cells [25]. In this study, ES/iPS cells cultured on each feeder layer with the ROCK inhibitor, Y-27632 or the Rho activator, LPA (also acts as an Erk1/2 activator), resulted in significantly increased Figure 3C or decreased Figure 3D AP-positive colonies. These results indicate that the Rho-ROCK signaling pathway regulates the undifferentiated state of ES/iPS cells in the presence and absence of LIF. Nevertheless, it is difficult to specify the E-cadherin signal from E-cadherin-expressing feeder cells to ES/iPS cells, because ES/iPS cells also express E-cadherin. At least, E-cadherin expression in feeder cells did not affect the Rho-ROCK signaling pathway in ES/iPS cells, although cell-cell interactions were enhanced between STO and ES/iPS cells. Thus, the Rho-ROCK signaling pathway does not seem to be the dominant mechanism for supporting the undifferentiated state of ES/iPS cells using STO/EL feeder. Taken together, our results suggest that STO/EL cells support undifferentiated culture of ES/iPS cells in LIF-free medium due to enhanced LIF/STAT3 signaling in response to LIF expression via forced physical contact by E-cadherin expression.

Interestingly, the effectiveness of feeder layers for supporting undifferentiated culture of ES/iPS cells in LIF-free medium varies among the three cell lines Figure 3A. iPS cells cultured on STO, STO/E and STO/L feeder layers formed a higher number of AP-positive colonies compared with those of the other two ES cell lines. The gene expression profiles of ES/iPS cells vary among cell lines due to variations in establishment procedures and culture history [26, 27], which may affect compatibility with various feeder layers. A MEF feeder layer is the most robust system for maintaining the undifferentiated state of ES/iPS cell cultures in the presence of LIF. In this study, the STO/EL feeder system even in LIF-free medium is compatible with the ES/iPS cell lines used for experimentation, and comparable to that of a MEF feeder system in the presence of LIF.

In conclusion, we demonstrate that E-cadherin and LIF gene transfer into STO cells improves the performance of a STO feeder layer for ES/iPS cell culture in LIF-free medium. The forced cell-cell interaction via E-cadherin between ES/iPS and feeder cells enhanced the cytokine signal through a paracrine mechanism. MEF isolation is a laborious and time-consuming process requiring the sacrifice of mice. Feeder-free culture systems for ES cell culture have been reported, but may cause chromosomal instability in ES cells [28,29]. Thus, although the long-term stability of ES/iPS cells using a STO/EL feeder layer and application to other pluripotent cell lines including human ES/iPS cells should be examined, the STO/EL feeder system may contribute toward medical and biological research using ES/iPS cells.

References

1. Passier R, van Laake LW, Mummery CL (2008) Stem-cell-based therapy and lessons from the heart. Nature 453: 322-329.

2. Okita K, Ichisaka T, Yamanaka S (2007) Generation of germline-competent induced pluripotent stem cells. Nature 448: 313-317.

3. Larue L, Antos C, Butz S, Huber O, Delmas V, et al. (1996) A role for cadherins in tissue formation. Development 122: 3185-3194.

4. Horie M, Ito A, Kiyohara T, Kawabe Y, Kamihira M (2010) E-cadherin gene-engineered feeder systems for supporting undifferentiated growth of mouse embryonic stem cells. J Biosci Bioeng 110: 582-587.

5. Ito A, Kiyohara T, Kawabe Y, Ijima H, Kamihira M (2008) Enhancement of cell function through heterotypic cell-cell interactions using E-cadherin-expressing NIH3T3 cells. J Biosci Bioeng 105: 679-682.

6. Nakamura H, Kimura T, Ogita K, Koyama S, Tsujie T, et al. (2004) Alteration of the timing of implantation by in vivo gene transfer: delay of implantation by suppression of nuclear factor kappaB activity and partial rescue by leukemia inhibitory factor. Biochem Biophys Res Commun 321: 886-892.

7. Dang SM, Kyba M, Perlingeiro R, Daley GQ, Zandstra PW (2002) Efficiency of embryoid body formation and hematopoietic development from embryonic stem cells in different culture systems. Biotechnol Bioeng 78: 442-453.

8. Shinkai M, Yanase M, Honda H, Wakabayashi T, Yoshida J, et al. (1996) Intracellular hyperthermia for cancer using magnetite cationic liposomes: in vitro study. Jpn J Cancer Res 87: 1179-1183.

9. Horie M, Ito A, Maki T, Kawabe Y, Kamihira M (2011) Magnetic separation of cells from developing embryoid bodies using magnetite cationic liposomes. J Biosci Bioeng 112: 184-187.

10. Evans MJ, Kaufman MH (1981) Establishment in culture of pluripotential cells from mouse embryos. Nature 292: 154-156.

11. Park Y, Lee SJ, Choi IY, Lee SR, Sung HJ, et al. (2010) The efficacy of human placenta as a source of the universal feeder in human and mouse pluripotent stem cell culture. Cell Reprogram 12: 315-328.

12. Liu T, Cheng W, Liu T, Guo L, Huang Q, et al. (2010) Human amniotic epithelial cell feeder layers maintain mouse embryonic stem cell pluripotency via epigenetic regulation of the c-Myc promoter. Acta Biochim Biophys Sin (Shanghai). 42: 109-115.

13. Kim J, Seandel M, Falciatori I, Wen D, Rafii S (2008) CD34+ testicular stromal cells support long-term expansion of embryonic and adult stem and progenitor cells. Stem Cells 26: 2516-2522.

14. Eiselleova L, Peterkova I, Neradil J, Slaninova I, Hampl A, et al. (2008) Comparative study of mouse and human feeder cells for human embryonic stem cells. Int J Dev Biol 52: 353-363.

15. Lim JW, Bodnar A (2002) Proteome analysis of conditioned medium from mouse embryonic fibroblast feeder layers which support the growth of human embryonic stem cells. Proteomics 2: 1187-1203.

16. Prowse AB, McQuade LR, Bryant KJ, Marcal H, Gray PP (2007) Identification of potential pluripotency determinants for human embryonic stem cells following proteomic analysis of human and mouse fibroblast conditioned media. J Proteome Res 6: 3796-3807.

17. Smith AG (2001) Embryo-derived stem cells: of mice and men. Annu Rev Cell Dev Biol 17: 435-462.

18. Nichols J, Chambers I, Smith A (1994) Derivation of germline competent embryonic stem cells with a combination of interleukin-6 and soluble interleukin-6 receptor. Exp Cell Res 215: 237-239.

19. Yoshida K, Chambers I, Nichols J, Smith A, Saito M, et al. (1994) Maintenance of the pluripotential phenotype of embryonic stem cells though direct activation of gp130 signalling pathways. Mech Dev 45: 163-171.

20. Buehr M, Meek S, Blair K, Yang J, Ure J, et al. (2008) Capture of authentic embryonic stem cells from rat blastocysts. Cell 135: 1287-1298.

21. Hanna J, Cheng AW, Saha K, Kim J, Lengner CJ, et al. (2010) Human embryonic stem cells with biological and epigenetic characteristics similar to those of mouse ESCs. Proc Natl Acad Sci U S A 107: 9222-9227.

22. McMahon AP, Bradley A (1990) The Wnt-1 (int-1) proto-oncogene is required for development of a large region of the mouse brain. Cell 62: 1073-1085.

23. Nagy A, Rossant J, Nagy R, Abramow-Newerly W, Roder JC (1993) Derivation of completely cell culture-derived mice from early-passage embryonic stem cells. Proc Natl Acad Sci U S A 90: 8424-8428.

24. Harb N, Archer TK, Sato N (2008) The Rho-Rock-Myosin signaling axis determines cell-cell integrity of self-renewing pluripotent stem cells. PLoS One 3: e3001.

25. Watanabe K, Ueno M, Kamiya D, Nishiyama A, Matsumura M, et al. (2007) A ROCK inhibitor permits survival of dissociated human embryonic stem cells. Nat Biotechnol 25: 681-686.

26. Chin MH, Pellegrini M, Plath K, Lowry WE (2010) Molecular analyses of human induced pluripotent stem cells and embryonic stem cells. Cell Stem Cell 7: 263-269.

27. Boué S, Paramonov I, Barrero MJ, Izpisúa Belmonte JC (2010) Analysis of human and mouse reprogramming of somatic cells to induced pluripotent stem cells. What is in the plate? PLoS One 5: e12664.

28. Draper JS, Smith K, Gokhale P, Moore HD, Maltby E, et al. (2004) Recurrent gain chromosomes 17q and 12 in cultured human embryonic stem cells. Nat Biotechnol 22: 53-54.

29. Catalina P, Montes R, Ligero G, Sanchez L, de la Cueva T, et al. (2008) Human ESCs predisposition to karyotypic instability: Is a matter of culture adaptation or differential vulnerability among hESC lines due to inherent properties?. Mol Cancer 7: 76.

28

Influence of Process Conditions on Drying by Atomization Pulp Umbu

Jackelinne de A Silva[1], Maria IS Maciel[1]*, Naíra P de Moura[1], Marcony E da S Júnior[1], Janaína V de Melo[2], Patrícia M Azoubel[3] and Enayde de A Melo[1]

[1]*Federal Rural University of Pernambuco, Dois Irmãos, 52171-900, Recife, Brazil*
[2]*Center of Strategic Technologies of the Northeast, Cidade Universitária, 50740-540, Recife, Brazil*
[3]*Federal University of Pernambuco, Cidade Universitária, 50740-521, Recife, Brazil*

Abstract

The present study evaluated the influence of variables involved in the drying process by atomization on the physico-chemical properties of umbu powder. The process was carried out using a laboratory scale atomizer with DE 15 maltodextrin as the carrier agent. Seventeen assays were performed according to a central composite rotational design. The independent variables were drying air temperature, mass feeding flow, and carrier agent concentration. The analyzed responses were water activity, moisture, hygroscopicity, process yield, and retention of phenolic compounds (RPC) in the final product. Drying air temperature negatively affected water activity and moisture content, i.e., higher applied temperatures led to lower water activity and moisture content in atomized umbu. However, the effects of the linear, quadratic, and interaction factors were not statistically significant over hygroscopicity and process yield at 95% level of statistical significance; therefore, it was not possible to generate a model. The RPC was influenced by the mass feeding flow and carrier agent concentration; the use of faster flows and higher concentrations of maltodextrin presented a final product with higher RPC. Based on the analysis of response surface graphs, an assay was selected and physically characterized: apparent density of 0.61 g/mL and percentage of solubility of 80.28%. Atomized umbu particles showed uniform size and formed numerous small pellets with spherical shape and predominantly rough surface.

Keywords: Dehydration; Experimental planning; Maltodextrin; Microstructure

Introduction

Atomization is a widely used technique by the food industry and consists in the transformation of liquid products to powdered products through exposure of the product to a current of hot air under high pressure. This type of drying process is usually used to ensure microbiological stability of products and avoid risk of biological degradation while reducing storage and transportation costs [1].

This process has been used in the production of fruit powder and reported in a series of studies [2-5] the goals of which were to establish the best drying conditions by optimizing and adapting the conditions to each type of fruit. Atomization drying represents an attractive alternative in the case of umbu because its consumption in *natura* is limited to the production region due to its high perishability.

In general, powdered fruits present many advantages and economic potentials over fruit in *natura,* such as reduced volume and/or weight, reduced packaging, easier transport and storage, and increased shelf life. However, the atomization of products with low molecular weights and high sugar content, such as fruits, presents problems of deposition and adhesion of particles, reducing the process yield [6,7].

Nevertheless, these problems can be minimized with the use of carrier agent additives. These agents are applied before the product is atomized and are responsible for the micro-encapsulation that protects sensitive constituents, such as pigments, compounds responsible for flavor and aroma, and bioactive compounds, such as phenolic compounds.

The physico-chemical properties of food powders obtained by atomization depend on variables such as the characteristics of the feeding liquid and drying air, atomizer type, and operating mechanism. Thus, optimization of the process is important in order to obtain products with improved nutritional and sensory characteristics and optimized yield [3].

Hence, the study of the atomizing process effects on the characteristics of powdered umbu is important because this product is an ingredient in processed foods and drying ensures its availability in times of scarcity. This study evaluated the effect of the drying temperature, mass feeding flow, and maltodextrin concentration on responses related to water activity, moisture, hygroscopicity, yield, and phenolic compounds contents in atomized umbu pulp.

Material and Methods

Materials

Umbu fruits were acquired from the Supply and Logistics Center of Pernambuco (Pernambuco, Brazil). Semi-ripe and ripe fruits were selected, washed in running water, sanitized, and pulped. Pulp was stored frozen (-22°C) in polyethylene bags. The pulp was thawed as needed prior to the atomizing process. Table 1 shows the physico-chemical and colorimetric composition of umbu pulp in *natura.*

Maltodextrin *MOR-REX ® 1914* (Corn Products, Mogi-Guaçu, Brazil) with DE 15 was used as the carrier agent.

Preparation of samples

The pulp was sifted to eliminate particles whose diameter was greater than the diameter of the atomizer nozzle to facilitate passage. Maltodextrin at concentrations calculated on the basis of the weight of

*Corresponding author: Maria IS Maciel, Federal Rural University of Pernambuco, Dois Irmãos, 52171-900, Recife, Brazil

Component	Average value	Method of analysis
Water activity	0.98 ± 0.00	water activity analyzer
Humidity (%)	86.65 ± 0.25	moisture titrator
Soluble solids (°Brix)	10.00 ± 1.00	refractometer
Titratable acidity (g citric acid)	3.05 ± 0.12	AOAC (2006)
pH	2.00 ± 0.05	pH meter
Proteins (%)	0.84 ± 0.34	AOAC (1990)
Lipids (%)	0.34 ± 0.01	Bligh and Dyer (1959)
Carbohydrates (%)	11.68 ± 0.27	difference
Phenolic compounds (mg/100 g)	158.78 ± 8.47	Wettasinghe and Shahidi (1999)
Ashes	0.49 ± 0.13	AOAC (1990)
Color		McGuire (1992)
L*	63.89 ± 5.01	
a*	-4.92 ± 0.58	
b*	38.25 ± 3.77	

Table 1: Physico-chemical and colorimetric composition of umbu pulp in *natura*.

the sieved pulp was added to the stirred pulp until complete dissolution. Water (approximately 50% v/v) was added before the solution entered the atomizer.

Spray drying

Spray drying process was performed in a laboratory scale spray dryer Labmaq model MSD 1.0 (Ribeirao Preto/SP, Brazil), with a 1.2 mm diameter nozzle. The mixture was fed into the chamber by a peristaltic pump, drying flow rate was 30 m^3/h air flows and compressor air pressure was 0.6 bar.

A rotatable central composite design was used to perform the tests for the microencapsulation of umbu pulp, considering three factors (independent variables): inlet air temperature (90-190°C), feed flow rate (0.2-1.0 L/h) and carrier agent concentration (10-30%, w/w). Five levels of each variable were chosen for the trials, including the central point and two axial points, giving a total of 17 combinations (Table 2). The responses (dependent variables) were: moisture content, water activity, hygroscopicity, phenolic compounds content and yield. Process yield was calculated as the ratio between the total solids content in atomized umbu and total solids content in the mixture.

The following polynomial equation was fitted to the data:

$y = \beta_0 + \beta_1 x_1 + \beta_2 x_2 + \beta_3 x_3 + \beta_{11} x_1^2 + \beta_{22} x_2^2 + \beta_{33} x_3^2 + \beta_{12} x_1 x_2 + \beta_{13} x_1 x_3 + \beta_{23} x_2 x_3$ (Equation 1)

Where β_0 are constant regressions coefficients; y is the response (moisture content, water activity, hygroscopicity, RPC and yield) and x_1, x_2 and x_3 are the coded independent variables (inlet air temperature, feed flow rate and carrier agent concentration, respectively).

The analysis of variance (ANOVA), test of lack of fit (F test), determination of regression coefficients, and the generation of response surfaces were carried out using Statistica 7.0 *software* (Stat Soft, Tulsa, USA).

Analytical methods

Atomized umbu was analyzed for moisture content, water activity, hygroscopicity, and phenolic compounds. Moreover, the assay that presented the best drying conditions selected from the graphs of surface response was physically characterized (apparent density, solubility, and morphology). Analyses were performed in triplicate and according to the procedures described below.

Water activity: A water activity analyzer (Aqualab model 4TE) was used to determine water activity at 25°C.

Moisture content: Moisture content was determined over the course of thirty minutes at 105°C using the Marte apparatus model IDSO. The results were expressed as percentage of moisture.

Hygroscopicity: Hygroscopicity was determined according to the modified methodology proposed by Cai and Corke [8]. Samples of approximately 1 g were placed in a hermetically sealed container containing a NaCl saturated solution (relative humidity 75.29%) at 25°C; these samples were weighed after one week, and the hygroscopicity was expressed as g of adsorbed moisture per 100 g of dry mass sample (g/100 g).

Phenolic compounds: Phenolic compounds were extracted with acetone according to the procedure described by Singleton et al. [9]. Total phenolic compounds content was measured with the Folin-Ciocalteau reagent (Merck) and determined by interpolating sample absorbance against a gallic acid calibration curve, according to the methodology described by Wettasinghe and Shahidi [10]. The results were converted to RPC terms using the dry mass (m.s.) of umbu pulp. Maltodextrin concentration was not used in the calculation.

Physical characterization of atomized umbu:

Apparent density: The apparent density (ρ_b) was measured following the procedure described in previous studies with modifications [11,12]. Approximately 2 g of atomized umbu was transferred to a 10 mL measuring cylinder. The powder was compressed by hitting the measuring cylinder on the bench for about 50 times. The $\rho_{(b)}$ was calculated by dividing the powder mass by the volume occupied on the measuring cylinder.

Solubility: Solubility was measured according to the methodology described by Cano-Chauca et al. [13].

Particle morphology: Particle morphology was determined by Scanning Electron Microscopy (SEM) in the Laboratory of Electron Microscopy and Microanalysis (Center of Strategic Technologies of the Northeast – CETENE). The samples were fixed in metallic containers for specimens (*stubs*) with a conventional conductive double-sided adhesive tape. Subsequently, samples were gold metalized at the covering

Assay	Temperature (°C)	Mass flow rate (l/h)	Carrier agent (%)
01	-1 (110)	-1 (0.36)	-1 (14)
02	+1 (170)	-1 (0.36)	-1 (14)
03	-1 (110)	+1 (0.84)	-1 (14)
04	+1 (170)	+1 (0.84)	-1 (14)
05	-1 (110)	-1 (0.36)	+1 (26)
06	+1 (170)	-1 (0.36)	+1 (26)
07	-1 (110)	+1 (0.84)	+1 (26)
08	+1 (170)	+1 (0.84)	+1 (26)
09	0 (140)	0 (0.60)	0 (20)
10	0 (140)	0 (0.60)	0 (20)
11	0 (140)	0 (0.60)	0 (20)
12	-1.68 (90)	0 (0.60)	0 (20)
13	+ 1.68 (190)	0 (0.60)	0 (20)
14	0 (140)	-1.68 (0.20)	0 (20)
15	0 (140)	+ 1.68 (1.00)	0 (20)
16	0 (140)	0 (0.60)	-1.68 (10)
17	0 (140)	0 (0.60)	+ 1.68 (30)

Table 2: Experimental planning for drying by atomization.

Assay	Water activity	Moisture (%)	Hygroscopicity (g/100 g)	Yield (%)	Retention of Phenolic Compounds (%)
1	0.19	3.25	18.22	12.29	35.55
2	0.13	2.36	17.07	12.18	33.33
3	0.25	5.62	19.76	14.90	35.55
4	0.10	2.38	16.89	20.49	33.33
5	0.24	4.20	10.23	14.96	44.44
6	0.09	2.01	12.98	11.79	40.00
7	0.21	4.70	17.94	17.88	48.88
8	0.17	3.86	13.67	08.88	60.00
9	0.13	3.02	14.85	09.21	44.44
10	0.12	3.89	13.85	10.07	44.44
11	0.12	3.15	17.28	15.39	40.00
12	0.25	5.59	15.42	09.85	33.33
13	0.10	2.09	15.16	13.01	33.33
14	0.11	2.34	19.44	15.98	57.77
15	0.19	4.18	13.53	07.34	68.88
16	0.11	4.35	14.77	14.79	28.88
17	0.13	2.86	14.65	10.68	57.77

Table 3: Water activity, moisture, process yield, hygroscopicity, and RPC in 17 assays based on the experimental planning.

Coefficients	Water activity	Moisture	Hygroscopicity	Yield	Retention of phenolic compounds
β_0	0.15	3.52	15.63	12.92	43.52
β_1	-0.09	-1.91	NS	NS	NS
β_2	0.03	1.14	NS	NS	-8.63
β_3	NS	NS	NS	NS	6.31
β_{11}	0.04	NS	NS	NS	15.25
β_{22}	0.02	NS	NS	NS	12.62
β_{33}	NS	NS	NS	NS	NS
β_{12}	NS	NS	NS	NS	NS
β_{13}	NS	NS	NS	NS	NS
β_{23}	NS	NS	NS	NS	NS
R^2	83.61	80.04	44.72	40.06	86.7
F_{cal}	15.78	28.07	-	-	17.44

Table 4: Regression coefficients encoded for the second-order polynomial equation, F and p, and values of the coefficients of determination (R^2).

rate of 10 nm thick, for 80 seconds, and 40 mA current. Samples were subsequently observed in a scanning electron microscope (FEI Quanta 200 FEG model, Netherlands) operating at 20 kV.

Results and Discussion

Analysis of response surface

Water activity, humidity, process yield, hygroscopicity, and RPC values are presented in Table 3.

The regression coefficients for the second order polynomial equations are presented in Table 4. The obtained equation was tested for adequacy and fitness by Analysis of Variance (ANOVA). Table 5 summarizes the results.

Water activity (Aw): The analysis of variance for the model as fitted showed significance ($p \leq 0.05$) and explained 83.61% of the variability in the water activity. Therefore, the model as fitted provides an approximation to the true system.

Figure 1 shows response surfaces generated through the proposed model, considering the central point of carrier agent concentration,

mass flow rate, and drying air temperature.

The increase in drying air temperature, regardless of the mass feeding flow and carrier agent concentration, led to product Aw reduction (Figure 1). Thus, greater temperature gradient between the atomized product and drying air leads to higher heat transfer and, consequently, higher evaporation of water from the product resulting in lower Solval et al. [14] and Fazaeli et al. [15] used spray drying technology and observed a decrease in the particles Aw with increasing drying air temperature for melons and blackberries, respectively. However, the increase in mass feeding flow, independent of the carrier agent concentration used, led to increased product Aw values (Figure 1). Therefore, faster processes results in shorter contact time between the product and the drying air, making the process of heat transfer less efficient.

Moisture: The regression was significant and explained 80.04% of the variability ($p \leq 0.05$). According to Figure 2, the drying air temperature was the variable that showed greater influence over the final moisture content of the particles; increasing this variable, independent of the mass feeding flow and concentration of the carrier agent values, there was a reduction in the product moisture content. Quek et al. [16] explained that in drying conditions with higher input temperatures, the heat transfer rate is increased and, consequently, particles with reduced moisture content are formed. Goula and Adamopoulos [4] also observed a decrease in particles moisture with increasing drying air temperature when using a new atomization technique in concentrated orange juice.

The moisture content in the obtained powders ranged from 2.01 to 5.62%. Values within this range (3.81 and 5.39%) are presented by Solval et al. [14]. The moisture content was significantly influenced by the drying air temperature and mass feeding flow.

The mass feeding flow in the mixture, presented in turn, a positive effect on moisture, i.e., the processes performed at higher mass feeding flows resulted in moister particles. However, when higher temperatures are used, the use of high mass feeding flow also resulted in powders with low moisture content, which confirmed that high temperature influences the drying process more than the mass feeding flow (Figure 2).

The results presented by Tonon et al. [3] when evaluating the influence of process conditions of spray drying on the physico-chemical properties of açaí powder also indicated that the powders moisture content decreased with increasing drying air temperatures and decreasing mass feeding flow; the temperature effect was greater than that of mass feeding flow.

Source	Water activity			Moisture			RPC*		
	DF	MS	F	DF	MS	F	DF	MS	F
Regression	4	0.01	15.78	2	8.47	28.07	4	412.46	17.44
Residual	12	0.00		14	0.30		12	23.64	
Lack of fit	10	0.00		12	0.31		10	27.05	
Pure error	2	0.00		2	0.22		2	6.57	
TOTAL	16			16			16		
R^2	0.83			0.80			0.86		

NS: Not significant
DF: degree of freedom
MS: mean square
*Significant at 5% level

Table 5: Analysis of variance for water activity, moisture and retention of phenolic compounds (RPC) in regression models of sonication treatment.

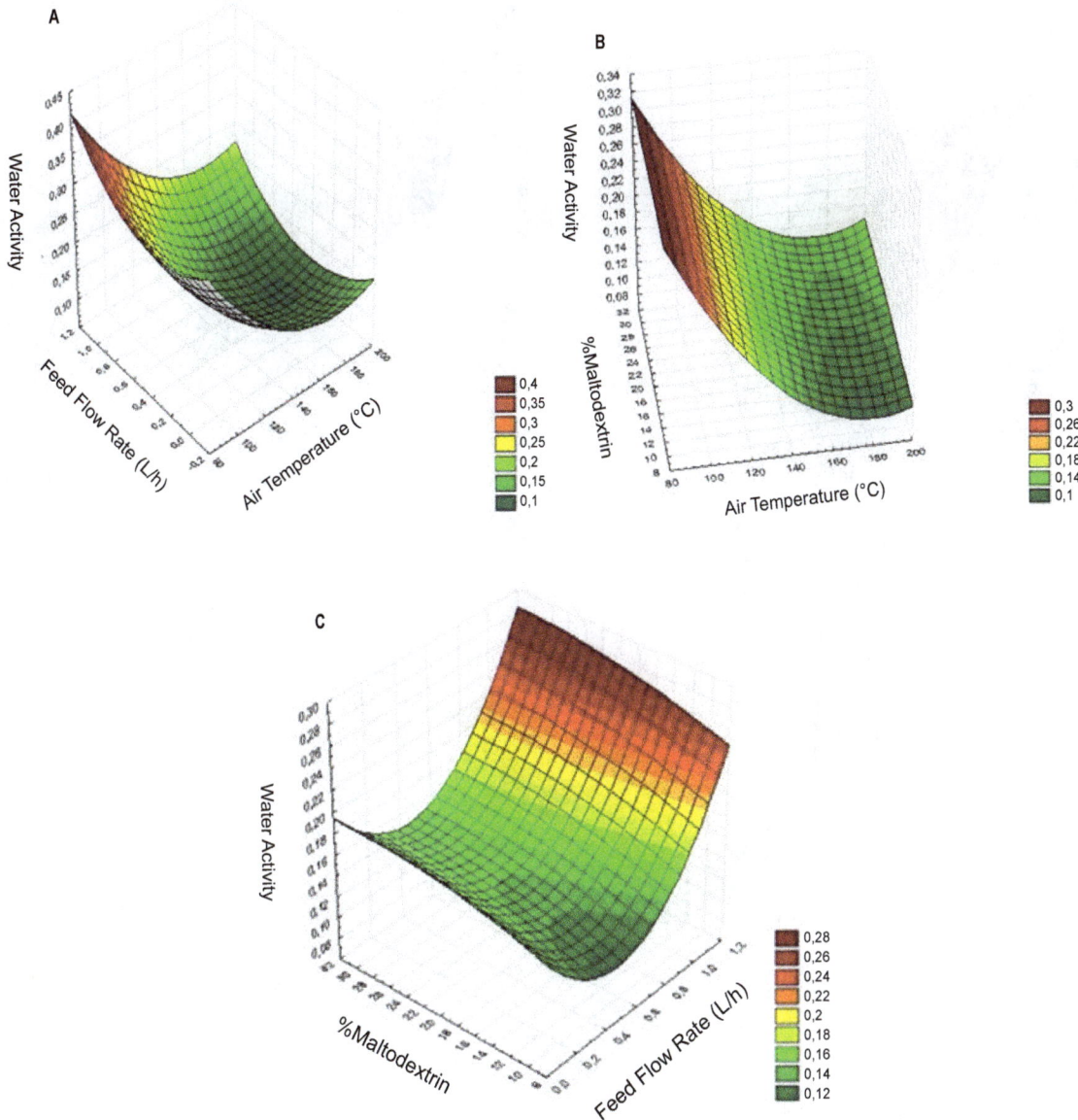

Figure 1: Response surface for water activity, (A) as a function of temperature x mass feeding flow, (B) as a function of temperature x carrier agent concentration, and (C) as a function of mass feeding flow x carrier agent concentration.

Hygroscopicity: Results regarding the effects of processing conditions on powder hygroscopicity were not statistically significant. Thus, the response surface methodology was not applied to evaluate the experimental data.

According to Teunou and Fitzpatrick [17], the adsorption of water by powdered food depends on the exposure time to high moisture conditions because water has to diffuse from the air into the food, resulting in greater powder cohesion and, consequently reducing the flow.

The hygroscopicity in atomized umbu ranged from 10.23 to 19.76 g/100 g (Table 3). It decreased with increasing carrier agent concentration, due to the protection characteristic of the maltodextrin. This same behavior was observed by Moreira et al. [18], using different concentrations of maltodextrin in the spray drying acerola pulp.

The process yield: The yield of atomized umbu ranged from 7.34 to 20.49% (Table 3), which are considerably below the values obtained by most studies evaluating yield in spary drying processes [3,15,19,20]. According to Goula and Adamopoulos [21,22], the low yields in this type of process can be explained by the thermoplastic nature of sugar and low molecular weight organic acid molecules causing adhesion of particles on the drying chamber wall and the formation of unwanted clusters in the transmission systems. Retention and incomplete recovery of particles occurred during the process of atomization: particles remained adhered inside the atomizer, confirming the reports from those authors.

Retention of phenolic compounds: The analysis of variance for the model as fitted showed significance ($p \leq 0.05$) and explained 86.70% of the variability in retention of phenolic content.

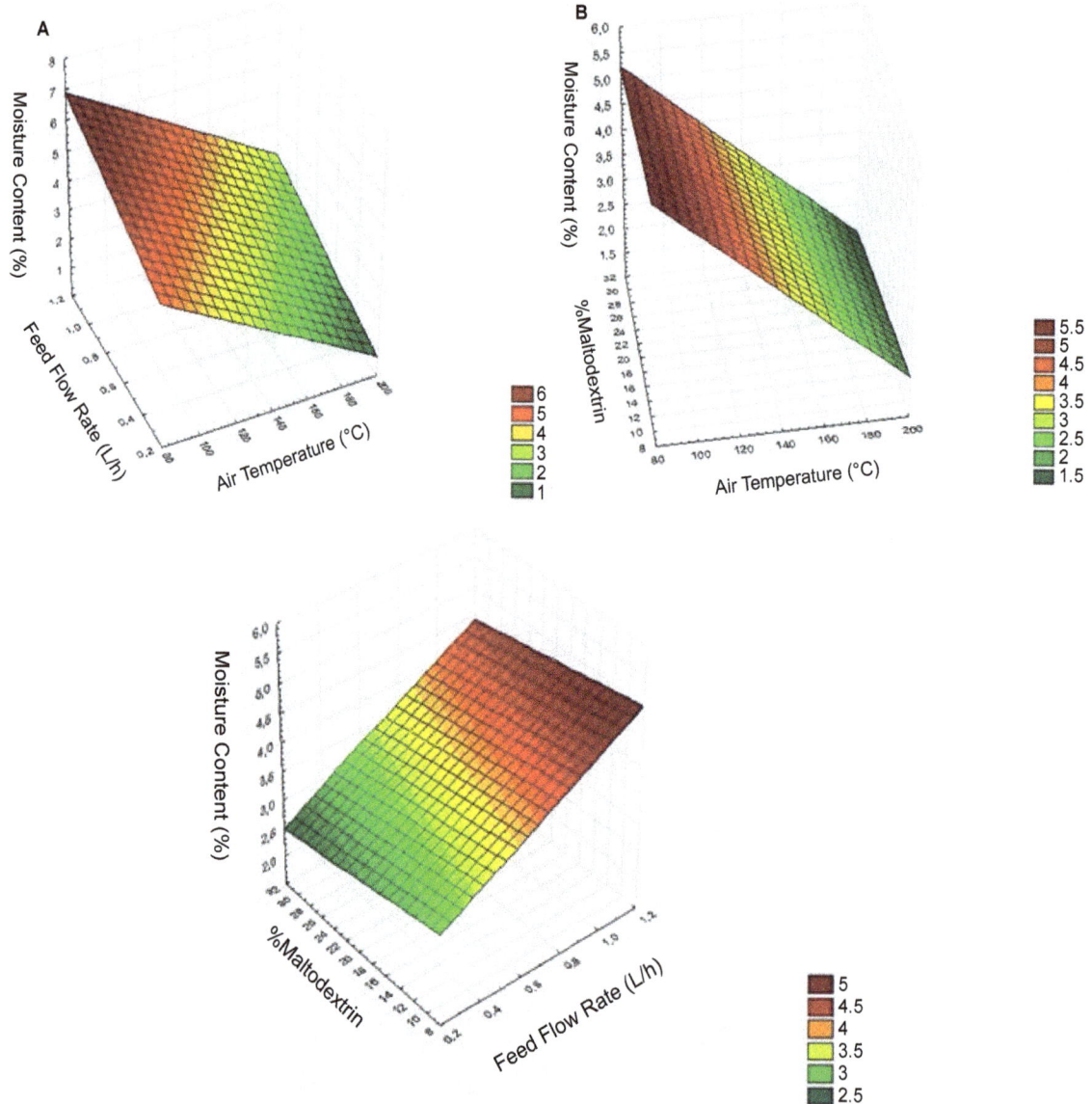

Figure 2: Response surface for moisture response, (A) as a function of temperature × mass feeding flow, (B) as a function of temperature × carrier agent concentration, and (C) as a function of mass feeding flow × carrier agent concentration.

Figure 3 presents the response surfaces generated through the proposed model, considering the central points of carrier agent concentration, mixture mass feeding flow, and drying air temperatures. Umbu powders produced under maximum mass feeding flow showed higher rates of RPC regardless of the temperature used. However, the use of moderate to low flows, under high temperatures, resulted in significant losses of phenolic compounds, which could be explained by the fact that longer particles remained in the drying chamber leads, having greater exposure to elevated temperatures and, consequently, an increase in loss of heat-sensitive compounds.

In addition, particles produced at lower temperatures have a tendency to agglomerate [16]. This agglomeration causes the particles to have reduced exposed surface and uneven exposure to same environmental conditions, protecting compounds against degradation. This phenomenon could be responsible for the behavior shown in

Figure 3A, where increased retentions can be observed at moderate to low temperatures as a function of lower mass feeding flow.

The variable carrier agent concentration had the greatest influence on the RPC. Increased maltodextrin concentration led to higher RPC in atomized umbu. Such phenomenon is predictable because, according to Desai and Park [23], carrier agents are able to seal and hold active materials within their structures during processing or storing, thereby promoting maximum protection against environmental conditions.

The results showed that an increase in both mass feeding flow and carrier agent concentration led to smaller losses of phenolic compounds (Figure 3).

Selection of the best drying condition

Models predicting responses in water activity, moisture, and RPC were generated by determining coefficients of regression and ANOVA.

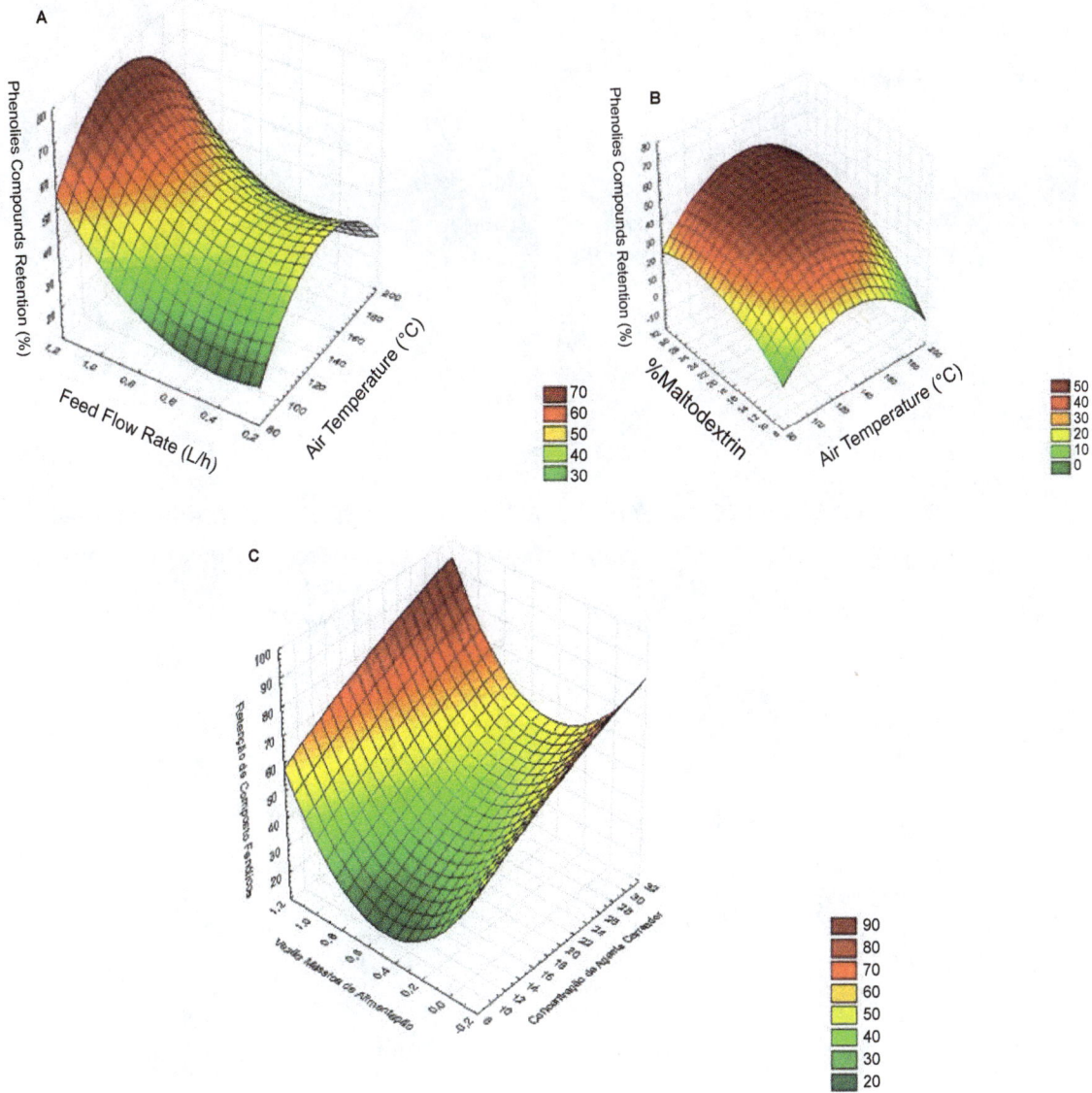

Figure 3: Response surface for the RPC, (A) as a function of temperature × mass feeding flow, (B) as a function of the carrier agent concentration t × temperature, and (C) as a function of mass feeding flow × carrier agent concentration.

The obtained models were not statistically significant for process yield and hygroscopicity responses. Additional studies are needed to optimize these properties in powdered umbu.

RPC was the main criterion used for the selection of the best drying condition because water activity and moisture content, determined in all samples, conferred chemical and microbiological stability to the product (Table 3); the hygroscopicity and process yield responses were not statistically significant.

The optimum set was chosen and an assay was performed at 133°C of air temperature, mass feeding flow of 0.91 L/h, and carrier agent concentration of 22%.

Validation of the obtained model

The best drying condition which was explained in above (selection of the best drying condition) was used for the validation of the model

Analysis	Experimental Values	Predicted Values	DR%
Water activity	0.14	0.16	9.13
Humidity (%)	5.51	5.58	1.25
RCF (%)	60.00	75.26	20.27

RD = Relative Deviation, RFC = Retention of Phenolic Compounds

Table 6: Experimental and predicted values for the analyses performed on powder produced under selected conditions based on the response surface graphs.

obtained from the experimental design. The experimental results and predicted values are presented in Table 6.

The experimental results for water activity and moisture were consistent with those predicted by the obtained model and showed deviations less than 10%. The RPC, however, showed a relatively higher deviation, indicating that the model's prediction ability for this response was not as efficient as the others.

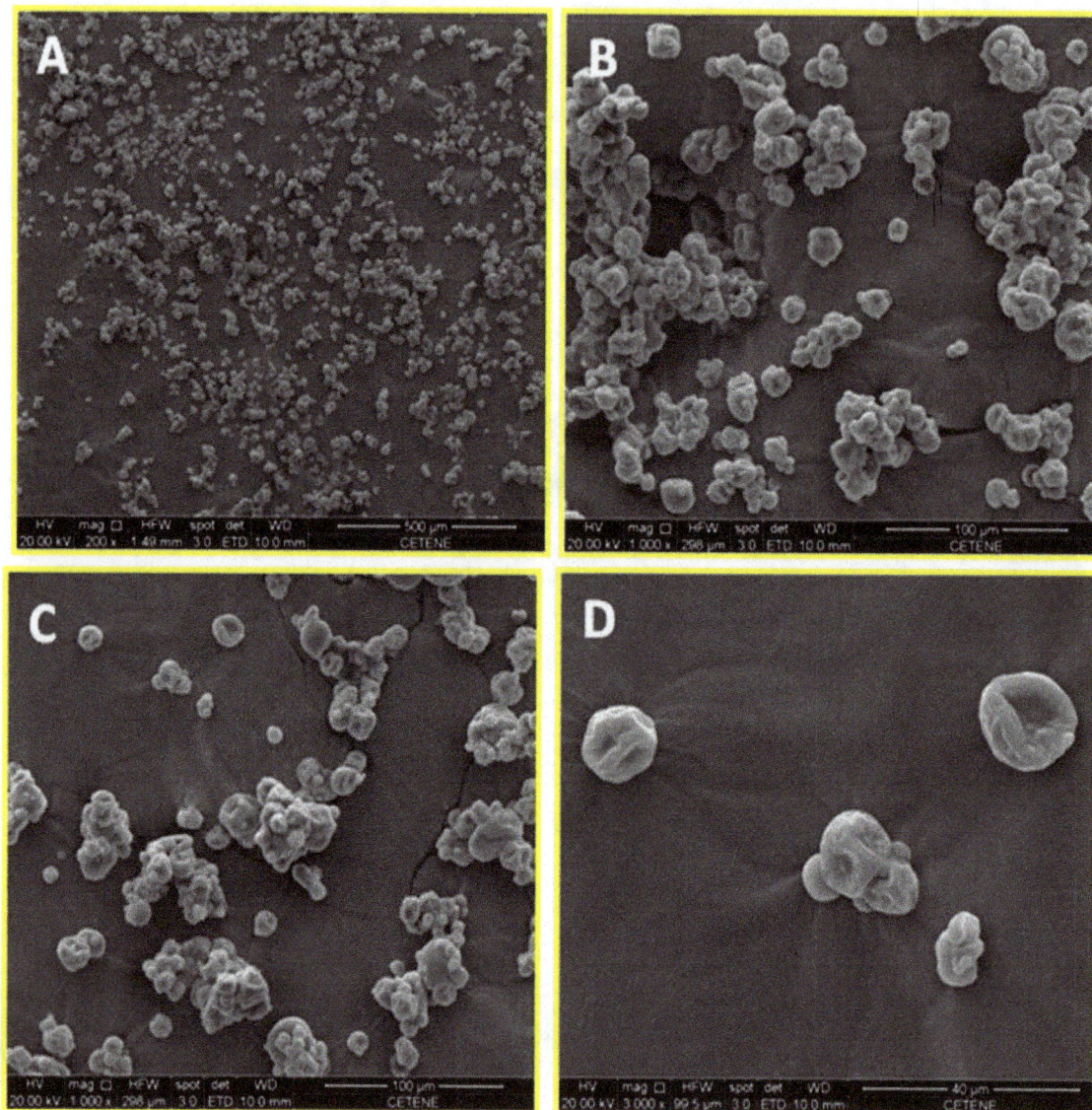

Figure 4: Electron Micrographs of umbu particles atomized at 133°C. A – Overview of the distribution of particles; B – Presence of agglomerated particles; C – Minority presence of particles with smooth surface; and D - Detail of isolated particles with rough surface.

Physical characterization of atomized umbu

Apparent density and solubility: The apparent density of atomized umbu was 0.61 g/mL, which is similar to that presented by Abadio et al. [2] in a study with atomized pineapple, and is within the range presented by Caparino et al. [12], in a study with atomized mangoes (from 0.40 to 0.80 g/mL). However, smaller values of apparent density were presented by Tonon et al. [3] in a study with atomized açaí (0.37 to 0.48 g/mL).

The volume of air present in powdered foods is inversely proportional to their density. Thus, according to Goula and Adamopoulos [24], a decrease in air volume causes an increase in apparent density. In addition, differences in drying conditions can influence the density of powdered food. Fazaeli et al. [15], for example, report density variations due to changes in drying temperature applied during atomization of blackberries.

The solubility of atomized umbu was 80.28%, which is similar

to data presented by Fazaeli et al. [15] and Abadio et al. [2]. The characteristics of the matter, drying conditions used, and type of carrier agent added to the atomized product directly influence the solubility of food powders. According to Cano-Chauca et al. [13], maltodextrin is widely used in the atomized drying processes because of its physical properties such as high solubility in water. Abadio et al. [2] reported that lower atomizing speeds were enough to obtain products with good solubility. Teunou and Fitzpatrick [17] and Goula and Adamopoulos [24] highlight that increased drying air temperature tends to increase the solubility of atomized food. According to Goula and Adamopoulos [21], these results from the effects that air temperature exerts on the residual moisture content in the powder; the lower the moisture content, the greater the solubility.

Particle Morphology: Particle surface analysis in atomized umbu was performed by scanning electron microscopy and the results are presented in Figure 4.

According to Figure 4, atomized umbu particles showed uniform size (A) and formed small and numerous clusters (B) (Figure 4). Spherical shaped particles with predominantly rough surface, which is the characteristic morphology of food powders produced by the atomizing process, were observed; smooth surface was observed in the minority of the particles (C and D).

Similar results were obtained by Tonon et al. [3] in a study with atomized açaí, mainly when using 20% maltodextrin at 138°C, conditions which are similar to those used in the present study. Alamilla-Beltrán et al. [25] observed that particles of maltodextrin produced under higher drying temperatures (170-200°C) presented a spherical and smooth surface format, while particles produced under lower temperature (110°C) presented a wrinkled appearance, similar to that of the atomized umbu particles produced at 133°C. Nijdam and Langrish [26] explain that when the drying temperature is high enough, the moisture is rapidly evaporated with a subsequent formation of a dry and hard envelope, avoiding emptying the particle. However, when the drying temperature is low, the envelope remains moist and supple for longer periods, allowing the particle to deflate and wither during cooling.

As for agglomeration, the behavior of particles produced by atomization is closely related to the origin and concentration of the carrier agent used during the process. Cano-Chauca et al. [13] showed that the treatment of atomized mangoes with only 12% maltodextrin resulted in the formation of larger, agglomerated, and amorphous particles. Fazaeli et al. [15] observed similar results when using 8% of maltodextrin in the process of drying blackberries by atomization. In the present study, particles with lesser degree of agglomeration where obtained using 22% of maltodextrin. Therefore, the higher concentrations of maltodextrin used during atomizing processes in fruits tend to form less agglomerated particles [27-30].

Conclusions

The factorial planning applied in this study did not allow optimizing the drying process of umbu by atomization. However, the analysis of response surface graphs for the RPC proved that the development of atomized umbu is viable by applying the following drying conditions: air temperature of 133°C, mass feeding flow of 0.91 L/h, and carrier agent concentration of 22%. The obtained results show that there are good prospects for using the production of umbu pulp in the development of atomized fruit. This could expand the economic prospects of small producers and the diversification of products derived from umbu powder. Future studies could contribute to the further optimization of this process.

Acknowledgments

The authors are thankful to CNPq and Capes for the financial support.

This manuscript was reviewed by a professional science editor and a native English-speaking editor to improve readability.

References

1. Gharsallaoui A, Roudaut G, Chambin O, Voilley A, Saurel R (2007) Applications of spray-drying in microencapsulation of food ingredients: An overview. Food Research International 40: 1107-1121.

2. Abadio FDB, Domingues AM, Borges SV, Oliveira VM (2004) Physical properties of powdered pineapple (Ananáscomosus) juice – effect of malt dextrin concentration and atomization speed. Journal of Food Engineering 64: 285-287.

3. Tonon RV, Brabet C, Hubinger MD (2008) Influence of process conditions on the physicochemical properties of açai (Euterpeoleraceae Mart.) powder produced by spray drying. Journal of Food Engineering 88: 411-418.

4. Goula AM, Adamopoulos KG (2010) A new technique for spray drying orange juice concentrate. Innovative Food Science and Emerging Technologies 11: 342-351.

5. Osorio C, Forero DP, Carriazo JG (2011) Characterization and performance assessment of guava (Psidiumguajava L.) microencapsulates obtained by spray-drying. Food Research International 44: 1174-1181.

6. Wang S, Langrish T (2009) A review of process simulations and the use of additives in spray drying. Food Research International 42: 13-25.

7. Turchiuli C, Gianfrancesco A, Palzer S, Dumoulin E (2011) Evolution of particle properties during spray drying in relation with stickiness and agglomeration control. Powder Technology 208: 433-440.

8. Cai YZ, Corke H (2000) Production and properties of spray-dried Amaranthus Betacyanin Pigments. Journal of Food Science 65: 1248-1252.

9. Singleton VI, Orthofer R, Lamucla-Reverentos RM (1999) Analysis of total phenol and other oxidation substrates and antioxidants by means of Folin-Ciocalteau. Methods in Enzymology 299: 152-178.

10. Wettasinghe M, Shahidi F (1999) Evening primrose meal: a source of natural antioxidants and scavenger of hydrogen peroxide and oxygen-derived free radicals. Journal Agriculture Food Chemistry 47: 1801-1812.

11. Barbosa-Cánovas GV, Juliano P (2005) Physical and chemical properties of food powders. In: Onwulata, C. (Ed.), Encapsulated and powdered foods.

12. Caparino OA, Tang J, Nindo CI, Sablani SS, Powers JR, et al. (2012) Effect of drying methods on the physical properties and microstructures of mango (Philippine 'Carabao' var.) powder. Journal of Food Engineering 111: 135-148.

13. Cano-Chauca M, Stringheta PC, Ramos AM, Cal-Vidal J (2005) Effect of the carriers on the microstructure of mango powder obtained by spray drying and its functional characterization. Innovative Food Science & Emerging Technologies 6: 420-428.

14. Solval KM, Sundararajan S, Alfaro L, Sathivel S (2012) Development of cantaloupe (Cucumismelo) juice powders using spray drying technology. LWT - Food Science and Technology 46: 287-293.

15. Fazaeli M, Emam-Djomeh Z, Ashtari AK, Omid M (2012) Effect of spray drying conditions and feed composition on the physical properties of black mulberry juice powder. Food and Bioproducts Processing 90: 667-675.

16. Quek SY, Chok NK, Swedlund P (2007) The physicochemical properties of spray-dried watermelon powders. Chemical Engineering and Processing 46: 386-392.

17. Teunou E, Fitzpatrick JJ (1999) Effect of relative humidity and temperature on food powder flowability. Journal of Food Engineering 42: 109-116.

18. Moreira GEG, Costa MGM, Souza ACR, Brito ES, Medeiros MFD, et al. (2009) Physical properties of spray dried acerola pomace extract as affected by temperature and drying AIDS. LWT - Food Science and Technology 42: 641-645.

19. Maury M, Murphy K, Kumar S, Shi L, Lee G (2005) Effects of process variables on the powder yield of spray-dried trehalose on a laboratory spray-dryer. European Journal of Pharmaceutics and Biopharmaceutics 59: 565-573.

20. Gallo L, Llabot JM, Allemandi D, Bucalá V, Piña J (2011) Influence of spray-drying operating conditions on Rhamnuspurshiana (Cáscarasagrada) extract powder physical properties. Powder Technology 208: 205-214.

21. Goula AM, Adamopoulos KG (2005) Spray drying of tomato pulp in dehumidified air: I. The effect on product recovery. Journal of Food Engineering 66: 25-34.

22. Goula AM, Adamopoulos KG (2006) Retention of ascorbic acid during drying of tomato halves and tomato pulp. Drying Technology 24: 57-64.

23. Desai KGH, Park HJ (2005) Recent developments in microencapsulation of food ingredients. Drying Technology 23: 1361-1394.

24. Goula AM, Adamopoulos KG (2008) Effect of Maltodextrin Addition during Spray Drying of Tomato Pulp in Dehumidified Air: II. Powder Properties. Drying Technology 26: 726-737.

25. Allamilla-Beltrán L, Chanona-Pérez JJ, Jiménez-Aparicio AR, Gutiérrez-López GF (2005) Description of morphological changes of particles along spray drying. Journal of Food Engineering, 67: 179-184.

26. Nijdam JJ, Langrish TAG (2006) The effect of surface composition on the

functional properties of milk powders. Journal of Food Engineering 77: 919-925.

27. Adhikari B, Howes T, Bhandari BR, Troung V (2004) Effect of addition of maltodextrin on drying kinetics and stickiness of sugar and acid-rich foods during convective drying: Experiments and modeling. Journal of Food Engineering 62: 53-68.

28. AOAC (2006) Official Methods of Analysis. 18th edtn, Association of Official Analytical Chemists, Gaithersburg, Maryland, USA.

29. AOAC (1990) Official Methods of Analysis. 14th edtn, Association of Official Analytical Chemists, Washington, DC, USA.

30. Barreto LS, Castro MS (2010) Boas práticas de manejo para o extrativismo sustentável do umbu, Embrapa Recursos Genéticos e Biotecnologia, Brasília.

29

Application of Home-Made Enzyme and Biosurfactant in the Anaerobic Treatment of Effluent with High Fat Content

Jaqueline do Nascimento Silva[1]*, Melissa Limoeiro Estrada Gutarra[1], Denise Maria Guimaraes Freire[2] and Magali Christe Cammarota[1]

[1]Department of Biochemical Engineering, School of Chemistry, Universidade Federal do Rio de Janeiro, Federal University of Rio de Janeiro, Cidade Universitária, Centro de Tecnologia, Bl. E, Sl. 203, Ilha do Fundão, 21941-909, Rio de Janeiro, Brasil.
[2]Department of Biochemistry, Institute of Chemistry, Universidade Federal do Rio de Janeiro, Federal University of Rio de Janeiro, Cidade Universitária, Centro de Tecnologia, Bl. A, Sl. 549, Ilha do Fundão, 21949-900, Rio de Janeiro, Brasil.

Abstract

The Solid State Fermentation (SSF) lipase production from *Penicillium simplicissimum* using an agro industrial residue and the biosurfactant from *Pseudomonas aeruginosa* were employed for a combined pre-treatment of wastewater from poultry processing industry. During the hydrolysis step of the wastewater fat better results were obtained at 34°C with biosurfactant 0.4% (v/v) and liquid enzyme preparation 6.2% (v/v) or at 46°C with biosurfactant 0.1% (v/v) and liquid enzyme preparation 3.8% (v/v). The pre-hydrolysis in two different conditions of enzyme and biosurfactant concentrations at 34°C and posterior anaerobic treatment allowed the COD removal and methane production, while the no hydrolyzed wastewater besides the COD removal, did not show methane production. Those results suggest the importance and efficiency of the pre-treatment with home-made lipases and biosurfactant during the high fat content wastewater pre-treatment.

Keywords: Anaerobic treatment; Biosurfactant; Fat; Lipase; Wastewater

Introduction

In Brazil, poultry slaughtering industry stands out as one of the most important industrial activities. Brazil is the third largest chicken's producer, reaching 11 million tons of chicken meat produced, with 3.6 million for export [1]. However, the poultry slaughtering generate high volumes of effluent containing biodegradable organic matter, lipids, proteins and cellulose. Lipids can represent over 67% of the COD particulate in slaughterhouses wastewater [2].

Nowadays, with demand for energy, the search for alternative energy sources is increasing. Thus, the anaerobic processes involving as effluents generated in the food industry become advantageous, since they allow the energy production in form of methane gas [3].

For an efficient anaerobic wastewater treatment containing high fat content, it becomes necessary a step of pre-hydrolysis in order to avoid operational problems. Enzymatic fats hydrolysis has been studied using commercial enzyme. However, the utilization of these enzymes involves the increase of treatment costs [4].

Several research groups evaluated the process of solid state fermentation (SSF) as a viable and economic alternative for the enzymes production, mainly due to the possibility of using agro industrial wastes as culture medium. Castilho et al. [5] demonstrated that lipase production SSF using the babassu cake as residue requires 78% less investment in comparison to the submerged fermentation. Lipases from *Penicillium simplicissimum* produced by SSF have high biotechnological potential, since they have higher production and yield [6].

In the environmental area, many studies have reported the use of biosurfactants to increase interaction water/oil, accelerate the degradation of several oils by microorganisms and promote the contaminated soils and waters bioremediation [7]. However, there are a few reports in the literature of the combined use of enzymes and biosurfactant to increase the treatment efficiency of wastewater from food industries [8].

Considering the need to promote the proper treatment of wastewater with high fat content, this paper aims the use of an extracellular lipase from *Penicillium simplicissimum* produced by SSF in babassu cake and a rhamnolipid produced by Pseudomonas *aeruginosa* PA1 and its application in wastewater pre-treatment from a poultry processing industry.

Material and Methods

Solid-state fermentation

The enzyme pool was produced by solid state fermentation of waste from babassu oil production by *Penicillium simplicissimum* microorganism in tray reactor containing 15 g of basal medium (babassu cake). After milling and supplementation with 6.25% (w/w) molasses, to a C:N ratio of 12.8:1, the cake was autoclaved at 121°C/15 min and inoculated with a spores suspension to an initial concentration of 107 spores/g [6]. After 48 h of growth conducted at 30°C and 95% humidity, a liquid enzyme preparation (LEP) was obtained by extraction with phosphate buffer pH 7.0 and 0.5% (w/v) Tween 80. The mixture was incubated in a rotary shaker at 25°C and 200 rpm for 40 minutes. The liquid fraction was then extracted by manual pressing and centrifugation. The supernatant (LEP) was used for the assays.

Biosurfarctant production

The biosurfactant was produced using *Pseudomonas aeruginosa*. After 7 days, the cell-free crude fermented broth was recovered from the fermented medium and characterized. The main characteristics of

*Corresponding author: Jaqueline do Nascimento, Department of Biochemical Engineering, School of Chemistry, Federal University of Rio de Janeiro, Cidade Universitária, Centro de Tecnologia, Bl. E, Sl. 203, Ilha do Fundão, 21941-909, Rio de Janeiro, Brasil

this cell-free fermented medium are: rhamnolipid concentration (4.7 g/L), surface tension (28mN/m), critical micelle concentration (CMC = 205 mg/L), emulsification index (61%) and chemical oxygen demand (COD = 20767 mg/L) [8]. It was stored at -20°C until use.

Collection and characterization of the effluent and sludge

The effluent was obtained from a poultry processing industry in the city of Rio de Janeiro (Rio de Janeiro - Brazil) at the treatment plant before the flotation step. After collection, properly preserved aliquots were taken for determining the characterization parameters and the remaining volume was stored in freezer until the moment of use (Table 1). The amounts of nitrogen and phosphorus in the effluent were enough to supply the necessities of the microorganisms in the anaerobic biodegradability tests. The sludge used in the anaerobic biodegradability tests was collected from a bioreactor UASB (upflow anaerobic sludge blanket) operating in this poultry processing industry, being characterized in terms of total volatile solids (14895 ± 1113mg/L) and oil and grease accumulated according standard procedures [9] (Table 1).

Pre-treatment with enzyme and biosurfactant

Firstly, a study was conducted in order to determine the time for enzymatic hydrolysis of fat present in the effluent (2034 mgO&F/L). Times of 4, 8 and 24h, using concentrations of 1 and 5% (v/v) LEP, at 400 rpm and 30ºC, were evaluated. These experiments were conducted with 100 mL of effluent in 500mL flasks. Then, the best conditions were set for hydrolysis employing an experimental design 2^k type, having as variables the temperature, LEP and biosurfactant concentrations. The actual and coded values of these variables are shown in Table 2. The experiments were conducted with 100 mL of raw wastewater in glass reactors with stirring (400 rpm) and temperature control (34ºC, 40ºC or 46ºC) for 8 h. The hydrolysis was monitored through free acid measurement performed by titration with 0.04M NaOH until pH 11 [8] (Table 2).

Variable	Average ± SD
pH	6.4 ± 0.1
T (ºC)	33
COD (mg/L)	8692 ± 1262
mgO&F/L	2403 ± 521
Total Solids (mg/L)	6717 ± 3049
Fixed Total Solids (mg/L)	1807 ± 151
Volatile Total Solids (mg/L)	4910 ± 2898
Total Nitrogen (mg/L)	434 ± 2
Total Phosphorus (mg/L)	6.5 ± 0.7

Average and standard deviations of three samples of collected effluent

Table 1: Characterization of the raw effluent from the poultry processing industry. The effluent was characterized in terms of pH, COD, suspended solids (total, fixed and volatile), total nitrogen, total phosphorus and oils and greases, according to standard procedures (APHA 2005).

Variable	Levels		
	-1	0	1
Biosurfactant (% v/v)	0.10	0.25	0.40
Enzyme pool (% v/v)	3.8	5.0	6.2
T (ºC)	34	40	46

Average and standard deviations of three samples of collected effluent

Table 2: Experimental design 2³ type, having as variables the temperature, LEP and biosurfactant concentrations. Conditions used in the pretreatment with enzyme pool and biosurfactant.

Anaerobic biodegradability tests

Anaerobic biodegradability tests were conducted with the raw and treated effluents with LEP and biosurfactant. All the tests were conducted in batch using six penicillin flasks of 100mL. The flasks were incubated at 30°C for up to 8 days with 90mL of a mixture of anaerobic sludge and effluent with pH adjusted to 7.0 ± 0.1. The effluent and sludge volumes were calculated in order to obtain an initial COD: sludge VSS ratio of 1:1 in the flasks. The biodegradability was assessed by measuring the COD removal efficiency and biogas production, performed by the piston displacement of 20 mL plastic syringes connected to the flasks. Aliquots for the initial COD determination were taken after pre-treatment and before contact with the anaerobic sludge. The final soluble COD was determined on the last day of the degradation test, after collecting the biogas for analysis in gas chromatograph. Specific methane production (L CH_4/g removed COD) was also used to assess the biodegradation process.

Analytical methods

The lipase activity was determined by colorimetric and titrimetric methods using p-nitrophenil laureate and olive oil as substrates, respectively [6], where one lipase unit was defined as the enzyme amount that causes the release of 1μmole of fatty acids per minute, under the assay conditions (titrimetric) and one unit of lipase activity is defined as the amount of enzyme which releases 1 μmole of p-nitrophenol under the assay conditions (colorimetric). Enzyme activity was expressed as units per gram of dry babassu cake. The rhamnolipid concentration, reported as rhamnose, was determined according to the method of Pham et al. [10]. The emulsification index was determined according to the method of Cooper and Goldenberg [11]. The determination of the crude extract surface tension (mN/m) was performed on an Aqua-Pi tensiometer (Kibron Inc., Helsinki) at 25°C, using the Du Noüy [12] method. The critical micelle concentration (CMC) was determined according to methodology described by Cooper et al. [13]. The biogas composition was determined in gas chromatograph VARIAN MICRO GC 4900 employing 10 m x 0.32 mm PPQ column, column temperature of 50°C, thermal conductivity detector (TCD) with temperature of 250°C, injector temperature of 80°C and helium as carrier gas. The other parameters used in the effluent and sludge characterization and in the monitoring of the anaerobic biodegradability tests were determined according to procedures described in Standard methods [9].

Results and Discussion

Determination of enzymatic hydrolysis pre treatment time

A preliminary study was conducted in order to determine the optimum pre hydrolysis time. We investigated the kinetic of hydrolysis with two different amounts of liquid enzyme preparation (LEP) (1 and 5% v/v) at 30°C, 400 rpm. The concentration of free acids was monitored after 4, 8 and 24h hydrolysis, as shown in Table 3. For both concentrations of LEP, the maximum free acids production was

Hydrolysis time (h)	1% LEP	5% LEP	Δ (1%)	Δ (5%)
0	30.5	30.5	0	0
4	31.2	31.5	0.7	1.0
8	34.4	41.3	3.9	10.8
24	28.7	33.7	--	3.2

Δ Variation between 4, 8 and 24 h and the zero time

Table 3: Free acid production (μmol/ml) for different concentrations of LEP (with 4.7 U/mL of lipase activity) and hydrolysis times (Standard deviation less than 5%).

obtained with 8 h hydrolysis. So, this time was selected for the next assays and the LEP concentration of 5% (v/v) was the basis for the experimental design (Table 3).

Valladão et al. [14] found maximum values of hydrolysis at 22 h (7.3 mol/ml) using 1% w/v (0.21 U/mL) lipase from *P. restrictum* in treating poultry slaughterhouse wastewater containing 1200mgO&F/L. In this work it was used substantially the same lipase activity (0.23 U/mL or 5% v/v) to hydrolyze almost twice as fat (2034 mgO&F/L), yielding 8.10mol/mL of free acids in only 8 h. These results show that the enzyme preparation produced by the fungus *P. simplicissimum* is most effective for hydrolysis of fats present in the effluent from a poultry processing industry.

Determination of optimum conditions for the pre-treatment

In order to evaluate the synergistic effect on the pretreatment step of the enzyme pool with the biosurfactant, an experimental design was carried out with effluent containing 2034 mgO&F/L. The free acids concentrations obtained in each condition evaluated are presented in Table 4.

The standard variables effects (t) and the significance probability test (p) were used to evaluate the effects of biosurfactant concentration, LEP concentration and temperature on the fat hydrolysis present in the effluent. Using a 10% confidence level (p<0.10) was observed that the LEP concentration, the biosurfactant concentration and interaction between biosurfactant concentration and temperature showed higher significant effect on the fat hydrolysis in the effluent (Table 5). The positive effect of the interaction between the variables biosurfactant and enzyme concentration prove the success of use of these products on the pretreatment when combined (Table 5).

Based on the results it was possible to construct an empirical model for quantification of free fatty acids released due to the LEP concentration, biosurfactant concentration and temperature (Equation 1) that includes the statistically significant and marginally significant variables, considering p<0.1.

$$FA = 8.48 + 1.38\,LEP - 1.26\,B - 0.97\,T + 1.01\,LEP \times B - 1.75\,B \times T \quad (Eq.\ 1)$$

Condition	LEP (% v/v)	BS (% v/v)	T (°C)	Free acids (µmol/ml)*
1	3.8	0.10	34	8.73
2	6.2	0.10	34	10.07
3	3.8	0.40	34	6.81
4	6.2	0.40	34	13.97
5	3.8	0.10	46	10.94
6	6.2	0.10	46	13.29
7	3.8	0.40	46	6.00
8	6.2	0.40	46	8.41
9	5.0	0.25	40	7.2
10	5.0	0.25	40	6.97
11	5.0	0.25	40	6.92

*Difference between sample and control after 8h. Standard deviation less than 5%

Table 4: Conditions (experimental design) and results of the pretreatment with enzyme pool (LEP) and biosurfactant (BS).

Variable	Effect	Standard Error	t	p
LEP concentration	2.75	1.00	2.73	0.0525
Biosurfactant concentration	-2.52	1.00	-2.51	0.0663
Temperature	-1.93	1.00	-1.91	0.1280
LEP conc x Biosurfactant conc.	2.04	1.00	2.01	0.1135
Biosurfactant conc. x Temperature	-3.52	1.00	-3.49	0.0251

Table 5: Standard variables effects (t) and the significance probability test (p).

Where: FA = free acids concentration (µmol/ml), LEP = liquid enzyme preparation concentration (coded values), T = Temperature (coded values), B = Biosurfactant concentration (coded values).

The model generated was considered predictive by analysis of variance (ANOVA) and it showed satisfactory determination coefficient (R^2=0.84) and F test value (7.71) greater than the critical value (2.14). This model allowed the construction of outline curves, which show the predicted values of free acids produced for each condition between studied ranges (Figure 1).

At Figure 1a, it was observed that at 34°C it was possible to obtain high free acids concentrations in the hydrolysis applying high LEP concentrations and biosurfactant (within the studied range). However, keeping the LEP concentration and biosurfactant in their highest levels, the concentration of free acid liberated in hydrolysis decreases with increasing temperature (Figure 1b). This could be explained by deactivation of the enzyme at higher temperature, but it is known that the enzyme preparation used in this work presents optimal activity at 50°C [6]. It was also observed in Figures 1c and Figures 1d that high amounts of free acids are obtained in high temperature employing low concentrations of biosurfactant and LEP. Increasing the temperature can lead to a greater solubility in oils and greases from wastewater, reducing the concentration of biosurfactant necessary to maintain the emulsion quality and consequently the access of the enzyme to the substrate, combined with a good enzyme activity at higher temperatures.

However, when comparing the hydrolysis conditions only with the effluent from a poultry processing industry, the free acids production obtained in the present studied with *P. simplissicimum* LEP (7 a 14µmol/mL) demonstrates that optimization of the enzymatic hydrolysis combined with biosurfactant had positive effects, producing better results in enzymatic hydrolysis step of the wastewater fat. Results indicate two optimum conditions of pretreatment (with a maximum hydrolysis 8h): (1) hydrolyzing the effluent to 34°C, which, thereafter, follow to one biological treatment with mesophilic microflora and (2) hydrolyzing the effluent at 46°C for further treatment with thermophilic microorganisms.

Anaerobic biodegradability tests

The anaerobic biodegradability tests were done after enzymatic hydrolysis optimization step, using the hydrolyzed obtained under two distinct conditions: in the first one, the hydrolyzed shows a high acids content (condition 1); in the second one, the hydrolyzed shows a not so high (medium) acids content (condition 2). Such strategy was made to combine good hydrolysis efficiency with a high specific methane production (SMP), once the free acids released during the enzymatic hydrolysis can be inhibitors to anaerobic biodegradability step [15]. Those conditions were selected fixing the temperature in 34°C, once it was used mesophilic microorganisms during the anaerobic biodegradability tests. Condition 1 had hydrolysis conditions at 34°C, biosurfactant 0.4% (v/v) and enzyme (LEP) 6.2% (v/v), reaching 14 µmol/mL of free acids; for condition 2 was used the same temperature and biosurfactant quantity, however with enzyme 3.8% (v/v), reaching nearly 8 µmol/mL of free acids. Table 6 shows the results in terms of COD and methane specific production (Table 6).

It's possible to see that for the hydrolyzed wastewater (conditions 1 and 2) it was possible to obtain methane as principal product during this step. Besides that, the reduction of initial fat content provides a high COD removal, proving the efficiency of enzymatic treatment. However, the specific methane production was higher for the wastewater with

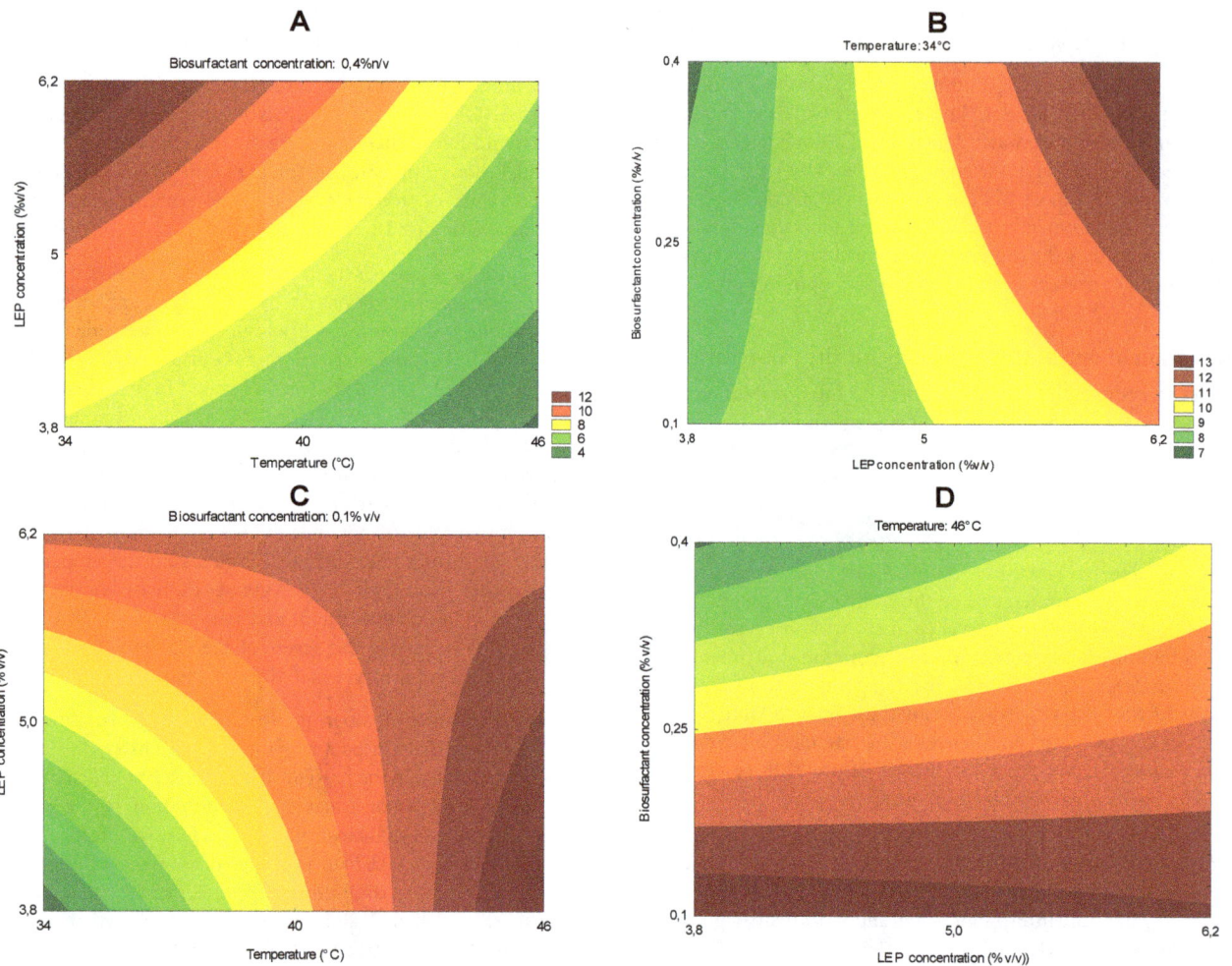

Figure 1: Contour curves for the hydrolysis of the effluent as a function of the variables biosurfactant concentration and LEP concentration to 34°C (a); LEP concentration and temperature for biosurfactant concentration of 0.4% v/v (b); concentration of LEP and temperature for the biosurfactant concentration of 0.1% v/v (c); and biosurfactant concentration and the concentration of LEP at 46°C (d).

Condition	Initial COD (mg/L)	Final\ COD (mg/L)	COD removal (%)	Biogas volume (mL)[a]	Methane (%)	Methane volume (mL)[a]	mL CH₄/g removed COD[a]
Control	3400	240	93	65	nd	nd	0
Condition 1	6147	412	93	79	80	63.2	122.4
Condition 2	4867	304	94	80	82	65.6	159.7

[a]30ºC, 1 atm, 8 days

Table 6: Anaerobic biodegradability tests with the addition of the enzyme pool and biosurfactant under different conditions (Standard deviation less than 5%).

lower acids content released (condition 2), once in this condition the wastewater may contain lower concentration of long chain fat acids, potential inhibitors of methanogen archaeas [15]. In condition control case (without enzymatic hydrolysis) there was COD reduction (93%) and gas productions (65 mL), however it was not possible to detect the methane production. The value of COD reduction in control is the same reached by hydrolyzed wastewater and the gas production was lower when compared with hydrolyzed wastewater. This indicate that part of the COD is removed by other metabolic pathways and/or by other microorganisms producing another by-products and gas as CO_2 (detected by chromatography), or H_2 and volatile acids (not determined). Other possibility is that part of COD removal was reached by adsorption of the fat content on the sludge or onto flasks wall used during the test.

While Valladão et al. [14] obtained 165 mL CH₄/g COD provided (at 34ºC) with effluent containing 1200 mgO&F/L hydrolyzed with 0.21 U/mL for 22 h, this work yielded 150 mLCH₄/g COD provided (at 34ºC) with wastewater containing 2034 mgO&F/L hydrolyzed with 0.19 U/mL and 0.4% biosurfactant for 8h. These results shows that higher concentrations of fat can be degrade, keeping the same enzymatic activity for shorter pre-hydrolysis time when adding a small quantity of biosurfactant. Probably the biosurfactant facilitates access of the enzyme to particles of fat.

Conclusions

The hydrolysis of wastewater from poultry processing industry with a home-made lipase showed high efficiency compared to other lipases.

The effect of biossurfactat alone was not evaluated in this study, but the use of it aimed to assist and minimize the use of the enzymatic preparation. After the optimization step, the use of the biosurfactant resulted in a reduction of 30% in the quantity of enzyme to release the same amount of fatty acids (8 μmol/mL). While an increase of about 35% in the quantity of enzyme preparation led to a 75% increase in the content of free fatty acids (14 μmol/mL) due the use of the biosurfactant. The enzymatic preparation and ramnolipid showed higher production of free fat acids in two different conditions of enzyme and biosurfactant concentration, and temperature due to variables interaction effect of biosurfactant and enzyme concentration, and temperature and biosurfactant concentration as showed in experimental design analysis. This result allows the pre-treatment to be conducted in 34 or 46ºC with efficiency and indicate a subsequent anaerobic treatment application with mesophilic or thermophilic microflora.

Is notable the favorable joint application between enzyme and biosurfactant also for methane production. The treatment shows high COD removal efficiency and methane production for a more concentrated effluent treatment with shorter time probable allowing operation of the system for long periods without problems caused by fat accumulation, reducing costs of the process.

Acknowledgments

This work was supported by the Brazilian state funding agency CNPq (Conselho Nacional ao Desenvolvimento Científico e Tecnológico) and PETROBRAS.

References

1. (2009) UBA (União Brasileira de Avicultura) Annual Report.

2. Sayed S, Zanden JVD, Wijffels R, Lettinga G (1988) Anaerobic degradation of the various fractions of slaughterhouse wastewater. Biol Waste 23: 117-142.

3. Speece RE (1983) Anaerobic Biotechnology for Industrial Wastewaters. Environ Sci Technol 17: 416–427.

4. Cammarota MC, Freire DMG (2006) A review on hydrolytic enzymes in the treatment of wastewaters with high oil and grease content. Bioresource Technol 97: 2195-2210.

5. Castilho LR, Polato CMS, Baruque EA, Sant'Anna Jr GL, Freire DMG (2000) Economic analysis of lipase production by Penicillium restrictum in solid state and submerged fermentations. Biochem Eng J 4: 239-247.

6. Gutarra MLE, Godoy MG, Maugeri F, Rodrigues MI, Freire DMG et al., (2009) Production of an acidic and thermostable lipase of the mesophilic fungus Penicillium simplicissimum by solid-state fermentation. Bioresource Technol 100: 5249–5254.

7. Millioli VS, Servulo EFLC, Sobral LGS, Carvalho DD (2009) Bioremediation of crude oil-bearing soil: evaluating the Effect of rhamnolipid addition to soil toxicity and to crude oil biodegradation efficiency. Global NEST J 11: 181-188.

8. Damasceno FRC, Cammarota MC, Freire DMG (2012) The combined use of a biosurfactant and an enzyme preparation to treat an effluent with a high fat content. Colloids and Surfaces B 95: 241-246.

9. APHA (American Public Health Association) (2005) American Water Works Association, Water Pollution Control Federation. Standard Methods for the Examination of Water and Wastewater. 18th edn, New York.

10. Pham TH, Webb JS, Rehm, BHA (2004) The role of polyhydroxyalkanoate biosynthesis by Pseudomonas aeruginosa in rhamnolipid and alginate production as well as stress tolerance and biofilm formation. Microbiol 150: 3405-3413.

11. Cooper DG, Goldenberg BG (1987) Surface-active agents from two Bacillus species. Appl Environ Microb 53: 224-229.

12. Du Noüy PL (1925) An interfacial tensiometer for universal use. J Gen Physiol 7: 625-632.

13. Cooper DG, Zajic JE, Gerson F (1979) Production of surface-active lipids by Corynebacterium lepus. Appl Environ Microb 37: 4-10.

14. Valladão ABG, Freire DMG, Cammarota MC (2007) Enzymatic pre-hydrolysis applied to the anaerobic treatment of effluents from poultry slaughterhouses. Int Biodeter Biodegr 60: 219-225.

15. Koster IW, Cramer A (1987) Inhibition of methanogenesis from acetate in granular sludge by long-chain fatty acids. Appl Environ Microb 53: 403-409.

Identification and Characterization of 1,4-β-D-Glucan Glucohydrolase for the Saccharification of Cellooligomers from *Paenibacillus* sp. HPL-001

Dal Rye Kim[1], Hee Kyung Lim[1] and In Taek Hwang[1,2]*

[1]*Biorefinery Research Group, Green Chemistry Division, Korea Research Institute of Chemical Technology, Yusoeong P.O. Box 107, Daejon 305-600, Republic of Korea*
[2]*Department of Green Chemistry and Environmental Biotechnology, University of Science & Technology, 217 Gajungro, Yuseong-gu, Daejon, 305-350, Republic of Korea*

Abstract

The 1,4-β-D-glucan glucohydrolase gene (Ggh) which was isolated from *Paenibacillus* sp. strain HPL-001 (KCTC11365BP) has been cloned and expressed in *Escherichia coli*, and efficiently purified using affinity column chromatography. Recombinant 1,4-β-D-glucan glucohydrolase (719aa, NCBI accession number KJ573391) was highly active at 40°C in pH 6.0 without significant changes to most salts tested, and exhibited K_m 0.893 mg/mL V_{max} 33.33 U/mg on *p*NPG, respectively. All soluble cellooligomers (e.g., cellobiose, cellotriose, cellotetraose, cellopentaose, and cellohexaose) had been virtually hydrolyzed to glucose by this enzyme. When compared with the commercialized-enzymes, the hydrolytic pattern to all substrates was almost same to the Celluclast 1.5L with about 1/3 strength, however hydrolytic pattern of Glucosidase and Almonds was significantly decreased with substrates increasing number of glucose polymer from cellotriose to cellohexaose. About 90% of the initial enzyme activity was maintained even after 10 consecutive recycle by the 11-carbon bridge and aldehyde-functionalized MCF-immobilization. 3-D structure of this enzyme has the ligand of cellobiose and cellulose binding in the center of niche according to the sequence information.

Keywords: 1,4-β-D-glucan glucohydrolase; Cellooligomer; Immobilization; Saccharification

Introduction

The conversion of lignocellulosic biomass to platform chemicals for fermentation and biorefinery typically involves a disruptive pretreatment process followed by enzyme-catalyzed hydrolysis of the cellulose and hemicelluloses to fermentable sugars [1]. The hydrolyzing process from cellulose to glucose is critical to the bio refinery of plant biomass for a wide variety of biotechnological and industrial application alternatives to fossil fuels [2].

The enzymatic hydrolysis involves the synergistic activity of endo-β-1,4-glucanases (EC3.2.1.4, EG), exoglucanases, including both cellobiohydrolases (EC 3.2.1.91, CBH) and β-glucosidases (EC 3.2.1.21, BGL). All three enzyme classes must be present in this system in order to produce glucose. These enzymes cooperate in the following manner: EG act randomly along the chain cellulose depolymerization, producing new attack sites for the CBH act as exo-enzymes, liberating cellobiose as their main product; and β-glucosidases, which are not regarded as legitimate cellulases, play an important role in the process, because it complete the process, which converts the intermediate cellobiose into two glucose molecules by the hydrolysis of β-glucosidic linkages, is a key rate-limiting enzyme in the cellulolytic process [3].

A high level of β-glucosidase is important to avoid the accumulation of cellobiose, which is a strong inhibitor of CBH to increase the productivity of saccharification and get a better product composition. The amount of BGL secreted from *T. reesei* is quite low compared to that of CBH and EG [4]. Usually, BGLs produced by other fungi must be added to a *T. reesei* enzyme mixture to increase the efficiency of the hydrolysis of cellulosic substrates. The competitive product inhibition of cellobiose can be overcome to some extent by addition of a surplus of β-glucosidase activity. The hydrolyzing enzyme mixture can be supplemented with additional soluble or immobilized β-glucosidase produced by another organism.

Recently, an excellent method for cellulose depolymerization using a solid catalyst in an ionic liquid has been reported [5]. The solid catalyst, it was possible to displace the role of EG, decreasing degree of polymerization (DP), over the course of hydrolysis for cellulosic substrates. Conclusively, identification of an enzyme for hydrolysis cellulose after depolymerization using a solid catalyst in an ionic liquid was essential to complete the alternative approach for cellulose saccharification. An enzyme 1,4-β-D-glucan glucohydrolase was identified from direct evolution [6], which showed the successive hydrolysis functions of cello-oligomers as substrates. This enzyme is capable of producing glucose from soluble cello-oligomers without requiring a CBH; this is different from the traditional model for cellulose digestion process of CBH and BGL, what is known about a fungal cellulase system. At this point, we decided to develop an alternative approach to hydrolyze cellulose to glucose with combining the starting step of cellulose depolymerization with solid catalyst in ionic liquid and the final step of glucose producing from soluble cellooligomers with this enzyme 1,4-β-D-glucan glucohydrolase, completely.

In this study, we suggested a 1,4-β-D-glucan glucohydrolase gene *Ggh* (PGDβ4, GenBank accession code: ZP_07902992.1) isolated from a *Paenibacillus* sp. HPL-001 (Korean Collection for Type Culture: KCTC11987BP) from an organic-rich, soil sample collected from a rearing farm of the wood-eating, oriental horned beetle, *Allomyrina dichotoma* (Linnaeus) in Okcheon County, Korea (ROK). The enzyme was purified and characterized for its activity and properties. Also, we immobilized this enzyme on aldehyde-, amine-, SH-, epoxy-functionalized meso-structured cellular foam silica (MCF) for practical applications.

***Corresponding author:** In Taek Hwang, Department of Green Chemistry and Environmental Biotechnology, University of Science & Technology, 217 Gajungro, Yuseong-gu, Daejon, 305-350, Republic of Korea*

Material and Methods

Gene selection from gDNA library

A strain of *Paenibacillus* sp. HPL-001 was isolated from an organic-rich soil, and obtained the whole genomic DNA from the strain. Size-fractionation was conducted in a 0.5% low-melting-point agarose gel after fragmentation by shearing with a nebulizer (Invitrogen). DNA fragments (around 5 kb) were collected and purified for library construction, and fragments were blunt-end repaired and dephosphorylated, and then ligated into pCB31 plasmid vector (MACROGEN Co., Korea). The packaged library was electro-porated into *E. coli* DH10B cells according to the manufacturer's instructions. *E. coli* transformants were selected on LB agar plates supplemented with kanamycin, a total of 1,536 clones were collected into sixteen 96-well plates containing 200 μl LB broth (Difco) in each well. After incubation for 24 h at 37°C, 25 μl glycerol (Sigma-Aldrich) was added to each well and mixed prior to storage at -80°C. Routine 1,4-β-D-glucan glucohydrolase activities were determined in duplicate using 100 mM 4-nitrophenyl-*p*-n-glucopyranoside (*p*NPG) as substrate. Reactions were performed in 96-well plate filled with 90 μl of 100 mM potassium phosphate buffer pH 7.0, 10 μl of *p*NPG solution and 100 μl of crude enzyme from each library clone [7]. Crude enzyme was harvested from destruction of the library cells with sonification after cultivation for 48 hrs at 37°C. Assays were conducted at 40°C water bath for 10 min. The amount of 4-nitrophenol (*p*NP) released was estimated by absorbance at 400 nm using a *p*NP standard curve. One unit (U) of 1,4-β-D-glucan glucohydrolase activity is defined as one μmol of *p*NP released mg protein^{-1} min^{-1}.

Activities against other 4-nitrophenyl derivatives were measured in the same way while activities on other substrates were determined by the release of reducing sugars.

DNA sequencing and expression of 1,4-β-D-glucan glucohydrolase in *E. coli*

The most glucosidase-active clone (PGDβ4, arbitrary named) was selected from the library screening, and the nucleotide sequence of the insert was determined by automated sequencing under BigDyeTM terminator cycling condition. The reacted product was purified using ethanol precipitation and run with Automatic Sequencer 3,730 × l (Applied Biosystems, Weiterstadt, Germany). Open Reading Frames (ORFs) from the sequence data of the insert in clone PGDβ4 were predicted using the ORF Finder (NCBI), taking ATG, GTG, and TTG as possible start condones. And also, homology searches for resulting seven ORFs were carried out by using the BLAST program in the GenBank database. All seven sets of primers fully-covering each ORF (ORF clone ID: PGDβ4 ORF1 ~ PGDβ4 ORF7) were designed, and amplified to construct transformants. The each PCR product was inserted into pGEM-T Easy plasmid vector and transformed to *E. coli* JM109 (Takara Bio Inc.), and the 1,4-β-D-glucan glucohydrolase activity of each transformant was examined with same method above mentioned. The most 1,4-β-D-glucan glucohydrolase-active subclone inserted with ORF1 was selected and the insert, which was modified with detaching Ribosomal Binding Site (RBS) from ORF1 sequence, was designated as *Ggh* gene. The gene, *Ggh* was inserted into the pIVEX GST fusion vector for the transformation into *E. coli* BL21 (Roche Applied Science) to produce the recombinant fusion protein.

Purification of recombinant 1,4-β-D-glucan glucohydrolase

The transformed *E. coli* with GST-fused 1,4-β-D-glucan glucohydrolase were grown overnight in 10 ml LB medium containing ampicillin (100 μg/ml) at 37°C and 200 rpm in a shaking incubator. After re-inoculation with 100 ml of new LB medium and grown to 0.6~0.7 at A595, the culture was induced with 1 mM IPTG and incubated at 18°C and 200 rpm for 18 hr longer. The induced bacteria were collected by centrifugation, suspended in 10 ml of ice-cold 1X phosphate-buffered saline (Sigma-Aldrich), repeated triple times. The cells from the final washing process was resuspended in the lysis buffer (pH 7.0, 200 mM Tris-HCl, 10 mM NaCl, 10 mM β-mercaptoethanol, 1 mM EDTA), and treated with sonic disruptor (CosmoBio Co., LTD). After cell disruption, the lysate was centrifuged at 10,000 × g for 20 min at 4°C, the supernatants was eluted through the GST binding resin column (Novagen, Madison WI, USA). The column was previously equilibrated with the washing buffer (pH 7.0, 50 mM Tris-HCl, 15 mM NaCl), then the lysate was applied to the equilibrated column with the flow rate at 10-15 cm/h. After washing the GST resin with 20 bed volumes of cold washing buffer, the fusion protein was eluted and fractionated with 10 bed volumes of freshly made elution buffer (pH 7.0, 20 mM Glutathione, 50 mM Tris-HCl, 15 mM NaCl). Active fractions were pooled and treated with Restriction Protease Factor Xa (Roche Applied Science), and eluted through *p*-aminobenzamidine-agarose column (Sigma-Aldrich) according to the manufacturer's instruction. The elute was precipitated with 70% ammonium sulfate, solubilized in phosphate-buffered saline, and dialyzed to concentrate at 4°C with a dialysis membrane (Spectra/Por CE, MWCO 10,000), then protein content was determined prior to store at -20°C. Sodium dodecyl sulfate-polyacrylamide gel electrophoresis (SDS-PAGE) on 15% polyacrylamide was performed to separate each protein, and the protein fractions mixed with denaturing agent were boiled for 5 min and applied to the gel. Proteins were visualized by Coomassie brilliant blue R 250 staining. The protein concentration was determined with Bradford reagent (Sigma-Aldrich) assay using bovine serum albumin as a standard.

Properties of recombinant 1,4-β-D-glucan glucohydrolase

The enzyme activity, recombinant 1,4-β-D-glucan glucohydrolase, was measured according to the method mentioned above. The amount of 4-nitrophenol (*p*NP) released was estimated by absorbance at 400 nm using a *p*NP standard curve. The optimal temperature and pH condition for the 1,4-β-D-glucan glucohydrolase activity of recombinant protein were examined in 96-well micro plates at various temperatures (ranging from 20 to 70°C) and pH conditions (ranging from pH 2 to 12). The kinetic parameters (Hanes-Woolf constant, Km and maximal reaction velocity, Vmax) were estimated by linear regression from double-reciprocal plots. Effect of metallic ions and other chemicals on the 1,4-β-D-glucan glucohydrolase activity was studied as described above at pH 7 with addition of 1 mM NaCl, LiCl, KCl, NH$_4$Cl, CaCl$_2$, MgCl$_2$, MnCl$_2$, CuSO$_4$, ZnSO$_4$, FeCl$_3$, CsCl$_2$, ethylenediamine tetraacetic acid (EDTA), 2-mercaptoethanol (2-ME), dithiothreitol (DTT), phenylmethane sulphonyl fluoride (PMSF), acetate, and furfural, respectively.

Immobilization of recombinant 1,4-β-D-glucan glucohydrolase

Meso-structured cellular foam silica (MCF) was prepared by the hydrothermal method that has been reported elsewhere. 1.62 g (0.279 mmol) of Pluronic P123 (triblock copolymer, Chemical formula: HO(CH$_2$CH$_2$O)$_2$O(CH$_2$CH(CH$_3$)O)$_7$O(CH$_2$CH$_2$O)$_2$OH, Molecular weight: 5,800) was transferred into 100 ml volume of polypropylene bottle and dissolved in a mixture of 33.33 g (1,852 mmol) of de-ionized water containing 0.8 g (13.32 mmol) of acetic acid. After clearly dissolved, the solution was heated up to 60°C and kept it for 1 hr in oil bath. Another solution was prepared with 2.67 g (11 mmol) of sodium silicate as a silica source in 33.33 g of water. The molar composition of the final solution was Na_2SiO_3: P123: H_2O: acetic acid = 1: 0.025: 336.4: 1.21. The latter solution was dropped into the solution, then aged at 60°C for 1h and reacted at 100°C for 12 h. After that, the bottle was cooled to room temperature.

White precipitated product was filtered with water, ethanol. The final product was obtained by heat treatment at 550°C for 6 hrs. For the amine-group grafting on MCF, as-prepared MCF sample (SBET = 780m^2/g) was degassed for 6 h at 150°C under vacuum of about 10^{-3} torr. 1.0 g of the MCF sample was suspended in 80 ml of toluene. 1.11 g (5 mmol) of (3-aminopropyl) triethoxylsilane was slowly dropped to the solution. After mixing for 5 min, the mixture was heated to 110°C and maintained at that temperature for 24 h under reflux condition. Final product of MCF-C_3-NH_2 was recovered by filtration with water and ethanol. In case of aldehyde grafting on MCF, 0.5 g amine functionalized MCF was added to 50 ml water and then 0.4 g (2 mmol) of glutaraldehyde (50 wt% in water) was injected to the mixture. After stirring at room temperature for 5 min, the mixture was heated to 60°C and maintained at that temperature for 24 h. The final product of MCF-C3-RC(=O)H was filtered with water and ethanol. Final products were aldehyde(3)-MCF, aldehyde(3)-SiO2, and aldehyde(11)-MCF. A proper degree of silylation was confirmed by FTIR spectra with Thermo Nicolet (US) 6700 FTIR spectrometer (data were not shown). Each sample was degassed for 6 h at 150°C under vacuum of about 10^{-3} torr in the degas port of the adsorption apparatus before being analyzed. Surface areas were calculated by the Brunauer-Emmett-Teller (BET) method. The pore size was calculated using Barrett-Joyner-Hatenda (BJH) model (data were not shown). Enzyme immobilization was conducted at pH 7, MCF or alkyl MCF (10 mg) was added to 300 µl buffer solution (50 mM of potassium phosphate, pH 7.0) containing 1,4-β-D-glucan glucohydrolase (0.5 mg/ml) in a 15-ml micro tube. The micro tube was shaken at 60 strokes per minute using a shaking incubator model SI-300 overnight at 4°C. The suspension was centrifuged at 4,000 rpm for 2 min and then the supernatant was taken from each micro tube. The solid (the MCF or alkyl-MCF with immobilized enzyme) was washed twice with 4 ml of buffer, centrifuged and the liquid removed. Repeated batch reaction was conducted with this enzyme immobilized on MCF using the same reaction mixture as for batch mode reaction of 1,4-β-D-glucan glucohydrolase activity assay after wash enzyme and filtering with membrane filter (ø 100 µm).

Analysis of hydrolytic products

Hydrolysis products were measured as substrates of cellobiose, cellotriose, cellotetraose, cellopentaose, and cellohexaose with 50 µg of purified 1,4-β-D-glucan glucohydrolase in 2 ml vial under constant temperature of 40°C, with a stirring rate of 200 rpm in a shaking incubator for time-based reaction. Each portion of 10 µl, cello-oligomer standard, and enzyme blank was analyzed after clean up with boiling 5 min and filtering with 0.2 µm syringe filter. Hydrolytic activity of 1,4-β-D-glucan glucohydrolase was also compared with commercialized-enzymes, Glucosidase (Lucigen), Almonds (Sigma), and Celluclast 1.5L (Novozyme).

Results and Discussion

Identification and expression of 1,4-β-D-glucan glucohydrolase in *E. coli*

Based on the screening of the gene library clones for their ability to exhibit the best 1,4-β-D-glucan glucohydrolase activity toward substrate 4-nitrophenyl-p-n-glucopyranoside (pNPG), the most glucosidase-active clone was selected and analyzed Open Reading

Frames (ORF) from the sequence data. Seven putative 1,4-β-D-glucan glucohydrolase-encoding genes (ORF 1-7) were detected by sequence similarity search with using the ORF Finder at the National Center for Biotechnology Information (NCBI) from the BLAST software (Table 1). All seven sets of genes were cloned into pIVEX GST fusion vector and transformed to *E. coli* BL21 (Roche Applied Science). Among seven genes a bacterial recombinant carrying ORF1 was selected as the most active plastid by measuring the amount of 4-nitrophenol (pNP) released in the glucohydrolase activity assay by absorbance at 400 nm using a pNP standard curve and designated as 1,4-β-D-glucan glucohydrolase. The gene sequence was submitted to GenBank, and assigned as Accession Number JF573391.

Purification and characterization of recombinant 1,4-β-D-glucan glucohydrolase

The final preparation of 1,4-β-D-glucan glucohydrolase gave a major single band on SDS-PAGE (Figure 1), with a molecular weight (MW) of 79 kDa, which appeared between the 66 and 97.4 kDa MW marker (Figure 1, lane 2). Also, the GST-fused *Ggh* protein of MW 105 kDa was observed between the 97.4 and 116 kDa MW markers (Figure 1, lane 1).

The optimal pH, showing maximal activity, was measured at a pH of 6.0, which was considered 100% activity, and retaining about 60-70% of its activity at a pH range of 5-7 under citrate buffer solution and potassium phosphate buffer solution, respectively (Figure 2A). The optimal temperature for this enzyme activity appeared to be 40°C (Figure 2B). This enzyme was stable for 40 min at 40°C; however, the activity was sharply decreased at 50°C after incubation for 10 minutes (Figure 3A). Most salts, such as NaCl, LiCl, KCl, NH_4Cl, $CaCl_2$, $MgCl_2$, $MnCl_2$, $FeCl_2$, and $CsCl_2$, did not significantly change the enzyme activity at 1 mM, however, $ZnSO_4$ inhibited 60% of the control (Figure 3B). Kinetic analysis of this enzyme with 4-nitrophenyl-β-D-glucopyranoside (pNPG) was performed at 40°C in pH 6.0. The Km value of 0.893 mg pNPG/ml and $Vmax$ of 33.333 was estimated by means of Hanes-Woolf equation (data were not shown) Hans-Woolf analysis of this enzyme kinetics with 4-nitrophenyl-p-n-glucopyranoside (pNPG) as substrate.

The predicted active sites of this enzyme through the 3D-strcture which is obtained from the hierarchical protein structure modeling approach [8], based on secondary-structure enhanced Profile-Profile threading Alignment (PPA) and the iterative implementation of the Threading ASSEmbly Refinement (TASSER) program shows binding site residues for cellulose and glucose of D88, F132, R146, K186, H187, R198, M230, F233, D265, W266, S405, M468 residues (Figure 4).

Enzyme immobilization

Many enzymes secreted by microorganisms are available on a large scale and there is no effect on their cost if they are used only once in a process. In addition, many more enzymes are such that they affect the cost and could not be economical if not reused. Therefore, reuse of enzymes led to the development of immobilization techniques. It involves the conversion of water soluble enzyme protein into a solid form of catalyst by several methods.

This enzyme was immobilized to aldehyde(3)-MCF, aldehyde(3)-SiO_2, and aldehyde(11)-functionalized mesostructured cellular foam with mesopores. Activity of this enzymes immobilized on the aldehyde(3)-MCF and aldehyde(3)-SiO_2 were maintained until recycle number of 3 (R3), however, decreased after R4 with same pattern regardless of immobilizing temperature at 4°C. After recycle number of 10, this enzyme activity was decreased to 30-40% of the beginning

ORF clone ID	Length		Blast P result	Identities (%)	Ref
	bp	aa			
PGD4 β4-ORF1	2160	719	glycoside hydrolase family 3 domain protein [*Paenibacillus* vortex V453] (721aa)	68	ZP_07902992.1
PGD4 β4-ORF2	2616	871	hypothetical protein GYMC10_5946 [*Geobacillus* sp. Y412MC10] (1125aa)	64	YP_003245958.1
PGD4 β4-ORF3	1476	491	putative beta-glucosidase [*Mycobacterium tuberculosis* K85] (296aa)	49	ZP_05770822.1
PGD4 β4-ORF4	501	166	RagB, SusD and hypothetical protein [*Bacteroides ovatus* SD CMC 3f] (578aa)	40	ZP_06616036.1
PGD4 β4-ORF5	1233	410	hypothetical protein PBAL39_16576 [*Pedobacter* sp. BAL39] (607aa)	42	ZP_01885563.1
PGD4 β4-ORF6	762	253	set domain containing protein [*Grosmannia clavigera* kw1407] (1619aa)	34	EFW99026.1
PGD4 β4-ORF7	678	225	ABC-type sugar transport system, periplasmic component [*Roseburia intestinalis* XB6B4] (580aa)	47	CBL11917.1

Table 1: BlastP analysis results and schematic representation of each ORFs for β-glucosidase production clone in GenBank

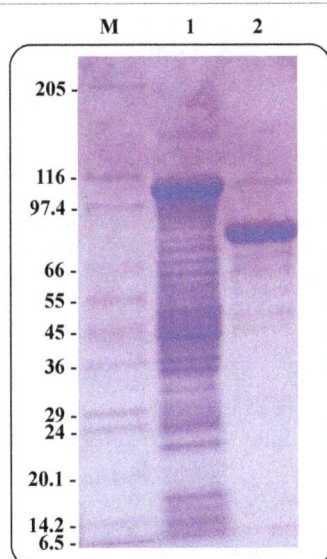

Figure 1: SDS PAGE analysis of the purified 1,4-β-D-glucan glucohydrolase (MW 79 kDa, lane 2) from cell lysate (lane 1) over-expressed with GST-fused 1,4-β-D-glucan glucohydrolase in E. coli BL21 (105 kDa, fused with GST 26 kDa and 1,4-β-D-glucan glucohydrolase 79 kDa), and molecular weight markers (lane M).

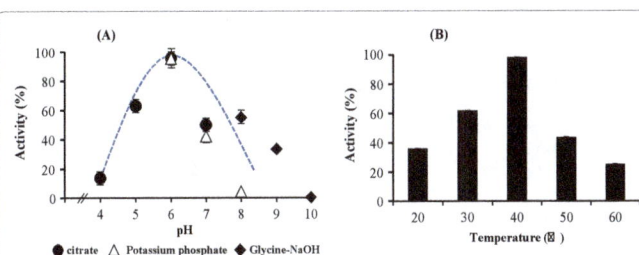

Figure 2: Optimal pH (A) and temperature (B) of 1,4-β-D-glucan glucohydrolase.

activity. Among them, about 90% of the initial enzyme activity was maintained even after 10 consecutive recycles by the aldehyde(11)-functionalized MCF-immobilized enzyme (Figure 5). The advantages of using immobilized enzymes are : (i) reuse (ii) continuous use (iii) less labor intensive (iv) saving in capital cost (v) minimum reaction time (vi) less chance of contamination in products, (vii) more stability (viii) improved process control and (ix) high enzyme : substrate ratio.

Analysis of hydrolytic products

The degradation profile of cellobiose by this enzyme, monitored through the HPLC analysis, revealed that the product was glucose only with the pattern of glucose increasing and cellobiose decreasing during the reaction time (Figure 6).

However, this enzyme might hydrolysis substrates longer than cellobiose, such as cellotriose, cellotetraose, cellopentaose, and cellohexaose, by this enzyme was monitored through the HPLC analysis, which revealed that the major hydrolysis product was glucose with smaller amounts of cellobiose. The amount of glucose was continuously increased during the entire reaction time and from the all substrates. At the beginning of reaction, substrate was hydrolyzed to glucose by the only this enzyme without cellobiohydrolase or exocellulase. This enzyme released detectable levels of glucose within 10 min, regardless of the oligosaccharide substrates. This enzyme was able to release glucose units progressively from each substrate. When incubated with cellotriose, both glucose and cellobiose were released within 10 min. As the incubation progressed, the cellobiose was hydrolyzed subsequently to glucose. The cellotriose was subsequently hydrolyzed to cellobiose, and the cellobiose to glucose, so that, after 360 min, virtually 82.0, 82.0, and 84.1% of the cellotetraose, cellopentaose, and cellohexaose had been reduced to glucose as the same manner, respectively. HPLC result shows that increasing in glucose and decreasing in all of these substrates via intermediates (Table 2).

Traditionally, three enzyme classes must be present in the cellulose hydrolysis system in order to produce glucose. According to the standard endo-, exo-synergy model, EG act randomly along the chain, producing new attack sites for the CBH act as exo-enzymes, liberating cellobiose as their main product; and β-glucosidases converts the intermediate cellobiose into two glucose molecules by the hydrolysis of β-glucosidic linkages [1-3]. The purified enzyme in this study showed not only successive hydrolysis functions of cellobiose but also soluble small cellooligomers as substrates. As the enzyme described here had a higher affinity for cellodextrin substrates longer than cellobiose than for cellobiose, it is suggested that it is a 1,4-β-D-glucan glucohydrolase (EC 3.2.1.74) [9-11].

When compared with the commercialized-enzymes and this enzyme, after 20 min reaction time, the hydrolytic pattern to all substrates was almost same to the Celluclast 1.5L (Novozyme) with about 1/3 strength, however hydrolytic pattern of Glucosidase (Lucigen) and Almonds (Sigma) was significantly decreased with substrates increasing number of glucose polymer from cellotriose to cellohexaose (Figure 7). The average activity of this enzyme for total substrate shows 1/2 of Glucosidase (Lucigen), 2 times of Almonds

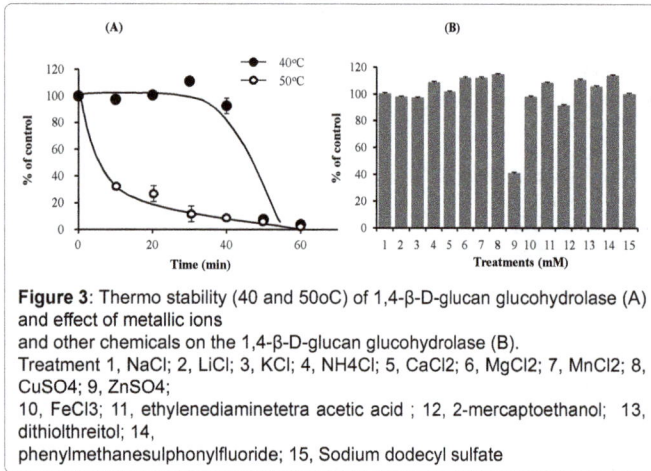

Figure 3: Thermo stability (40 and 50oC) of 1,4-β-D-glucan glucohydrolase (A) and effect of metallic ions
and other chemicals on the 1,4-β-D-glucan glucohydrolase (B).
Treatment 1, NaCl; 2, LiCl; 3, KCl; 4, NH4Cl; 5, CaCl2; 6, MgCl2; 7, MnCl2; 8, CuSO4; 9, ZnSO4;
10, FeCl3; 11, ethylenediaminetetra acetic acid ; 12, 2-mercaptoethanol; 13, dithiolthreitol; 14,
phenylmethanesulphonylfluoride; 15, Sodium dodecyl sulfate

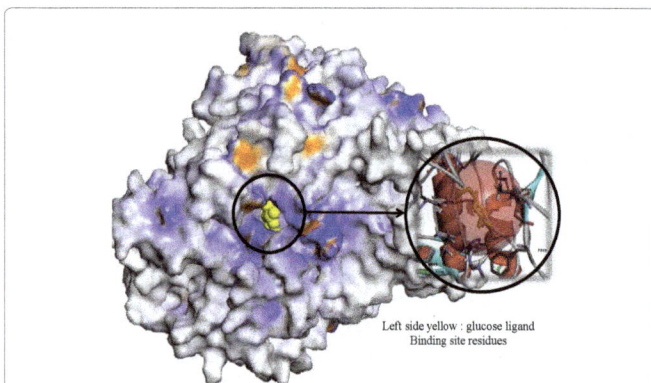

Figure 4: 1,4-β-D-glucan glucohydrolase homology model structure and active sites. Left side yellow : glucose ligand, Binding site residues : D88, F132, R146, K186, H187, R198, M230, F233, D265, W266, S405, M468. Nucleotide and deduced amino acid sequences were analyzed with CLC Free Workbench, Ver. 3.2.1 (CLC bio A/S, www.clcbio.com). Related sequences were obtained from database searches (SwissPort and GenBank). The biomolecular 3D structure of 1,4-β-D-glucan glucohydrolase was predicted with a deduced amino acid sequence as a homology model structure [8].

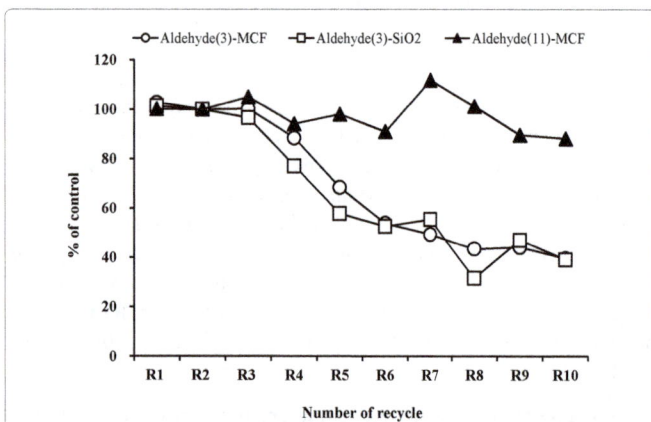

Figure 5: Enzyme activity of 1,4-β-D-glucan glucohydrolase according to the recycle number after surface functionalized MCF-immobilization.

(Sigma), and 1/3 of Cellulclast 1.5L (Novozyme).

Although classified as a glucohydrolase, the enzyme purified from *G. natalis* was similar in size to β-glucosidases found in other arthropods, such as the larvae of the moth *Erinnyis ello* [12] and the rice weevil *Sitophilus oryzae* [13]. The β-glucosidase from *S. oryzae*, like the glucohydrolase from *G. natalis*, was active towards both cellobiose and 4-methylumbelliferyl β-D-glucopyranoside. It is therefore possible that the enzyme found in *S. oryzae* is similar to that from *G. natalis* and, therefore, might also be a glucohydrolase. Glucohydrolase activity has been noted in other insects, such as the larvae of the cardinal beetle *Pyrochroa coccinea* [14] and the termite *Coptotermes lacteus* [15], but was typically considered to be a minor component of cellulose digestion [16]. These enzymes have therefore largely been ignored when discussing cellulolytic systems in invertebrates.

The mechanical disruption of cellulose during pretreatment increases the efficiency of enzymatic breakdown by increasing the surface area of cellulose exposed to the enzymes within the digestive juice and might further reduce the need for a cellobiohydrolase by improving access to glucose residues normally buried within crystalline regions of cellulose fibers. This is in contrast to fungal and bacterial systems, which lack a mechanical mill, relying instead on the presence of a cellobiohydrolase and its synergistic activity with the endo-β-1,4-glucanase to digest crystalline cellulose efficiently [17].

Cellulose solubility decreases drastically with increasing DP due to intermolecular hydrogen bonds. Cellodextrins with DP from 2-6 are soluble in water [18,19], while cellodextrins from 7-13 or longer are somewhat soluble in hot water [8]. A glucan of DP = 30 already represents the polymer "cellulose" in its structure and properties [19]. The DP of cellulosic substrates varies greatly, from 100 to 15,000, depending on substrate origin and preparation.

The change in DP over the course of hydrolysis for cellulosic substrates is determined by the relative proportion of exo-and endo-acting activities and cellulose properties. Exoglucanases act on chain ends, and thus decrease DP only incrementally. Endoglucanases act on interior portions of the chain and thus rapidly decrease DP [20,21]. Exoglucanase has been found to have a marked preference for substrates with lower DP [5], as would be expected given the greater availability of chain ends with decreasing DP. It is well known that endoglucanase activity leads to an increase in chain ends without resulting in appreciable solubilization [22]. We know of no indication in the literature that the rate of chain end creation by endoglucanase is impacted by substrate DP.

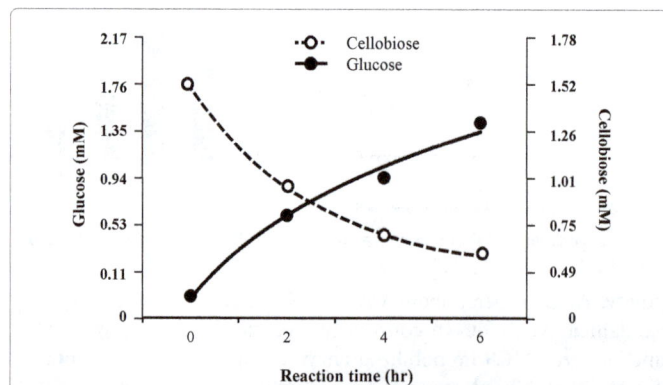

Figure 6: HPLC analysis results of the substrate cellobiose and product glucose by 1,4-β-D-glucan glucohydrolase with the time course of the reaction. HPLC (Younglin Instrument Acme 9000, Korea) analysis was conducted under conditions of Shodex Asahipak column NH2P-50 4E, fluent of water/CH3CN (1:3), flow rate of 1 ml/min., detector of RI (refractive index), column temperature of 30oC.

Reaction (min)	Cellobiose		Cellotriose		Cellotetraose		Cellopentose		Cellohexose	
	Glu (mmol)	Con (%)	Glu (mmol)	Con (%)	Glu (mmol)	Con (%)	Glu (mmol)	Con (%)	Glu (mmol)	Con (%)
0	0	0	0	0	0	0	0	0	0	0
60	0.61	44.4	0.92	55.8	1.06	60.2	1.20	50.7	1.25	59.0
120	0.74	57.6	1.04	63.9	1.20	68.0	1.50	67.0	1.89	78.2
180	0.82	64.7	1.09	70.2	1.38	75.1	1.63	75.0	1.99	82.9
360	0.98	79.8	1.22	79.6	1.50	82.0	1.79	82.0	2.13	84.1

˙Glu: Glucose; Con: Conversion

Table 2: Enzymatic glucose producing efficiency of a new 1,4-β-D-glucan glucohydrolase PGDβ4 according to the different glucose number of substrates

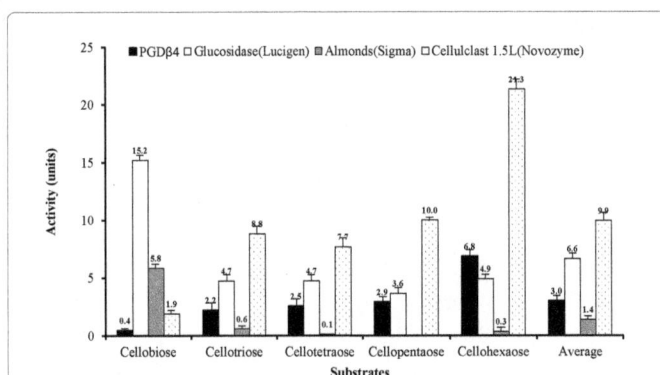

Figure 7: Comparison of hydrolytic activity with 1,4-β-D-glucan glucohydrolase and commercialized-enzymes using different number of glucose polymer substrates.

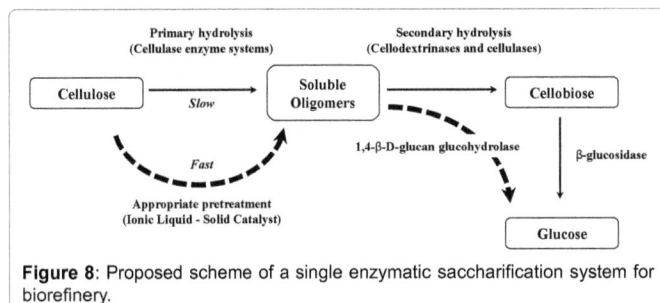

Figure 8: Proposed scheme of a single enzymatic saccharification system for biorefinery.

At this point, if the depolymerization of cellulose proceeds progressively by chemical hydrolysis instead of endo-β-1,4-glucanases, resulting in the formation of soluble oligosaccharides, cellulose fragments ideally suited for further processing by glucohydrolase single enzymatic hydrolysis can easily be isolated.

Conclusion

We identified a new 1,4-β-D-glucan glucohydrolase gene from common bacteria *Paenibacillus* sp. non-extremophile [6], which was cloned and expressed in *Escherichia coli* and efficiently purified using affinity column chromatography, which showed the successive hydrolysis functions of cello-oligomers virtually reduced to glucose. It is suggested that the single enzymatic saccharification system for bio refinery using this enzyme might be expected, if it will be connected to appropriate pretreatment technology, instead of the traditional concept using a three-enzyme system for cellulose hydrolysis (Figure 8). We think this paper will be of interest to those who involved in biorefinery process as well as in enzymatic hydrolysis.

Acknowledgement

This study was supported by a grant 10035574 funded by the Ministry of Knowledge Economy, Republic of Korea.

References

1. Wilson DB (2009) Cellulases and biofuels. Curr Opin Biotechnol 20: 295-299.

2. Konur O (2012) The scientometric evaluation of the research on the production of bioenergy from biomass. Biomass and Bioenergy 47: 504-515.

3. Zhang YH, Lynd LR (2004) Toward an aggregated understanding of enzymatic hydrolysis of cellulose: noncomplexed cellulase systems. Biotechnol Bioeng 88: 797-824.

4. Mach RL, Seiboth B, Myasnikov A, Gonzalez R, Strauss J, et al. (1995) The bgl1 gene of *Trichoderma reesei* QM9414 encodes an extracellular, cellulose-inducible β-glucosidase involved in cellulase induction by sophorose. Mol Microbiol 16: 687-697.

5. Rinaldi R, Palkovits R, Schüth F (2008) Depolymerization of cellulose using solid catalysts in ionic liquids. Angew Chem Int Ed Engl 47: 8047-8050.

6. McCarthy JK, Uzelac A, Davis DF, Eveleigh DE (2004) Improved catalytic efficiency and active site modification of 1,4-beta-D-glucan glucohydrolase A from *Thermotoga neapolitana* by directed evolution. J Biol Chem 279: 11495-11502.

7. Zhang YH, Lynd LR (2003) Cellodextrin preparation by mixed-acid hydrolysis and chromatographic separation. Anal Biochem 322: 225-232.

8. Zhang Y (2007) Template-based modeling and free modeling by I-TASSER in CASP7. Proteins 69 Suppl 8: 108-117.

9. Henrissat B, Davies G (1997) Structural and sequence-based classification of glycoside hydrolases. Curr Opin Struct Biol 7: 637-644.

10. Yernool DA, McCarthy JK, Eveleigh DE, Bok JD (2000) Cloning and characterization of the glucooligosaccharide catabolic pathway β-glucan glucohydrolase and cellobiose phosphorylase in the marine hyperthermophile *Thermotoganea politana*. J Bacteriol 182: 5172-5179.

11. Allardyce BJ, Linton SM, Saborowski R (2010) The last piece in the cellulase puzzle: the characterisation of beta-glucosidase from the herbivorous gecarcinid land crab *Gecarcoidea natalis*. J Exp Biol 213: 2950-2957.

12. Santos CD, Terra WR (1985) Physical properties, substrate specificities and a probable mechanism for a β-D-glucosidase (cellobiase) from midgut cells of the cassava hornworm (*Erinnyis ello*). Biochim Biophys Acta 831: 179-185.

13. Baker JE, Woo SM (1992) β-glucosidases in the rice weevil, *Sitophilus oryzae*: purification, properties, and activity levels in wheat-and legume-feeding strains. Insect Biochem Mol Biol 22: 495-504.

14. Chararas C, Eberhard R, Courtois JE, Petek F (1983) Purification of three cellulases from the xylophageous larvae of *Ergates faber* (Coleoptera: Cerambycidae). Insect Biochem 3: 213-218.

15. Hogan ME, Schulz MW, Slaytor M, Czolij RT, O'Brien RW (1988) Components of termite and protozoal cellulases from the lower termite, *Coptotermes lacteus* Froggatt. Insect Biochem 18: 45-51.

16. Scrivener AM, Slaytor M (1994) Properties of the endogenous cellulase from *Panesthia cribrata* Saussure and purification of major endo-β-1,4-glucanase components. Insect Biochem Mol Biol 24: 223-231.

17. Watanabe H, Tokuda G (2010) Cellulolytic systems in insects. Annu Rev Entomol 55: 609-632.

18. Pereira AN, Mobedshahi M, Ladisch MR (1988) Preparation of cello-dextrins. Methods Enzymol 160: 26-43.

19. Klemm D, Philipp B, Heinze T, Heinze U, Wagenknecht W (1998) Comprehensive cellulose chemistry. I. Fundamentals and analytical methods. Weinheim. Wiley-VCH.

20. Kleman-Leyer KM, Siika-Aho M, Teeri TT, Kirk TK (1996) The Cellulases Endoglucanase I and Cellobiohydrolase II of *Trichoderma reesei* Act Synergistically To Solubilize Native Cotton Cellulose but Not To Decrease Its Molecular Size. Appl Environ Microbiol 62: 2883-2887.

21. Srisodsuk M, Kleman-Leyer K, Keränen S, Kirk TK, Teeri TT (1998) Modes of action on cotton and bacterial cellulose of a homologous endoglucanase-exoglucanase pair from *Trichoderma reesei*. Eur J Biochem 251: 885-892.

22. Seo YK, Suryanarayana I, Hwang YK, Shin N, Ahn DC, et al. (2009) Swift Synthesis of Hierarchically Ordered Mesocellular Mesoporous Silica by Microwave-Assisted Hydrothermal Method. J of Nanosci Nanotech 8: 3995-3998.

31

Presence of Low Concentrations of Acetic Acid Improves Fermentations using *Saccharomyces cerevisiae*

Greetham D*

School of Biosciences, University of Nottingham, Loughborough, Leics, LE12 5RD, UK

Abstract

Fermentation of sugars released from lignocellulosic biomass (LCMs) is potentially a sustainable option for the production of bioethanol. LCMs release fermentable hexose sugars and the currently non-fermentable pentose sugars; ethanol yield from lignocellulosic residues is dependent on the efficient conversion of available sugars to ethanol. One of the challenges facing the commercial application for the conversion of lignocellulosic material to ethanol is the presence of inhibitors released by the breakdown of plant cell walls.

Presence of acetic acid is an inevitable side-effect for the release of fermentable sugars from the deconstruction of plant cell walls, increasing temperatures used for the pre-treatment process releases acetic acid from the lignin component of the plant cell wall. Using phenotypic microarray analysis revealed that low concentrations (20 mM) acetic acid augmented metabolic output in yeast for an initial period, however, assays at higher concentrations (>50 mM) reduced metabolic output.

Fermentations in the presence of acetic acid where characterized by an improved fermentation efficiency in assays containing 20 mM acetic acid compared with control conditions, however, efficiency was reduced in assays using 50 mM acetic acid. Yeast cells in the presence of 20 mM acetic acid produced less glycerol, and produced more ATP when compared with control conditions or in the presence of 50 mM acetic acid.

Keywords: Acetic acid; Yeast; Microarrays; Fermentation; Glycerol; ATP

Introduction

Short-chain weak organic acids are potent inhibitors of microbial growth and are widely applied as preservatives in food and beverages. Short-chain organic acids also occur as inhibitory compounds in industrial fermentation processes, for example the detrimental effect of acetic acid and on the production of bioethanol from lignocellulosic material in a fermentation using *Saccharomyces cerevisiae* [1].

Acetic acid is produced by the deacetylation of xylan during pre-treatment [2] as well as a by-product of bacterial contamination and a minor product of yeast fermentation [3]. The toxicity of acetic acid and other weak organic acids is pH dependent, as it is the un-dissociated form which passively enters the yeast cell [4]. Un-dissociated acetic acid that diffuses through the cell membrane will become dissociated intracellularly [5], the degree of dissociation will depend on the cytosolic pH. In order to maintain a constant intracellular pH, protons are transported across the cell membrane through the activity of ATPases [5]. This results in an increase in ATP consumption and addition of acetate to a media has been shown to lower biomass produced [6].

Acetic acid also stimulates Programmed Cell Death (PCD) in yeast cells through a mitochondria specific caspase cascade [7]. This appears to be separate from weak acids causing anion accumulation due to acidification of the cytoplasm through passive diffusion of acetic acid through the cell membranes.

Saccharomyces cerevisiae is currently used for the production of bioethanol; Pre-treatment of lignocellulose to release constituent sugars results in the formation of aromatic and acidic compounds such as acetic acid, formic acid, furfural, Hydroxy-Methyl Furfural (HMF), levulinic acid and vanillin [8] that are detrimental to the growth of *S. cerevisiae*. In addition, fermentations carried out within bioreactors generate additional difficulties, such as osmotic stress due to high sugar levels, elevated heat and increasing ethanol concentrations [9-

11]. Acetic acid is ubiquitous in hydrolysates where hemicellulose and components of the plant cell wall have acetyl groups which can undergo hydrolysis [12-14]. The precise mode of action for many of the inhibitors has yet to be fully determined [15]. Weak acid stress is induced when acetic, formic or levulinic acid is liberated from LCMs, they inhibit yeast fermentations reducing both growth and ethanol production.

Weak acids effect fermentation profiles where at low concentrations weak acids improve fermentation rates with increased ethanol yield, weak acids at low concentrations are believed to stimulate ATP production [16], and under anaerobic conditions ethanol is produced [17]. However, at high concentrations the beneficial stimulation of ATP production is overtaken by the acid stimulating the cell to increase ATPase activity.

Strain selection for the production of ethanol from LCM derived sugars has traditionally involved the use of several assays based on cell growth and division, maintenance of viability in stress tests and fermentation analyses [18,19]. Whilst very useful, these approaches are time consuming and interpretations can be subjective [20]. The Phenotypic Microarray (PM) developed by Bochner and colleagues, provides an analogous two-dimensional array technology for

---

*****Corresponding author:** Greetham D, School of Biosciences, University of Nottingham, Loughborough, Leics, LE12 5RD, UK

simultaneous analysis of live yeast cell populations in a 96-well micro titre plate format [21,22]. Use of be-spoke PM plates have been described previously [1], and how metabolic output relates to growth and production of ethanol [1,23]. In this present work, the effect of acetic acid on *Saccharomyces cerevisiae* NCYC2592 on metabolic output and conversion of sugar into ethanol has been assessed and correlated with acetic acid concentrations in the medium.

Material and Methods

Yeast strain and growth conditions

S. cerevisiae NCYC 2592 (www.ncyc.co.uk) was maintained on YPD containing agar containing 10 g/L yeast extract, 20 g/L peptone, 20 g/L glucose, and 20 g/L agar.

Phenotypic microarray analysis

Biolog growth medium was prepared using 0.67% (w/V) minimal medium (YNB- Yeast Nitrogen Base) supplemented with mixture of 6% (w/v) glucose, 2.6 µL of yeast nutrient supplement mixture (NS×48- 24 mM Adenine-HCl, 4.8 mM L-histidine HCl monohydrate, 48 mM L-leucine, 24 mM L-lysine-HCl, 12 mM L-methionine, 12 mM L-tryptophan and 14.4 mM uracil), and 0.2 µl of dye D (Biolog, USA). Final volume was made up to 30 µL using Reverse Osmosis (RO) sterile distilled water and aliquoted into individual wells with varying concentrations of acetic acid or levulinic acid (both prepared as 1M stock solutions) as required.

The inhibitory effects of pH was measured via Biolog by adjusting media containing 6% glucose, 2.8 % YNB to pH 5 with phosphoric acid, acetic acid was then added and the pH again adjusted using either phosphoric acid or NaOH.

Strains were prepared for inoculation and prepared for the PM assay plates as described previously [1], the plates were then placed in the OmniLog reader and incubated for 96 h at 30°C.

The OmniLog reader reads the plates at 15 min intervals, converting the pixel density in each well to a signal value reflecting cell growth and dye conversion. After completion of the run, the signal data was compiled and exported from the Biolog software and compiled using Microsoft® Excel. In all cases, a minimum of three replicate PM assay per plate were conducted, and the average of the signal values was used. To ensure that dye reduction was not occurring in the absence of growth, all PM plates were carefully examined following each run.

Effect of acetic acid on logarithmic metabolic output was determined by calculating the time required to double maximal output when cells were in logarithmic phase of metabolic activity. Exit from lag phase was determined by determining when metabolic activity was above 10 redox signal intensity units as wells containing media but no cells can produce a metabolic signal up to 10. Data representative of triplicate wells run on the same plate.

Budding index

10 µL of yeast cells were spotted onto 1 mL of YPD agar on a microscope slide containing acetic acid and single or clustered cells counted at x 20 magnification. Viable cells will start budding and become clumps of cells over time; dead cells remain as single cells. All slides were kept at 30°C for 42 hours with cell counts occurring after 18 and 42 hours respectively. All experiments were done in triplicate.

Confirmation of phenotypic microarray results using mini fermentation vessels

Fermentations were conducted in 180 mL mini-Fermentation Vessels (FV). Cryopreserved yeast colonies were streaked onto YPD plates and incubated at 30°C for 48 hrs. Colonies of *S. cerevisiae* NCYC2592 were used to inoculate 20 mL of YPD broth and incubated in an orbital shaker at 30°C for 24 hrs. These were then transferred to 200 mL of YPD and grown for 48 hrs in a 500 mL conical flask shaking at 30°C. Cells were harvested and washed three times with sterile RO water and then re-suspended in 5 mL of sterile water. For control conditions, 1.5×10^7 cells.mL^{-1} were inoculated in 99.6 mL of medium containing 8% glucose, 2% peptone, 1% yeast extract with 0.4 mL RO water. For stress conditions, 1.5×10^7 cells.mL^{-1} were incubated in 99.6 mL of medium containing 8% glucose, 2% peptone, 1% yeast extract with 0-50 mM acetic acid. Volumes of media were adjusted to account for the addition of the inhibitory compounds (0-400 µL) to ensure that all fermentations began with the same carbon load.

Anaerobic conditions were prepared using a sealed butyl plug (Fisher, Loughborough, UK) and aluminium caps (Fisher Scientific). A hypodermic needle attached with a Bunsen valve was purged through rubber septum to facilitate the release of CO_2. All experiments were performed in triplicate and weight loss was measured at each time point. Mini-fermentations were conducted at 30°C, with orbital shaking at 200 rpm.

Determination of glucose, acetic acid, glycerol and ethanol concentrations from fermentation experiments via HPLC

Glucose, acetic acid, glycerol and ethanol were quantified by HPLC. The HPLC system included a Jasco AS-2055 Intelligent auto sampler (Jasco, Tokyo, Japan) and a Jasco PU-1580 Intelligent pump (Jasco). The chromatographic separation was performed on a Rezex ROA H+ organic acid column, 5 µm, 7.8mm × 300 mm, (Phenomenex, Macclesfield, UK) at ambient temperature. The mobile phase was 0.005N H_2SO_4 with a flow rate of 0.5 mL/min. For detection a Jasco RI-2031 Intelligent refractive index detector (Jasco) was employed. Data acquisition was via the Azur software (version 4.6.0.0, Datalys, St Martin D'heres, France) and concentrations were determined by peak area comparison with injections of authentic standards. The injected volume was 10 µl and analysis was completed in 28 minutes. All chemicals used were analytical grade (>95% purity, Sigma-Aldrich, UK).

ATP concentration

Determination of ATP was using a ATP assay kit (ab8335, Abcam, UK), yeast cell pellets (106 cells/mL) taken during the fermentation were then broken using a MagNA lyser (Roche Applied Science, UK), cells were subjected to vigorous shaking/vortexing via the MagNa lyser for 1 min and repeated five times at a speed of 7,000 rpm while temperature was kept as low as practicable. ATP concentrations were determined using a using a Tecan (Mannedorf, Switzerland) Infinite M200 Pro plate reader at 570 nm.

Statistical analysis

Data derived from phenotypic microarrays was analysed for analysis of variance (ANOVA) using ezANOVA (http://www.cabiatl.com/mricro/ezanova), a free for use online statistical program with statistical significance signified by use of $*$, $* = 0.05\%$ significant, $** = 0.01\%$ significant and $***$ 0.001% significant.

Results

Presence of acetic acid influences metabolic output in S. cerevisiae

Presence of acetic acid (0-100 mM) on metabolic output was assessed with a reduction in metabolic output observed at 75 mM acetic acid and no metabolic output observed in an assay containing 100 mM acetic acid (Figure 1A). Low concentrations of acetic acid (<20 mM) had little or no effect on metabolic output when compared with the control (control defined as absence of acetic acid); however, assays containing low concentrations of acetic acid outperformed the unstressed control for the first 8 hours of the experiment (p=0.034) (Figures 1B and 1C). Augmentation of metabolic output observed for acetic acid was not observed for other weak acids, metabolic output in the presence of levulinic acid failed to show an early augmentation at low concentrations of levulinic acid (p=0.54) (Figure 1D). pH for all experiments was adjusted to pH 5 following the addition of acetic or levulinic acid with the addition of NaOH or phosphoric acid as appropriate.

Using metabolic output data and determining when the cell exits lag phase and enters logarithmic phase of metabolic output allows us to investigate how long a cell takes to overcome acetic acid stress. Plotting acetic acid concentration against entry into logarithmic metabolic output revealed that increasing concentrations of acetic acid increased the length of time the yeast takes to enter into log phase of metabolic output (Figure 2A). Presence of acetic acid also reduced maximal rates of metabolic output when compared with unstressed controls (Figure 2B).

Presence of acetic acid slows conversion of metabolic output into cell mass

Measuring yeast growth in the presence of acetic acid shows a correlation between concentration of acetic acid and growth. Presence of low concentrations of acetic acid (10-25 mM) had little or no impact on growth, indeed growth in the presence of 25 mM acetic acid was improved with unstressed controls (Figure 3A). Increasing acetic acid to 50 mM slowed growth, characterised by a longer lag phase and a delay of entry into the exponential growth phase (Figure 3A). Comparing growth (OD600) and metabolic output (redox signal intensity) revealed that under control conditions there was no difference between rates of growth and metabolic output (Figure 3B), however, in the presence of 50 mM acetic acid there is a delay between metabolic output and cellular growth (Figure 3C).

An assessment of the number of budding cells after 18 and 42 hours exposure (these time points were chosen because after 18 hours yeast are principally in logarithmic phase of metabolic output but after 42 hours have reached stationary phase of metabolic output (Figure 1A)) observed a reduction in budding cells in the presence of acetic acid after 18 hours (10-50 mM) when compared with control conditions. However, after 42 hours in the presence of 10-30 mM the number of budding index had returned to unstressed control conditions (Figure 2F).

There was no increase in acetic acid toxicity at pH 4 or pH 7

Toxicity of acetic acid is closely related to the pH of the media as the concentrations of un-dissociated form increases as the pH of the media decreases [4], initial studies were performed at pH 5, however we also assayed for the toxicity of low concentrations (10-25 mM) at pH 4 and pH 7. Assays revealed that there was no inhibition caused by the presence of acetic acid at either pHs when compared with controls

in which the pH had been set using phosphoric acid (Figures 4A and 4B), indicating that the presence of low concentrations of acetic acid was not inhibitory at pH 4-7 when compared with assays just looking at the effect of pH.

Presence of 20 mM acetic acid improves ethanol production during fermentation

Presence of 20 mM acetic acid on ethanol and glycerol production was assessed and compared with unstressed control conditions during fermentations. Ethanol production in the presence of acetic acid was higher (43.58 ± 0.53 g/L) than under control conditions (41.04 ± 0.46) (Table 1), the theoretical maxima for glucose to ethanol conversion is 0.51 g/L [24] with an improved fermentation efficiency 93.53% compared with 88.82% (Table 1). Assessment of glycerol revealed that there was a reduction in glycerol production in the presence of acetic acid (2.78 ± 0.03 g/L) compared with control conditions (4.19 ± 0.045 g/L) (Table 1). Fermentations in the presence of 50 mM acetic acid where characterised by a reduced ethanol production (34.45 ± 0.4 g/L), reduced fermentation efficiency 78.31% and a reduced glycerol production (1.78 ± 0.2 g/L) (Table 1).

ATP levels increased in the presence of low concentrations of acetic acid

ATP concentrations have been shown to be increased at relatively low concentrations of acetic acid; we measured ATP concentrations in the presence of acetic acid throughout fermentations. ATP concentrations under control conditions at the start of the fermentation was determined to be 0.04 ± 0.003 mM and increased for the first eight hours of the fermentation to a peak of 0.09 ± 0.004 mM before decreasing to 0.06 ± 0.001 mM for the remainder of the fermentation (Table 2). Addition of 20 mM acetic acid stimulated ATP production for the first 8 hours of the fermentation (0.15 ± 0.04 mM) subsequently there was no increase in ATP production observed for the duration of

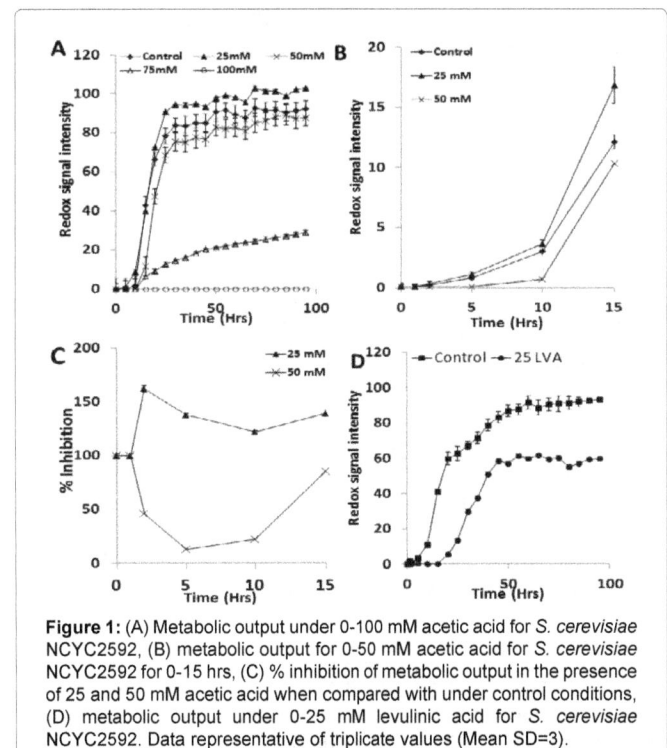

Figure 1: (A) Metabolic output under 0-100 mM acetic acid for S. cerevisiae NCYC2592, (B) metabolic output for 0-50 mM acetic acid for S. cerevisiae NCYC2592 for 0-15 hrs, (C) % inhibition of metabolic output in the presence of 25 and 50 mM acetic acid when compared with under control conditions, (D) metabolic output under 0-25 mM levulinic acid for S. cerevisiae NCYC2592. Data representative of triplicate values (Mean SD=3).

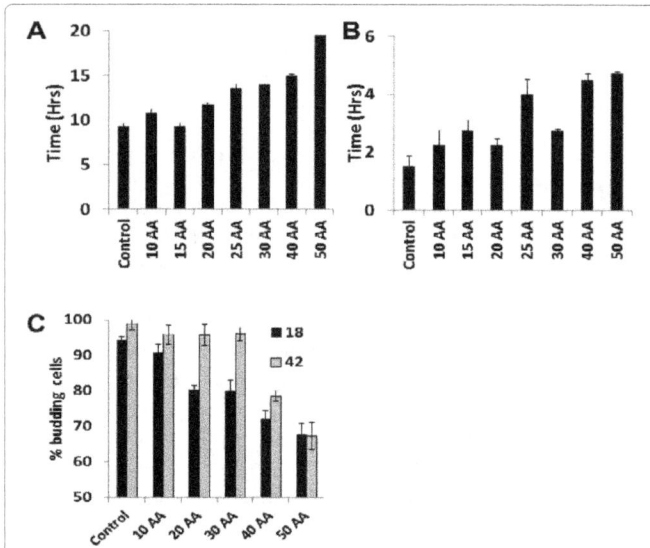

Figure 2: (A) effect of acetic acid (0-50 mM) on exit of lag phase of metabolic output, (B) effect of acetic acid (0-50 mM) on maximal metabolic output, (D) comparison of yeast growth and metabolic output under control conditions, (E) comparison of growth and metabolic output in the presence of 50 mM acetic acid and (C) budding index in the presence of acetic acid (0-50 mM) after 18 and 42 hours exposure respectively. Data representative of triplicate values (Mean SD=3).

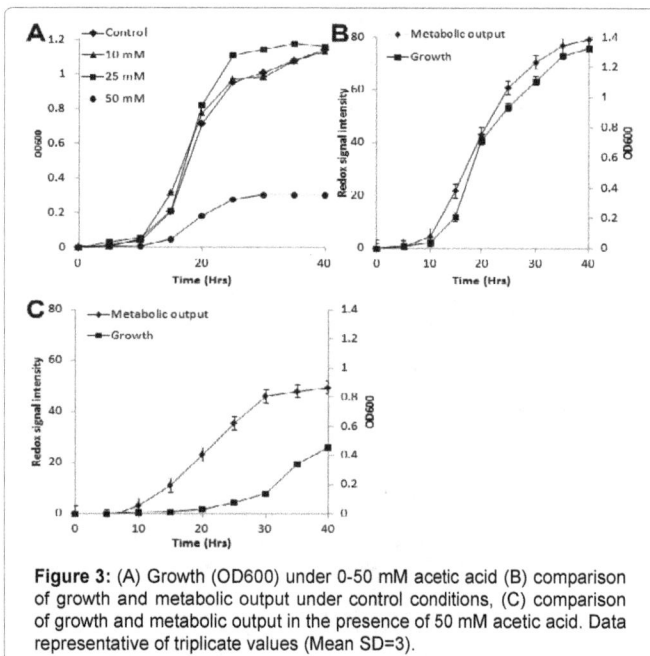

Figure 3: (A) Growth (OD600) under 0-50 mM acetic acid (B) comparison of growth and metabolic output under control conditions, (C) comparison of growth and metabolic output in the presence of 50 mM acetic acid. Data representative of triplicate values (Mean SD=3).

the fermentation (p=0.08) (Table 2). Addition of 50 mM acetic acid was characterised by no increase in ATP production for the first 10 hours of the fermentation before an increase was observed (Table 2). ATP concentrations in fermentations in the presence of 50 mM acetic acid did not increase above that observed under control conditions for any time point measured during the fermentation (p=0.91) (Table 2).

Discussion

Presence of acetic acid is an unavoidable consequence of the pre-treatment of lignocellulosic material as lignocellulosic material contains hemicellulose which reinforces the plant cell wall [25]. Regardless of

the lignocellulosic material, a structural component of hemicellulose is that some of the pyranose subunits are substituted for acetyl groups [26], upon treatment for the liberation of monomeric sugars acetic acid is generated from the degradation of acetylated sugars with concentrations present at 1-10 g/L [27].

Presence of acetic acid at higher concentrations is inhibitory, however, assays at sub-lethal concentrations (10-20 mM) improves metabolic output, rates of fermentation and ethanol production. Acetic acid remains unchanged throughout the fermentation, however, at lower concentrations there is lower glycerol production and higher ATP levels within the cell when compared with unstressed conditions. Increasing acetic acid concentrations is characterised by a slower rate of fermentation and a reduced conversion of glucose into ethanol with increased glycerol production and reduced ATP production.

Presence of acetic acid (20-80 mM) has been shown to induce Programmed Cell Death (PCD) [28], either involving mitochondria (intrinsic pathway) or a pathway involving cytosolic caspases called the extrinsic pathway [29,30]. The toxicity of acetic acid is pH dependent, acetic acid in its un-dissociated form diffuses through the cell membrane and dissociation is dependent on cytosolic pH, however, at low acetic acid concentrations (<20 mM) there was no increase in toxicity in the presence of acetic acid compared with pH adjusted assays.

Fermentations in the presence of 20 mM acetic acid were characterised by an increase in fermentation efficiency, a reduction in glycerol production and an increase in ATP production. An increase in ATP production and a reduction in glycerol production has been

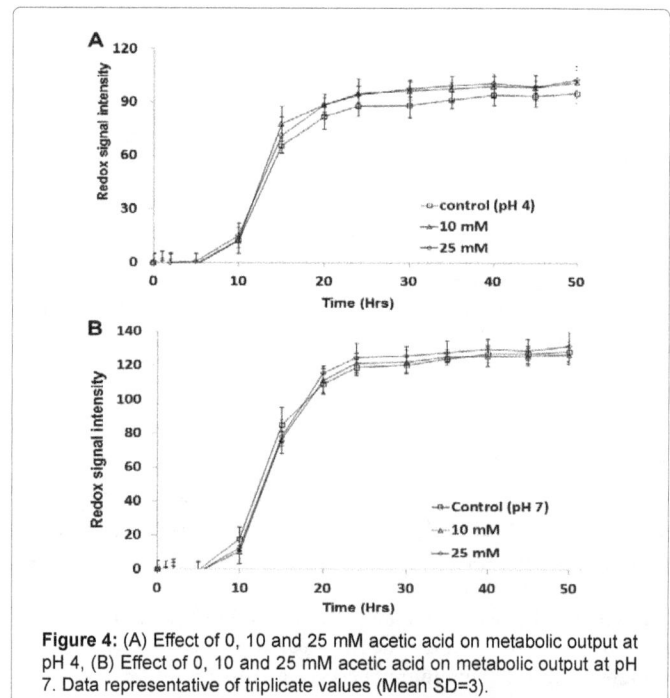

Figure 4: (A) Effect of 0, 10 and 25 mM acetic acid on metabolic output at pH 4, (B) Effect of 0, 10 and 25 mM acetic acid on metabolic output at pH 7. Data representative of triplicate values (Mean SD=3).

Acetic acid (mM)	Ethanol (g/L)	Glycerol (g/L)	Fermentation Efficiency (%)
0	41.04 ± 0.46	4.19 ± 0.045	88.82
20	43.58 ± 0.53	2.78 ± 0.03	93.53
50	34.45 ± 0.4	1.78 ± 0.2	78.31

Table 1: Ethanol, glycerol production (g/L) for *S. cerevisiae* NCYC2592 under control conditions (no acetic acid), 20 mM and 50 mM acetic acid. Fermentation efficiencies (%) were calculated using ethanol production-0.51 g/L theoretical maxima.

Time (Hrs)	ATP concentration (mM)		
	Control	20 mM acetic acid	50 mM acetic acid
0	0.04 ± 0.003	0.04 ± 0.003	0.04 ± 0.003
2	0.06 ± 0.004	0.08 ± 0.007	0.05 ± 0.001
4	0.08 ± 0.001	0.09 ±0.008	0.05 ± 0.001
6	0.09 ± 0.006	0.10 ± 0.009	0.05 ± 0.002
8	0.09 ± 0.004	0.12 ± 0.004	0.05 ± 0.004
10	0.08 ± 0.001	0.15 ± 0.001	0.07 ± 0.003
12	0.07 ± 0.005	0.15 ± 0.007	0.06 ± 0.001

Table 2: ATP concentrations for *S. cerevisiae* NCYC2592 at selected time points during fermentation in the presence of acetic acid (0-50 mM). Data representative of triplicate values (Mean SD=3).

shown previously for yeast cells under acetic acid stress [16], glycerol is one of the main by-products in an ethanol fermentation and may account for 5% of the available carbon [31], reducing glycerol has been shown to increase ethanol production in an ethanol fermentation [32]. ATP levels for yeast cells under higher concentrations of acetic acid have been shown to be reduced along with inhibition of nutrient uptake when compared with controls [33].

Results here have revealed that presence of relatively low concentrations of acetic acid improve ethanoic fermentations with concurrent reduced accumulation of glycerol and increased ATP production, however, at higher concentrations of acetic acid these effects are reversed.

Acknowledgements

The research reported here was supported (in full or in part) by the Biotechnology and Biological Sciences Research Council (BBSRC) Sustainable Bioenergy Centre (BSBEC), under the programme for 'Lignocellulosic Conversion to Ethanol' (LACE) [Grant Ref: BB/G01616X/1]. This is a large interdisciplinary programme and the views expressed in this paper are those of the authors alone, and do not necessarily reflect the views of the collaborators or the policies of the funding bodies. This project is part financed by the European Regional Development Fund project EMX05568.

References

1. Greetham D, Wimalasena T, Kerruish DW, Brindley S, Ibbett RN, et al. (2014) Development of a phenotypic assay for characterisation of ethanologenic yeast strain sensitivity to inhibitors released from lignocellulosic feedstocks. J Ind Microbiol Biotechnol 41: 931-945.

2. Palmqvist E, Hahn-Hägerdal B (2000) Fermentation of lignocellulosic hydrolysates. II: inhibitors and mechanisms of inhibition. Bioresource Technology 74: 25-33.

3. Thomas KC, Hynes SH, Ingledew WM (2002) Influence of medium buffering capacity on inhibition of Saccharomyces cerevisiae growth by acetic and lactic acids. Appl Environ Microbiol 68: 1616-1623.

4. Atkins P, Julio de Paula (2002) Chemical equilbrium. Oxford University Press, 7: 222-251.

5. Mollapour M, Shepherd A, Piper PW (2008) Novel stress responses facilitate Saccharomyces cerevisiae growth in the presence of the monocarboxylate preservatives. Yeast 25: 169-177.

6. Verduyn C, Postma E, Scheffers WA, van Dijken JP (1990) Energetics of Saccharomyces cerevisiae in anaerobic glucose-limited chemostat cultures. J Gen Microbiol 136: 405-412.

7. Madeo F, Herker E, Wissing S, Jungwirth H, Eisenberg T, et al. (2004) Apoptosis in yeast. Curr Opin Microbiol 7: 655-660.

8. Tomás-Pejó E, Oliva JM, Ballesteros M, Olsson L (2008) Comparison of SHF and SSF processes from steam-exploded wheat straw for ethanol production by xylose-fermenting and robust glucose-fermenting Saccharomyces cerevisiae strains. Biotechnol Bioeng 100: 1122-1131.

9. Casey GP, Ingledew WM (1986) Ethanol tolerance in yeasts. Crit Rev Microbiol 13: 219-280.

10. Aslankoohi E, Zhu B, Rezaei MN, Voordeckers K, De Maeyer D, et al. (2013) Dynamics of the Saccharomyces cerevisiae transcriptome during bread dough fermentation. Appl Environ Microbiol 79: 7325-7333.

11. Beltran G, Torija MJ, Novo M, Ferrer N, Poblet M, et al. (2002) Analysis of yeast populations during alcoholic fermentation: a six year follow-up study. Syst Appl Microbiol 25: 287-293.

12. Taherzadeh MJ, Karimi K (2008) Pretreatment of lignocellulosic wastes to improve ethanol and biogas production: a review. Int J Mol Sci 9: 1621-1651.

13. Zhang J, Zhang WX, Wu ZY, Yang J, Liu YH, et al. (2013) A comparison of different dilute solution explosions pretreatment for conversion of distillers' grains into ethanol. Prep Biochem Biotechnol 43: 1-21.

14. Cantarella M, Cantarella L, Gallifuoco A, Spera A, Alfani F (2004) Effect of inhibitors released during steam-explosion treatment of poplar wood on subsequent enzymatic hydrolysis and SSF. Biotechnol Prog 20: 200-206.

15. Mira NP, Teixeira MC, Sá-Correia I (2010) Adaptive response and tolerance to weak acids in Saccharomyces cerevisiae: a genome-wide view. OMICS 14: 525-540.

16. Pampulha ME, Loureiro-Dias MC (2000) Energetics of the effect of acetic acid on growth of Saccharomyces cerevisiae. FEMS Microbiol Lett 184: 69-72.

17. Taherzadeh MJ, Lidén G, Gustafsson L, Niklasson C (1996) The effects of pantothenate deficiency and acetate addition on anaerobic batch fermentation of glucose by Saccharomyces cerevisiae. Appl Microbiol Biotechnol 46: 176-182.

18. Attfield PV, Bell PJ (2006) Use of population genetics to derive nonrecombinant Saccharomyces cerevisiae strains that grow using xylose as a sole carbon source. FEMS Yeast Res 6: 862-868.

19. Watanabe I, Ando A, Nakamura T (2012) Characterization of Candida sp. NY7122, a novel pentose-fermenting soil yeast. J Ind Microbiol Biotechnol 39: 307-315.

20. Deák T (1993) Simplified techniques for identifying foodborne yeasts. Int J Food Microbiol 19: 15-26.

21. Bochner BR, Gadzinski P, Panomitros E (2001) Phenotype microarrays for high-throughput phenotypic testing and assay of gene function. Genome Res 11: 1246-1255.

22. Bochner BR (2003) New technologies to assess genotype-phenotype relationships. Nat Rev Genet 4: 309-314.

23. Wimalasena TT, Greetham D, Marvin ME, Liti G, Chandelia Y, et al. (2014) Phenotypic characterisation of Saccharomyces spp. yeast for tolerance to stresses encountered during fermentation of lignocellulosic residues to produce bioethanol. Microb Cell Fact 13: 47.

24. Krishnan MS, Ho NW, Tsao GT (1999) Fermentation kinetics of ethanol production from glucose and xylose by recombinant Saccharomyces 1400(pLNH33). Appl Biochem Biotechnol 77-79: 373-88.

25. Scheller HV, Ulvskov P (2010) Hemicelluloses. Annu Rev Plant Biol 61: 263-289.

26. Sjostrom B (1993) Wood polysaccharides. Academic Press, USA, 2: 51-70.

27. Mills TY, Sandoval NR, Gill RT (2009) Cellulosic hydrolysate toxicity and tolerance mechanisms in Escherichia coli. Biotechnol Biofuels 2: 26.

28. Ludovico P, Sousa MJ, Silva MT, Leão C, Côrte-Real M (2001) Saccharomyces cerevisiae commits to a programmed cell death process in response to acetic acid. Microbiology 147: 2409-2415.

29. Hengartner MO (2000) The biochemistry of apoptosis. Nature 407: 770-776.

30. Matsuyama S, Reed JC (2000) Mitochondria-dependent apoptosis and cellular pH regulation. Cell Death Differ 7: 1155-1165.

31. (1977) Reaction-products of yeast fermentations. Process Biochemistry 12: 19-21.

32. Pagliardini J, Hubmann G, Alfenore S, Nevoigt E, Bideaux C, et al. (2013) The metabolic costs of improving ethanol yield by reducing glycerol formation capacity under anaerobic conditions in Saccharomyces cerevisiae. Microb Cell Fact 12: 29.

33. Ding J, Bierma J, Smith MR, Poliner E, Wolfe C, et al. (2013) Acetic acid inhibits nutrient uptake in Saccharomyces cerevisiae: auxotrophy confounds the use of yeast deletion libraries for strain improvement. Appl Microbiol Biotechnol 97: 7405-7416.

32

Reduction of Toxic Heavy Metals in Traditional Asian Herbs By Decoction Preparation

Seung–Hoon Lee[1,2]*, In-Jun Wee[1,2] and Chang-Ho Park[3]

[1]Department of Acupuncture and Oriental Medicine, Healing Hand Healthcare Center, 3400 W. 6th St. Suite 305, Los Angeles, CA 90020, USA
[2]Department of Oriental Medicine and Society, Acupuncture Network, 430 32nd St. Suite 100, Newport Beach, CA 92663, USA
[3]Department of Chemical Engineering, College of Engineering, Kyung Hee University, 1 Seocheon-dong, Giheung-gu, Yongin-si, Gyeonggi-do, 446-701, Republic of Korea

Abstract

Heavy metal contents in traditional Asian herbs need to be monitored closely because of potential health risks of heavy metals at high concentrations. Total amount of toxic heavy metals (lead, copper, cadmium, chromium, mercury and arsenic) in each herb listed in the recipe of Ssanghwatang (one of the most popular herbal drinks in Korea) was 22.6-42.1 mg/kg as determined by Inductively Coupled Plasma - Atomic Emission Spectrometry (ICP-AES). These levels were notable because they were comparable to or above the guideline (30 mg/kg) set by Korean Food and Drug Administration (KFDA). However, herbal tea prepared by decoction preparation (a process of extracting medicinal components from the herbs by boiling them in water) contained only 6.35-12.2 % of the original toxic heavy metals in the herbs. Instead the remainder of toxic heavy metals was found in the herbal residue and the hempen cloth filter. The result suggests that drinking herbal tea is a much safer way of taking beneficial nutrients from the herbs as compared with consuming the whole herbs.

Keywords: Herbal tea; Ssanghwatang; Heavy metals; KFDA guideline; Decoction: ICP-AES

Introduction

The world market size of traditional Asian herbs amount to 200-250 billion US $ per year, among which 30% is consumed in North America [1,2]. In spite of the medical efficacy of herbal treatment a wider use of herbs has been limited due to public concerns on contamination by heavy metals. Herbs, like other plants, absorb and accumulate heavy metals from soil during growth. Therefore, herbs may contain too high level of heavy metals if they are grown on soil contaminated by heavy metals coming from mining or manufacturing industry, the use of synthetic products containing heavy metals (pesticides, insecticides, paints, and batteries, etc.), and the land application of industrial wastes or domestic sludge [3,4]. In agricultural area pesticides and fertilizers are known to be the main sources of heavy metal pollution [5].

Heavy metals are defined as metals with a density higher 5 g/cm³. However, based on the solubility at physiological conditions the number of heavy metals available for living cells and of importance for organism and ecosystem are limited to 17 [6,7]. Among these some heavy metals such as iron (Fe), molybdenum (Mo), and manganese (Mn) are important as micronutrients, but nickel (Ni), copper (Cu), vanadium (V), cobalt (Co), tungsten (W), and chromium (Cr) are toxic at high concentrations. Arsenic (As), mercury (Hg), cadmium (Cd), and lead (Pb) have no known function as nutrients and seem to be toxic to plants and micro-organisms [8,9]. These heavy metals are transported from soil to plants through plant cell wall and membranes, and plants are able to accumulate heavy metals to certain level in their tissues [10,11]. These toxic heavy metals will eventually be transferred to human body through food web [12-14].

When the concentration of heavy metals in human body reaches a certain high level, it will cause various acute and chronic disorders (Table 1) [15]. Chronic exposure to heavy metals was linked to the development of various diseases [16-18]. To resolve health related concerns, we must monitor the level of heavy metals in the herbs before the herbs are applied for treatment. The management and regulation of herb distribution systems have been established and are currently in

effect in East Asian countries including South Korea, and the level of heavy metals and pesticide residues are controlled for both domestic and imported traditional Asian herbs. Nonetheless, when the media reports on heavy metal contamination of medicinal herbs, the public becomes anxious and hesitant to select herbal treatment for their illnesses.

In this paper we determined the levels of heavy metals (lead, copper, cadmium, chromium, mercury and arsenic) in nine herbs used for the preparation of Ssanghwatang. Compared with the guideline of Korean Food and Drug Administration [19] the results were noteworthy. When we applied decoction preparation (a process of boiling herbs in water in order to extract medicinal components from the herbs), the level of heavy metals in the herbal decoction (tea) was lowered far below the KFDA guideline. This was because most of the metals in the herbs were left in the residue and hempen cloth filter, but not extracted during decoction process.

Materials and Methods

Sources of traditional Asian herbs

Ssanghwatang is one of the most popular traditional herbal drinks available in Korea. In the name "Ssang" means a couple, "hwa" means a harmony, and "tang" means decoction, and it is said to be originally prepared by a court lady for a king after she slept with him at night. As the name stands for, it is known to be effective for both men and

*Corresponding author: Seung–Hoon Lee, Healing Hand Healthcare Center, 3400 W. 6th St. Suite 305, Los Angeles, CA 90020, USA

Heavy Metal	Acute	Chronic	Toxic Concentration
Lead	Nausea, vomiting, encephalopathy (headache, seizures, ataxia, obtundation)	Encephalopathy, anemia, abdominal pain, nephropathy, foot-drop/ wrist-drop	Pediatric: symptoms or [Pb] ≥45 μ/dL (blood); Adult: symptoms or [Pb] ≥70 μ/dL
Copper	Blue vomitus, GI irritation/ hemorrhage, hemolysis, MODS (ingested); MFF (inhaled)	Vineyard sprayer's lung (inhaled); Wilson disease (hepatic and basal ganglia degeneration)	Normal excretion: 25 μg/24 hours (urine)
Cadmium	Pneumonitis (oxide fumes)	Proteinuria, lung cancer, osteomalacia	Proteinuria and/or ≥15 μg/g creatinine
Chromium	GI hemorrhage, hemolysis, acute renal failure (Cr^{6+} ingestion)	Pulmonary fibrosis, lung cancer (inhalation)	No clear reference standard
Mercury	Elemental (inhaled): fever, vomiting, diarrhea, ALI; Inorganic salts (ingestion): caustic gastroenteritis	Nausea, metallic taste, gingivo-stomatitis, tremor, neurasthenia, nephrotic syndrome; hypersensitivity (Pink disease)	Background exposure "normal" limits: 10 μg/L (whole blood); 20 μg/L (24 hours urine)
Arsenic	Nausea, vomiting, "rice-water" diarrhea, encephalopathy, MODS, LoQTS, painful neuropathy	Diabetes, hypopigmentation/ hyperkeratosis, cancer: lung, bladder, skin, encephalopathy	24 hours urine: ≥50 μg/L urine, or 100 μg/g creatinine

MODS, multi-organ dysfunction syndrome; MFF, metal fume fever; GI, gastrointestinal; LoQTS, long QT syndrome and a rare inborn heart condition; ALI, acute lung injury.

Table 1: Typical presentation of poisoning by six heavy metals.

Heavy metals	Wave length (nm)	Estimated detection limit (μg/L)	Alternate wave length (nm)	Calibration concentration (mg/L)	Upper Limit concentration (mg/L)
Lead (Pb^{2+})	220.35	40	217	10.0	100
Copper (Cu^{2+})	324.75	6	219.96	1.0	50
Cadmium (Cd^{2+})	226.50	4	214.44	2.0	50
Chromium (Cr^{6+})	267.72	7	206.15	5.0	50
Mercury (Hg^{2+})	253.70	2	184.91	1.0	50
Arsenic (As^{2+})	193.70	50	189.04	10.0	100

Table 2: Operating conditions of inductively coupled plasma - atomic emission spectrometer (ICP-AES).

women in relieving physical fatigue, boosting vigor and energy after a sexual intercourse, and controlling cold sweat.

Traditional Asian herbs listed in the recipe of Ssangwhatang [20] were obtained from Oriental Medical Center of Kyung Hee University (Seoul, Republic of Korea), and identified by traditional Asian herbal medicinal experts. The herbs were originated from different locations in East Asia: *Paeoniae Radix Alba* (Dae-Jeon-Si, Republic of Korea), *Zizyphi Fructus* (Yeong-Cheon-Si, Republic of Korea), *Cnidii Rhizoma* (Yeong-Deok-Gun, Republic of Korea), *Rehmanniae Radix Preparata* (China), *Angelicae Gigantis Radix* (Cheong-Yang-Gun, Republic of Korea), *Astragali Radix* (Je-Cheon-Si, Republic of Korea), *Glycyrrhizae Radix* (China), *Cinnamomi Cortex Spissus* (China) and *Zingiberis Rhizoma Crudus* (China).

Heavy metal analysis

Decoction is a process of boiling herbs in water, and extracting medicinal components from the herbs. When herbs are processed by decoction, we obtain the herbal decoction (tea) together with herbal residue and used hempen cloth filter. Therefore, heavy metals originally present in the herbs will be distributed in the herbal decoction, residue, and hempen cloth filter. The contents of lead (Pb^{2+}), copper (Cu^{2+}), cadmium (Cd^{2+}), chromium (Cr^{6+}), mercury (Hg^{2+}) and arsenic (As^{2+}) in herbal decoction, herbal residue, and hempen cloth filter were determined by inductively coupled plasma - atomic emission spectrometer (ICP-AES) (Shimazu Co., Japan). The decoction experiments were performed in triplicate, and the same sample from each experiment was analyzed twice. The detailed procedure followed literature information [21,22]. Operating conditions of the ICP-AES are shown in Table 2. In the analysis standard solution of heavy metal (1000 mg/L, Showa Co., Japan) was used with appropriated dilution.

Preparation of herbal decoction

One gram of oven-dried herb was wrapped in one gram of an oven-dried hempen cloth filter, and was boiled for 3 hours in a flask with 300 mL of deionized water. The herbal residue and hempen cloth filter were oven-dried for 24 hours in a porcelain mortar at 100°C before dry weight analysis. The dried herbal residue and hempen cloth filter were milled separately in a pestle and subsequently digested by acids (HNO_3 and HCl) in a flask according to the wet digestion method (U.S. E.P.A. Standard Method 3030 F) [23-26] (Figure 1). The herbal decoction was oven-dried for 72 hours in a porcelain mortar at 100°C before it was weighed and analyzed by the wet digestion method (Figure 1). The con-

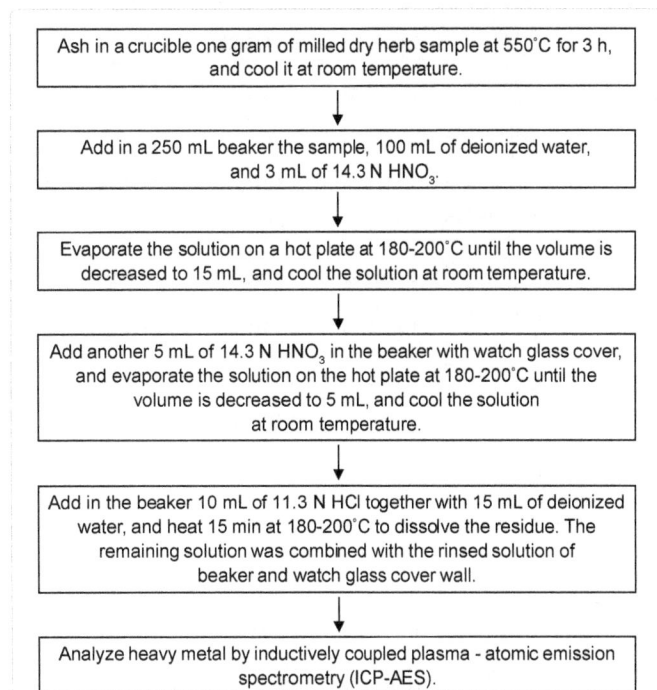

Ash in a crucible one gram of milled dry herb sample at 550°C for 3 h, and cool it at room temperature.

↓

Add in a 250 mL beaker the sample, 100 mL of deionized water, and 3 mL of 14.3 N HNO_3.

↓

Evaporate the solution on a hot plate at 180-200°C until the volume is decreased to 15 mL, and cool the solution at room temperature.

↓

Add another 5 mL of 14.3 N HNO_3 in the beaker with watch glass cover, and evaporate the solution on the hot plate at 180-200°C until the volume is decreased to 5 mL, and cool the solution at room temperature.

↓

Add in the beaker 10 mL of 11.3 N HCl together with 15 mL of deionized water, and heat 15 min at 180-200°C to dissolve the residue. The remaining solution was combined with the rinsed solution of beaker and watch glass cover wall.

↓

Analyze heavy metal by inductively coupled plasma - atomic emission spectrometry (ICP-AES).

Figure 1: Procedure of wet digestion and heavy metal analysis (U.S. E.P.A. Standard Method 3030 F and ICP-AES).

Heavy metal	Sample	Rehmanniae Radix Preparata	Zizyphi Fructus	Cnidii Rhizoma	Cinnamomi Cortex Spissus	Astragali Radix	Paeoniae Radix Alba	Glycyrrhizae Radix	Angelicae Gigantis Radix	Zingiberis Rhizoma Crudus
Lead (Pb^{2+}) (mg/kg) ±error	Herbal decoction	0.16±0.03	0.30±0.09	0.24±0.06	1.15±0.14	0.69±0.05	0.93±0.03	0.61±0.10	0.11±0.01	0.12±0.03
	Herbal residue	3.13±0.14	2.96±0.44	1.42±0.20	2.19±0.09	3.64±0.34	6.53±0.74	7.40±0.56	4.19±0.53	5.96±0.36
	Hempen cloth filter	0.59±0.12	0.42±0.07	0.13±0.22	0.16±0.04	0.23±0.03	0.43±0.01	0.68±0.04	0.11±0.02	1.38±0.17
	Total level	3.88±0.23	3.68±0.42	1.79±0.28	3.49±0.18	4.56±0.33	7.89±0.72	8.69±0.53	4.41±0.52	7.46±0.20
	% in decoction	4.12	8.15	13.4	33.0	15.1	11.8	7.02	2.49	1.61
Copper (Cu^{2+}) (mg/kg) ±error	Herbal decoction	0.23±0.03	0.34±0.10	0.56±0.32	0.51±0.09	0.87±0.07	0.50±0.09	0.59±0.04	0.65±0.06	0.43±0.06
	Herbal residue	2.18±0.17	7.27±0.86	6.93±0.59	7.11±1.02	4.11±1.06	7.43±0.27	3.91±0.14	8.24±0.76	2.45±0.13
	Hempen cloth filter	0.58±0.08	0.36±0.14	0.71±0.03	1.42±0.03	1.72±0.39	0.55±0.07	0.71±0.05	0.54±0.06	0.59±0.04
	Total level	2.99±0.10	7.97±0.84	8.20±0.61	9.04±1.09	6.70±1.22	8.48±0.41	5.21±0.22	9.43±0.77	3.47±0.10
	% in decoction	7.69	4.27	6.80	5.64	13.0	5.90	11.3	6.89	12.4
Cadmium (Cd^{2+}) (mg/kg) ±error	Herbal decoction	0.07±0.02	0.05±0.02	0.09±0.02	0.08±0.01	0.08±0.01	0.05±0.01	0.04±0.01	0.10±0.03	0.09±0.03
	Herbal residue	1.25±0.12	0.28±0.09	0.55±0.08	0.68±0.18	0.39±0.07	0.84±0.12	0.75±0.04	2.11±0.05	4.64±0.37
	Hempen cloth filter	0.11±0.05	0.11±0.04	0.06±0.01	0.10±0.04	0.25±0.10	0.20±0.11	0.14±0.02	0.07±0.01	0.22±0.04
	Total level	1.43±0.10	0.44±0.05	0.70±0.11	0.87±0.20	0.72±0.16	1.09±0.23	0.93±0.04	2.28±0.03	4.95±0.43
	% in decoction	4.90	11.4	12.9	9.20	11.1	4.59	4.30	4.39	1.82
Chromium (Cr^{6+}) (mg/kg) ±error	Herbal decoction	1.09±0.02	1.83±0.17	1.16±0.11	0.94±0.02	0.85±0.04	1.05±0.05	0.51±0.01	0.70±0.05	1.65±0.23
	Herbal residue	6.48±0.20	4.46±0.31	6.10±0.51	5.64±0.31	5.93±0.26	4.00±0.76	6.69±0.29	7.74±0.18	12.7±1.58
	Hempen cloth filter	0.45±0.10	0.91±0.37	0.71±0.31	1.48±0.07	1.67±0.22	1.73±0.32	2.19±0.25	1.84±0.32	1.94±0.06
	Total level	8.02±0.12	7.20±0.67	7.97±0.38	8.06±0.27	8.45±0.13	6.78±0.54	9.39±0.13	10.3±0.12	16.3±1.43
	% in decoction	13.6	25.4	14.6	11.7	10.1	15.5	5.43	6.80	10.1
Mercury (Hg^{2+}) (mg/kg) ±error	Herbal decoction	0.00±0.01	0.14±0.02	0.20±0.03	0.00±0.00	0.17±0.01	0.09±0.01	0.00±0.00	0.21±0.02	0.12±0.03
	Herbal residue	1.61±0.07	2.33±0.43	3.21±0.31	1.45±0.17	3.16±0.52	2.27±0.13	3.72±0.15	4.60±0.53	2.11±0.08
	Hempen cloth filter	1.49±0.07	0.11±0.02	0.10±0.04	0.16±0.03	0.67±0.05	0.02±0.01	1.34±0.28	0.01±0.01	0.10±0.01
	Total level	3.10±0.14	2.58±0.41	3.51±0.28	1.61±0.20	4.00±0.50	2.38±0.12	5.06±0.13	4.82±0.54	2.33±0.09
	% in decoction	0.00	5.43	5.70	0.00	4.25	3.78	0.00	4.36	5.15
Arsenic (As^{2+}) (mg/kg) ±error	Herbal decoction	0.42±0.02	0.51±0.03	0.58±0.05	0.34±0.02	0.59±0.01	0.59±0.05	0.38±0.12	0.67±0.07	1.40±0.13
	Herbal residue	1.29±0.03	3.03±0.07	3.21±0.22	5.45±0.16	3.56±0.11	2.56±0.31	1.86±0.03	6.05±0.06	4.82±0.12
	Hempen cloth filter	1.45±0.21	0.62±0.22	1.39±0.15	1.32±0.16	1.89±0.11	0.87±0.05	0.18±0.03	0.49±0.05	1.44±0.14
	Total level	3.16±0.18	4.16±0.21	5.18±0.29	7.11±0.25	6.04±0.19	4.02±0.37	2.41±0.12	7.21±0.05	7.66±0.29
	% in decoction	13.3	12.3	11.2	4.78	9.77	14.7	15.4	9.29	18.3
Total heavy metals in herbal decoction in mg/kg (%)		1.97 (8.73)	3.17 (12.18)	2.83 (10.35)	3.02 (10.01)	3.25 (10.67)	3.21 (10.48)	2.12 (6.69)	2.44 (6.35)	3.81 (9.04)
Total heavy metals in herbal residue in mg/kg		15.94 (70.59)	20.33 (78.10)	21.42 (78.32)	22.52 (74.62)	20.79 (68.23)	23.63 (77.12)	24.33 (76.77)	32.93 (85.69)	32.68 (77.51)
Total heavy metals in hempen cloth filter in mg/kg		4.67 (20.68)	2.53 (9.72)	3.1 (11.33)	4.64 (15.37)	6.43 (21.10)	3.80 (12.40)	5.24 (16.54)	3.06 (7.96)	5.67 (13.46)
Sum of heavy metals in decoction, residue and hempen cloth filter in mg/kg		22.58 (100)	26.03 (100)	27.35 (100)	30.18 (100)	30.47 (100)	30.64 (100)	31.69 (100)	38.43 (100)	42.16 (100)

Table 3: Distribution of heavy metals in the herbal decoction (tea), the herbal residue, and the hempen cloth filter.

tents of heavy metals in herbal decoction, herbal residue, and hempen cloth filter were determined by ICP-AES.

Heavy metal contents in the unused hempen cloth filter

Because hempen cloth is also from a plant source, we need to determine the initial level of heavy metals in the unused hempen cloth filter. We calculated a net gain of heavy metals in hempen cloth filter by subtracting the initial level from the final level of used hempen cloth filter. The unused hempen cloth filter was oven-dried for 24 hours in a porcelain mortar at 100°C. One gram of oven-dried hempen cloth was milled by a pestle, and followed procedures in Figure 1. The grey white ash was obtained at the completion of the ashing procedure. The ICP-AES was used to analyze the concentration of the heavy metals in the clear digested solution. The initial levels of heavy metals in one gram of dry unused hempen cloth filter were 0.41 mg/kg (lead), 1.03 mg/kg (copper), 0.13 mg/kg (cadmium), 1.74 mg/kg (chromium), 1.38 mg/kg (mercury) and 1.50 mg/kg (arsenic), respectively.

Calculation of heavy metal percentage in herbal decoction

Percentage of heavy metal in the herbal decoction was obtained as follows;

$$\text{Percentage of heavy in the herbal decoction} = \frac{M(B)}{M(A)} \times 100 \qquad (1)$$

Where,

M (A): The sum of heavy metals in herbal decoction, herbal residue, and a gain of heavy metals in hempen cloth filter

M (B): Mass of heavy metals in herbal decoction

Results and Discussion

Total heavy metal levels for each traditional Asian herb

Total amount of toxic heavy metals (Pb^{2+}, Cu^{2+}, Cd^{2+}, Cr^{6+}, Hg^{2+} and As^{2+}) in traditional Asian herbs listed in the recipe of Ssanghwatang varied from 22.6 to 42.1 mg/kg, and the value for each herb was as follows (in mg/kg): *Rehmanniae Radix Preparata* (22.6), *Zizyphi Fructus* (26.0), *Cnidii Rhizoma* (27.4), *Cinnamomi Cortex Spissus* (30.2), *Astragali Radix* (30.5), *Paeoniae Radix Alba* (30.6), *Glycyrrhizae Radix* (31.7), *Angelicae Gigantis Radix* (38.4), and *Zingiberis Rhizoma Crudus* (42.2) (Table 3). This level of heavy metals was close to or above the guideline (30 mg/kg) set by Korean Food and Drug Administration. Especially, the heavy metal contents of *Angelicae Gigantis Radix* (38.4 mg/kg) and *Zingiberis Rhizoma Crudus* (42.2 mg/kg) were 1.28 and 1.40 times higher than the guideline of KFDA.

Individual heavy metal level in each traditional Asian herb

Heavy metal contents for each herb are presented in Table 3. Lead

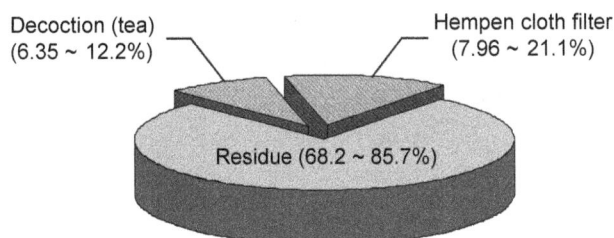

Figure 2: Heavy metal distribution after decoction of medicinal herbs (in the decoction: 6.35–12.2%, in the hempen cloth filter: 7.96–21.1%, in the residue: 68.2–85.7%). Exact percentages vary depending on the kind of heavy metal and the herb. Refer to Table 3 for detailed data.

(Pb^{2+}) content in the herbs ranged from 1.79 (*Cnidii Rhizoma*) to 8.69 mg/kg (*Glycyrrhizae Radix*). Copper (Cu^{2+}) content was from 2.99 (*Rehmanniae Radix Preparata*) to 9.43 mg/kg (*Angelicae Gigantis Radix*). Cadmium (Cd^{2+}) content varied from 0.44 (*Zizyphi Fructus*) to 4.95 mg/kg (*Zingiberis Rhizoma Crudus*). Chromium (Cr^{6+}) content ranged from 6.78 (*Paeoniae Radix Alba*) to 16.3 mg/kg (*Zingiberis Rhizoma Crudus*). The level of mercury (Hg^{2+}) was between 1.61 (*Cinnamomi Cortex Spissus*) and 5.06 mg/kg (*Glycyrrhizae Radix*). Arsenic (As^{2+}) content ranged from 2.41 (*Glycyrrhizae Radix*) to 7.66 mg/kg (*Zingiberis Rhizoma Crudus*).

In the case of *Zingiberis Rhizoma Crudus* the levels of cadmium (4.95 mg/kg) and chromium (16.3 mg/kg) were noticeably higher compared with other herbs. This result concurs with a literature report that cadmium and chromium were easily taken up by root of plants and transported to different plant parts [5]. Total heavy metal level was also highest for *Zingiberis Rhizoma Crudus* probably because this root type of herb was directly influenced by soil pollution [3].

Heavy metal levels in each herbal decoction

Herbal decoction (tea) can be prepared from traditional Asian herbs by decoction process: wrapping herbs in hempen cloth and steeping them for a few hours in boiling water [20, 27-29]. Unlike the unprocessed herb the level of heavy metals in the herbal decoction was much less than the guidelines set by KFDA. The percentage (contents) of heavy metals ($Pb2+$, $Cu2+$, $Cd2+$, $Cr6+$, $Hg2+$ and $As2+$) in the herbal decoction (tea) as defined by equation (1) was 8.73% (1.97 mg/kg) of the total contents for *Rehmanniae Radix Preparata*. And the percentage (contents) for other herbs were 12.2% (3.17 mg/kg) for *Zizyphi Fructus*, 10.4% (2.83 mg/kg) for *Cnidii Rhizoma*, 10.0% (3.02 mg/kg) for *Cinnamomi Cortex Spissus*, 10.7% (3.25 mg/kg) for *Astragali Radix*, 10.5% (3.21 mg/kg) for *Paeoniae Radix Alba*, 6.69% (2.12 mg/kg) for *Glycyrrhizae Radix*, 6.35% (2.44 mg/kg) for *Angelicae Gigantis Radix*, and 9.04% (3.81 mg/kg) for *Zingiberis Rhizoma Crudus* (Table 3). Our results indicate that decoction preparation leaves most of (87.8-93.7%) of heavy metals in the residue and hempen cloth, and the problem of heavy metals in traditional Asian herbs can be solved efficiently by applying decoction preparation. Also our results suggest that in order to prepare powder form of herbal medicine we better use decoction process followed by drying rather than washing procedure. In our previous work washing procedure removed 17.8-36.7% of heavy metals from the herbs [30], which was much lower than the decoction process (87.8-93.7%). Decoction process may increase the cost of powder preparation due to a large evaporation load during drying process. However, preparation of healthier herbal medicine (tea) with less heavy metal content is definitely a more important issue to be addressed.

Conclusion

Total amount of toxic heavy metals (Pb^{2+}, Cu^{2+}, Cd^{2+}, Cr^{6+}, Hg^{2+} and As^{2+}) in each traditional Asian herbs listed in the recipe of Ssanghwatang are notable because they are comparable to or above the guideline (30 mg/kg) set by Korean Food and Drug Administration (KFDA). However, after decoction process most of the heavy metals were present in the herbal residue and used hempen cloth filter, and the percentage of heavy metals in the herbal decoction (tea) was only 6.35-12.2% of the total amount, which was far below the guideline set by KFDA. This result suggests that public concerns on the level of heavy metals in traditional Asian herbs can be resolved efficiently by applying decoction process (a process of extracting medicinal components from the herbs by boiling them in water).

Acknowledgment

This work was supported by a grant from the Kyung Hee University in 2010.

References

1. Shin HK, Lee KK, Hwang DS (2007) A study on the current status and prospect of CAM world market. Korea Institute of Oriental Medicine, Seoul.

2. Kim JN, Lee YS, Park SY, Lee YC, Choi YW, et al. (2007) Studies on establishing of good supplying practice for standardized goods of herbal material. Korean Food and Drug Administration, Seoul.

3. Brady NC, Weil RR (1999) The nature and properties of soils. Prentice Hall, Saddle River.

4. Ogundiran MB, Osibanjo O (2008) Heavy metal concentration in soils and accumulation in plants growing in a deserted slag dump site in Nigeria. Afr J Biotechnol 7: 3053-3060.

5. Jarvis SC, Jones LHP, Hopper MJ (1976) Cadmium uptake from solution by plants and its transport from roots to shoots. Plant AND Soil 44: 179-191.

6. Weast RC (1989) CRC handbook of chemistry and physics, (69th edn). CRC Press Inc, Boca Raton.

7. Al-Yousuf MH, El-Shahawi MS, Al-Ghais SM (2000) Trace metals in liver, skin and muscle of Lethrinus lentjan fish species in relation to body length and sex. Sci Total Environ 256: 87-94.

8. Breckle CW (1991) Growth under heavy metals. In: Waisel Y, Eshel A, Kafkafi U (eds) Plant roots: the hidden half. Marcel Dekker, New York.

9. Nies DH (1999) Microbial heavy metal resistance. Appl Microbiol Biotechnol 51: 730-750.

10. Samecka-Cymerman A, Kempers AJ (2004) Toxic metals in aquatic plants surviving in surface water polluted by copper mining industry. Ecotoxicol Environ Saf 59: 64-69.

11. Prasad MNV, Sajwan KS, Naidu R (2006) Trace elements in the environment biogeochemistry, biotechnology and bioremediation. Tylor & Francis Group LLC, London.

12. Rai UN, Tripathi RD, Vajpayee P, Pandey N, Ali MB, et al (2003) Cadmium accumulation and its phytoxicity in Potamogeton pectinatus L. (Potamogetonaceae). Bull Environ Contam Toxicol 70: 566-575.

13. Tessier A, Buffle J, Campbell PGC (1994) Uptake of trace metals by aquatic organisms. In: Buffle J, De Vitre RR (edn) Chemical and biological regulation of aquatic system. CRC Press Inc, Boca Raton.

14. Yang H, Rose NL, Battarbee RW (2002) Distribution of some trace metals in Lochnagar, a Scottish mountain lake ecosystem and its catchment. Sci Total Environ 285: 197-208.

15. Soghoian S, Sinert RH (2009) Heavy Metal Toxicity. Medscape reference publishing web.

16. Prozialeck WC, Edwards JR, Nebert DW, Woods JM, Barchowsky A, et al. (2008) The vascular system as a target of metal toxicity. Toxicol Sci 102: 207-218.

17. Alam SK, Brim MS, Carmody GA, Parauka FM (2000) Concentrations of heavy and trace metals in muscle and blood of juvenile Gulf sturgeon (Acipenser oxyrinchus desotoi) from the Suwannee River. J Environ Sci Health 35: 645-660.

18. Tripathi RM, Raghunath R, Sastry VN, Krishnamoorthy TM (1999) Daily intake of heavy metals by infants through milk and milk products. Sci Total Environ 227: 229-235.

19. Director of Korean Food and Drug Administration (2010) The specifications and test methods for contaminants and residues in herbal medicines. Korean Food and Drug Administration, Seoul.

20. Choi TS, An DG (1998) Korean tonics. The Open Books Co, Seoul.

21. Kho BS, Park KJ, Kim HK, Eun YA, Kang EJ, et al. (1998) Studies on security of quality control and safety in traditional medicine herbs. Korea Institute of Oriental Medicine, Seoul.

22. Ha YD, Lee IS (1990) Investigation of heavy metal contents in Ganoderma lucidum (Fr.) karst. J Kor Soc Food Sci Nut 19: 187-193.

23. Eaton AD, Clesceri LS, Rice EW, Greenberg AE, Franson MAH (2005) Standard methods for the examination of water and wastewater. Amer Public Health Assn, Washington DC.

24. Hwang HW, Chung DH, Oh SY, Moon YS, Park MK (2006) A study on heavy metal contents in herbal medicine. Proceedings Kor Environ Sci Soc Conf 15: 547-549.

25. Ayari F, Hamdi H, Jedidi N, Gharbi N, Kossai R (2010) Heavy metal distribution in soil and plant in municipal solid waste compost amended plots. Int J Environ Sci Technol 7: 465-472.

26. Ernst E (2002) Toxic heavy metals and undeclared drugs in Asian herbal medicines. Trends Pharmacol Sci 23: 136-139.

27. Bensky D, Barolet R (2009) Chinese herbal medicine: formulas and strategies. Eastland Press, Seattle, WA.

28. Zuo YF, Zhu ZB, Huang YZ, Tao JW, Li ZG (2002) Science of prescriptions. Publishing House of Shanghai University of Traditional Chinese Medicine, Shanghai.

29. Long ZX, Li QG, Liu ZW (2005) Formulas of traditional Chinese medicine. Academy Press, Beijing.

30. Lee SH, Choi HY, Park CH (2003) Determination of heavy metal contents in oriental medical materials and the effect of washing. Korean Journal of Biotechnology and Bioengineering 18: 90-93.

Permissions

The contributors of this book come from diverse backgrounds, making this book a truly international effort. This book will bring forth new frontiers with its revolutionizing research information and detailed analysis of the nascent developments around the world.

We would like to thank all the contributing authors for lending their expertise to make the book truly unique. They have played a crucial role in the development of this book. Without their invaluable contributions this book wouldn't have been possible. They have made vital efforts to compile up to date information on the varied aspects of this subject to make this book a valuable addition to the collection of many professionals and students.

This book was conceptualized with the vision of imparting up-to-date information and advanced data in this field. To ensure the same, a matchless editorial board was set up. Every individual on the board went through rigorous rounds of assessment to prove their worth. After which they invested a large part of their time researching and compiling the most relevant data for our readers.

The editorial board has been involved in producing this book since its inception. They have spent rigorous hours researching and exploring the diverse topics which have resulted in the successful publishing of this book. They have passed on their knowledge of decades through this book. To expedite this challenging task, the publisher supported the team at every step. A small team of assistant editors was also appointed to further simplify the editing procedure and attain best results for the readers.

Apart from the editorial board, the designing team has also invested a significant amount of their time in understanding the subject and creating the most relevant covers. They scrutinized every image to scout for the most suitable representation of the subject and create an appropriate cover for the book.

The publishing team has been an ardent support to the editorial, designing and production team. Their endless efforts to recruit the best for this project, has resulted in the accomplishment of this book. They are a veteran in the field of academics and their pool of knowledge is as vast as their experience in printing. Their expertise and guidance has proved useful at every step. Their uncompromising quality standards have made this book an exceptional effort. Their encouragement from time to time has been an inspiration for everyone.

The publisher and the editorial board hope that this book will prove to be a valuable piece of knowledge for researchers, students, practitioners and scholars across the globe.

List of Contributors

Bernadette E. Teleky
Department of Mechanical Engineering, Technical University of Cluj-Napoca, Romania

Mugur C. Bălan
Department of Mechanical Engineering, Technical University of Cluj-Napoca, Romania

Sharma Alok
Food Technology (Food Process Engineering), SHIATS, Allahabad, India

Genitha Immanuel
Department of Food Process Engineering, SHIATS, Allahabad, India

Adipa Chongsuksantikul
Department of Chemical Engineering, Tokyo Institute of Technology, Japan

Kazuhiro Asami
Department of Chemical Engineering, Tokyo Institute of Technology, Japan

Shiro Yoshikawa
Department of Chemical Engineering, Tokyo Institute of Technology, Japan

Kazuhisa Ohtaguchi
Department of Chemical Engineering, Tokyo Institute of Technology, Japan

Walter David Obregón
Research Laboratory Vegetable Proteins (LIPROVE), Faculty of Sciences, National University of La Plata (UNLP), 47 and 115 s / N, La Plata (B1900AVW), Argentina

José Sebastián Cisneros
Research Laboratory Vegetable Proteins (LIPROVE), Faculty of Sciences, National University of La Plata (UNLP), 47 and 115 s / N, La Plata (B1900AVW), Argentina

Florencia Ceccacci
Research Laboratory Vegetable Proteins (LIPROVE), Faculty of Sciences, National University of La Plata (UNLP), 47 and 115 s / N, La Plata (B1900AVW), Argentina

Evelina Quiroga
Membranes and Biomaterials Laboratory, Institute of Applied Physics (INFAP) -CONICET, National University of San Luis (UNSL), Almirante Brown 907, San Luis (D5700HHW), Argentina

R. F. Branco
Federal Technological University of Paraná - Campus Pato Branco, Coordination Chemistry, Research Group in bioprocesses and food technology, Fraron, Pato Branco, Paraná, Brazil

J. C. Santos
University of Sao Paulo - School of Engineering of Lorena, Department of Biotechnology, Group of Applied Microbiology and Bioprocess,Road of the city soccer field PO Box 116, Postal code 12602810, Lorena, SP, Brazil

S.S. Silva
University of Sao Paulo - School of Engineering of Lorena, Department of Biotechnology, Group of Applied Microbiology and Bioprocess,Road of the city soccer field PO Box 116, Postal code 12602810, Lorena, SP, Brazil

Swati Hegde
B.V.B College of Engineering & Technology, Hubli-580031, Karnataka, India

Gururaj Bhadri
B.V.B College of Engineering & Technology, Hubli-580031, Karnataka, India

Kavita Narsapur
B.V.B College of Engineering & Technology, Hubli-580031, Karnataka, India

Shanta Koppal
B.V.B College of Engineering & Technology, Hubli-580031, Karnataka, India

Princy Oswal
B.V.B College of Engineering & Technology, Hubli-580031, Karnataka, India

Naina Turmuri
B.V.B College of Engineering & Technology, Hubli-580031, Karnataka, India

Veena Jumnal
B.V.B College of Engineering & Technology, Hubli-580031, Karnataka, India

Basavaraj Hungund
B.V.B College of Engineering & Technology, Hubli-580031, Karnataka, India

Mohamed E Selim
Agricultural Botany Department, Faculty of Agriculture, Menoufiya University, Egypt

EZ Khalifa
Agricultural Botany Department, Faculty of Agriculture, Menoufiya University, Egypt

GA Amer
Agricultural Botany Department, Faculty of Agriculture, Menoufiya University, Egypt

AA Ely-kafrawy
Plant Pathology research institute, Agriculture research center, Gizza, Egypt

Nehad A El-Gammal
Plant Pathology research institute, Agriculture research center, Gizza, Egypt

Sudhish Mishra
Translational Science and Molecular Medicine, Michigan State University, Grand Rapids, MI 48202, USA

A.Mandal
Edward Waters College, Biology Program, Jacksonville, FL 32209, USA

RC Gupta
Translational Science and Molecular Medicine, Michigan State University, Grand Rapids, MI 48202, USA

PK Mandal
Edward Waters College, Biology Program, Jacksonville, FL 32209, USA

Fatemeh Ajalloueian
Department of Textile Engineering, Center of Excellence in Applied Nanotechnology, Isfahan University of Technology, Isfahan, 84156-83111, Iran
Polymer Chemistry, Department of Chemistry, Ångström Laboratory, SciLife Lab, Uppsala University, 751 21 Uppsala, Sweden

Moa Fransson
Department of Immunology, Genetics and Pathology, Uppsala University, 751 85 Uppsala, Sweden

Hossein Tavanai
Department of Textile Engineering, Center of Excellence in Applied Nanotechnology, Isfahan University of Technology, Isfahan, 84156-83111, Iran

Mohammad Massumi
Department of Physiology, University of Toronto, Toronto, ON M5S 1A8, Canada

Jöns Hilborn
Polymer Chemistry, Department of Chemistry, Ångström Laboratory, SciLife Lab, Uppsala University, 751 21 Uppsala, Sweden

Katarina Leblanc
Department of Clinical Immunology and Transfusion Medicine, Karolinska Institutet and Hematology Center at Karolinska University Hospital, Stockholm, Sweden

Ayyoob Arpanaei
Department of Industrial and Environmental Biotechnology, National Institute of Genetic Engineering and Biotechnology, Tehran, 14965-161, Iran

Peetra U Magnusson
Department of Immunology, Genetics and Pathology, Uppsala University, 751 85 Uppsala, Sweden

Aravindan Rajendran
Biochemical Engineering Laboratory, Department of Chemical Engineering, Annamalai University, Annamalai Nagar– 608 002, Tamil Nadu, India

Viruthagiri Thangavelu
Biochemical Engineering Laboratory, Department of Chemical Engineering, Annamalai University, Annamalai Nagar– 608 002, Tamil Nadu, India

Chun-Hsiang Huang
Industrial Enzymes National Engineering Laboratory, Tianjin Institute of Industrial Biotechnology, Chinese Academy of Sciences, Tianjin 300308, China

Zhen Zhu
Industrial Enzymes National Engineering Laboratory, Tianjin Institute of Industrial Biotechnology, Chinese Academy of Sciences, Tianjin 300308, China

Ya-Shan Cheng
Genozyme Biotechnology Inc., Taipei 106, Taiwan
AsiaPac Biotechnology Co., Ltd., Dongguan 523808, China

Hsiu-Chien Chan
Industrial Enzymes National Engineering Laboratory, Tianjin Institute of Industrial Biotechnology, Chinese Academy of Sciences, Tianjin 300308, China

Tzu-Ping Ko
Institute of Biological Chemistry, Taiwan

Chun-Chi Chen
Industrial Enzymes National Engineering Laboratory, Tianjin Institute of Industrial Biotechnology, Chinese Academy of Sciences, Tianjin 300308, China

Iren Wang
Institute of Biological Chemistry, Taiwan

Meng-Ru Ho
Institute of Biological Chemistry, Taiwan

Shang-Te Danny Hsu
Institute of Biological Chemistry, Taiwan
Institute of Biochemical Sciences, Taiwan

Yi-Fang Zeng
Institute of Biotechnology, Taiwan

Yu-Ning Huang
Institute of Biotechnology, Taiwan

Je-Ruei Liu
Agricultural Biotechnology Research Center, Academia Sinica, Taipei 115, Taiwan
Institute of Biotechnology, Taiwan
Department of Animal Science and Technology, National Taiwan University, Taipei 106, Taiwan

Rey-Ting Guo
Industrial Enzymes National Engineering Laboratory, Tianjin Institute of Industrial Biotechnology, Chinese Academy of Sciences, Tianjin 300308, China

EA Echiegu
Department of Agricultural and Bioresources Engineering, University of Nigeria, Nsukka, Enugu State, Nigeria

AE Ghaly
Department of Process Engineering and Applied Sciences, Dalhousie University, Halifax, Nova Scotia, Canada

VV Ramakrishnan
Department of Process Engineering and Applied Sciences, Dalhousie University, Halifax, Nova Scotia, Canada

AN Pathaka
Research Dean and Head of Institution, Amity Institute of Biotechnology, Amity University Rajasthan, India

Pranav H Nakhate
Biochemical Engineering Scholar Jaipur, Amity Institute of Biotechnology, Amity University Rajasthan, India

Nuvula Ashok Kumar
Centre for Biotechnology, JNTUH, Hyderabad, India

Korla Lakshmana Rao
Centre for Biotechnology, JNTUH, Hyderabad, India

Archana Giri
Centre for Biotechnology, JNTUH, Hyderabad, India

Komath Uma Devi
Centre for Biotechnology, JNTUH, Hyderabad, India

Mukama Omar
Key Laboratory of Carbohydrate Chemistry and Biotechnology, School of Biotechnology, Jiangnan University, Wuxi 214122, P.R. China
Department of Applied Biology, College of Science and Technology, Avenue de l'armée, P.O. Box: 3900 Kigali, Rwanda

Ndikubwimana Jean de Dieu
Key Laboratory of Carbohydrate Chemistry and Biotechnology, School of Biotechnology, Jiangnan University, Wuxi 214122, P.R. China

Muhammad Shamoon
The synergistic Innovation Center of Food Safety and Nutrition, State Key Laboratory of Food Science and Technology, Jiangnan University, Wuxi 214122, P.R. China

Byong H Lee
Department of Food Science/Agricultural Chemistry, McGill University, Saint-Anne-de-Bellevue, Qc H9X3V9, Canada

Blaz Gerbec
Department of Chemical, Biochemical and Environmental Engineering, Faculty of Chemistry and Chemical Technology, University of Ljubljana, Askerceva 5, SI-1115 Ljubljana, Slovenia

Eva Tavčar
Department of Pharmaceutical Biotechnology, Faculty of Pharmacy, University of Ljubljana, Aškerčeva 7, SI-1115 Ljubljana, Slovenia

Andrej Gregori
Institute for Natural Science (Zavod za naravoslovje), Ulica Bratov Ucakar 108, SI-1000 Ljubljana; MycoMedica d.o.o. Podkoren 72, SI-4280 Kranjska Gora, Slovenia

Samo Kreft
Department of Pharmaceutical Biotechnology, Faculty of Pharmacy, University of Ljubljana, Aškerčeva 7, SI-1115 Ljubljana, Slovenia

Marin Berovic
Department of Chemical, Biochemical and Environmental Engineering, Faculty of Chemistry and Chemical Technology, University of Ljubljana, Askerceva 5, SI-1115 Ljubljana, Slovenia

Swaran Nandini
King Lab Burnett School of Biomedical Sciences, College of Medicine, UCF, USA

KE Nandini
Plant & Microbial Technology (PMT) Group, Biotechnology Department, Jaypee Institute of Information Technology (JIIT), Noida, U.P., India

S. Krishna Sundari
Plant & Microbial Technology (PMT) Group, Biotechnology Department, Jaypee Institute of Information Technology (JIIT), Noida, U.P., India

Farrukh Jamal
Department of Biochemistry, Dr. Ram Manohar Lohia Avadh University, Faizabad-224001, U.P., India

Tanvi Goel
Division of Life Sciences, Research Centre, Nehru Gram Bharti University, Jhunsi, Allahabad-221505, U.P., India

Laura Bulgariu
Technical University Gheorghe Asachi of Iaşi, Faculty of Chemical Engineering and Environmental Protection, Department of Environmental Engineering and Management, D. Mangeron Street, 71A, 700050-Iaşi, Romania

Dumitru Bulgariu
Alexandru Ioan Cuza University of Iaşi, Faculty of Geography and Geology, Department of Geology, Carol I Street, no. 20A, 700506-Iaşi, Romania
Romanian Academy, Filial of Iaşi, Collective of Geography, Carol I Street, no. 18, 700506-Iaşi, Romania

S. M. Hosseini
Department of Food Science and Technology, Shaheed Beheshti University, M.C., Tehran, Iran

K. Khosravi-Darani
Department of Food Technology Research, National Nutrition and Food Technology Research Institute, Shaheed Beheshti University, M.C., P. O. Box: 19395-4741, Tehran, Iran

Farrukh Jamal
Department of Biochemistry, Dr. Ram Manohar Lohia Avadh University, Faizabad-224001, U.P, India

Sangram Singh
Department of Biochemistry, Dr. Ram Manohar Lohia Avadh University, Faizabad-224001, U.P, India

Sadiya Khatoon
Department of Biochemistry, Dr. Ram Manohar Lohia Avadh University, Faizabad-224001, U.P, India

Sudhir Mehrotra
Department of Biochemistry, University of Lucknow, Lucknow, U.P, India

Hee Kyung Lim
Nanocatalysis Center, Green Chemistry Division, Korea Research Institute of Chemical Technology, Yuseong PO Box 107, Daejon 305-600, Republic of Korea

No-Joong Park
Nanocatalysis Center, Green Chemistry Division, Korea Research Institute of Chemical Technology, Yuseong PO Box 107, Daejon 305-600, Republic of Korea

Young Kyu Hwang
Nanocatalysis Center, Green Chemistry Division, Korea Research Institute of Chemical Technology, Yuseong PO Box 107, Daejon 305-600, Republic of Korea
Department of Green Chemistry and Environmental Biotechnology, University of Science and Technology, 217 Gajungro, Yuseong- gu, Daejon, 305-350, Republic of Korea

Kee-In Lee
Nanocatalysis Center, Green Chemistry Division, Korea Research Institute of Chemical Technology, Yuseong PO Box 107, Daejon 305-600, Republic of Korea
Department of Green Chemistry and Environmental Biotechnology, University of Science and Technology, 217 Gajungro, Yuseong- gu, Daejon, 305-350, Republic of Korea

In Taek Hwang
Nanocatalysis Center, Green Chemistry Division, Korea Research Institute of Chemical Technology, Yuseong PO Box 107, Daejon 305-600, Republic of Korea
Department of Green Chemistry and Environmental Biotechnology, University of Science and Technology, 217 Gajungro, Yuseong- gu, Daejon, 305-350, Republic of Korea

Emilia Furia
Department of Chemistry and Chemical Technologies, University of Calabria, 87036 Arcavacata di Rende (CS), Italy

Anna Napoli
Department of Chemistry and Chemical Technologies, University of Calabria, 87036 Arcavacata di Rende (CS), Italy

Antonio Tagarelli
Department of Chemistry and Chemical Technologies, University of Calabria, 87036 Arcavacata di Rende (CS), Italy

Giovanni Sindona
Department of Chemistry and Chemical Technologies, University of Calabria, 87036 Arcavacata di Rende (CS), Italy

Duygu Karaalp
Ege University, Graduate School of Natural and Applied Sciences, Biotechnology Department, Izmir,Turkey

Kubra Arslan
Ege University, Faculty of Engineering, Bioengineering Department, Izmir, Turkey

Nuri Azbar
Ege University, Faculty of Engineering, Bioengineering Department, Izmir, Turkey
Ege University, Center for Enviromental Studies, Izmir, Turkey

Kishore Kumar Gopalakrishnan
Research Scholar, Department of Chemical & Process Engineering, University of Canterbury, New Zealand

Swaminathan Detchanamurthy
Research Scholar, Department of Chemical & Process Engineering, University of Canterbury, New Zealand

Emad Malaekah
School of Electrical and Computer Engineering, RMIT University, Australia

SobhanSalari Shahrbabaki
School of Electrical and Computer Engineering, RMIT University, Australia

Dean Cvetkovic
School of Electrical and Computer Engineering, RMIT University, Australia

Masanobu Horie
Department of Chemical Engineering, Faculty of Engineering, Kyushu University, 744 Motooka, Nishi-ku, Fukuoka 819-0395, Japan

Akira Ito, Yoshinori Kawabe
Department of Chemical Engineering, Faculty of Engineering, Kyushu University, 744 Motooka, Nishi-ku, Fukuoka 819-0395, Japan

Masamichi Kamihira
Department of Chemical Engineering, Faculty of Engineering, Kyushu University, 744 Motooka, Nishi-ku, Fukuoka 819-0395, Japan

Jackelinne de A Silva
Federal Rural University of Pernambuco, Dois Irmãos, 52171-900, Recife, Brazil

Maria IS Maciel
Federal Rural University of Pernambuco, Dois Irmãos, 52171-900, Recife, Brazil

Naíra P de Moura
Federal Rural University of Pernambuco, Dois Irmãos, 52171-900, Recife, Brazil

Marcony E da S Júnior
Federal Rural University of Pernambuco, Dois Irmãos, 52171-900, Recife, Brazil

Janaína V de Melo
Center of Strategic Technologies of the Northeast, Cidade Universitária, 50740-540, Recife, Brazil

Patrícia M Azoubel
Federal University of Pernambuco, Cidade Universitária, 50740-521, Recife, Brazil

Enayde de A Melo
Federal Rural University of Pernambuco, Dois Irmãos, 52171-900, Recife, Brazil

Jaqueline do Nascimento Silva
Department of Biochemical Engineering, School of Chemistry, Universidade Federal do Rio de Janeiro, Federal University of Rio de Janeiro, Cidade Universitária, Centro de Tecnologia, Bl. E, Sl. 203, Ilha do Fundão, 21941-909, Rio de Janeiro, Brasil

Melissa Limoeiro Estrada Gutarra
Department of Biochemical Engineering, School of Chemistry, Universidade Federal do Rio de Janeiro, Federal University of Rio de Janeiro, Cidade Universitária, Centro de Tecnologia, Bl. E, Sl. 203, Ilha do Fundão, 21941-909, Rio de Janeiro, Brasil

Denise Maria Guimaraes Freire
Department of Biochemistry, Institute of Chemistry, Universidade Federal do Rio de Janeiro, Federal University of Rio de Janeiro, Cidade Universitária, Centro de Tecnologia, Bl. A, Sl. 549, Ilha do Fundão, 21949-900, Rio de Janeiro, Brasil

Magali Christe Cammarota
Department of Biochemical Engineering, School of Chemistry, Universidade Federal do Rio de Janeiro, Federal University of Rio de Janeiro, Cidade Universitária, Centro de Tecnologia, Bl. E, Sl. 203, Ilha do Fundão, 21941-909, Rio de Janeiro, Brasil

Dal Rye Kim
Biorefinery Research Group, Green Chemistry Division, Korea Research Institute of Chemical Technology, Yusoeong P.O. Box 107, Daejon 305-600, Republic of Korea

Hee Kyung Lim
Biorefinery Research Group, Green Chemistry Division, Korea Research Institute of Chemical Technology, Yusoeong P.O. Box 107, Daejon 305-600, Republic of Korea

In Taek Hwang
Biorefinery Research Group, Green Chemistry Division, Korea Research Institute of Chemical Technology, Yusoeong P.O. Box 107, Daejon 305-600, Republic of Korea
Department of Green Chemistry and Environmental Biotechnology, University of Science & Technology, 217 Gajungro, Yuseong-gu, Daejon, 305-350, Republic of Korea

D. Greetham
School of Biosciences, University of Nottingham, Loughborough, Leics, LE12 5RD, UK

Seung-Hoon Lee
Department of Acupuncture and Oriental Medicine, Healing Hand Healthcare Center, 3400 W. 6th St. Suite 305, Los Angeles, CA 90020, USA
Department of Oriental Medicine and Society, Acupuncture Network, 430 32nd St. Suite 100, Newport Beach, CA 92663, USA

In-Jun Wee
Department of Acupuncture and Oriental Medicine, Healing Hand Healthcare Center, 3400 W. 6th St. Suite 305, Los Angeles, CA 90020, USA
Department of Oriental Medicine and Society, Acupuncture Network, 430 32nd St. Suite 100, Newport Beach, CA 92663, USA

Chang-Ho Park
Department of Chemical Engineering, College of Engineering, Kyung Hee University, 1 Seocheon-dong, Giheung-gu, Yongin-si, Gyeonggi-do, 446-701, Republic of Korea

www.ingramcontent.com/pod-product-compliance
Lightning Source LLC
Chambersburg PA
CBHW080533200326
41458CB00012B/4420